XXIVth Conference of the Danubian Countries on Hydrological Forecasting and Hydrological Bases of Water Management 2008

IOP Conference Series: Earth and Environmental Science
Volume 4

Bled, Slovenia
2 - 4 June 2008

Editors:

Mitja Brilly
Ognjen Bonacci
Peter Hans Nachtnebel
Jan Szolgay
Gabor Balint

ISBN: 978-1-5108-0206-3
ISSN: 1755-1307

Printed from e-media with permission by:

Curran Associates, Inc.
57 Morehouse Lane
Red Hook, NY 12571

CURRAN ASSOCIATES INC. proceedings.com

Some format issues inherent in the e-media version may also appear in this print version.

Printed by Curran Associates, Inc. (2015)

For permission requests, please contact the Institute of Physics at the address below.

Institute of Physics
Dirac House, Temple Back
Bristol BS1 6BE UK

Phone: 44 1 17 929 7481
Fax: 44 1 17 920 0979

techtracking@iop.org

Additional copies of this publication are available from:

Curran Associates, Inc.
57 Morehouse Lane
Red Hook, NY 12571 USA
Phone: 845-758-0400
Fax: 845-758-2634
Email: curran@proceedings.com
Web: www.proceedings.com

Table of contents

Volume 4

XXIVTH CONFERENCE OF THE DANUBIAN COUNTRIES ON THE HYDROLOGICAL FORECASTING AND HYDROLOGICAL BASES OF WATER MANAGEMENT

2–4 June 2008, Bled, Slovenia

Accepted papers received: 7 November 2008
Published online: 10 December 2008

HYDROLOGICAL FORECASTING

012001

Online updating procedures for a real-time hydrological forecasting system B Kahl and H P Nachtnebel pg. 1

012002

Assessment of evapotranspiration and soil water content in the Kysuca River basin (Slovakia) using a rainfall-runoff model Romana Košková, Soňa Němečková and Zuzana Sitková pg. 9

012003

Forecasting analysis of runoff for reservoir regulation of dams and weirs in terms of hydro power plant operation Mariana Maradjieva and Nikola Nikolov pg. 23

012004

The implementation of the HBV model on the Sava River Basin Miha Primožič, Mira Kobold and Mitja Brilly pg. 36

012005

Continuous hydrological modelling in the context of real time flood forecasting in alpine Danube tributary catchments Ph Stanzel, B Kahl, U Haberl, M Herrnegger and H P Nachtnebel pg. 46

HYDRO-METEOROLOGICAL EXTREMES, FLOODS AND DROUGHTS

012006

The necessity of flood risk maps on Timiş River Geogr Cătălin Aldescu pg. 53

012007

Analyzing temporal changes in maximum runoff volume series of the Danube River Dana Halmova, Pavla Pekarova, Jan Pekar and Milan Onderka pg. 60

012008

Hydrological analysis of high waters and flash floods occurred in September 2007 in Slovenia M Kobold, M Sušnik, M Robič, F Ulaga and B Lalić pg. 68

012009

Precipitation intensity probability distribution modelling for hydrological and construction design purposes Georgy Koshinchanov and Dobri Dimitrov pg. 80

012010

Groundwater drought in different geological conditions A Machlica and M Stojkovova pg. 92

012011

Low flow analysis of the lower Drava River T Mijuskovic-Svetinovic and S Maricic pg. 101

012012

T-year maximum discharges on water courses in Slovakia J Podolinská and H Šipikalová pg. 112

012013

Multiple-coincidence of flood waves on the main river and its tributaries S Prohaska, A Ilic and B Majkic pg. 118

012014

Quality management for discharge measurements in Bavaria Carmen Roth pg. 126

012015

The survey of ecologically acceptable flows in Slovenia Nataša Smolar-Žvanut and Darko Burja pg. 132

012016

Modelling snow accumulation and snow melt in a continuous hydrological model for real-time flood forecasting Ph Stanzel, U Haberl and H P Nachtnebel pg. 139

012017

The main characteristics of the high water registered in the River Basin Bega in February 1999 Niculae Iulian Teodorescu pg. 146

012018

The minimum flow in the upstream basin of the Timis River Niculae Iulian Teodorescu pg. 153

GLOBAL CLIMATE CHANGE AND HYDROLOGICAL PROCESSES

012019

Proving the ecosystem value through hydrological modelling W Dorner, K Spachinger, M Porter and R Metzka pg. 161

012020

Water balance of Slovenia 1971 – 2000 P Frantar, M Dolinar and B Kurnik pg. 172

012021

Analysis of interrelation between water quality and hydrologic conditions on a small karst catchment area of sinking watercourse Trbuhovica V Hinić, J Rubinić, I Vučković, I Ružić, A Gržetić, G Volf, M Ljubotina and N Kvas pg. 181

012022

Hydrological scenarios of future seasonal runoff distribution in Central Slovakia K Hlavcová, J Szolgay, S Kohnová and G Bálint pg. 189

012023

Estimation of climate change impact on water resources by using Bilan water balance model Stanislav Horáček, Ladislav Kašpárek and Oldřich Novický pg. 198

012024

Slovenian experience in applicability of remote sensing data in hydrology A Horvat, A Vidmar, S Petan and M Brilly pg. 205

012025

The influence of Arctic and North Atlantic Oscillation on precipitation regime in Serbia Gordana Jovanović, Irini Reljin and Branimir Reljin pg. 217

012026

The effect of flow regulation on the distribution and dynamics of channel geomorphic units (CGUs) and implications for Marble Trout (*Salmo marmoratus*) spawning habitat in the Soča River, Slovenia Ian Maddock, Nataša Smolar-Žvanut and Graham Hill pg. 223

012027

Climate Change and water resources: Scenarios of low-flow conditions in the Upper Danube River Basin W Mauser, T Marke and S Stoeber pg. 233

012028

Detection of changes in flow variability of the upper Danube between 1876-2006 Pavla Pekarova, Peter Skoda, Pavol Miklanek, Dana Halmova and Jan Pekar pg. 244

012029

Influence of vegetation cover on air and soil temperatures in the Šumava Mts. (Czech Republic) M Tesař, M Šír, M Krejča and J Váchal pg. 253

012030

Trends of river discharges in Slovenia F Ulaga, M Kobold and P Frantar pg. 260

WATER MANAGEMENT

012031

Integrated modelling and the impacts of water management on land use W Dorner, K Spachinger and R Metzka pg. 273

012032

Influence of spreading urbanization in flood areas on flood damage in Slovenia B Komac, K Natek and M Zorn pg. 284

012033

Hydro power plants on the Middle Sava River section A Kryžanowski, A Horvat and M Brilly pg. 293

012034

Temperature and oxygen conditions peculiarities of the cooling ponds of nuclear and thermal power plants N Mostova pg. 298

012035

River archaeology – a new tool for historical hydrology Attila J Tóth pg. 302

012036

The UN Convention on International Watercourses and integrated water management: A bridge built Vasiliki-Maria Tzatzaki pg. 310

012037

Public or private water management: Experience from different European Countries Johann Wackerbauer pg. 315

FLOODS, MORPHOLOGICAL PROCESSES, EROSION, SEDIMENT TRANSPORT AND SEDIMENTATION

012038

Complex hydro- and sediment dynamics survey of two critical reaches on the Hungarian part of river Danube Sándor Baranya, László Goda, János Józsa and László Rákóczi pg. 324

012039

Torrent classification – Base of rational management of erosive regions Zoran Gavrilovic, Milutin Stefanovic, Irina Milovanovic, Jelena Cotric and Mileta Milojevic pg. 337

012040

Erosion control works and the intensity of soil erosion in the upper part of the river Toplica drainage basin S Kostadinov, N Dragović, M Zlatić and M Todosijević pg. 345

012041

Response of conservation measures from small cultivated watersheds, concerning runoff and erosion, under the impact of extreme rainfall events N Popa pg. 356

012042

Model based estimation of sediment erosion in groyne fields along the River Elbe Sandra Prohaska, Thomas Jancke and Bernhard Westrich pg. 361

012043

Flood Risk and Flood hazard maps – Visualisation of hydrological risks Karl Spachinger, Wolfgang Dorner, Rudolf Metzka, Kamal Serrhini and Sven Fuchs pg. 372

012044

Variability of bed sediments in channel network of Rye Island (Slovakia) Y Velísková and R Dulovičová pg. 389

012045

Interrill soil erosion on flysch soil under different types of land use in Slovenian Istria M Zorn and S Petan pg. 395

DEVELOPMENTS IN HYDROLOGY

012046

Automatic calibration of the Watbal model A Cotar and M Brilly pg. 404

012047

Flood warning level forecasting for ungauged catchments by means of a combined API storage concept T Lehmann and H Holzmann pg. 412

012048

WaterNet: The NASA water cycle solutions network – Danubian regional applications Dave Matthews, Mitja Brilly, Mira Kobold, Mark Zagar and Paul Houser pg. 422

012049

Hydrological and seasonal export mechanisms for nitrate transport from a forested catchment S Rusjan, M Mikoš and M Brilly pg. 441

012050

Measuring and modelling forest transpiration M Šír, J Čermák, N Nadezhdina, J Pražák and M Tesař pg. 449

012051

The experimental watersheds in Slovenia M Šraj, S Rusjan, S Petan, A Vidmar, M Mikoš, L Globevnik and M Brilly pg. 454

IOP Conf. Series: Earth and Environmental Science **4** (2008) 012001 doi:10.1088/1755-1307/4/1/012001

Online updating procedures for a real-time hydrological forecasting system

B Kahl and H P Nachtnebel

Institute of Water Management, Hydrology and Hydraulic Engineering, University of Natural Resources and Applied Life Sciences, Muthgasse 18, 1190 Vienna, Austria

E-mail: bianca.kahl@boku.ac.at

Abstract. Rainfall-runoff-models can explain major parts of the natural runoff pattern but never simulate the observed hydrograph exactly. Reasons for errors are various sources of uncertainties embedded in the model forecasting system. Errors are due to measurement errors, the selected time period for calibration and validation, the parametric uncertainty and the model imprecision. In on-line forecasting systems forecasted input data is used which additionally generates a major uncertainty for the hydrological forecasting system. Techniques for partially compensating these uncertainties are investigated in the recent study in a medium sized catchment in the Austrian part of the Danube basin. The catchment area is about 1000 km². The forecasting system consists of a semi-distributed continuous rainfall-runoff model that uses quantitative precipitation and temperature forecasts. To provide adequate system states at the beginning of the forecasting period continuous simulation is required, especially in winter. In this study two online updating methods are used and combined for enhancing the runoff forecasts. The first method is used for updating the system states at the beginning of the forecasting period by changing the precipitation input. The second method is an autoregressive error model, which is used to eliminate systematic errors in the model output. In combination those two methods work together well as each method is more effective in different runoff situations.

1. Introduction

When forecasting runoff, sources of errors are numerous. Errors are due to measurement errors, the selected time period for calibration and validation, the parametric uncertainty and the model imprecision. In early warning systems forecasted input data is used for extending lead time. This additionally generates a major uncertainty for the hydrological forecasting system. Therefore the simulated and forecasted hydrographs will never fit perfectly to the measurements. To partially compensate these uncertainties different techniques have been elaborated that try to reduce the different sources of errors in the system. A main target of most methods is analyzing and minimizing errors in historic simulation. Errors in the recent past can influence the forecast negatively because of the inertia of the hydrologic system.

Continuous modelling can identify wrong tendencies in system states and is of specific importance in alpine regions, where snow melt is a major source of flood events. System states are an important input for a continuous rainfall-runoff-model and contribute a lot to forecasted hydrographs. Although optimized system states give a better fitting in historic simulations it is not assured that the forecast

IOP Conf. Series: Earth and Environmental Science **4** (2008) 012001 doi:10.1088/1755-1307/4/1/012001

connects directly to the observed hydrograph. For a proper visualization still an output correction is needed.

Two updating methods have been tested in the Traisen basin, an alpine tributary to the Danube with an area of about 1000 km². The catchment is prone to flash floods especially from late spring to early autumn and regularly shows snow melt events in early spring. Within the early warning system hydrological modelling is operated with the continuous, semi-distributed rainfall-runoff model COSERO, developed by the Institute of Water Management, Hydrology and Hydraulic Engineering by [1] and [2] among many others. It accounts for processes of snow accumulation and melt, interception, evapotranspiration, infiltration, soil storage, runoff generation and routing. Separation of runoff into fast surface runoff, inter flow and base flow is calculated by means of a cascade of linear and non-linear reservoirs, following the design of the HBV model [3]. Spatial discretization of the catchment relies on the division of the watersheds into subbasins and subsequently into hydrologic response units (HRUs) based on available spatial information on subcatchment boundaries, soil types and land cover, and 200-400 m elevation bands.

2. Forecasting and Updating

In operational systems with a continuous model a good historic simulation is needed to build up reliability into the hydrologic model. It also shows that hydrological processes are simulated in a proper way. Error sources are reduced in good simulations and make the updating process more efficient. A good calibration therefore is required. In this study meteorological input raster data from the Austrian Central Institute for Meteorology and Geodynamics (ZAMG) was used for calibrating and validating the model. A period from 2002 until 2006 was chosen. Precipitation and temperature analyses with a spatial resolution of 1x1 km and a temporal resolution of 15 minutes incorporate point ground measurements and radar information [4] were available. Operationally quantitative precipitation and temperature forecasts with the same source as the analysis data is used. The forecasts consist of two components. The first component is an observation-based extrapolation (nowcasting) of the precipitation field (INCA – [4]). The second component is a weighted mean of the forecasted fields of the ALADIN and ECMWF numerical weather prediction models.

Online, each forecasting sequence in the hydrological model starts with the re-calculation of the last 24 to 48 hours, using the above described ZAMG precipitation and temperature analysis grids. Thus the system states at the beginning of the prediction period are determined. For the hydrological forecast quantitative 48 hour forecasts of precipitation and temperature are fed into the hydrological model. As the uncertainty of forecasted rainfall can be high, ensemble precipitation forecasts are used to estimate meteorological uncertainty of the system. The forecasting sequences are repeated each 15 minutes. Results and evaluation of ensemble forecasts are shown in more detail in another contribution to the conference [5].

Depending on the method, the correction is applied at different points in the time flow of a forecasting system. Two major groups can be classified (Figure 1): corrections in the historic simulation (updating) and in the forecasted simulation (output correction). The two time periods are divided by t_0 (date of the forecast). Updating can further be divided into permanent and temporal modifications. Permanent updating intervenes in the long term water balance and influences the system states slowly with time. Temporal updating can take stronger influence on the system and react faster to the recent situation since resulting system states are not stored in the system for a long time. This contribution concentrates on a temporal updating method and a method for output correction. The schematic use of the presented methods can be seen in Figure 1. The first method starts with correcting the input of the last 48 to 24 hours and results in updated system states at t_0 (denominated simulation with updating in Figure 1). The correction of the output (denominated forecast with output correction in Figure 1) is done with an autoregressive (AR) model. Both updating methods are presented in the following passages in more detail.

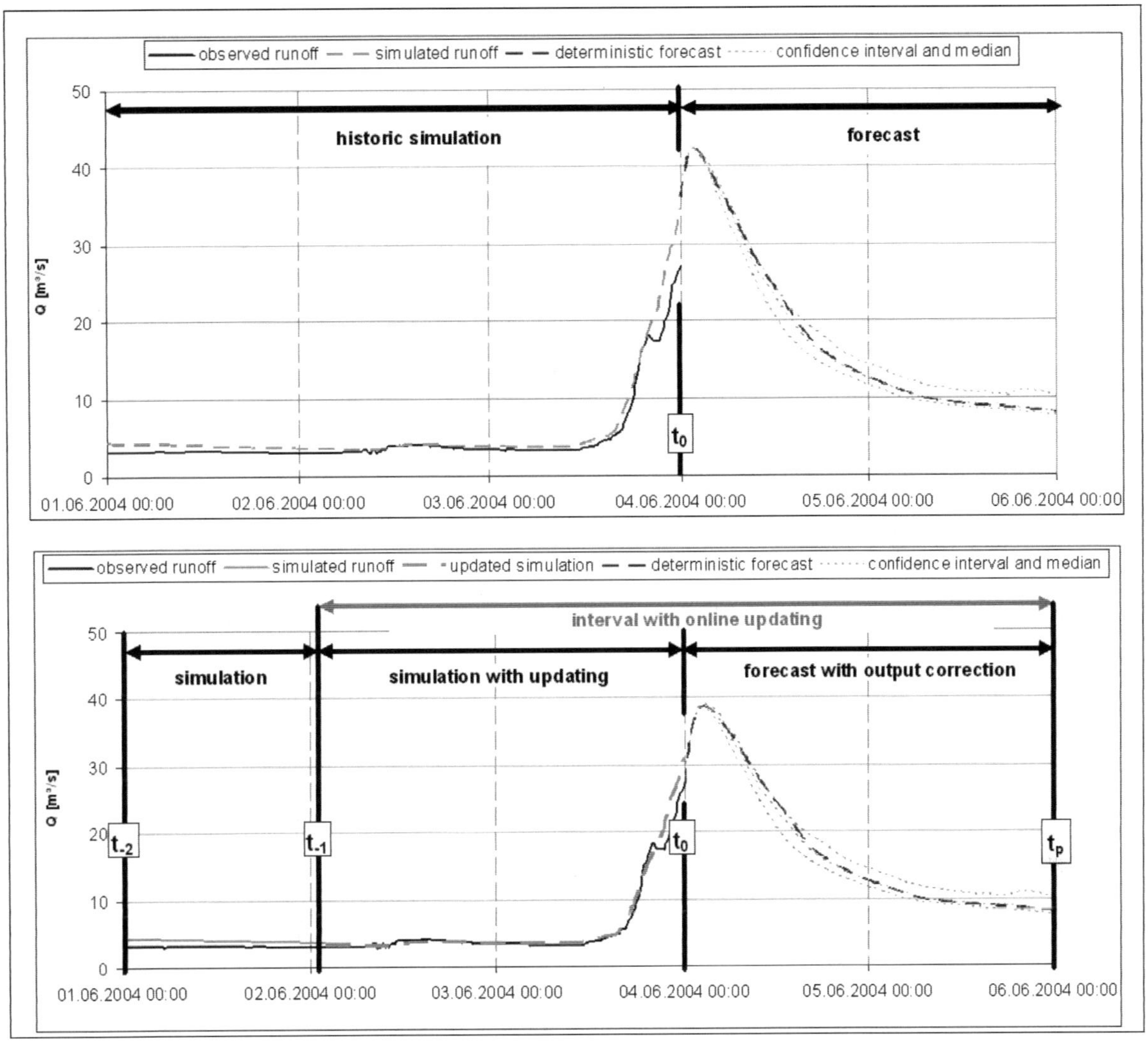

Figure 1. Time flow in a forecasting system (top) Application and time flow of online correction methods used in this paper (bottom)

3. Updating system states with the input

A good forecast needs an optimal state of the system at t_0 (date of forecast). Therefore recalculating the recent history until the beginning of the forecast with some kind of updating mode is essential for a forecasting system. As sources of error are various it is hard to separate the error of the output to all of its sources. Assigning the error of the output to only one source can be problematically and often has negative effects on the water balance. But solving the problems associated with multiple updating [6] is still complex and needs more investigations. The choice of the right updating mode must always be started with analysing the mean sources of errors in the system. In the recent study the precipitation input with its quantity and spatial distribution was seen as the mayor uncertainty in the system. Changing precipitation input therefore seems reasonable. And this is the point were the updating of the system states with the correcting the input attaches. The correction of the input has the primary target to improve system states in a way that the resulting runoff fits better to the measurements. In this study the updated system states are only valid for a short time period and do not influence the long-term water balance of the hydrological model.

3.1. Method

The main target of this updating method is to decrease the deviation of simulated to observed runoff in the recent past of the forecasting run. Therefore a spatial linearly varying correction factor is applied to the precipitation raster. The parameters of the correction plane are adapted with an optimization algorithm. The objective function (OF) is dependent on the deviation of simulated and observed runoff and obtained to reach a minimum. The application flow is as following:

1. rainfall-runoff-modelling without correction
2. calculation of start value of OF
3. adaptation of correction plane
4. re-modelling with corrected precipitation
5. re-calculation of OF and comparison to best value of OF
6. repetition of point 3 to 5 until an optimum of the Of is reached
7. rainfall-runoff-modelling with optimum of corrected precipitation
8. output of updated system states
9. start of forecast with updated system states at t_0

Changing the precipitation input happens with a correction factor. The factor can vary spatially and every raster element has its own correction factor. For keeping the factor related to neighbouring elements a linear relationship in form of a correction plane is defined. Another reason for working with a correction plane is that it is defined by only three parameters. The fewer parameters have to be optimized the more efficient an optimization routine can work. The correction plane (*KFAK*) is calculated with (1). And the corrected precipitation (*PCOR*) forms from precipitation input (*P*) for each time step (*t*) and *KFAK* with equation (2).

$$KFAK_{(nrow,ncol)} = A + B \cdot nrow + C \cdot ncol \tag{1}$$

$$PCOR_{(nrow,ncol,t)} = P_{(nrow,ncol,t)} \cdot KFAK_{(nrow,ncol)} \tag{2}$$

In equations (1) and (2) *nrow* denominates the row number and *ncol* the column number of the raster. The letters *A*, *B* and *C* are the parameters of the correction plane. Parameter *A* is responsible for the vertical shift. The parameters *B* and *C* effect the inclination of the correction plane. If there is no correction applied the correction plane is horizontal with the parameters: A=1, B=C=0.

In Figure 2 a scheme of the input correction with a correction plane is shown. The parameters of the shown correction plane are: A=-1.1167, B=0.037, C=-0.0072. That means a decrease of the precipitation in the north-eastern part of the raster and more precipitation for the south-western part of the catchment.

The optimization of the parameters for the correction plane is done with a Rosenbrock-method [7]. The method of Rosenbrock is similar to the method of steepest gradients and works faster in finding an optimum than the discrete step method. A drawback of the Rosenbrock method (and all other gradient methods) is that it is dependent on the initial values of the parameters. This means that it can occur that the optimization program finds a parameters set that is located at a local optimum and not a global optimum.

Essential for the result of the optimization is the objective function. Due to the fact that simulated runoff needs to be fit to more than one gauging station all deviations have to be taken into account in the objective function. Deviations of simulated runoff to the corresponding measurements should be minimized. Considering the different magnitude of discharge within the catchment, values of deviation get standardizes with the one year flood. The standardization effects that all gauging stations have the same weight in the objective function. The objective function (*OF*) for one gauging station is shown in (3). Where ε is representing the deviation between simulated and observed runoff and HQ_1 is the discharge value of the one year flood at this gauging station. The sum includes all observed and simulated time steps.

$$OF = \sum_{t=1}^{n} \left(\frac{\varepsilon}{HQ_1} \right)^2 \tag{3}$$

The definite OF is a combination of equation (3) for every observed gauging stations in the updated time period. In that way precipitation distribution is corrected considering all areas of the catchment with the same weight.

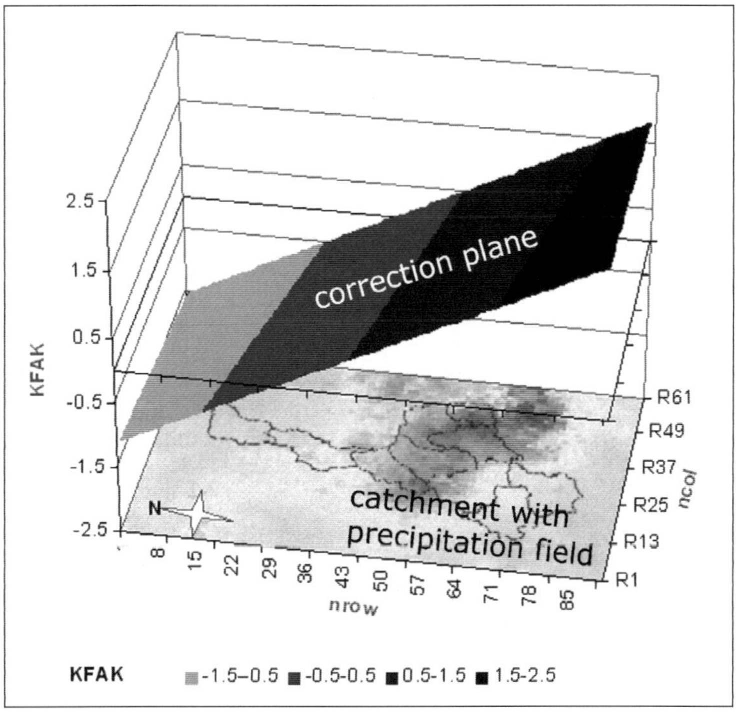

Figure 2. Scheme of input correction. The precipitation field gets multiplied with a correction plane (KFAK).

3.2. Application

This method was developed for correcting system states especially before and during flood events but it is not efficient in medium and low flow situations. Updating the system states with this method of input correction is only useful with occurring rainfall. Otherwise no correction of precipitation can be made with a factor.

The method had been tested for several flood events in post-processing for the whole event. Also real time operation was simulated by applying the method to different time periods before and during events. For testing the method only meteorological analysis data were used but no forecasts. Analysis data were assumed to be the perfect forecast to test only the method of the hydrological updating. In Figure 3 the effectivity of the method is shown in post processing for a whole flood event (dark grey solid line). This is only for theoretical analysing how the method works on different types of events. For operational use only the effect on the system states and the forecast during an event is of special interest. Two different forecast dates are also shown in Figure 3, where the vertical lines marked with t_1 and t_2 are the start dates of the forecast simulations with updated system states. Until t_1 and t_2 the updating mode was used, after this dates the model was running in forecast mode.

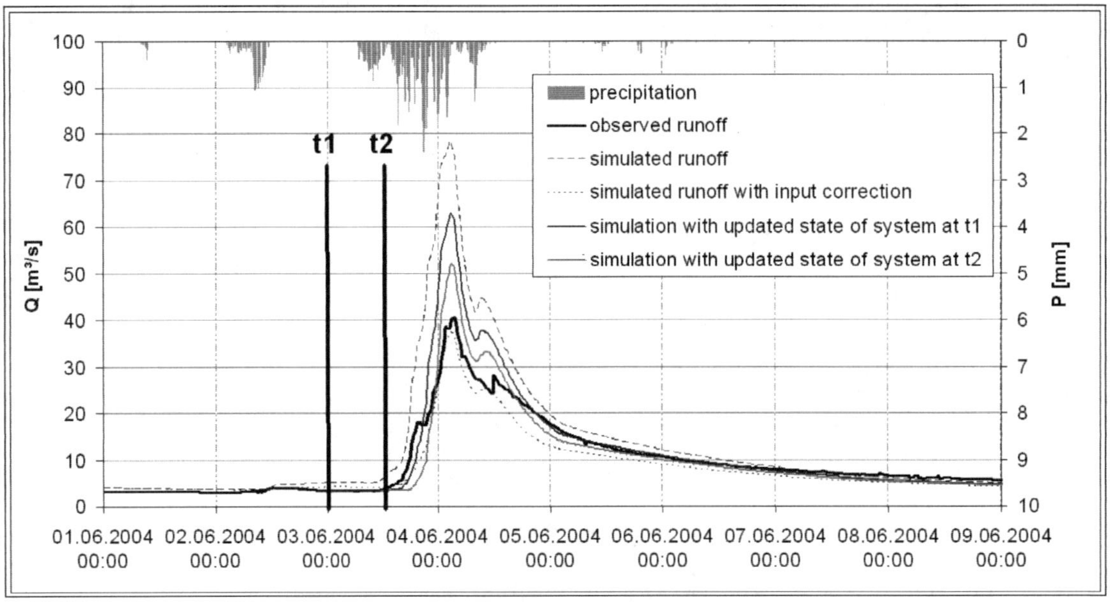

Figure 3. Input correction applied to a whole event in post-processing and simulating forecasting mode for dates t_1 and t_2.

As seen in Figure 3 updated system states show a positive effect on the following forecast run. Results of the updating procedure show a better fitting of the forecasts to the measurement. The closer the flood event comes the better the fitting gets. A reason is that already more precipitation has come to discharge and can be corrected with the updating procedure.

4. Output correction with an AR-model

Correcting the output of a model means correcting an error without dealing with the sources of the error. However, output correction can be very effective in simulating the forecast errors. In discharge modelling future errors of the output are often depended on already occurred errors when comparing simulations with observed data from the past. These systematic errors can easily be eliminated with a simple autoregressive method applied to the output of the model. Especially in the first few hours of forecasting this method is useful to build a connection between observed and forecasted hydrograph.

4.1. Method

Based on the most widely used single autoregressive (AR) model a method with implicit discharge thresholds is applied. The prospective error is dependent on the last observed error and the corresponding discharge level and is calculated with equation (4).

$$\varepsilon(t) = \varepsilon(t_0) \cdot R_Q^t + \varphi \tag{4}$$

Where $\varepsilon(t)$ is the estimate of the prospected error, $\varepsilon(t_0)$ is the simulation error at t_0 or the last observed error and R_Q is the regression coefficient. Here R_Q is dependent on the corresponding discharge. Analysing different discharge situations have shown a stronger error-autocorrelation at low and medium flow than in high flow situations.

IOP Conf. Series: Earth and Environmental Science **4** (2008) 012001 doi:10.1088/1755-1307/4/1/012001

4.2. Application

Calculation of the regression coefficient is done for every discharge class separately. Discharge is divided into eight classes geared on statistical flood events for each river gauge. Calibrating the AR-model resulted in values for the coefficient R_Q from 1 for low flow situation, which means a parallel dislocation, to 0.9 during high flood events. The regression coefficient decreases linearly with the discharge class. If no observation is available for the last few hours, R_Q is set to 0 and no output correction is calculated.

In Figure 4 an application example of updating with the AR-threshold-model at a flood even is shown. Obviously at higher discharge corrected simulation show only at the beginning of a forecast a strong influence of past errors but returns fast to the simulation with time. At low flows corrected simulations are longer influenced by the past error and are almost shifted parallel through the point of the last observation.

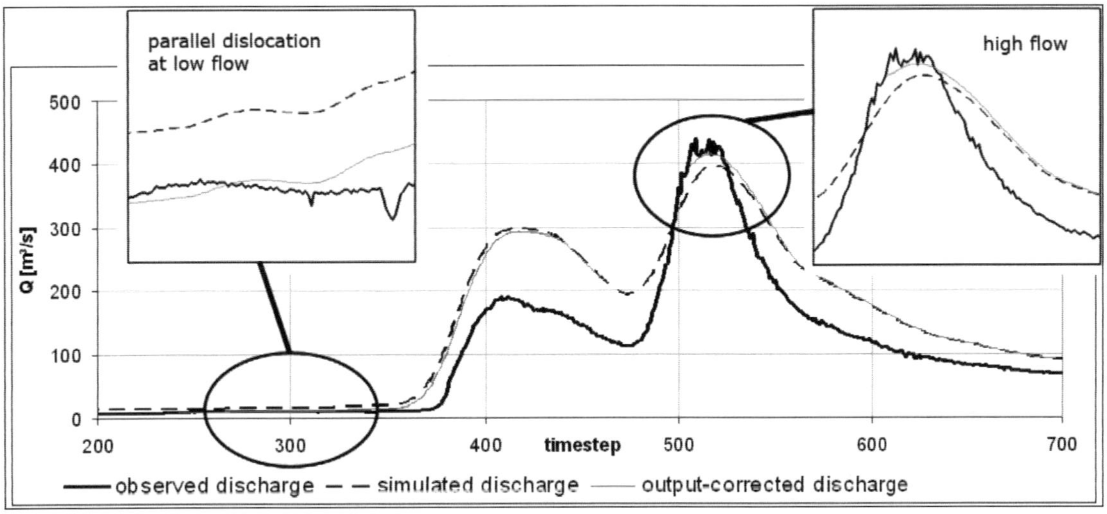

Figure 4. Examples of AR corrected forecasts with discharge level dependent coefficiant

5. Conclusions

In combination the two tested methods work well since both act at different magnitudes of discharge. Applying only the updating method gave unsatisfying results at low flows because this method only works in time periods with precipitation. When working only with the method of output correction to little adjustment happened during flood events because the major source of error in these situations, the precipitation input, is not corrected. However, past experiences have shown that in real-time flow forecasting, the simple autoregressive (AR) model is very efficient in simulating the forecast errors from the substantive model operating in the simulation mode [8]. Analysing data showed a dependency of regression coefficient with magnitude of discharge. This results where applied in the AR-threshold-model. In low flow situations a parallel dislocation of the simulation through the last observed value can be obtained since hydrologic dynamic is simulated well with the model. In flood situations the regression coefficient (R) can be reduced because time shifts between simulation and observations could eliminate too much information when R is too high. The updating method for adjusting the input with a correction plane is a simple method and seems a good basis for further investigations. Advantages and adaptations are still needed especially when simulating snow that brings delays in the hydrologic response of the system.

The study shows also the need for multiple updating since one method cannot deal with all sources of errors.

Acknowledgements
The presented work has been carried out in the framework of a project funded by the provincial government of Lower Austria. Data was provided by this institution and the Central Institute for Meteorology and Geodynamics (ZAMG)

References
[1] Kling H 2002 *Developement of Tools for a Semi-Distributed Runoff Model* (Vienna: master thesis at IWHW/BOKU)
[2] Nachtnebel H P, Lettl W and Baumung S 1993 *Abflussprognosemodell für das Einzugsgebiet der Enns und Steyr - Endbericht im Auftrag der EKW* (Vienna: Institut für Wasserwirtschaft, Hydrologie u. konstruktiven Wasserbau)
[3] Bergström S 1995 The HBV-Model in *Computer Models of Watershed hydrology ISBN 0-918334-91-8* Singh V P (Colorado: Water Resources Publications) 443-476
[4] Haiden T, Kann A, Stadlbacher K, Steinheimer M and Wittmann C 2007 *Integrated Nowcasting through Comprehensive Analysis (INCA) – System overview* (Vienna: ZAMG)
[5] Stanzel P, Kahl B, Haberl U, Herrnegger M and Nachtnebel H P 2008 Continuous hydrological modelling in the context of real time flood forecasting in alpine Danube tributary catchments *IOP Conference Series: Earth and Environmental Science - XXIVTH CONFERENCE OF THE DANUBIAN COUNTRIES ON THE HYDROLOGICAL FORECASTING AND HYDROLOGICAL BASES OF WATER MANAGEMENT*
[6] Messal H E E 2000 *Rückkopplungen und Rückwirkungen in der hydrologischen Modellierung am Beispiel von kontinuierlichen Niederschlags-Abfluß-Simulationen und Hochwasservorhersagen* (Berlin: PhD thesis at Technischen Universität Berlin) Mitteilungen aus dem Institut für Wasserbau und Wasserwirtschaft – Technische Universität Berlin 135
[7] Rosenbrock H H 1960 An Automated Method for finding the Greatest or Least Value of a Function *The Computer Journal* **3**
[8] Xiong L, O'Connor K 2002 Comparison of four updating models for real-time river flow forecasting *Hydrological Sciences Journal* **47** 621-639

IOP Conf. Series: Earth and Environmental Science **4** (2008) 012002 doi:10.1088/1755-1307/4/1/012002

Assessment of evapotranspiration and soil water content in the Kysuca River basin (Slovakia) using a rainfall-runoff model

Romana Košková[1], Soňa Němečková[1], and Zuzana Sitková[2]

[1] The Institute of Hydrodynamics, AS CR, v.v.i. Pod Patankou 5, 166 12 Prague, Czech Republic

[2] The National Forest Centre T. G. Masaryka st. 22, 960 92 Zvolen, Slovakia

E-mail: koskova@ih.cas.cz, nemeckova@ih.cas.cz, sitkova@nlcsk.org

Abstract. Rainfall-runoff modelling is a commonly-used tool for analysis of various aspects of hydrological research. An application of the distributed hydrological model brings the benefits of assessment to the spatially distributed results. Description of the spatial distribution of soil and vegetation types is an essential task and substantially influences the accuracy of the results. The Kysuca River basin has been chosen for an evaluation of soil water content and evapotranspiration by a distributed hydrological model, SWIM (Soil and Water Integrated Model). The primary interest in this analysis was motivated by assessment of forest health conditions in the area and finding possible causes of forest deterioration. Evapotranspiration and soil water content in two selected soil layers in each hydrotop have been evaluated in a simulation period, 1995-1999. A set of maps each representing an individual day of the simulation period has been obtained as a result of the simulation for each variable mentioned above. The simulation has also been carried out for 2007, when the results for selected hydrotops were statistically processed and compared with the available observed data measured during the field survey. Nevertheless, the comparison is affected by the fact that the tensometers have not yet been calibrated to the local conditions. The resulting uncertainties had to be taken into account. The main trends of the simulated and measured soil water content have been evaluated and assessed as satisfactory.

1. Introduction

Rainfall-runoff modelling is a commonly-used tool for analysis of various aspects in hydrological research, recently very often related to global climate change. This article deals with modelling of evapotranspiration and soil water content in different layers in the mesoscale Kysuca River basin (1016 km^2). Soil water conditions are one of the main factors determining conditions of vegetation cover. The simulations were carried out as part of a larger study about the role of different aspects (biotic, abiotic and antropogenic factors) in the health of vegetation cover and its revitalisation. It is typical of the Czech and Slovak Republic that mountains are covered by pine-woods which are not the original ecosystems. The aim of this article is to describe the soil water conditions and evapotranspiration in the mentioned river basin using SWIM (Soil and Water Integrated Model).

2. Hydrological model SWIM

The SWIM model has been developed in the PIK (Potsdam Institute for Climatic Research, Germany). Two different models, SWAT and MATSALU, have been combined into one unified system known as SWIM [1]. SWIM is a continuous-time distributed simulation watershed model. It was developed to predict the effects of the alternative management decisions on water, sediment, and chemical yields with reasonable accuracy for the ungauged rural basins. It was motivated by water resources management in densely populated agricultural areas (water pollution problem), arid and semi-arid regions (water scarcity), and mountainous and loess regions (erosion problem). The other motivation is an ongoing climate change and land use/land cover change. Development of water resources in the conditions of global change requires an understanding and adequate (mesoscale and large-scale) representation in the models of the basic hydrological and related processes, i.e. the river basins of hundreds, thousands or tens of thousands of square kilometres. The SWIM model was developed in order to provide a comprehensive GIS-based tool for hydrological and water quality modelling in the mesoscale and large river basins, which can be parameterised with regionally available information. The model was mainly intended for use in Europe and temperate zones, though its application in other regions is possible as well. The model integrates hydrology, vegetation, erosion and nutrient dynamics at the basin scale. SWIM has a three-tier disaggregation scheme, 'basin – sub-basins – hydrotopes', and is coupled with the Geographic Information System GRASS [2]. Model test and validation were performed sequentially for hydrology, crop growth, nitrogen and erosion in a number of the mesoscale basins in the German part of the Elbe drainage basin. A comprehensive scheme of spatial disaggregation into sub-basins and hydrotopes combined with reasonable restriction on a sub-basin area allows assessment of water resources and water quality with SWIM in the mesoscale river basins to be performed. The hydrological model is based on water balance in several layers of soil column that includes precipitation, evapotranspiration, percolation, surface runoff and subsurface runoff (figure 1).

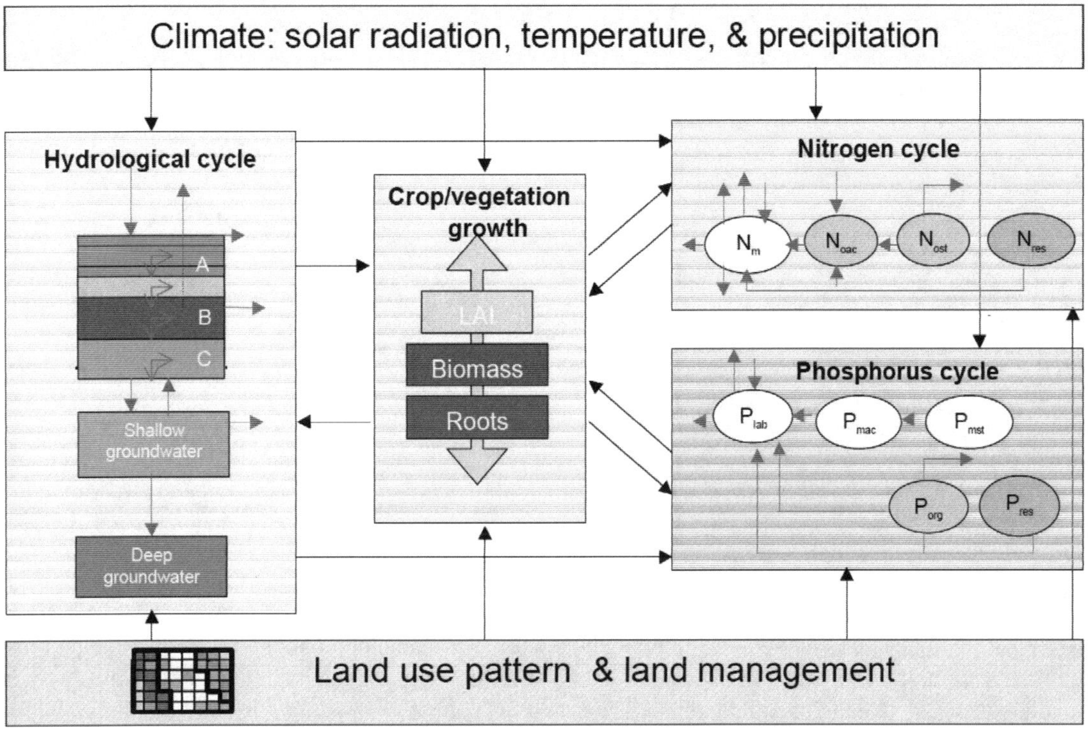

Figure 1. Flow chart of the SWIM model [1]

2.1. Evapotranspiration calculation

An expression of evapotranspiration in the model is based on the calculation of evaporation from soils and transpiration by plants separately, according to Ritchie [3]. The plant transpiration is calculated with the value of potential evapotranspiration and *LAI* (leaf area index). If soil water is limited, plant water transpiration is reduced.

$$EP = \frac{EO.LAI}{3} \text{ if } 0 \leq LAI \leq 3 \quad \text{ or } EP = EO \text{ if } LAI \geq 3 \tag{1}$$

EO the potential evapotranspiration [mm.d^{-1}],
EP the plant transpiration [mm.d^{-1}],
LAI the area of plant leaves relative to a soil surface area.

Potential soil evaporation from a soil surface *ESO* [mm.d^{-1}] is simulated by an exponential function of the leaf area index according to the equation of Richardson and Ritchie [4]:

$$ESO = EO.\exp(-0.4.LAI) \tag{2}$$

Actual evaporation from a soil surface is calculated in two stages. In the first stage only the energy available at the surface limits the actual soil evaporation, which is equal to the potential soil evaporation in this case. When the accumulated soil evaporation exceeds the first-stage threshold (equal to 6 mm), the second stage begins after the following expression:

$$ES = 3.5.(\sqrt{TST} - \sqrt{TST - 1}) \tag{3}$$

ES soil evaporation from a soil surface for day *t* [mm.d^{-1}],
TST number of days since stage-two evaporation began.

Actual soil water evaporation is estimated on the basis of Priestley-Taylor method from the top 30 cm of soil and snow cover, if any [5]. If the water content of snow cover is greater than or equal to *ES*, soil evaporation comes from the snow cover. If *ES* exceeds the water content of the snow cover, water is removed from the upper soil layers if available.

Potential evapotranspiration is estimated using this method or that of Penman-Monteith [6], according to data availability. The first method requires only solar radiation, air temperature and elevation as inputs. The latter requires solar radiation, air temperature, wind speed and relative humidity as inputs. In this study, the Penman-Monteith method has been used.

The SWIM model provides simulation of vegetation growth in a year using the EPIC method according to Williams [7]. In this approach the *LAI* is simulated as a function of heat unit and biomass in a year [1].

2.2. Soil water content calculation

The computation of soil water content in the hydrological sub-model in SWIM is based on the following water balance equation:

$$SW_{(t+1)} = SW_{(t)} + PRECIP - Q - ET - PERC - SSF \tag{4}$$

SW(t) the soil water content in the day *t*,
PRECIP the precipitation,
Q the surface runoff,
ET the evapotranspiration,
PERC the percolation,
SSF the subsurface flow.

The surface runoff volume is estimated by using a modification of the Soil Conservation Service (SCS) curve number technique [8, 9]. The method of evapotranspiration computation is described briefly below. The Subsurface Flow is modelled according to Sloan and colleagues [10].

The computation of the soil moisture in a soil profile is done with daily steps for each soil layer in the soil profile. A storage routing technique is used [11]. The method is based on evaluation of daily percolation in the soil layer and the soil water content at the beginning and end of the day. The percolation from the bottom soil layer is treated as recharge to the shallow aquifer. The storage routing technique is based on the equation:

$$SW_{(t+1)} = SW_{(t)}.\exp\left(\frac{-\Delta t}{TT}\right) \tag{5}$$

SW(t+1); SW(t) the soil water contents at the beginning and end of the day [mm],
Δt the time interval (24 h),
TT the travel time through layer [*h*].

The computation follows by determination of the percolation by subtracting SW_t from SW_{t+1} in each soil layer:

$$PERC = SW\left[1 - \exp\left(\frac{-\Delta t}{TT}\right)\right] \tag{6}$$

PERC the percolation rate in each soil layer in [$mm.d^{-1}$],
TT the travel time is calculated for each soil layer with the linear storage equation:

$$TT = \frac{SW - FC}{HC} \tag{7}$$

HC the hydraulic conductivity in [$mm.h^{-1}$]
FC the field capacity water content for layer [mm].

The hydraulic conductivity is given by the equation:

$$HC = SC.\left(\frac{SW}{UL}\right)^{\beta} \tag{8}$$

SC the saturated conductivity for layer in [$mm.h^{-1}$],
UL the water content at saturation in [$mm.mm^{-1}$],
β the shape parameter given by the expression:

$$\beta = \frac{-2.655}{\log_{10}\left(\frac{FC}{UL}\right)} \qquad (9)$$

Water flow through a soil layer may occur until the lower layer is not saturated. If the layer below the layer being considered is saturated, no flow can occur. In this case the percolation is further corrected.

3. The Kysuca River basin

3.1. The river basin description

The Kysuca River basin is located in north-western Slovakia and covers an area of 1016 km^2. The Kysuca River reaches the Vah River near the town Zilina and its length is 66.3 km. The region is rather mountainous with low population density. A large part of the river basin (654 km^2) is protected within a natural reservation, the Kysuce. About half of the basin is covered by forest (45 % - coniferous; 12 % - mixed; 43 % - deciduous). The extent of original fir and beech forest is limited. Dominant ecosystems are monocultures of spruce wood of antropogenic origin, which are less stable than the natural ones.

The geology is determined by the Carpathian flysh formation. It is a specific formation consisting of alternating layers of hard sandstones and soft shales. Because of these conditions there is a lack of groundwater storage. Therefore the spring discharges are low and unsteady. The most extended soil type in the basin is cambisol.

3.2. Input data for modelling

The distributed model processes a huge amount of data, firstly spatial data describing the river basin, and secondly the time series of climatic data and discharge at the outlet of the river basin as well.

Four layers of spatial data for the SWIM model are required: a river network and sub-basins, digital elevation model, soil types and land cover. The hydrological model SWIM simulates water conditions in several layers of a soil column that is why detailed soil data are required. Nine soil types (four agricultural and five forest) and eleven vegetation types of land cover in the Kysuca River basin were defined. The soil types were derived according to their physical characteristics from the Slovakian land valuation database (BPEJ) in the case of agricultural soils and from the database of forest land resources in the case of the forest soils (figure 2, figure 3 and figure 8).

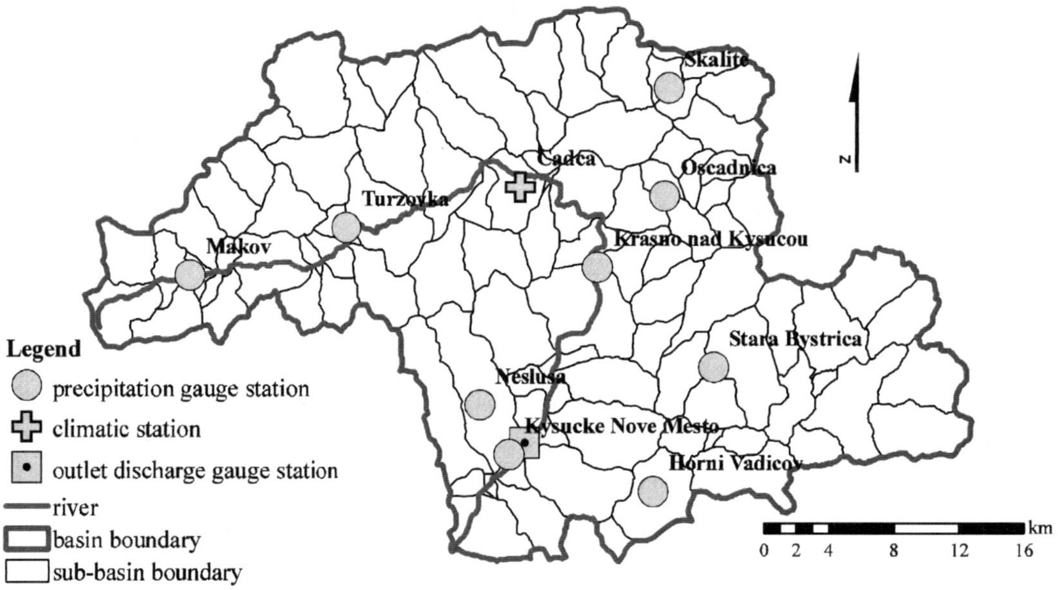

Figure 2. Sub-basins and gauge stations in the Kysuca River basin

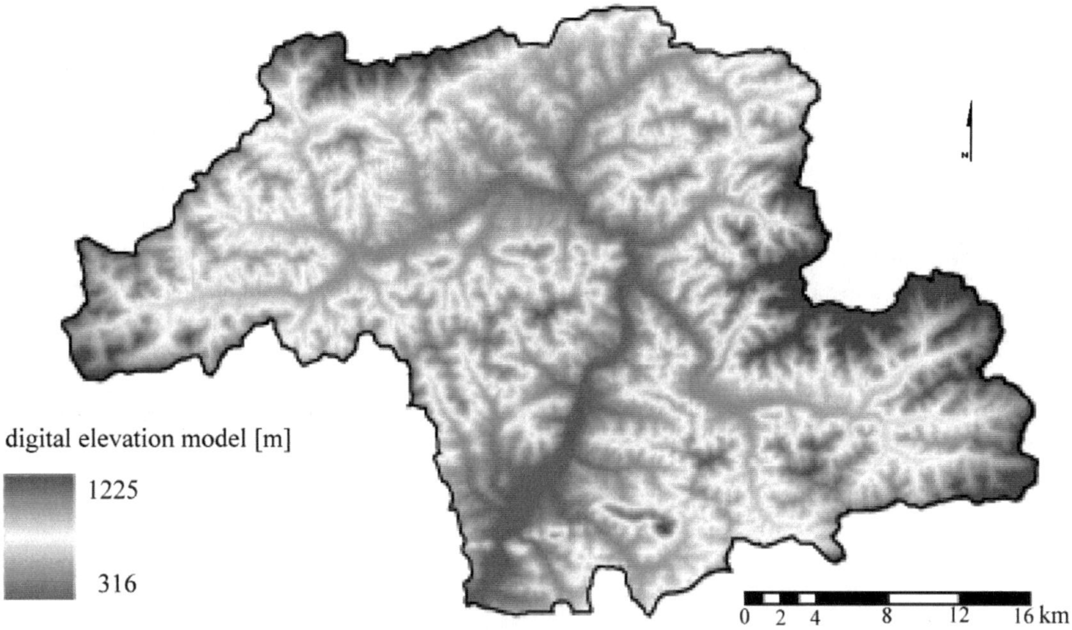

Figure 3. Digital elevation model of the Kysuca River basin

These geographic data were handled by GIS technology. Finally all the layers were converted into raster forms and subsequently into a text format that the SWIM required as input. During the GIS process hydrotops were defined by overlaying the four input layers of spatial data.

The climate data necessary to drive the model are precipitation, minimum, maximum and average temperature and solar radiation. The model requires the input of the meteorological data mentioned above for each sub-basin in a daily step. Nevertheless, there are only nine precipitation gauge stations providing rainfall data and one climatic station (Čadca) providing temperature, radiation and air humidity measurement, as well as rainfall data. Therefore it was necessary to interpolate the variables over sub-basins.

The model was calibrated and validated taking into account discharge at the outlet (Kysucké Nové Mesto) of the Kysuca basin.

4. Simulations and results

4.1. Calibration of the model

The model was calibrated and validated between 1995 and 1999. The simulated and observed discharges at the outlet gauge station were compared (figure 4). Nash Sutcliffe efficiency was 0.82 per calibration period and 0.80 per validation period. The water balance in the calibration period was 5 % and in the validation period 6 %. The output of the model could be described as satisfactory.

The model represents low flows better. There are some differences in high flows, especially in extreme peaks. The discrepancy could be connected with the problems of soil characteristic determination for the model. Field works have proved there are a huge number of rock fragments (skeleton) in the soil, which has been not included in the model soil parameterisation. There is no model parameter that can describe the amount of these fragments in the soil. The model could therefore underestimate high flow and overestimate soil water capacity owing to the overestimation of the real soil depth.

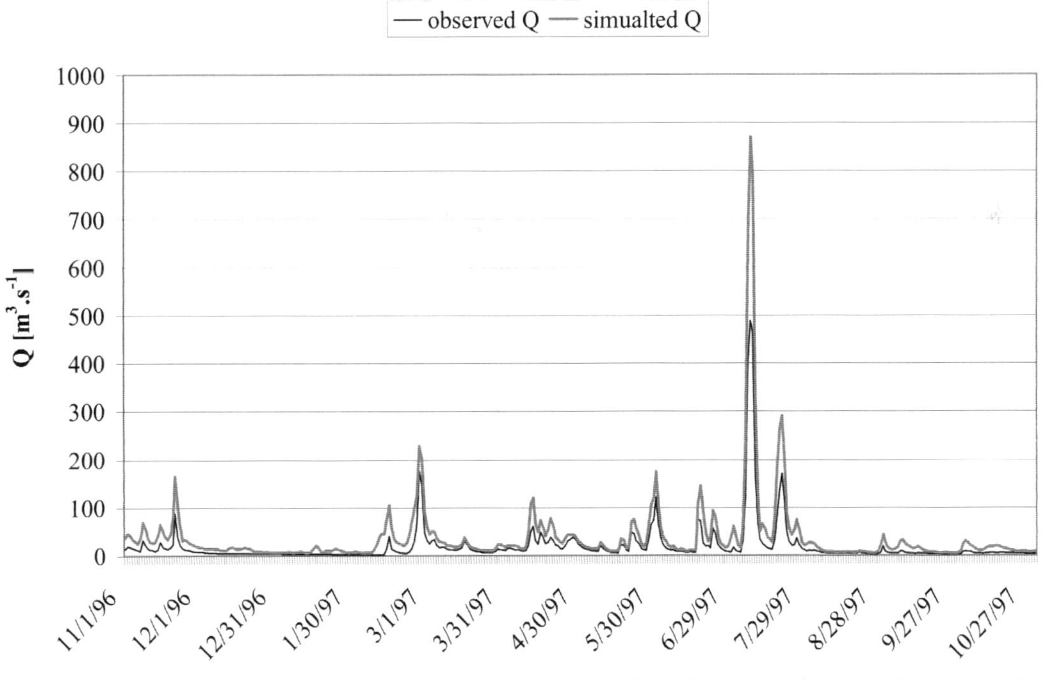

Figure 4. Comparison of simulated and observed discharge in daily step at the Kysucke Nove Mesto

4.2. Soil water content simulations

Soil water content is not a regularly measured physical quantity. Nevertheless, it is one of the limiting factors of the growth and health of vegetation cover. The main task is to assess the accuracy of the SWIM model for soil moisture simulation. The simulation should help to explain the relation between them in any given area during a selected time period. If it is proved, it will be possible to simulate soil water content and other quantities under changing conditions (climate change or vegetation cover change.)

IOP Conf. Series: Earth and Environmental Science **4** (2008) 012002 doi:10.1088/1755-1307/4/1/012002

The plausibility of soil water simulation results was evaluated by comparison with observed values of the volumetric water content using statistical methods. The monitoring of the soil moisture was established in May 2007 in eight localities within the Kysuca River basin (see figure 5 or figure 6). The observations were realised by probes positioned in two depths (15 and 45 cm). They measured continuously, which means every hour one value was recorded. In addition, selected meteorological elements (temperature, precipitation, radiation and air humidity) were observed close to the localities of the measurement.

The soil water content was simulated in two out of five soil layers, namely in the second and the third, in accordance with the depths of the observations. The model gives results of soil water content in mm of a water column in each hydrotop. There are 2255 hydrotops defined in the basin. In order to compare results with the measured data in [%] they had to be recalculated. Two maps, each representing an average water content of one layer simulated within the period 1995 to 1999, were created to display the result of the simulation (figure 5). The behaviour of soil moisture within one year is a subject for further analysis.

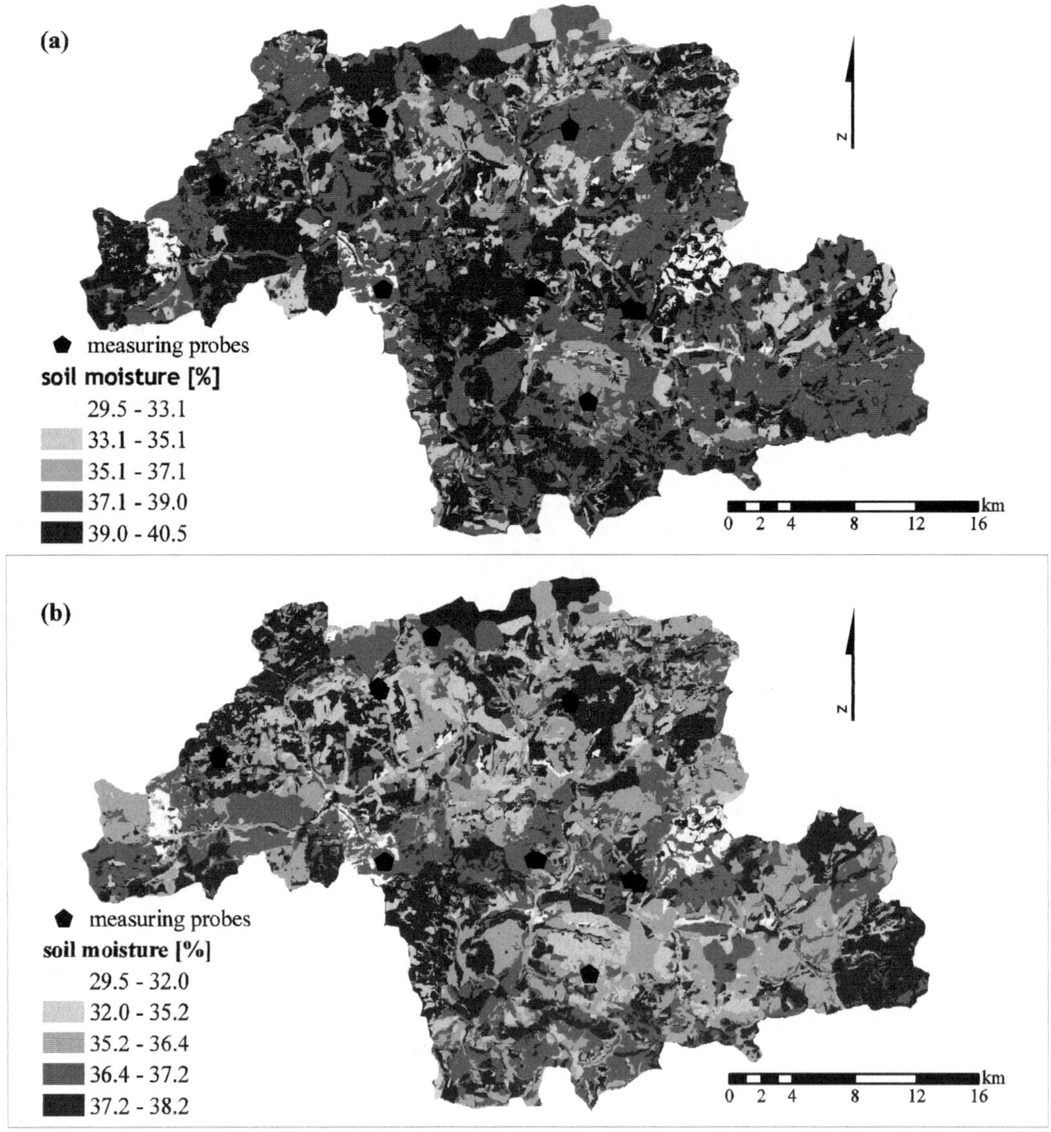

IOP Conf. Series: Earth and Environmental Science 4 (2008) 012002 doi:10.1088/1755-1307/4/1/012002

Figure 5. Map of the simulated average annual soil water content [%] in 15 cm (a) and 45 cm (b) depth

Besides the assessment of the behaviour of soil moisture in a year the comparison of simulated and observed data was accomplished (figure 6, figure 7). The observation was carried out by three probes A, B and C at each location. The pictures show the first results of comparison for the period June-October 2007 (growing season). The year 2007 is the only year of measurement. Moreover the comparison is affected by the fact that the tensometers have not been calibrated to the local conditions yet. The resulting uncertainties have to be taken into account. It has therefore not been possible to compare the absolute values precisely, but the trends of the simulated and measured soil water content could be evaluated as satisfactory.

In the upper zone the results of simulation are very good; in the lower zone they are not quite so good. The divergence of the trend is possibly caused by high ground water levels. Simulated values present conditions in the whole layer with some thickness on the contrary measurement just in the point of an exact depth. Another factor could be the absence of information about soil skeletons among the model inputs, as mentioned above.

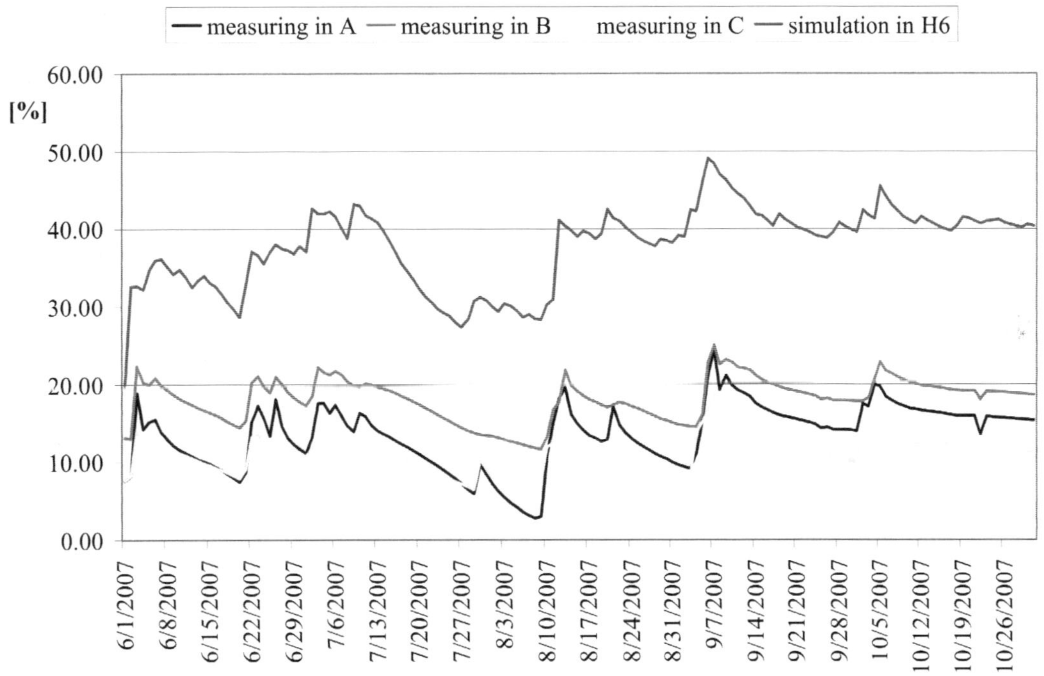

Figure 6. Comparison of observed values of the average daily volumetric water content in the locality 'Jelitov' and simulated values in the matching hydrotop (H6) at a depth of 15 cm (June-October 2007 - growing season)

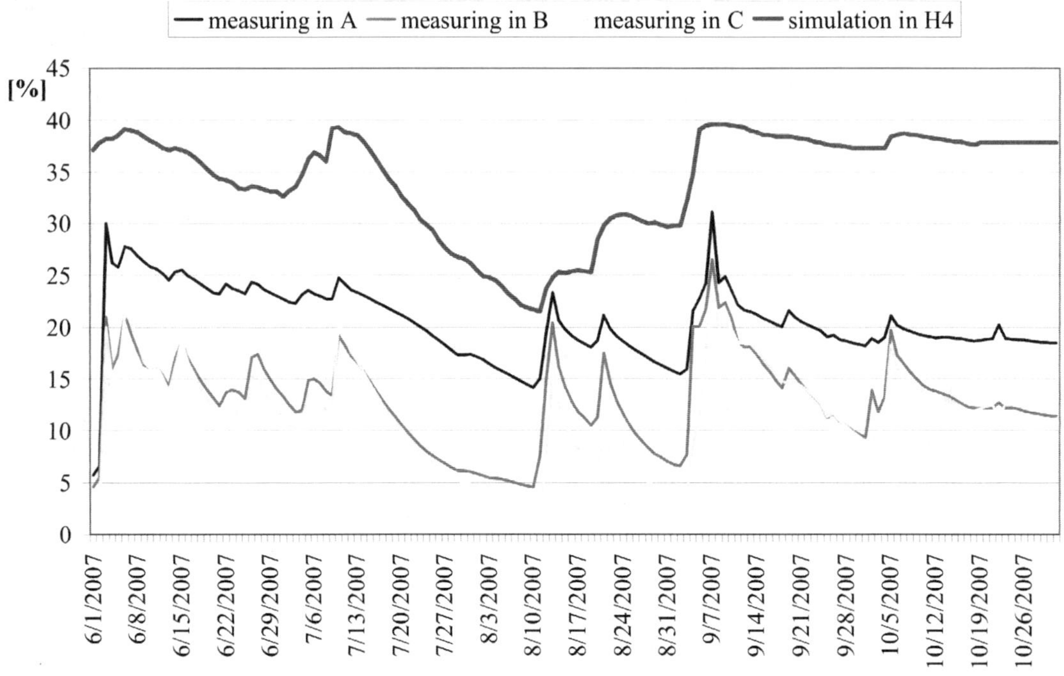

Figure 7. Comparison of observed values of the average daily volumetric water content in the locality 'Zborov' and simulated values in the matching hydrotop (H4) at a depth of 45 cm (June-October 2007 -growing season)

4.3. Evapotranspiration simulations

Actual evapotranspiration was simulated in each hydrotop for the period 1995-1999. The average annual values of actual evapotranspiration were obtained in mm of a water column. Then it was analysed how the results spatially correspond to data elements entering the model (namely the geographical data about soils and land use - see maps in figure 8). The results generally correspond to those of Tomlain [11], who states that the annual average evapotranspiration in the region of north-western Slovakia varies from less than 400 mm to more than 450 mm depending on elevation above sea level. Actual evapotranspiration increases with increasing amounts of precipitation up to a certain level. After that it decreases owing to lack of energy caused by increasing cloud amounts and relative air humidity, rising numbers of days with snow cover and fall in air and soil temperatures.

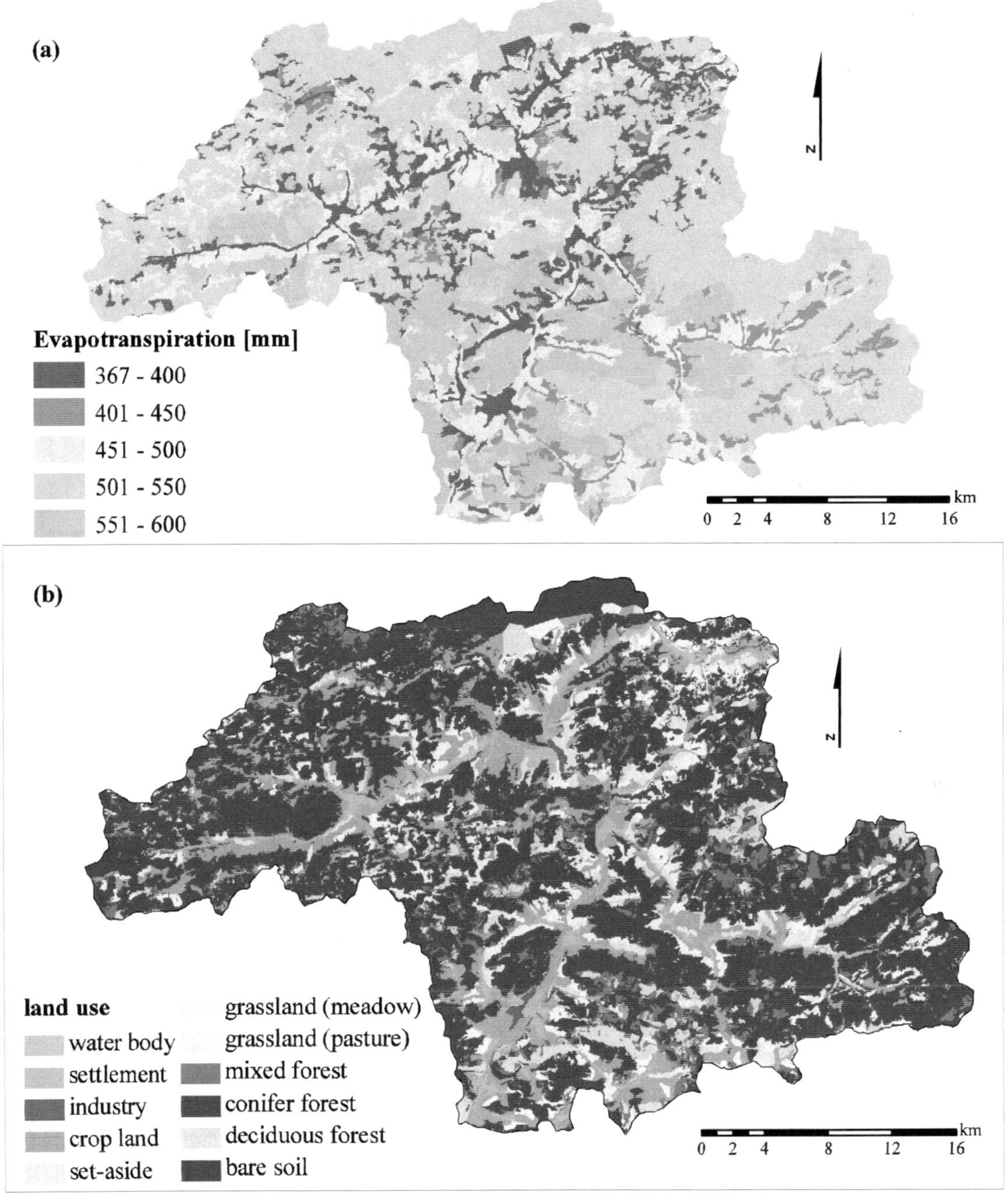
(a)
N
Evapotranspiration [mm]
367 - 400
401 - 450
451 - 500
501 - 550
551 - 600
km
0 2 4 8 12 16
(b)
N
land use
water body
settlement
industry
crop land
set-aside
grassland (meadow)
grassland (pasture)
mixed forest
conifer forest
deciduous forest
bare soil
km
0 2 4 8 12 16

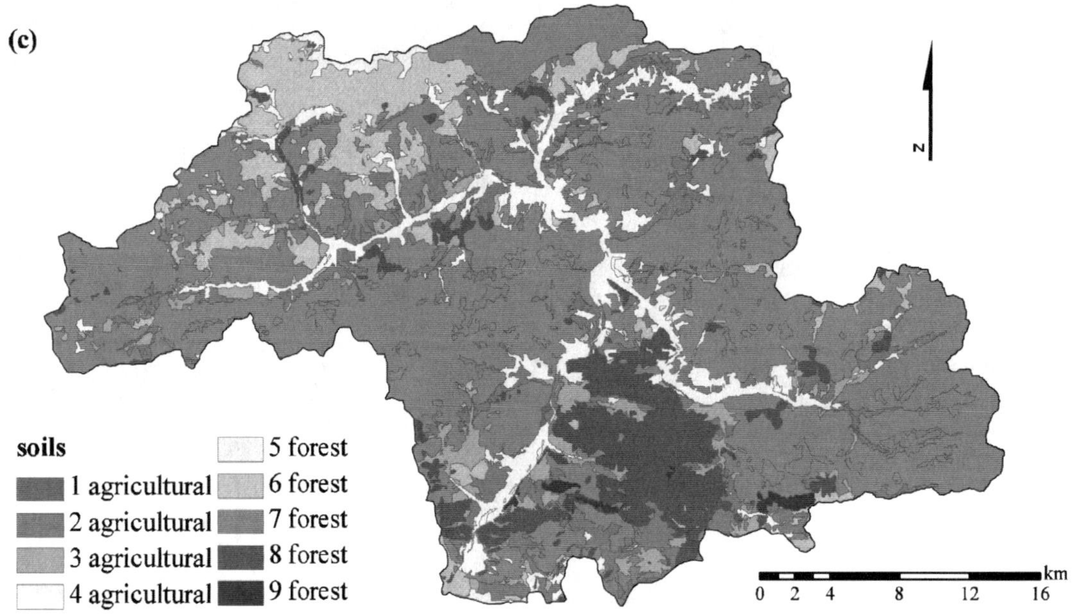

Figure 8. Comparison of the map of the average annual evapotranspiration (a) with the map of land use (b) and the soil map (c)

For an evaluation of the percentage of hydrotops in each category of land use and each soil group, see figure 9 and figure 10.

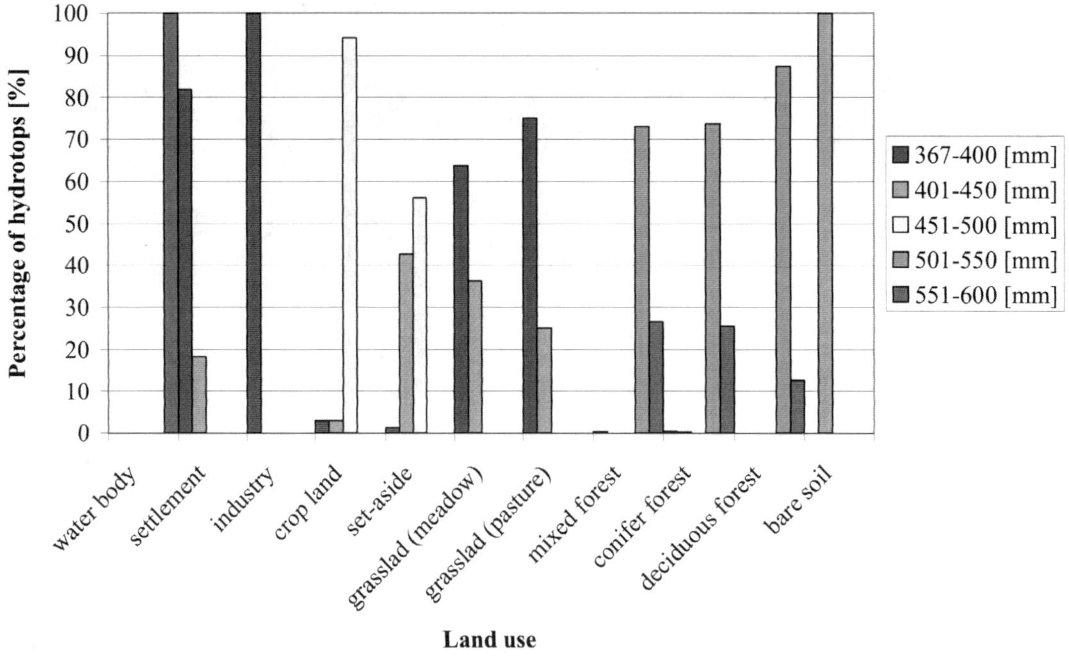

Figure 9. Percentage of hydrotops in each type of land use in categories of the annual average evapotranspiration

The figures have verified that values of evapotranspiration depend considerably on the type of land use as was expected. The values of evapotranspiration increase according to density of vegetation and the highest ones are reached over forested areas and naturally water bodies.

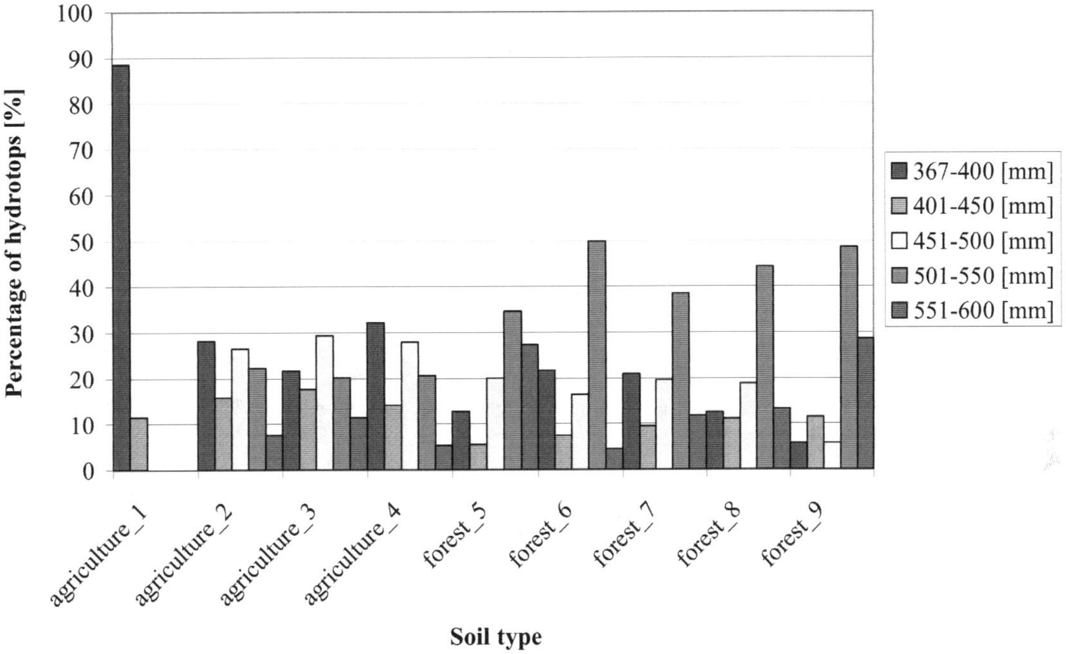

Figure 10. Percentage of hydrotops of each soil in categories of the annual average evapotranspiration

The effect of the characteristics of individual soil types on the evapotranspiration does not seem to be significant. More important is the presence or absence of forest in the hydrotop. The very high percentage of the lowest evapotranspiration category in the type 'agriculture_1' is probably caused by the fact that it is very rich in sand (85 %) and by the very small area it takes over (small number of hydrotops).

5. Conclusion

It appears that the SWIM model is a good tool for analysis of hydrological conditions in a river basin. Apart from simulation of discharge it is possible to simulate other variables in various time and spatial steps. The outputs can be visualised by use of further statistical software and GIS.

The pleasing ability of the SWIM model to simulate rainfall-runoff processes in the Kysuca River basin was proved by the simulation of discharge at the outlet. It can be assumed that other simulated variables, soil water content and evapotranspiration in this study, are also computed, satisfactorily. This assumption is supported by the establishment of expected relations between evapotranspiration and land cover.

To improve the results of the simulations, the calibration of the sensors in the field should help. Including information about the amount of rock fragments in the soil column in the SWIM model will be attempted.

Obtained results of simulation can then be processed in the further study on the health of vegetation cover and its revitalisation.

Acknowledgements
The study was supported by the research grant GA AS CR KJB300600602 and the Institutional Research Plan AV0Z20600510.

References
[1] Krysanova V, Müller – Wohlfeil D I and Becker A 1998 Development and test of spatially distributed hydrological/water quality model for mesoscale watersheds *Ecological Modelling* **106** 216-289
[2] Neteler M and Mitasova H 2004 *Open Source GIS - A GRASS GIS Approach* vol 689 (Boston, Dordrecht, London: Kluwer Academic Publishers) p.464
[3] Ritchie J T 1972 A model for predicting evaporation from a row crop with incomplete cover. *Water Resources Res.* **8** 1204-13
[4] Richardson C W and Ritchie J T 1973 Soil water balance for small watersheds *Trans. ASAE* **16(1)** 72-77
[5] Priestley C H B, Taylor R J 1972 On the assessment of surface heat flux and evaporation using large scale parameters *Monthly Weather Review* **100** 81-92
[6] Monteith J L 1965 Evaporation and environment *Symp. Soc. Exp. Biol.* **19** 205-234
[7] Williams J R, Renard K G and Dyke P T 1984 EPIC – a new model for assessing erosion´s effect on soil productivity *Journal of Soil and Water Conservation* **38(5)** 381-383
[8] USDA, Soil Conservation Service 1972 National Engineering Handbook, Section 4, Hydrology (Washington D.C.)
[9] Arnold J G, Williams J R, Nicks A D and Sammons N B 1990 *SWRRB – A Basin Scale Simulation Model for Soil and Water Resources Management* (Texas A&M University Press, College Station, TX.) p 255
[10] Sloan P G, Morre I D, Coltharp G B and Eigel J D 1983 *Modeling Surface and Subsurface Stormflow on Steeply-sloping Forested Watershed* (*Water Resources Inst. Report 142).* (Lexington: Univ. Kentucky)
[11] Tomlain J 2003 Potential and actual evapotranspiration on the territory of Slovakia for a period 1951 to 2000 *Evaporation and evapotranspiration: Proc. Int. Research Workshop (Brno, CZ, 23 March 2005)* ed Rožnovsky J and Litschmann T [in Slovak]

IOP Conf. Series: Earth and Environmental Science **4** (2008) 012003 doi:10.1088/1755-1307/4/1/012003

Forecasting analysis of runoff for reservoir regulation of dams and weirs in terms of hydro power plant operation

Mariana Maradjieva[1], Nikola Nikolov[2]

[1] University of Architecture Civil Engineering and Geodesy, Sofia, Bulgaria
[2] University of Architecture Civil Engineering and Geodesy, Sofia, Bulgaria

E-mail: m_marajieva@abv.bg

Abstract. In order to meet the needs of Hydropower Plant (HPP) production new algorithms and software were developed for daily, seasonal, annual and long-term control of the runoff for the design of dam and weirs. This control is carried out for monitored periods from 20 to 50 years. The control depends on economic considerations, namely that the accepted probability of required water power is 90%, i.e. concerning the runoff and in this way for the useful volume of water dams. The research is accomplished by a design with the observations.
First the hydrometric stations are selected at the available analogy with the building project and then the correlative connection is found assessed by general and true correlative coefficients. The transferring to the project of the observations for the average annual and average monthly water discharges is made with the coefficient of the analogy. The theoretical probability curves are chosen with a minimum dispersion. By the last curves the average monthly distributions are settled with probability from 2% to 90% by statistical method.
During the investigated period of the regulation the volumes of discharge, overflow and shortage are calculated as and the determination for the accepted volume of the reservoir if the normative probability of the need is executed. As well the power output of the HPP and its participation in the coverage of the charge diagram on the peak load, under peak load, daily and nightly part are determined in separate observed or forecasting periods. The upper problems about the design and the operation of HPP, water output, reservoir volume and coverage of the charge diagram are solved by iterations. Practical examples are given for the runoff and for the time forecasting system.

1. Introduction

The design of the building project demands a determination in the separate years and average power output of the HPP and the monthly share for the consumer need in peak load, under peak, day and night. The HPP is rational to share first of all for peak load but therefore the regulated volume is necessary to be created. For that purpose the monthly production is examined for 50 years in order to establish the probabilities and monthly distributions in the interval from 2% to 90% [1]. The choice of the years with their probability in this interval becomes with the help of equal area below the differential curve of probability density function in the field of the final water discharges with probabilities 2% and 90% (Fig.1). Further for the observed water discharge in 12 months period and for the average monthly water discharge (Table 1) the probability density curves are found. Then by the 12 theoretical probability curves the monthly distributions for 50 years between 2% and 90% are established. Namely with these statistical curves, through an algorithm and software, the annual and monthly power output, the hour coverage of the charge diagram, the useful volume with standard probability are examined.

When the observed water discharges are with duration from 20 to 50 years the upper investigations about the distributions can be implemented directly by observations but their accuracy is smaller than the statistical distributions. If the observations are shorter than 20 years, the investigation must be implemented with statistical characteristics or forecasting analysis during 50 years.

The choice of the probability density (Fig.1) and the theoretical distributive curve (Fig.2) are computed by observations of the dam A - Table 1.

First the hydrometric stations are selected at an available analogy of the dam A and then the correlative dependence is found assessed by general and true correlative coefficients.

The transferring to the dam A for the average annual and average monthly water discharges becomes with the coefficient of analogy by means of the data given in Table 1.The option of the probability density function and the theoretical curve of distribution among a set of functions is given by minimum dispersion. The accepted probabilities are obtained in accordance to economic considerations for a minimum - 90% and 2% for a maximum average annual water discharge (Fig.1, Fig.2). The last probability 2% is found from the correspondence of the full long term cycle of the runoff of 50 years.

2. Monthly distribution

The 12 monthly distributions throughout the years is established by two methods:

a) Statistical characteristics in the course of 50 years.

b) Observations within a period from 20 to 50 years.

The monthly observations are ordered in descending series for 12 monthly observations and are calculated by empirical probabilities with the next formula $p = \dfrac{m - 0.25}{n + 0.50}$. The probability density of the separate 12 months and the theoretical curves of probability are computed and selected according to a minimum dispersion among the functions: Pearson III, II and I type, three parameter gamma, gamma or lognormal, exponential and Weibul. The chosen theoretical curve for the separate 12 months can be one type or other.

The monthly productions are examined for 50 years by the established probability monthly distributions in the interval from 2% to 90%. The realized test shows that the function Pearson III has a minimum dispersion. The parameters are given below.

Test. Pearson III type with minimum dispersion for 12 months (Fig.1).

The mode of the function is:

$$P_o = \frac{n}{l_1} \frac{q^{q+1}}{e^q . \Gamma_{(q+1)}} \tag{1}$$

Probability density $P_{(x_o)}$ for x_o:

$$P_{(x_o)} = P_o (1 + \frac{x_o}{l_1})^q . e^{-\frac{q . x_o}{l_1}} \tag{2}$$

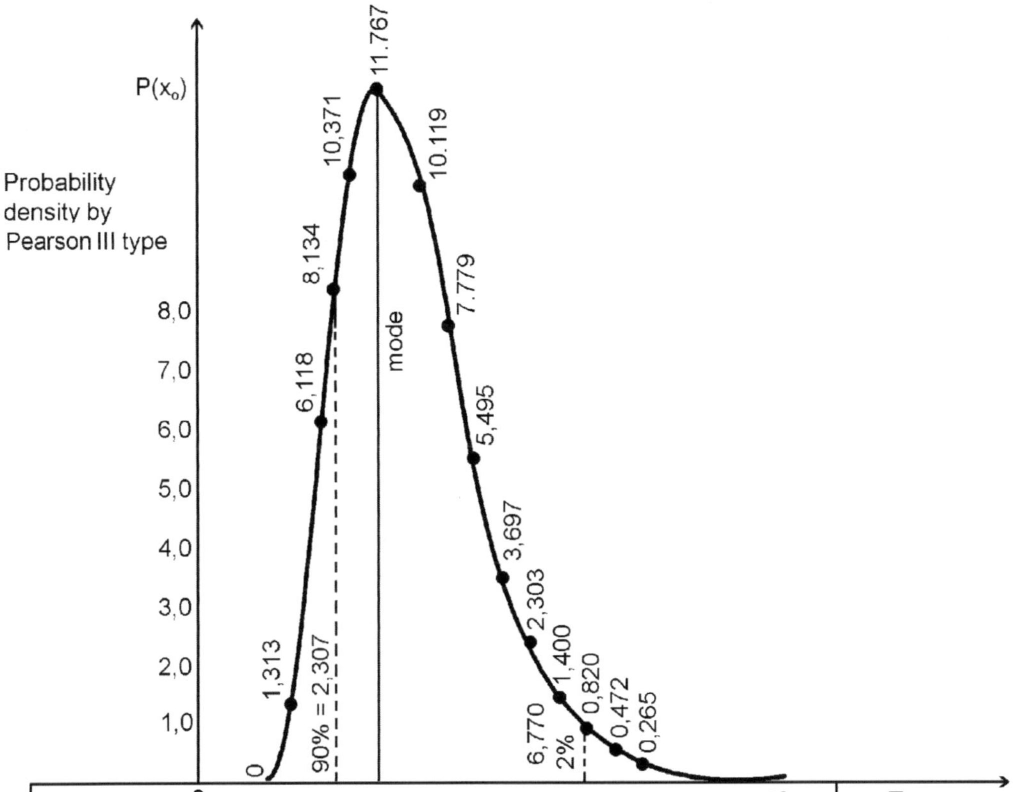

$x = \bar{Q}_A$	0	1,154	1,645	2,145	2,645	3,145	3,803	4,305	4,805	5,305	5,805	6,305	6,805	7,305	7,805	8,305	8,805	9,305	9,805	10,305
x_0 [m$_3$/sec]	-3,145	-1,991	-1,50	-1,00	-0,50	0	0,658	1,160	1,660	2,160	2,660	3,160	3,660	4,160	4,660	5,160	5,660	6,160	6,660	7,160
$k = \bar{Q}_A/\bar{Q}_0$		0,303	0,433	0,564	0,696		1	1,132	1,263	1,395	1,526	1,658	1,789	1,921	2,052	2,184	2,315	2,447	2,578	2,709

Figure 1. Differential curve of probability density – distribution of average annual water discharge $\overline{Q}_A$ by dam A

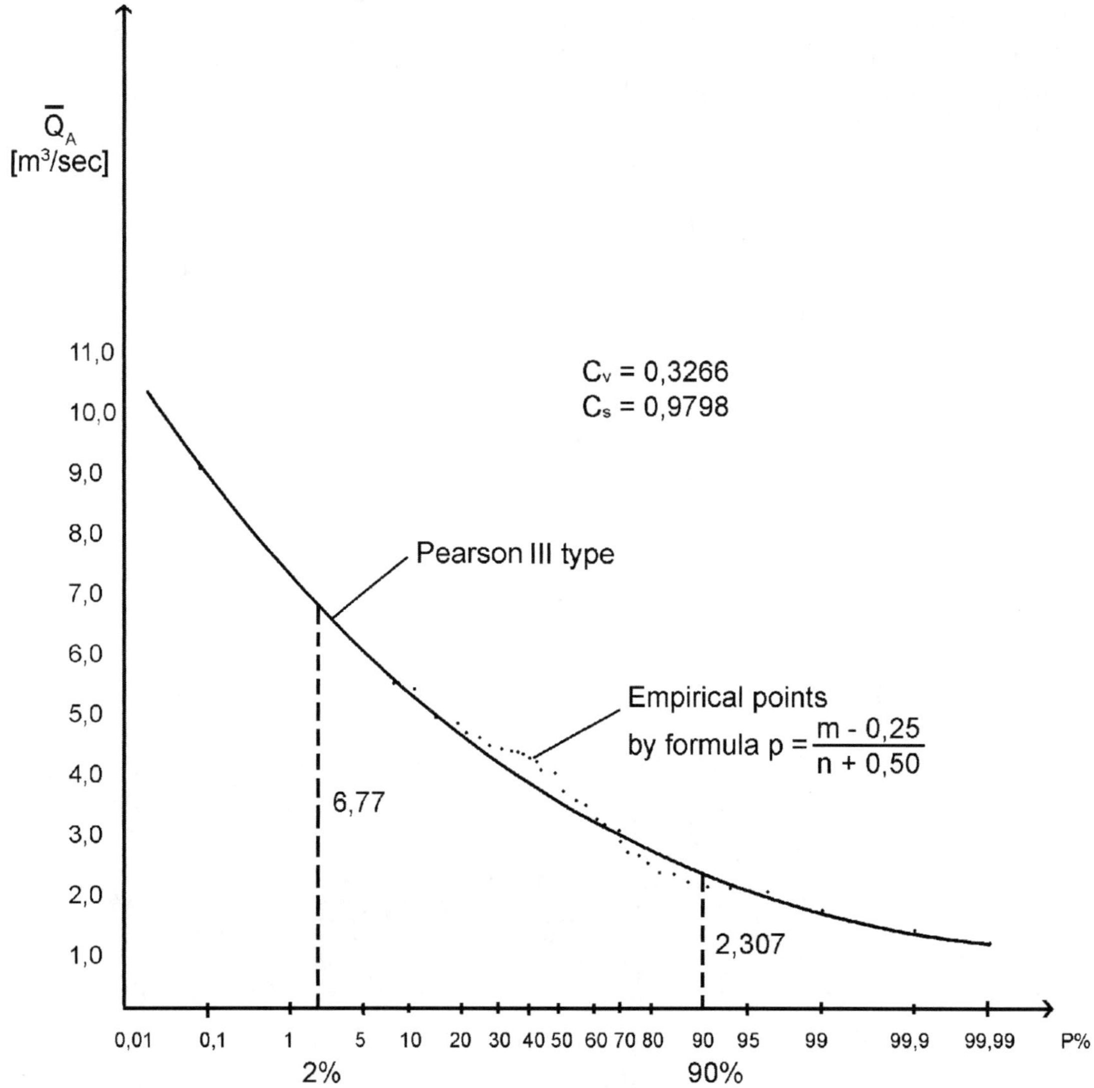

Figure 2. Empirical probability and theoretical curve of Pearson III type for average annual water discharge $\overline{Q}_A$ by dam A

2.1. Computation of the density and the water discharge distribution function

The parameters of Pearson III type function are estimated as follow. The number of years is n=33, q=3.340, l_1=1.99.

$\hat{x}$=3.145 is abscissa of the mode; $x_o = x - \hat{x}$ is abscissa about the mode (Fig.1); (3)

For 2% from Fig. 2 x =6.77, x_o=6.77-3.145 = 3.625;

For 90% from Fig. 2 x=2.307, x_o=2.307-3.145 = -0.838;

$P_o = 11.677$, $P^{2\%}_{(3.625)} = 0.856$ (Fig.1), $P^{90\%}_{(-0.838)} = 8.134$ (Fig.1)

Or $-0.838 \le x_{oi} \le 3.625$ (4)

$$P_{(x_{0i})} \le P^{\max}_{(x_{oo})} = 11.767 \text{, for } x_o = 0 \quad (5)$$

The area under the differential curve of the probability density from 2% to 90% is presented in Fig.1 and the sum is equal to:

$$\sum_{p=2\%}^{90\%} f_p = \sum_{p=0.01}^{99.99\%} f_p - \left(\sum_{p=0.01\%}^{2\%} f_p + \sum_{p=0.01\%}^{99.99\%} f_p - \sum_{0.01\%}^{90\%} f_p \right) = \quad (6)$$

=32.8-(0.65605+32.80-29.5224) = 28.866

The equal area below the differential curve of a probability density for 50 years in the field of the final water discharge from $\overline{Q}_{2\%} = x = 6.77 m^3/s$ to $\overline{Q}_{90\%} = x = 2.307 m^3/s$ is:

$$f = \frac{\sum_{p=2\%}^{90\%} f_p}{50} = \frac{28.866}{50} = 0.577 \quad (7)$$

The probability $P_{(x_{oi})}$ and the water discharge $\overline{Q}_i$ for 50 years from 2% to 90% are found with the theoretical probability density (Fig.1) by means of the chain system below. The above mentioned model can be used to calculate the probabilities and the water discharges for 50 years by the scheme:

$BOUNDARY\ CONDITIONS:$

$f = 0.577 : P_{(x_{01})} = 0.856 : P_1 = 2\% : \overline{Q_{A1}} = x_1 = 6.77$

$ITERATIVE\ SYSTEM:$

$DO\, k = 2{,}50$

$$\Delta x_k : x_{0k} : P_{(x_{0k})} = P_0 \left(1 + \frac{x_{0k}}{l_1}\right)^q e^{\frac{-q x_{ok}}{l_1}}$$

$$f = \Delta x_k \frac{P_{(x_{0k-1})} + P_{(x_{0k})}}{2}$$

$$P_k\% = \frac{\sum_{p=0.01}^{2\%} f_p + \sum_{1}^{k-1} f}{\sum_{p=0.01}^{99.99} f_p} x100$$

$$\overline{Q_k} = x_k = \hat{x} + x_{0k}$$

$END\ DO$

After determination of the above mentioned probabilities and water discharges the monthly distributions are settled. The monthly distributions $Q_{i,j}$ are improved, if an inadmissible difference has got among the years and months i.e. $\overline{Q}_{i,j}$ and $\overline{Q}_i$.

3. Reservoir regulation of the runoff

By means of the regulation of runoff it will be verified if the foreseen useful volume of the dam A for seasonal regulation will secure always the built-up capacity N_v and the maximum working water discharge Q_v.

From the Table 1 the average many years water discharge is $\overline{Q}_o$=3.803 m^3/s or the volume is W_o = 3.803x31.536.10^6=72.75x10^6 m^3. The season regulated volume is taken approximately 30%

from $W_{90\%}$ i.e. 22.10^6 m^3, Q_v=3x$\overline{Q}_o$=11.5 m^3/s and for H_{net}=100 m, N_v=10000 kW. The following symbols are used.

3.1. Symbols for the algorithm:

$W^r_{i,j}$ -river feeder in dam A during month j=I, ….,XII and in observed years i=1,2,….33;

t^h -diurnal hours of discharge to HPP for a peak coverage of the load;
winter 5h(I, II, XI,XII), spring and autumn 4h(III,IV,V,X), summer 3h(VI,VII,VIII,IX);

$W^h_{i,j} = Q_v\ t^h\, 3600\, t_{m.days}$ -useful volume in peak coverage during t^h of month j and year i;

$t_{m.days}$ -monthly days and nights;

$W^d_{\max}$ -max useful volume of the dam;

$W^d_{i,j}$ -useful volume of the dam during month j and year i;

$W^{of}_{i,j}$ -overflow volume water during month j and year i;

$W^{cs}_{i,j}$ -consumption from the HPP during month j and year i;

$W^{sh}_{i,j}$ -shortage from water volume during month j and year i;

$E_{i,j} = N_v t_{i,j} t_{m.days}$ -monthly output water power during month j and year i;

$E_i = \sum_{j=I}^{XII} E_{i,j}$ -annual output water power in year i;

$\overline{E}_i = \frac{1}{n}\sum_{i=1}^{n}\sum_{j=I}^{XII} E_{i,j}$ – average annual output in i=n years;

$t_{m.days}$ = 31 (I, III, V, VII, VIII, X, XII), 30 (IV, VI, IX, XI), 28 (II) –number of monthly days;

h_{add} -additional day hours of discharge to the HPP, which will take part along t^h for a coverage of the charge diagram;

$t_{i,j}$ - night and day hours of a discharge to the HPP.

3.2. Criteria of Regulation [2]

If $W^d_{\max} \le 0.30 W^r_{90\%}$ - season regulation is; $0.30\overline{W}_o \langle W^d_{\max} \le 0.50\overline{W}_o$ -annual one is; and when $0.50\overline{W}_o \langle W^d_{\max} \le 2\overline{W}_o$ - long term period regulation is.

Table 1. Dam A -values of monthly, annual and long term period of water discharges [m^3/s]

Year	I	II	III	IV	V	VI	VII	VIII	IX	X	XI	XII	AnnQ	Max Q
1	1.997	2.266	7.741	10.91	16.27	10.54	2.811	2.774	2.740	2.534	1.757	1.128	5.290	45.30
2	1.073	0.795	1.877	5.354	3.949	2.710	0.999	1.054	0.805	1.248	1.794	3.866	2.127	15.81
3	2.589	2.238	1.822	8.138	14.51	13.99	4.411	1.766	0.869	1.748	1.655	2.867	4.735	159.0
4														
…………														
…………														
28														
29	2.312	3.357	2.710	4.624	4.430	1.914	0.980	2.220	1.036	1.064	2.635	2.192	2.451	33.66
30	1.794	1.424	2.784	5.216	10.03	5.586	1.156	0.777	0.610	1.258	1.479	1.267	2.784	41.34
31	2.016	1.239	6.104	5.197	4.383	1.692	0.832	0.712	0/814	0.758	0.731	1.128	2.145	49.75
32	1.304	1.359	1.748	7.953	6.325	2.136	0.832	0.897	0.638	0.767	1.507	2.007	2.284	15.07
33	1.239	1.618	2.321	5.771	2.367	4.606	2.858	1.119	0.684	0.629	0.832	0.832	2.062	73.61
Sum	65.93	67.41	115.2	243.3	399.8	233.2	93.10	47.52	41.79	54.44	66.68	72.53	125.4	

IOP Conf. Series: Earth and Environmental Science **4** (2008) 012003 doi:10.1088/1755-1307/4/1/012003

Qo	1.998	2.043	3.492	7.374	12.11	7.067	2.821	1.440	1.267	1.650	2.021	2.198	3.803	

The chart of a runoff regulation is presented in Fig. 3.

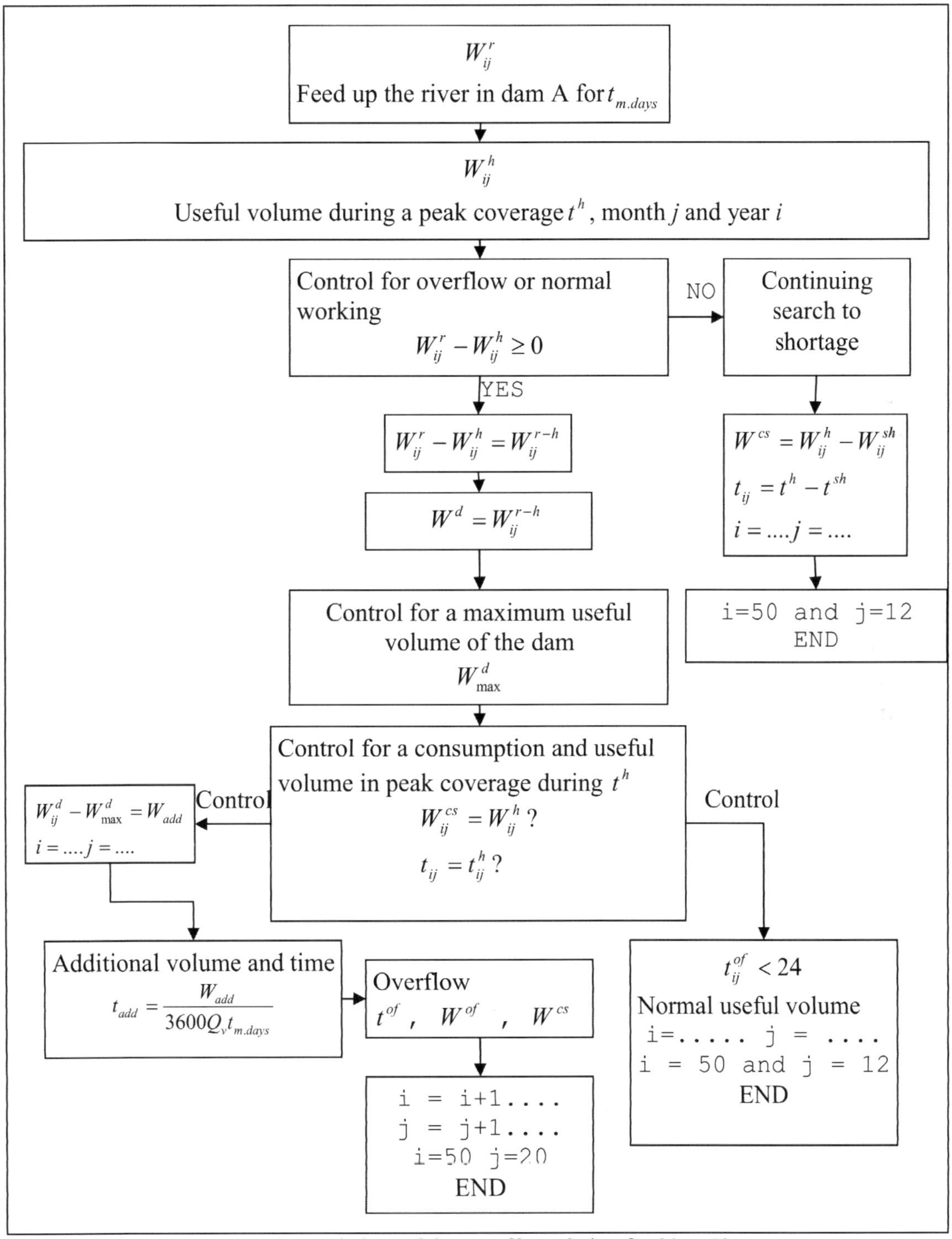

Figure 3. General chart of the runoff regulation for 20 to 50 years

This algorithm is applied for two methods:

a) statistics;

b) observations

The calculated result with the upper algorithm b) is presented in Table 2 and Table 3 by means of a month observations for 33 years given in Table1 and a seasonal regulation according to Fig.4.

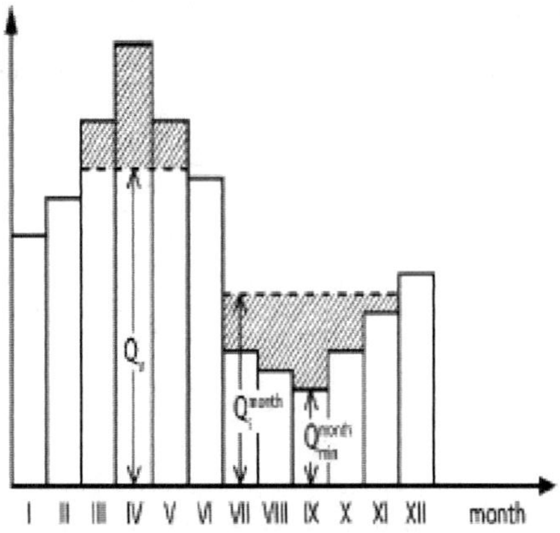

Figure 4. Seasonal regulation

Table 2. Regulation of a runoff: feeder, discharge, shortage, overflows of dam A

I	Quantity	Measures	j=I	j=II	j=III	j=IV	j=V	j=VI	j=VII	VIII	j=IX	j=X	j=XI	XII	Sum
1	$W_{1,j}^{r}$	$10^6 m^3$	5.349	5.482	20.73	28.28	43.59	27.32	7.529	7.430	7.102	6.787	4.554	3.021	167.1
1	t^h	hour	5	5	4	4	4	3	3	3	3	4	5	5	
1	$W_{1,j}^{h}$	$10^6 m^3$	6.417	5.796	5.134	4.968	5.134	3.726	3.850	3.850	3.726	5.134	6.210	6.417	
1	t_{add}	hour				13.62	29.96	19.02	2.867	2.789	2.718	1.288			
1	$t_{1,j}$	hour	4.168	4.729	4	17.62	24	22.00	5.867	5.789	5.718	5.288	5	5	
1	$W_{1,j}^{cs}$	$10^6 m^3$	5.349	5.482	5.134	21.88	30.80	27.32	7.529	7.430	7.102	6.787	6.210	6.417	137.4
1	$W_{1,j}^{d}$	$10^6 m^3$			15.59	22	22	22	22	22	22	22	20.34	16.94	16.94
1	t^{of}	hour					9.967								
1	$W_{1,j}^{of}$	$10^6 m^3$					12.79								12.79
1	$t_{1,j}^{sh}$	hour	0.832	0.271											
1	$W_{1,j}^{sh}$	$10^6 m^3$	1.068	0.314											-1.38
2	$W_{2,j}^{r}$	$10^6 m^3$	2.874	1.923	5.027	13.87	10.23	7.258	2.676	2.823	2.086	3.343	4.650	10.35	67.12
2	t^h	hour	5	5	4	4	4	3	3	3	3	4	5	5	
2	$W_{2,j}^{h}$	$10^6 m^3$	6.417	5.796	5.134	4.968	5.134	3.726	3.850	3.850	3.726	5.134	6.210	6.417	
2	t_{add}	hour					1.119	2.844							
2	$t_{2,j}$	hour	5	5	4	4	5.119	5.844	3	3	3	4	5	5	
2	$W_{2,j}^{cs}$	$10^6 m^3$	6.417	5.796	5.134	4.968	6.570	7.258	3.850	3.850	3.726	5.134	6.210	6.417	65.33
2	$W_{2,j}^{d}$	$10^6 m^3$	13.40	9.532	9.425	18.33	22	22	20.82	19.79	18.15	16.36	14.80	18.74	18.74
2	t^{of}	hour													
2	$W_{1,j}^{of}$	$10^6 m^3$													
2	$t_{1,j}^{sh}$	hour													

2	$W^{sh}_{1,j}$	$10^6 m^3$													

3...

....

33...

Table 3. The monthly share in the coverage of the charge diagram and output of the HPP

I	Quantity	measure	j=I	j=II	j=III	j=IV	j=V	j=VI	VII	VIII	j=IX	j=X	j=XI	j=XII	Sum
1	$t_{1,j}$	hour	4.168	4.729	4	17.62	24	22	5.867	5.789	5.718	5.288	5	5	
1	$E_{1,j}$	10^6 kwh	1.296	1.324	1.240	5.286	7.440	6.600	1.819	1.795	1.715	1.639	1.500	1.550	33.20
2	$t_{2,j}$	hour	5	5	4	4	5.119	5.844	3	3	3	4	5	5	
2	$E_{2,j}$	10^6 kwh	1.550	1.400	1.240	1.200	1.587	1.753	0.930	0.930	0.900	1.240	1.500	1.550	15.78

3...

....

33...

3.3. Results

Type of the coverage of the charge diagram and the output of HPP

1) Peak load: winter 5h, spring 4h, summer 3h,
   Sub-peak: winter 5h – 10h, spring 4h – 8h, summer 3h – 6h;
2) Daily: winter 10h – 16h, spring 8h – 16h, summer 6h - 16h;
3) Daily/nightly: winter16h - 24h, spring 16h = 24h, summer 16h – 24 h.

4. Forecasting analysis

In the previous subsection 2.1 the water discharges for 50 years and the density functions were found by a chain system on the basis of the observed data. New approaches can be used for the runoff regulation and for the operation of HPP applying the forecast analysis. Two methods are possible: time series analysis and advanced series one. The both methods could be classified according three categories:

- Quantitative
- Qualitative
- Unpredictable

Quantitative forecasting is applied when the information about the past is available. The following three steps are used:

1-preliminary analysis; 2-choice and fit the model; 3-evaluation of forecasting.

The first step can be carried out by analysis the trend, seasonality, cycle, and random variations by means of exploratory analysis and fitted load data. Several methods are available for the realization of the second step as moving average, exponential smoothing, Holt-Winter linear algorithm, regression and finally advance series analysis of the type autoregressive model (AV), autoregressive moving average (ARMA), autoregressive integrated moving average (ARIMA),etc. Since a quantitative forecasting varies widely and could be fall easy the third step is obligatory.

4.1. *Preliminary analysis of trend in time series*

Smooth curves are chosen commonly to describe and average fluctuations of original data. The least square method (LSM) and R-Squared value (RSV) are applied to fit the best curve. The exponential form is commonly used:

$$Y = Ce^{DX} \tag{8}$$

Where Y, X are dependent and independent variables, C and D are parameters to be computed.

Many mathematical functions exist using various forms of LSM or by Moving Average MA. The best curve, estimated by RSV, is called a coefficient of determination. It is an indicator whose value ranges in the interval from nought to unit and shows how closely the trend values correspond to the actual data. In general this indicator is defined as:

$$R^2 = \frac{\sum_{1}^{N}(Y' - Y_{avg})^2}{\sum_{1}^{N}(Y - Y_{avg})^2} \tag{9}$$

Where Y and Y' are forecasting variable and its estimated results respectively, Y_{avg} is the calibration period.

After estimation of parameters in formula (8) the de-trended series are computed by dividing the original data with the trend component:

$$Y_t / T_t = S_t E_t \tag{10}$$

Where Y_t is the time series value (actual data), S_t is the seasonal component, T_t is the trend-cycle component and E_t is the random component at time t.

The random component is computed by dividing the seasonal one from (10) and the final formula has the form:

$$Y_t / (T_t S_t) = E_t \tag{11}$$

The next table shows the seasonal components in the series represented by the data for the last period of five years – 1979 to 1983 for water volumes in a river (load data). The last bold numbers in the final column are known as seasonal indices.

Table 4. Seasonal components (indices) of the series for the last five years

Years / Months		1979	1980	1981	1982	1983	Average seasonal components (indices)
Jan	seasonal	0.94	0.84	0.93	0.96	0.96	**0.98**
	random	0.96	0.86	0.95	0.98	0.98	
Feb	seasonal	0.98	0.91	0.95	1.00	0.95	**0.981**
	random	1.00	0.93	0.97	1.02	0.97	
Mar	seasonal	1.08	1.01	1.12	1.06	1.13	**1.03**
	random	1.05	0.98	1.08	1.03	1.10	
Apr	seasonal	1.05	0.86	1.01	0.98	1.11	**1.062**
	random	0.99	0.91	0.96	0.93	1.05	
May	seasonal	0.97	0.88	1.06	0.98	1.07	**1.101**
	random	0.88	0.80	0.96	0.84	0.98	
Jun	seasonal	1.07	1.03	1.08	1.19	1.13	**1.152**
	random	0.93	0.90	0.94	1.01	0.98	
Jul	seasonal	1.169	1.17	1.19	1.21	1.21	**1.120**
	random	1.040	1.04	1.07	1.08	1.08	
Aug	seasonal	1.15	1.15	1.18	1.26	1.23	**1.061**
	random	1.09	1.09	1.12	1.19	1.17	
Sep	seasonal	1.04	1.06	1.08	1.07	1.06	**0.941**
	random	1.12	1.13	1.15	1.15	1.14	
Oct	seasonal	0.90	0.88	0.94	0.97	0.94	**0.923**
	random	0.99	0.96	1.03	1.06	1.02	

Nov	seasonal	0.781	0.751	0.84	0.86	0.79	**0.861**
	random	0.90	0.87	0.97	1.00	0.91	
Dec	seasonal	0.88	0.91	0.94	0.96	0.87	**0.900**
	random	0.97	1.01	1.05	1.07	0.97	

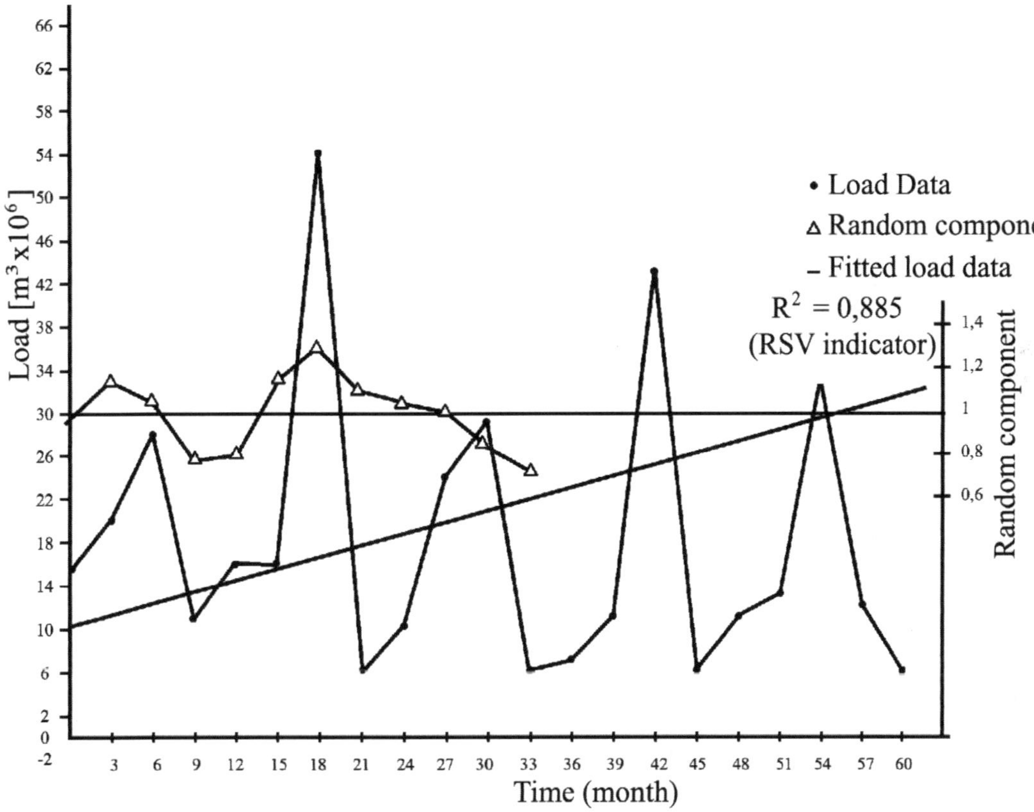

Figure 5. Time series plot for the last five years

Graphical plot of the original load, fitted data and random components are presented in the above figure with the root square indicator RSV.

4.2. Choice and Fit the Model

Holt-Winter method was tested because it is based on three smoothing equations that account the level of the series, trend and seasonal components. Two applications of this method are used as a rule-multiplicative method and additional one. The first model is preferable when the seasonal fluctuations vary in the level of the series and it is chosen in this work with mean square error (MSE) by the formula:

$$MSE = \sqrt{\frac{\sum_{month} (Q_{demand} - Q_{estimated})}{N}} \qquad (12)$$

The following test gives an idea for the monthly consumptions of HPP (Table2) over the past 24 months and a forecasting two months ahead- table 5.

Table 5. Forecasting by Holt-Winter multiplicative method for two months ahead

Month	Demand [m^3.10^6]	Level [m^3.10^6]	Trend [m^{3}s^{-1}]	Seasonal	Fitted Demand [m^3.10^6]	Random
1	3,49	4,31	1,30	0,96	3,45	0,04
2	3,28	4,28	1,36	1,00	3,80	-0,52
3	4,68	4,70	1,75	1,06	4,10	0,58
4	20,61	5,65	7,95	0,98	21,62	-1,01
5	16,94	5,40	6,32	0,98	17,25	-0,21
6	5,54	4,80	2,14	1,19	6,02	-0,48
7	2,16	3,85	0,83	1,21	2,21	-0,05
8	2,40	3,90	0,89	1,26	2,45	-0,05
9	1,65	2,15	0,64	1,07	1,50	1,15
10	2,05	3,70	0,77	0,97	2,15	-0,10
11	3,91	3,90	1,51	0,86	4,10	-0,19
12	5,37	5,40	2,00	0,87	5,27	0,10
13	3,32	3,25	1,24	0,96	3,45	0,13
14	3,91	3,95	1,62	0,95	3,80	0,11
15	6,22	4,90	3,32	1,13	6,12	0,10
16	14,96	5,20	5,72	1,13	13,90	1,06
17	6,34	4,83	2,37	1,07	6,50	-0,96
18	11,94	5,13	4,61	1,13	12,04	-0,10
19	7,65	4,95	2,85	1,21	7,30	0,35
20	2,99	3,50	1,11	1,23	2,50	0,49
21	1,77	2,26	0,68	1,06	1,87	-0,10
22	1,68	2,25	0,63	0,94	1,88	-0,20
23	2,16	3,85	0,83	0,79	2,36	-0,20
24	2,16	3,85	0,83	0,87	2,36	1,49
-		α = 0,83 β = 0,075 γ = 0 MSE = 25%				
-		Forecasting 2 months ahead				
months	Demand [m^3.10^6]	95% Confidence Upper Limits		95% Confidence Lower Limits		
25	3,21	3,85		2,95		
26	4,15	4,99		3,53		

4.3. Evaluation of forecasting by advance series analysis

Advance analysis includes other variables to describe the behaviour of the series using multiple time data. The new variables will capture additional effects for prediction and this comprehensive model depends on adequacy of calibration and residual uncertainty. In simulation the reservoir process, the principle of continuity is used to balance precipitation, storage, evapotranspiration and runoff written by (13). This conceptual model takes into consideration the nonlinearity, uncertainty factors, etc. [3]:

$$\text{Precipitation} - \text{Actual} \quad \text{Evapotranspiration} \pm \text{Storage} = \text{Runoff} \tag{13}$$

An Auto Regressive process of order p, AR(p) contains p number of parameters and can be expressed as:

$$x_t = \phi_1 x_{t-1} + \phi_2 x_{t-2} + \ldots + x_{t-p} + a_t \tag{14}$$

Where $\phi_1 \ldots$ are the AR parameters which describe the effect of a unit change in $x_{t-1} \ldots$ on x_t and which needs to be estimated; a_t is known as error or series white noise (Fig.6). A process with autoregressive and a moving average term is called an autoregressive moving process (ARMA) and it is represented by equation:

$$x_t = \phi_1 x_{t-1} + a_t - \theta_1 a_{t-1} \quad (15)$$

The higher order $\text{ARMA}(p,q)$ is expressed as

$$x_t = \phi_1 x_{t-1} + \phi_2 x_{t-2} + ...\phi_p x_{t-p} + a_t - \theta_1 a_{t-1} - ...\theta_q a_{t-q} \quad (16)$$

Where θ_1 is a moving average parameter; p refers to the number of autoregressive parameters and q - the number of moving average parameters (Fig.7).

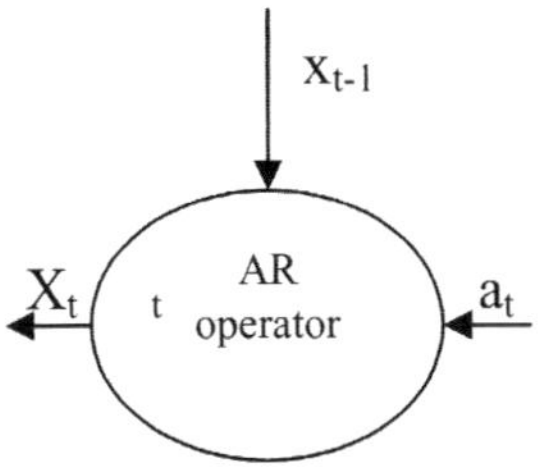

Figure 6. Scheme of AR

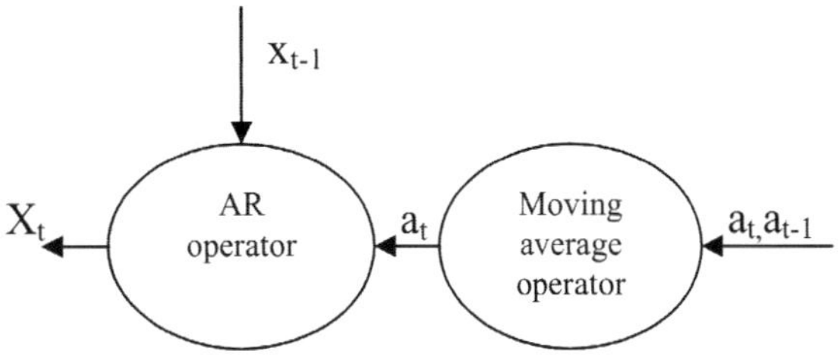

Figure 7. Scheme of ARMA

5. Conclusion

The seasonal, annual and long term period of regulations of the runoff can be investigated with the help of the developed algorithm and software. The 12 monthly distributions in the individual years are established by statistical characteristics in the interval of probabilities among 2%-90%. Examples for the density and distribution functions are solved as well as a test for seasonal regulation of runoff with the monthly share in the coverage of the charge diagram and the output of HPP. Finally a forecasting analysis of runoff is estimated using the observed data.

References

[1] Nikolov N 2007 Reservoir regulation of runoff in dams, weirs or cascades with a view to hydro power plant operation *XXXII IAHR congress* (Venice, Italy)

[2] Nikolov N, Maradjieva M, Chikalanov A and Emilov G 2005 Computer modeling and Design of Cost-Effective Pump Lines Systems for Different Thickness, Materials and Fluid *International Conference "Waterpower" XIV* (USA, Austin)

[3] Friedel M 2006 *Predictive stream flow uncertainty in relation to calibration-constraint information, model complexity, and model bias* Int. JRBM vol.4 N2

The implementation of the HBV Model on the Sava River Basin

Miha Primožič[1], Mira Kobold[1,2], Mitja Brilly[1]

[1]University of Ljubljana, Faculty of Civil Engineering, Jamova 2, Ljubljana, Slovenia,
[2]Environmental Agency of the Republic of Slovenia, Vojkova 1b, Ljubljana, Slovenia

E-mail: mbrilly@fgg.uni-lj.si; mira.kobold@gov.si

Abstract. The Swedish HBV model was used for modelling the Sava River discharge and modelling of snow cover over the Sava river watershed in Slovenia. The Sava River is the largest river in Slovenia and tributary of the Danube River, contributing to its largest runoff. It covers more than half of the territory of Slovenia, namely 10700 km^2. The watershed is heterogeneous, mountainous in the upper parts and plain in the middle reach. There are also some karstic regions. The floods are caused by heavy rainfall in headwater mountain areas, especially in autumn. Some tributary flows can rise more than a hundred times in such events. For the purpose of flood forecasting the HBV model was set up for the whole watershed using the time step of 24 hours to calibrate the set of model parameters, which then used for recalibration of the model with time step of 1 hour. The watershed was divided to 26 subcatchments. Model input data are precipitation, potential evapotranspiration, and in case of climates with temperatures below zero, temperature data. Measured discharges are needed to calibrate, verify and up-date the model. Satisfactory calibration of the model was achieved, although the topography has strong influence on the meteorological happening in the catchement, especially in the upper stream of the Sava river, and small number of available raingauges.
Keywords: HBV, modelling, runoff, Sava, precipitation, evaporation, watershed.

1. Introduction

Following the good results using the Swedish HBV model for the simulation of runoff for the Savinja river basin [1], the model was used for modelling the Sava River discharge and modelling of snow cover over the Sava river watershed in Slovenia. The Sava River is the largest river in Slovenia and tributary of the Danube River, contributing to its largest runoff. It is created by two rivers Sava Bohinjka (31 km long) and Sava Dolinka (45 km long) which join together near town of Lesce. The head part of the Sava River basin is located in Slovenia and before it reaches Danube it is 990 km long. On the way to Belgrade it flows through three European countries: Slovenia, Croatia and Serbia. At the start of its journey the Sava river has an average discharge of about 45 m^3/s (near town of Lesce), just after exiting Republic of Slovenia it has a moderate discharge of 255 m^3/s and before it joins Danube in Belgrade the discharge rises to amazing 1.722 m^3/s. Sava river watershed covers more than half of the territory of Slovenia, namely 10.700 km^2. The upper part of the basin is mountainous with altitudes up to 2800 meters. The altitudes of the plain area, in the middle reach are between 100 and 400 meters. The floods, usually flash floods, are caused by heavy rainfall in headwater mountain areas, especially in autumn. Some tributary flows can rise more than a hundred times in such events.

2. Methods

2.1. HBV model

The HBV model was originally developed at the Swedish Meteorological and Hydrological Institute (SMHI) in Norrköping, Sweden[2]. After twenty years, the HBV model has become a standard tool for runoff simulations in the Nordic countries, and the number of applications in other countries is growing. The HBV approach has since the early days proved flexible and robust in solving water resource problems and applications. It is characterised as a conceptual model. The aim of the first operational applications of the HBV model was hydrological forecasting. Since then, the field of applications has widened and covers today real-time forecasting, control of data quality, extension of runoff records and filling in of gaps, design floods, synoptic water balance mapping, water balance studies, simulations of the effects of a changing climate and simulations of groundwater response.

The model consists of several fundamental hydrological routines, including a snow routine based on a degree-day relation, and a soil moisture routine that accounts for soil field capacity and changes in soil moisture storage due to rainfall/snow melt and evapotranspiration. The runoff generation routine transforms water from the soil moisture zone to runoff.

The snow routine of the model controls snow accumulation and melting. It works separately for each elevation and vegetation zone. The snow routine is based on a simple degree-day relation. The precipitation is assumed to accumulate as snow when the air temperature drops below a threshold value which is usually close to 0°C. Melting starts with temperatures above the threshold. The threshold temperature is normally used to decide whether the precipitation falls as rain or snow, but it is possible to have different thresholds. It can also be extended to an interval and within this interval precipitation is assumed to be a mix of rain and snow (decreasing linearly from 100 % snow at the lower end to 0 % at the upper end). The snow pack is assumed to retain melted water as long as the amount does not exceed a certain fraction of the snow. Then, runoff is generated. When temperature decreases below threshhold temperature, this water refreezes. The temperature is an important parameter in snow routine and it is altitude corrected by applying altitude correction parameter.

The soil moisture accounting routine computes an index of the wetness of the entire basin and integrates interception and soil moisture storage. It is the main part controlling runoff formation. This routine is based on the three main parameters BETA, LP and FC. FC is the maximum soil moisture storage in the basin. BETA controls the increase in soil moisture storage from each millimetre of rainfall or snow melt at a given soil moisture deficit. LP is a soil moisture value above which evapotranspiration reaches its potential value, which means that it controls the shape of the reduction curve for potential evapotranspiration. At soil moisture values below LP, the actual evapotranspiration will be reduced. The parameter LP is given as a fraction of FC. The effect of the curve determining runoff generation is that the response is gradually increasing with decreasing wetness.

The runoff generation routine is the response function which transforms excess water from the soil moisture zone to runoff. It also includes the effect of direct precipitation and evaporation on a part which represents lakes, rivers and other wet areas. The function consists of one upper, non-linear, and one lower, linear, reservoir. These are the origins of the quick and slow runoff components of the hydrograph. The yield from the soil moisture zone, i.e. the effective precipitation, will be added to the storage in the upper reservoir. As long as there is water in the upper reservoir, water will percolate to the lower reservoir. At high yield from the soil, percolation is not sufficient to keep the upper reservoir empty, and the generated discharge will have a contribution directly from the upper reservoir, which represents drainage, through more superficial channels. The lower reservoir, on the other hand, represents the groundwater storage of the catchment contributing to the base flow. In the latest version of the model (HBV-96) [3,4], the recession is modelled by a function corresponding to a continuously increasing recession coefficient. Each one of the subbasins has individual soil moisture accounting procedures and response functions. The runoff is generated independently from each one of the subbasins and is then routed through a transformation function in order to get a proper shape of the hydrograph.

The flexible structure of the IHMS/HBV system allows the model to make necessary sub-divisions with respect to different climate zones, land-use, density of the hydrometeorological network etc. The HBV can be applied to catchments of virtually any size, from less than 1 km^2 to several hundred thousand km^2. Larger river watershed is divided into subbasins, which are also its primary hydrological units. The model is set up separately for each subbasin. Subbasins are linked together and the outflow from the upstream ones is routed through the downstream ones. The SMHI version of the HBV model is usually run with 24 hour (day) time steps, but it is possible to use shorter time steps down to 1 (one) hour, if higher resolution data is available.

Model input data is precipitation, values of potential evapotranspiration, e.g. standard monthly values and in case of climates with temperatures below zero, temperature data. The specified input potential evapotranspiration values can be corrected for altitude by using the lapse parameter which works similar to precipitation altitude correction parameter but decreases high altitude values instead of increasing them. If no evapotranspiration data is present, the values can be calculated directly from temperature data.

Measured stream-flow or reservoir inflow is needed to calibrate, verify and up-date the model. Standard model outputs are discharge/ streamflow/reservoir inflow, additionally the basin average temperature, precipitation, evapotranspitation, soil-moisture content and snow-pack are calculated. Most of those values can also, optional, be presented for all land-use types in all elevation zones.

As is the case in all conceptual hydrological models, the HBV model can be sensitive to calibration of parameters, and more or less the same result can be obtained using different parameter-sets. The model is usually calibrated by a manual procedure with set of model are changed during calibration process until an acceptable agreement with observations is obtained. The basins are computed in the order of the streamflow, so downstream basins require computed data from the upstream ones. As a result, the calibration cannot be made in the order of the subbasins the user wants. It has to follow the natural stream and flow of the river.

The number of parameters normally used in the model is in the order of 20 - 25. While 5 of them are in most cases set to standard values, some 10 are very important to calibrate. The importance of a long enough period of historical data to allow calibration, and then verification on an independent period must be taken into account when using this kind of models. Three main criteria of fit are used: visual inspection of the computed and observed hydrographs, Nash/Sutcliffe criterion R^2 and inspection of accumulated error. R^2 efficiency criterion was introduced by Nash and Sutcliffe (1970) and is commonly used in hydrological modelling. R^2 has a value of 1,0 if the simulation and the observations agree completely and 0 if the model does not perform any better than the mean value of the runoff record, but in practice values between 0,8 and 0,95 [5] can be achieved if the quality of observed data is good. Negative values can be the result of poor model performance or poor data. Accumulated error is another important indicator of performance. A perfect match between recorded and computed runoff would result in a straight line with a constant value of zero.

2.2. Implementation of the HBV model on the Sava river basin

The division of the Sava river basin was made to 26 watersheds with area ranging from 0,26 km^2 to 1.019,85 km^2 and further to elevation zones. The upper part of the basin is mountainous and subbasins in that area were divided in up to 5 (five) elevation zones. Subbasins in the plain area of the Sava basin, where altitudes reach maximum 400 meters, have only 1 (one) elevation zone. Each elevation zone was then divided in to two parts according to land use, so called vegetation zones (forest and field).

IOP Conf. Series: Earth and Environmental Science 4 (2008) 012004 doi:10.1088/1755-1307/4/1/012004

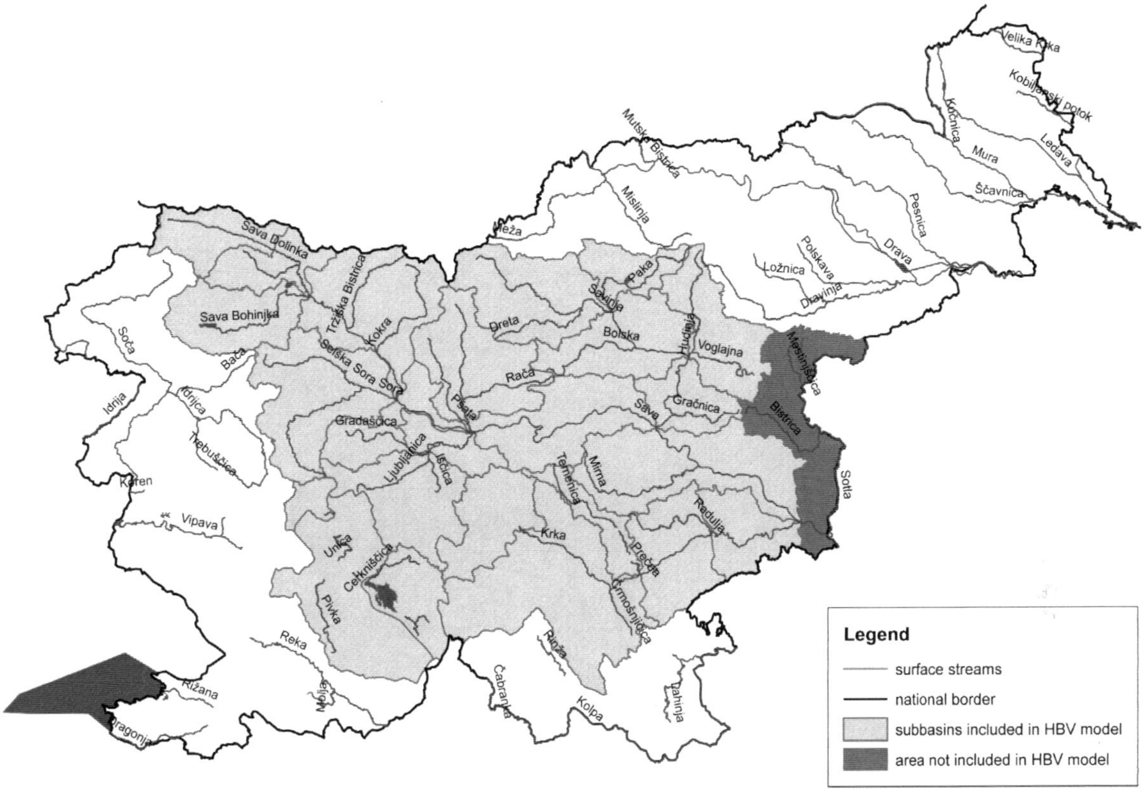

Figure 1. Slovenia, Sava river basin (grey) divided in to 26 watersheds (coloured in light grey) and thc arca which is not included in the modelling. (coloured in dark grey).

First the model with 24 hour (1 day) time step was set up. For this model the data was collected for the period from 1 January 1990 to 31 December 2006. Data from 1990 to 1999 was used for calibration to determine the values of a number of free parameters and data from 2000 to 2006 was used for validating the model. After the model with 24 hour time step was calibrated, it was recalibrated with time step of 1 hour. From the set of nearly 30 parameters, six parameters for each subbasin was needed to be recalibrated. The remaining parameters kept the values from the 24 hour time step model. Data for model with 1 hour time step was collected for the period from 1 January 1998 to 31 December 1999. Data was used only for calibrating the model and because of the huge amount of missing data no validation period was chosen. In spite of some difficulties with incomplete data sets, effects of the hydropower plants and wrong measurments, calibration of both models was successful.

Except for the discharge stations, missing station data sets in both models were substituted by the data sets from nearby station and corrected by correlation factor.

The effects of hydropower plants are best seen on figure 3 in 1 hour time step model. It takes one or two hours between the moment the hydropower plant takes the water from the Sava river and the moment it releases it. This explains the daily and especially hourly fluctuation that we can notice on the hydrographs (figure 3, figure 4). The model is unable to predict this behaviour of the discharge, unless the data of regulation schedule is fed into the model. This was not the aim of the present work. Consequently, the model gives lower R^2 values for 1-hour model (table 1), especially for upper part of the Sava river wateshed where hydropower plants are located.

A number of 7 temperature and 5 evapotranspiration stations was used in both models. In the model with 1 hour time step from 53 (used in 24 hour model) precipitation stations only 16 precipitation stations were used providing high resolution data, the rest of0 37 are classical raingauge station with daily precipitation data. Weight of each station was estimated by the Thiessen method. The results are snown for 3 discharge measuring stations.

All the data used during calibration and validation of the model (precipitation, temperature, discharge and evapotranspiration) were obtained from the station network of the Environmental Agency of the Republic of Slovenia (ARSO).

3. Results and discussion

Results of calibration for two models for the Sava watershed with daily and hourly time steps can be presented in the form of hydrographs for each subbasin as well for the whole watershed, but because of great number of subbasins (26) the results are presented only for four different drainage areas as seen in Figure 4.

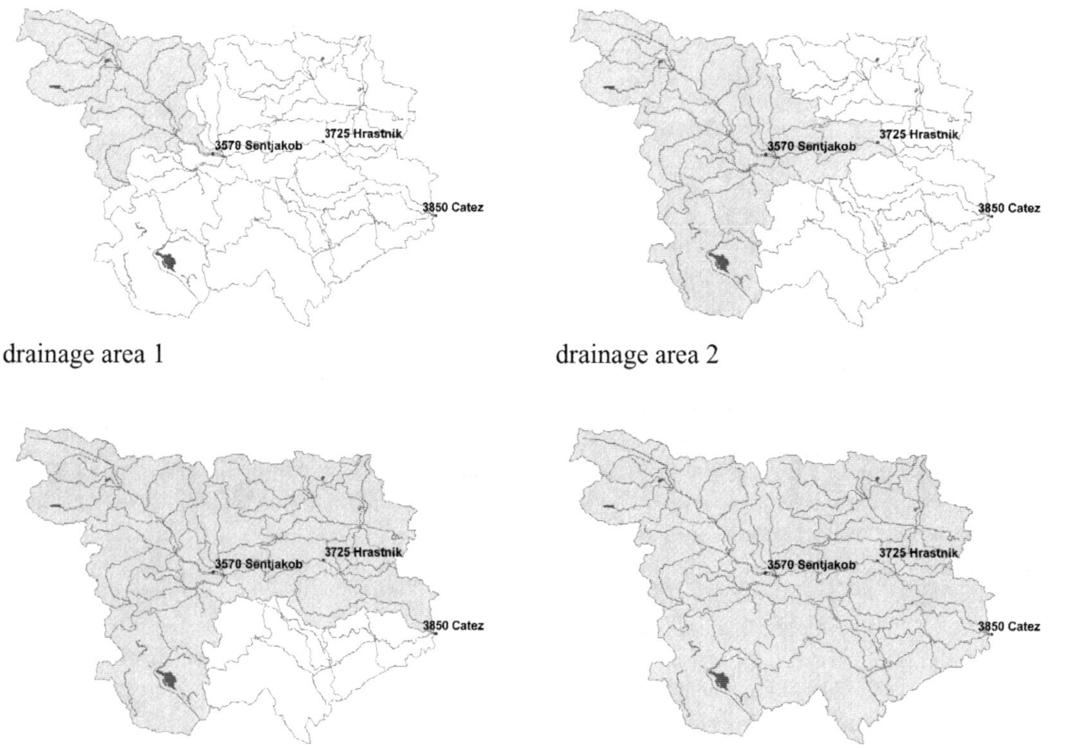

Figure 2. Modelled Sava river basin, divided in to 26 watersheds, with marked 4 different drainage area (coloured in cyan), for which results in form of hydrographs are presented.

Hydrographs for the 24-hour model, for different drainage areas are presented in Figure 3. Due to better presentation the results of calibration are not given for the whole period, but only from January 1998 to December 1999 and contain data from the flood in October 1998 (peaks seen in Figure 3). Hydrographs for 1-hour model, for the same drainage areas, are presented for the period from February to June 1998.

Drainage area 1

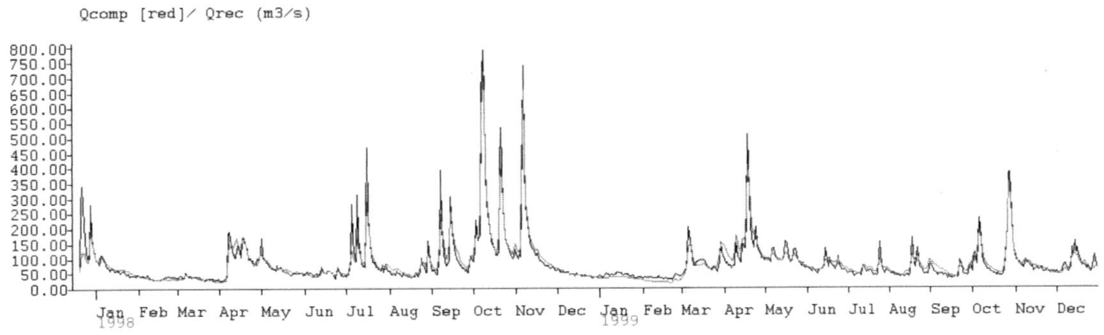

Drainage area 2

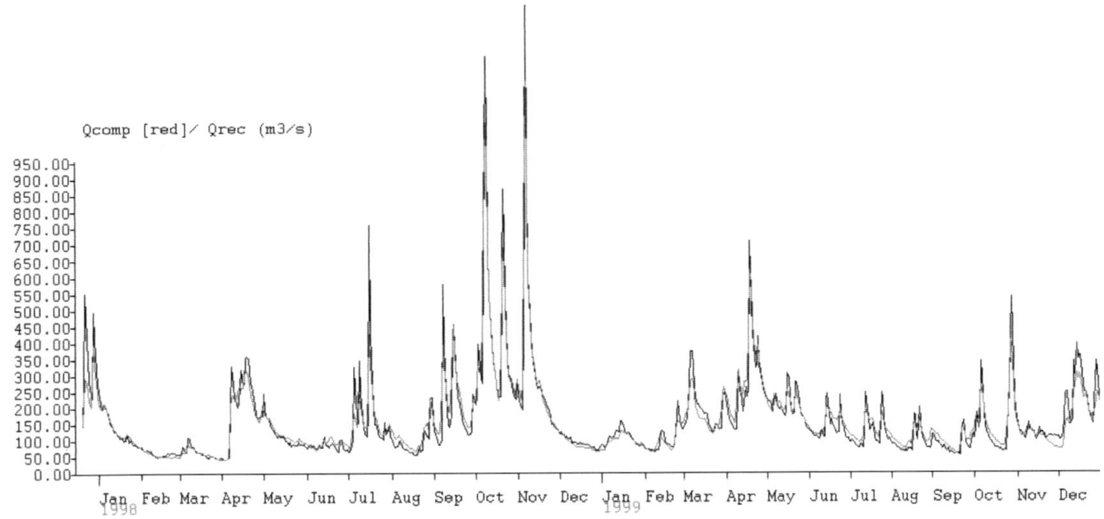

Drainage area 3

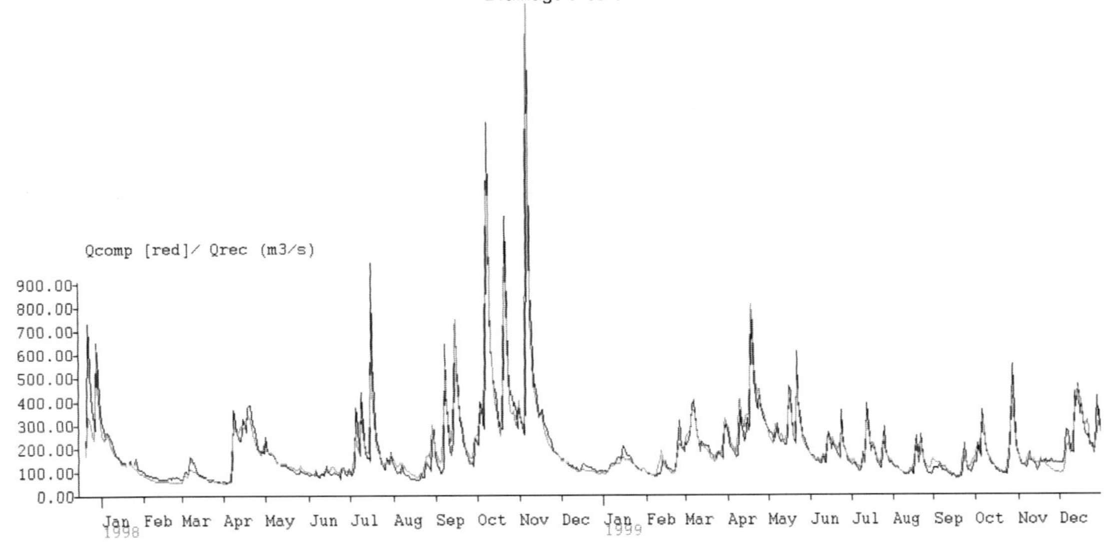

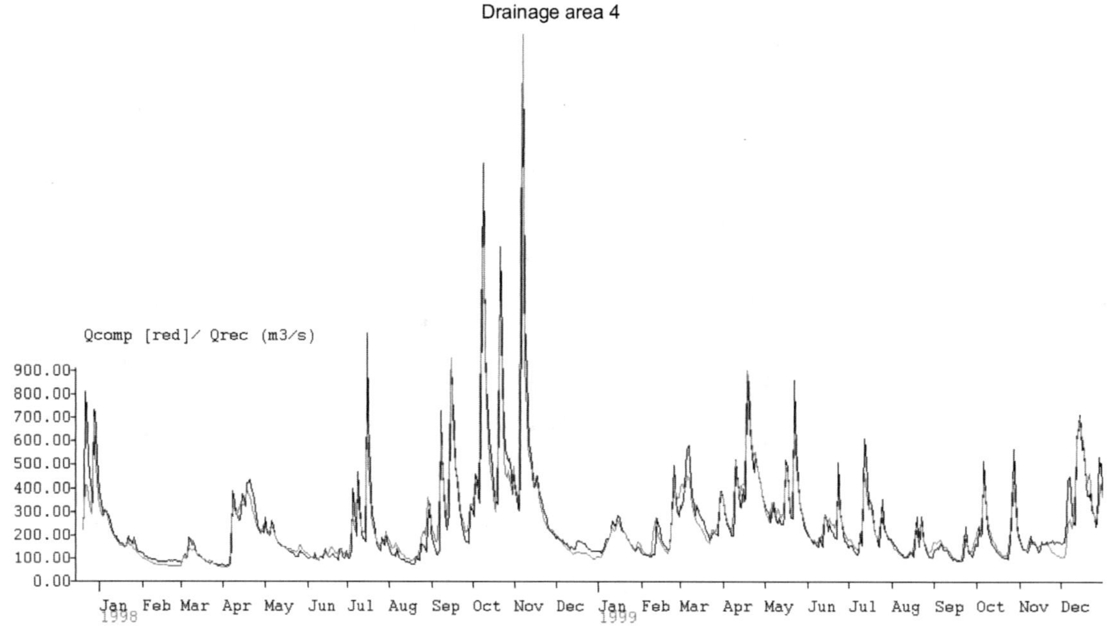

Figure 3. The comparison of recorded (Qrec) and computed (Qcomp [red]) discharges for four different drainage areas (Figure 2) from the 24-hour model.

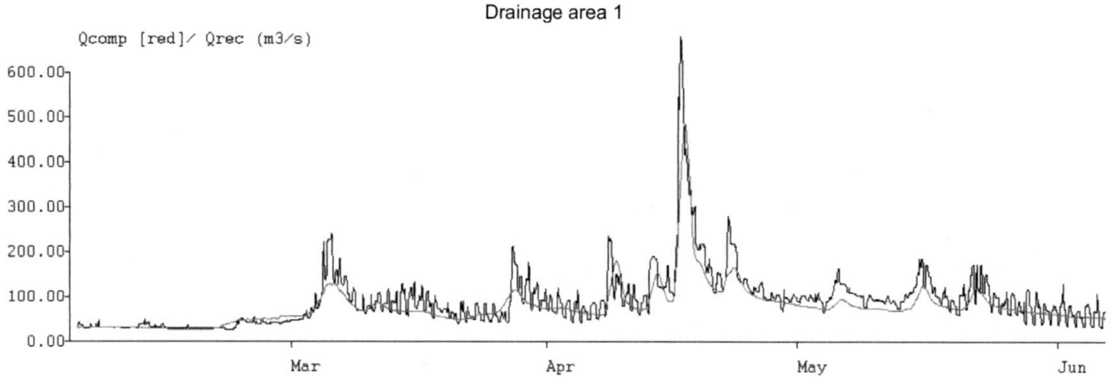

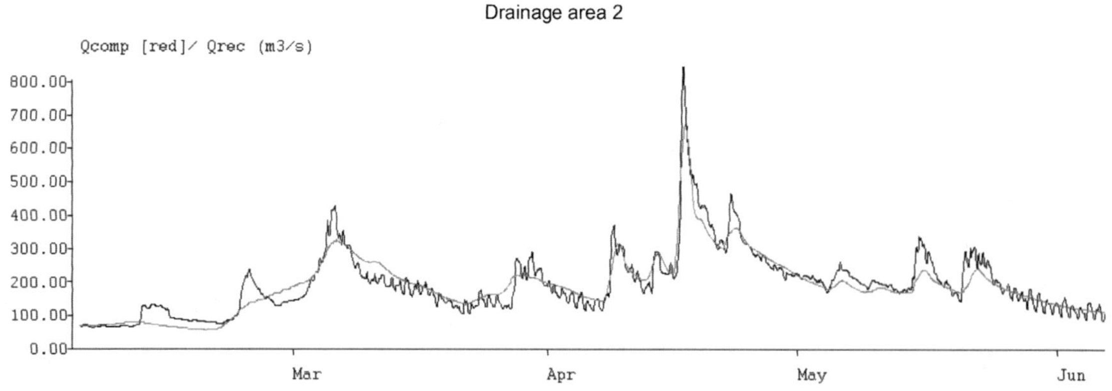

IOP Conf. Series: Earth and Environmental Science 4 (2008) 012004 doi:10.1088/1755-1307/4/1/012004

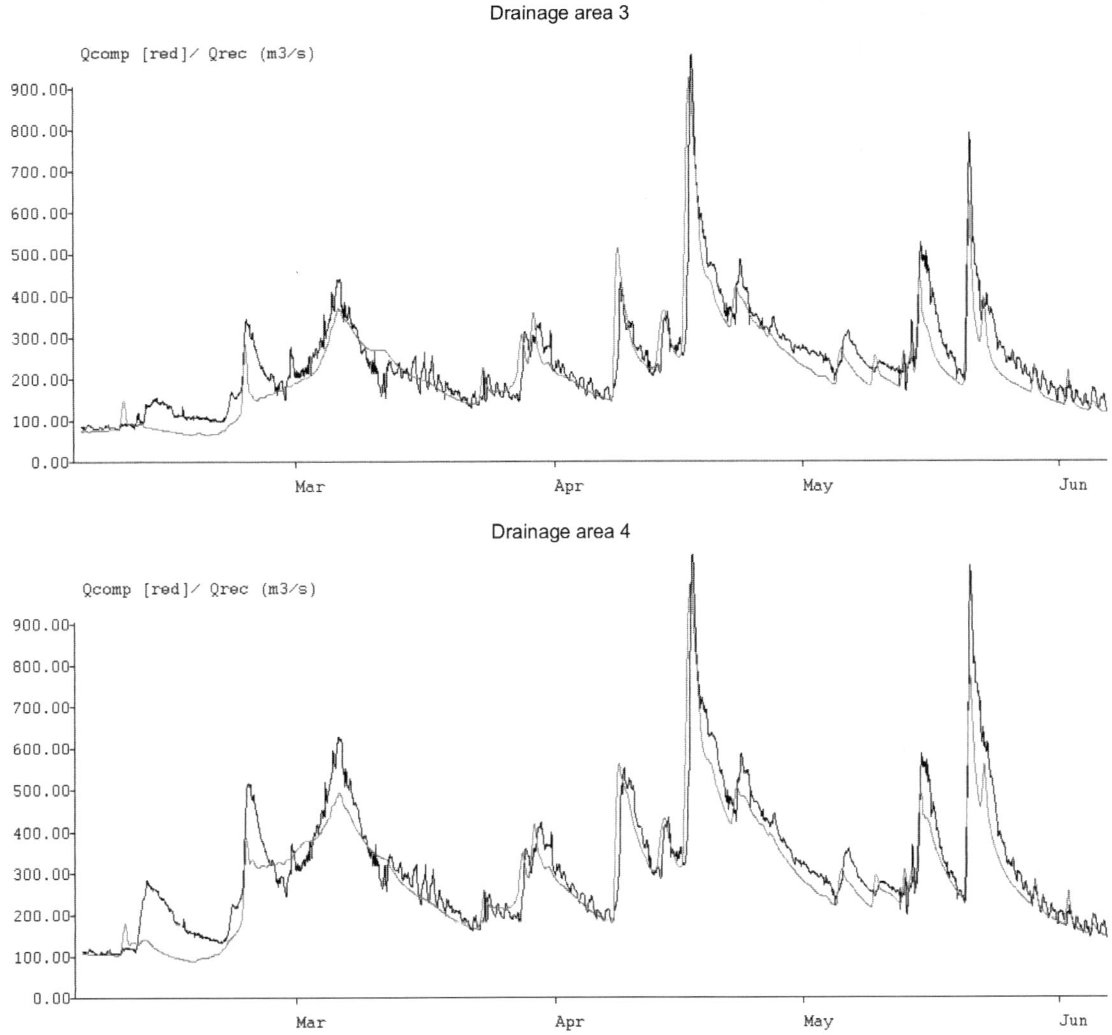

Figure 4. The comparison of recorded (Qrec) and computed (Qcomp [red]) discharges for four different drainage areas (Figure 2) from the 1-hour model.

The Nash/Sutcliffe criterion R^2, ranging from 0 to 1, was calculated for each subbasin (a value 1 represents the perfect performance). The result of present calibration is ranging from 0,644 to 0,898. It shows very good agreement between the recorded and computed discharges. The results of verification of the 24-hour model showed that the error was just slightly higher comparing with the calibration period with R^2 between 0,87 and 0,90.

Table 1. Comparison of R^2 values for both models

	24 hour timestep model		1 hour timestep model
	Calibration period	Validation period	Calibration period
Drainage area	R^2	R^2	R^2
1	0,89825	0,87905	0,64379
2	0,89749	0,90322	0,87061
3	0,86652	0,86942	0,83800
4	0,88410	0,87368	0,86132

Analysis indicated that the lack of input precipitation data caused the poor result of calibration for some subbasins. To make the model more accurate, the number of precipitation and other stations should be increased (especially for 1-hour time step model). However, the additional stations are possible to add after calibration process, but the model has then to be recalibrated in order to get useful results.

4. Conclusion and further use of the model

The results of calibration of two models for the Sava river basin with two different time steps, day and hour, are presented in this article. The model with timestep of one day was initially set up for easier designing (calibrating) of the 1-hour time step model (only 5 calibration parameters needed to be recalibrated), although the model can also be used for further studying of calibration parameters, water balance studies, simulations of the effects of a changing climate, filling in the missing input data sets, etc. The model with time step of one hour is much more suitable for real time flood forecasting or designing than the 24-hour timestep model. Only 16 recording raingauges used in one hour timestep model on the catchment of 10.098,49 km^2 and its closeness is not sufficient for accurate estimation of areal precipitation. Additional input is needed to get even more reliable river runoff output. The number of raingauges is namely crucial for the estimation of areal precipitation. Small number of raingauges leads to the deviation of runoff from observed values. Analyses show great deviations especially by small amount of precipitation on the catchment when there is storm and usually local precipitation [1]. The analysis of number of raingauges on the areal estimation of precipitation on Savinja river subbasin [1] has shown, that relative error can exceed 100 % in the case of small amount of precipitation. The relative error is smaller by precipitation causing high waters and floods and can reach 50 %. The correction of precipitation in the HBV model for some high water events showed that simulated runoff fits with measured one well by the proper precipitation input [1]. Satisfactory calibration of the model was achieved, although the topography of the catchement has strong influence on the meteorological happening, especially in the upper stream of the Sava river, and small number of included raingauges.

The model with one hour time step was further used for simulating discharges in more than 17 different rainfall scenarious, including scenarious with summer and autumn rainfall, heavy rainfall with moving velocity of 60 km/h, rainfall with intermediate dry period (example: 30 % probable maximum precipitation + 3 day dry period + 100 % probable maximum precipitation) and scenarious with snowmelt in connection with heavy summer rainfall.

References

[1] Kobold M 2007 *Vpliv napake ocene padavin na napako napovedi odtoka pri napovedovanju poplav* Doktorska disertacija. (Ljubljana: Univerza v Ljubljani, Fakulteta za gradbeništvo in geodezijo) p 139

[2] Bergström S 2002 *The HBV story in Sweden* (Sweden, Norrköping: Swedish Meteorological and Hydrological Institute) p 4

[3] Kobold M 2004 *IHMS-HBV 96 navodila za operativno delo* (Ljubljana: Hidrometeorološki zavod Republike Slovenije) p 7

[4] Lindström G, Johansson B, Persson M, Gardelin M and Bergström S 1997 Development and test of the distributed HBV-96 hydrological model *Journal of Hydrology* **201** 272-88 Brilly M and Šraj M 2000 *Osnove hidrologije* (Ljubljana: Univerza v Ljubljani, Fakulteta za gradbeništvo in geodezijo) p 234

[5] HMS 1999 *Integrated Hydrological Modelling System* Manual, Version 4.5. (Norrköping, Sweden: Swedish Meteorological and Hydrological Institute)

[6] Isaie D 2007 *Hydrological measurements and modeling - 2nd year Internship REA (2006-2007)* (France, Grenoble: Polytechnic Institute of Grenoble, Department of Hydraulics and the Environment) p 69

[7] Kobold M, Zgonc A and Sušnik M 2007 Nezanesljivost padavinskih meritev in napovedi pri

modeliranju hudourniških poplav = Uncertainty of precipitation measurements and predictions in flash flood modeling *Acta hydrotechnica* **23, 39**

[8] http://ksh.fgg.uni-lj.si/ksh/acta/ (07. 05. 2007).

[9] Primožič M 2007 *Umerjanje HBV modela za porečje reke Save v Sloveniji = Calibration of the HBV model for Sava watershed in Slovenia* Diplomska naloga. (Ljubljana: Univerza v Ljubljani, Fakulteta za gradbeništvo in geodezijo) p 81

[10] Reihan A, Kovalenko O 2000 Experience of an application of the HBV model for runoffcomputation in Estonia. *Proceedings of the Symposium dedicated to the 40th Anniversary of Institute of the Environmental Engineering at Tallinn Technical University*: pp 126-134

[11] Wood S J, Jones D A and Moore R J 2000 Accuracy of rainfall measurement for scales of hydrological interest *Hydrology and Earth system Sciences* **4(4)** 531-543

Continuous hydrological modelling in the context of real time flood forecasting in Alpine Danube tributary catchments

Ph Stanzel, B Kahl, U Haberl, M Herrnegger and H P Nachtnebel

Institute of Water Management, Hydrology and Hydraulic Engineering, University of Natural Resources and Applied Life Sciences, Muthgasse 18, 1190 Vienna, Austria

E-mail: philipp.stanzel@boku.ac.at

Abstract. A hydrological modelling framework applied within operational flood forecasting systems in three alpine Danube tributary basins, Traisen, Salzach and Enns, is presented. A continuous, semi-distributed rainfall-runoff model, accounting for the main hydrological processes of snow accumulation and melt, interception, evapotranspiration, infiltration, runoff generation and routing is set up. Spatial discretization relies on the division of watersheds into subbasins and subsequently into hydrologic response units based on spatial information on soil types, land cover and elevation bands. The hydrological models are calibrated with meteorological ground measurements and with meteorological analyses incorporating radar information. Operationally, each forecasting sequence starts with the re-calculation of the last 24 to 48 hours. Errors between simulated and observed runoff are minimized by optimizing a correction factor for the input to provide improved system states. For the hydrological forecast quantitative 48 or 72 hour forecast grids of temperature and precipitation – deterministic and probabilistic – are used as input. The forecasted hydrograph is corrected with an autoregressive model. The forecasting sequences are repeated each 15 minutes. First evaluations of resulting hydrological forecasts are presented and reliability of forecasts with different lead times is discussed.

1. Introduction

Devastating effects of extreme flood events in the recent past showed that in addition to structural protection the prediction of floods with early warning systems is needed to reduce damage. Rainfall-runoff models are at the core of predicting catchment response in the imminent future. This contribution outlines a hydrological modelling framework currently applied in newly developed flood forecasting systems in three alpine Danube tributary basins, Traisen, Salzach and Enns.

The described approach meets a number of specific requirements. A continuous simulation model is applied in order to provide adequate system states for all flood situations, which is of specific importance in alpine regions, where snow melt can contribute to floods. The heterogeneity of basin features and meteorological characteristics requires high spatial and temporal resolution of model structure and input data. Influences of hydropower production demand the incorporation of reservoir management modules. As alpine catchments have short response times, meteorological forecasts as modelling input are needed in order to achieve forecast lead times of 48 to 72 hours. Including ensemble forecasts contributes to taking uncertainty associated with forecasts into account.

2. Hydrological simulation

2.1. Methods

The applied model COSERO is a continuous, semi-distributed rainfall-runoff model developed by the Institute of Water Management, Hydrology and Hydraulic Engineering ([1], [2], [3], among others). It accounts for processes of snow accumulation and melt, interception, evapotranspiration, infiltration, soil storage, runoff generation and routing. Separation of runoff into fast surface runoff, inter flow and base flow is calculated by means of a cascade of linear and non-linear reservoirs, following the design of the HBV model [4].

Spatial discretization relies on the division of the watersheds into subbasins and subsequently into hydrologic response units (HRUs) based on available spatial information on subcatchment boundaries, soil types, land cover, and 200 m elevation bands.

The model allows redistribution of locally generated runoff in order to consider influences of karst cross-catchment groundwater flow [5] and diversions in the context of hydropower use. The effects of operation of reservoirs for power production and flood retention are considered in reservoir modules that try to display the operation based on predefined rules. In many cases, however, rules can hardly be defined, as short term power market transactions govern reservoir management.

In the Salzach basin COSERO is combined with a glacier melt module [6]. In the Salzach and Enns basins runoff simulated by the hydrological model is further transformed in a hydrodynamic model that also takes effects of river power plants into account [7].

Initial sets of parameters are estimated for each HRU, partly based on physical features of the zone, partly conceptual parameters that are estimated based on prior applications of the model. The initial estimates are calibrated and validated with historic time series of precipitation and temperature. The calibration procedure involves manual calibration until a satisfying representation of seasonal patterns is reached and optimization of the few parameters governing peak discharge.

2.2. Application

The Traisen, Enns and Salzach basins are located in the north-eastern Alps (see figure 1). The main attributes of the three basins and of their representation in the simulation model are given in Table 1. Enns and Salzach basins are larger than the Traisen basin, and reach to higher altitudes. They also have considerable glaciers in their catchments. In relation to the basin areas the spatial discretization is similar in all three basins, with the finest model structure in the Salzach basin, where the presented flood forecasting model replaces an existing with rather coarse spatial resolution [8].

Table 1. Basin attributes

	Salzach		Enns		Traisen	
Area (km²)	6700		6100		900	
Elevation range (m a.s.l.)	300 - 3800		300 - 3000		200 - 1800	
Nr. of subbasins / area range (km²)	82	2 - 308	41	3 - 651	10	40 - 125
Nr. of HRUs / mean size (km²)	1861	3.6	891	6.9	63	14.5

Meteorological input data for calibration and validation stem from two sources. Point measurements of daily precipitation and air temperature of ten years were regionalized using external drift kriging for precipitation and a linear regression model for temperature. For a shorter period – the period of the storm event in August 2002 and the time from 2003 – the Central Institute for Meteorology and Geodynamics (ZAMG) provided precipitation and temperature analyses derived with comparable algorithms as during online operation. These analyses, with a spatial resolution of a 1x1 km grid and a temporal resolution of 15 minutes or one hour, incorporate point ground measurements and radar information [9].

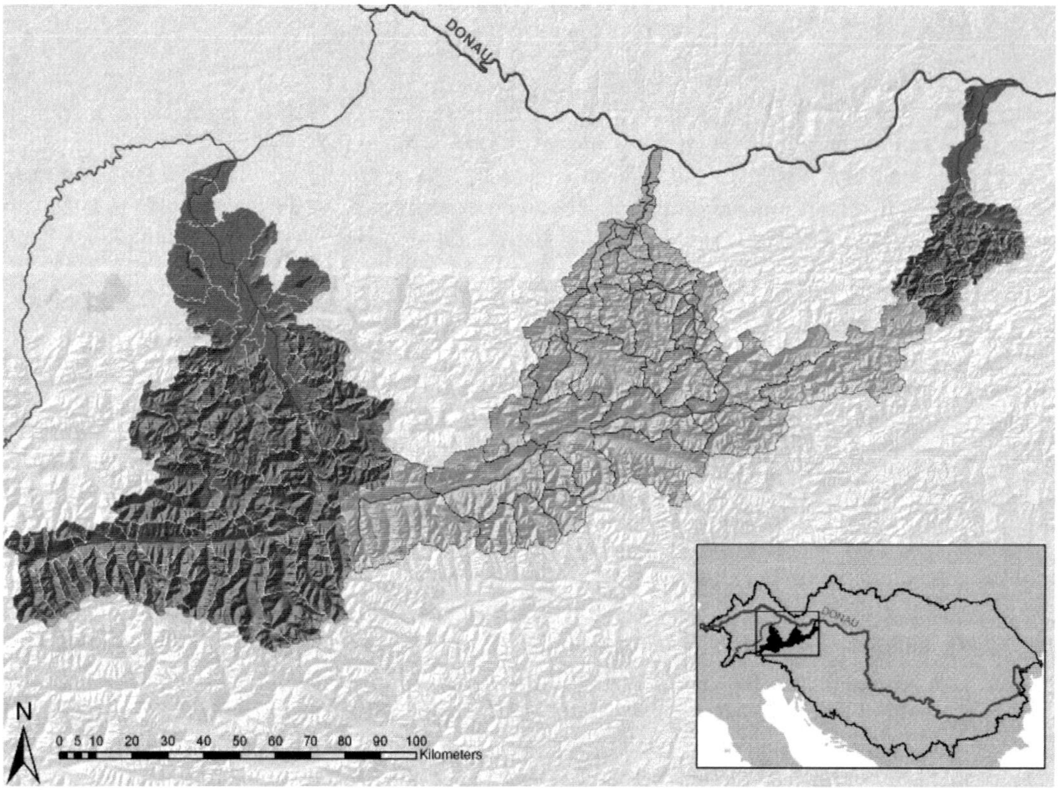

Figure 1. Map of the Salzach, Enns and Traisen basins (from west to east)

Figure 2 shows simulation results from the calibration period for the three basin outlet gauges. For the Enns, they are based on daily time steps and point measurement input, for Traisen and Salzach the input data were ZAMG analysis grids with 15 minutes and hourly time steps, respectively. The models generally show good performances, in the snow melt period as well as during summer flood events.

Only for the Traisen basin the calibration process has been completed so far. An evaluation of three flood events – which are of main interest in this context – in the validation period yielded an average Nash-Sutcliffe coefficient of model efficiency of 0.90 at the basin outlet gauge Windpassing. The mean relative peak error for these three events was 11 %, the mean peak time error was 1 hour.

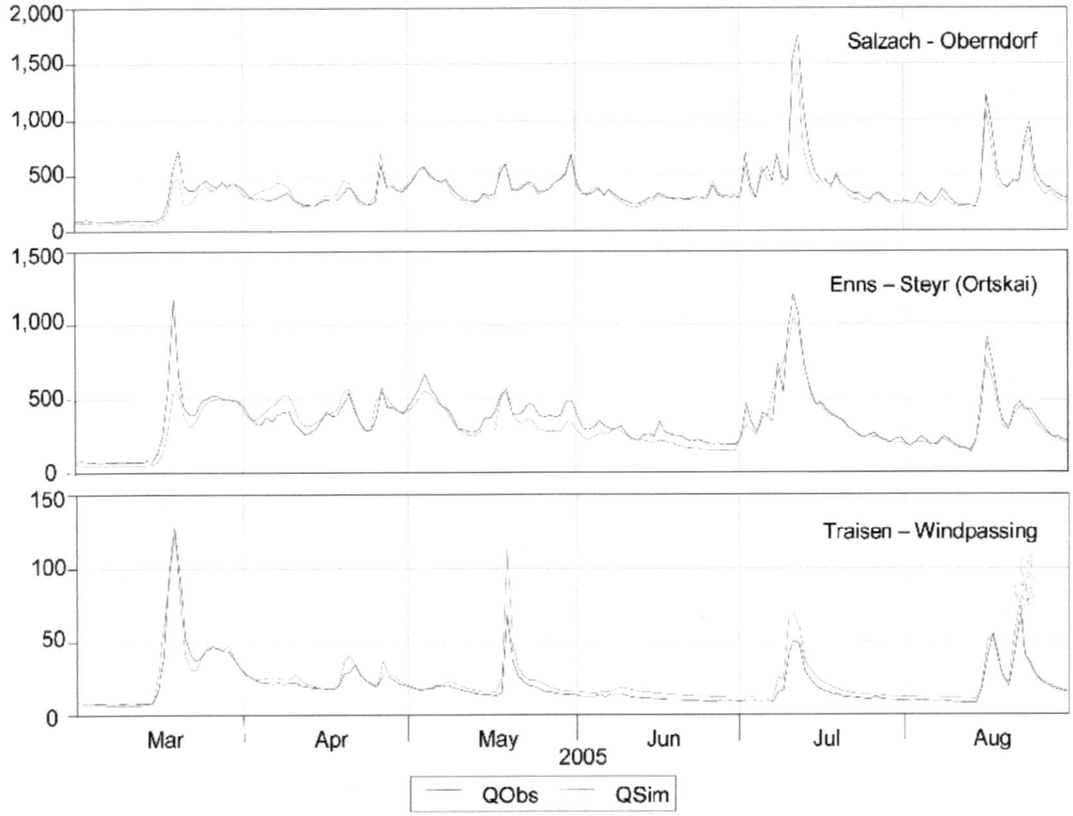

Figure 2. Simulation results (March to August 2005)

3. Real-time runoff forecast

3.1. Methods

Operationally, each forecasting sequence starts with the re-calculation of the last 24 to 48 hours, using the above described ZAMG precipitation and temperature analysis grids based on ground and radar measurements as input. Thus the system states at the beginning of the prediction period are determined. For the hydrological forecast quantitative 48 to 72 hour forecasts of precipitation and temperature are fed into the hydrological model. As the uncertainty of forecasted rainfall can be high, ensemble precipitation forecasts are used to estimate forecasting errors. The resulting hydrological ensemble is analysed statistically and the display of the 80 % confidence interval shows the uncertainty due to the meteorological forecast. The forecasting sequences are repeated each 15 minutes.

Model uncertainties are considered by applying online correction methods. In the presented framework, a combination of two online correction methods is used (see figure 3). The first method is applied in the recent history from t0 (date of forecast). It is a method to provide improved system states for the forecast at t0. Errors between simulated and observed runoff are evaluated and minimized by optimizing a correction factor for the input (precipitation and/or temperature). This is denominated updating and gives new system states at t0 for the start of the next forecast run. The second method (denominated output correction in figure 3) is an autoregressive (AR) model with eight discharge classes used for correcting the forecasted hydrograph. The online updating procedures are described in detail in [10].

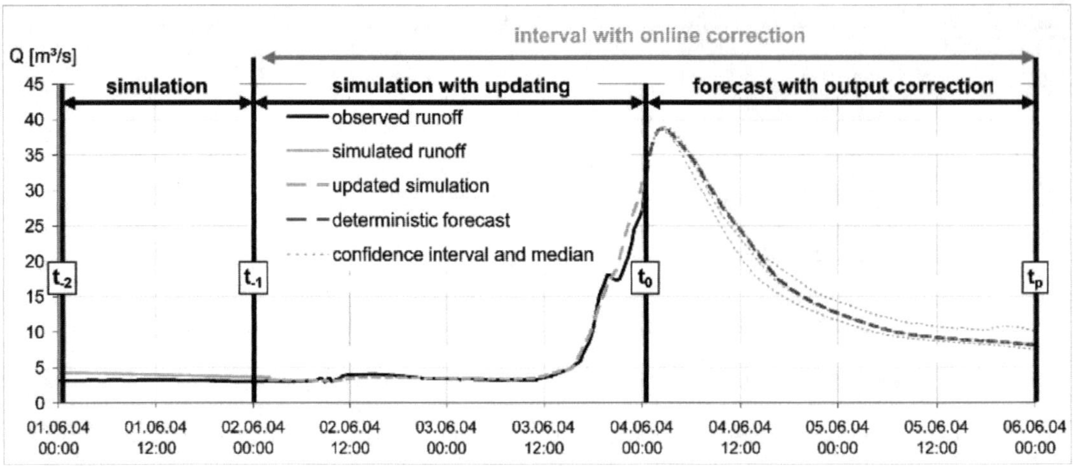

Figure 3. Online correction methods

Every 24 hours a special updating mode is started with meteorological re-analyses as input. These re-analyses can incorporate more input data than those transmitted in real-time every 15 minutes, which can be of importance in flood periods, when gauges transmit incomplete or inaccurate data.

3.2. Results

In the presented studies one deterministic forecast (resolution 1x1 km) and ensemble precipitation forecasts (10x10 km) from ZAMG are used as input data. These ensemble forecasts consist of two components: the first component is an observation-based extrapolation (nowcasting) of the precipitation field, the second component is a weighted mean of the forecasted fields of the ALADIN and ECMWF numerical weather prediction models. Ensembles are only provided by the ECMWF model [9].

Results from operational forecasting are so far available for the Traisen basin only. Reliability of forecasted runoff simulations for different forecasting time steps (τ hours ahead) was approached with a multiple-step-ahead forecasting technique. A multi-step-ahead hydrograph is built from the forecasted hydrographs at each time step of the event where only the value for τ hours ahead is taken out of every forecast. Figure 4 shows some multi-step-ahead hydrographs for one flood event in the Traisen basin. The graphs clearly depict the decrease of the forecast accuracy and the increase of the spread of the confidence interval with longer lead times.

The evaluation of similar analyses is shown in figure 5. The evaluation of the simulated runoff forecast (Qfc) is performed against two different reference hydrographs: the observed hydrograph (Qobs) and the best-possible simulation (Qref, which is the simulated hydrograph with observed rainfall data). The lines show the decrease of the simulation performance with increasing lead time (mean values for three flood events in the Traisen basin).

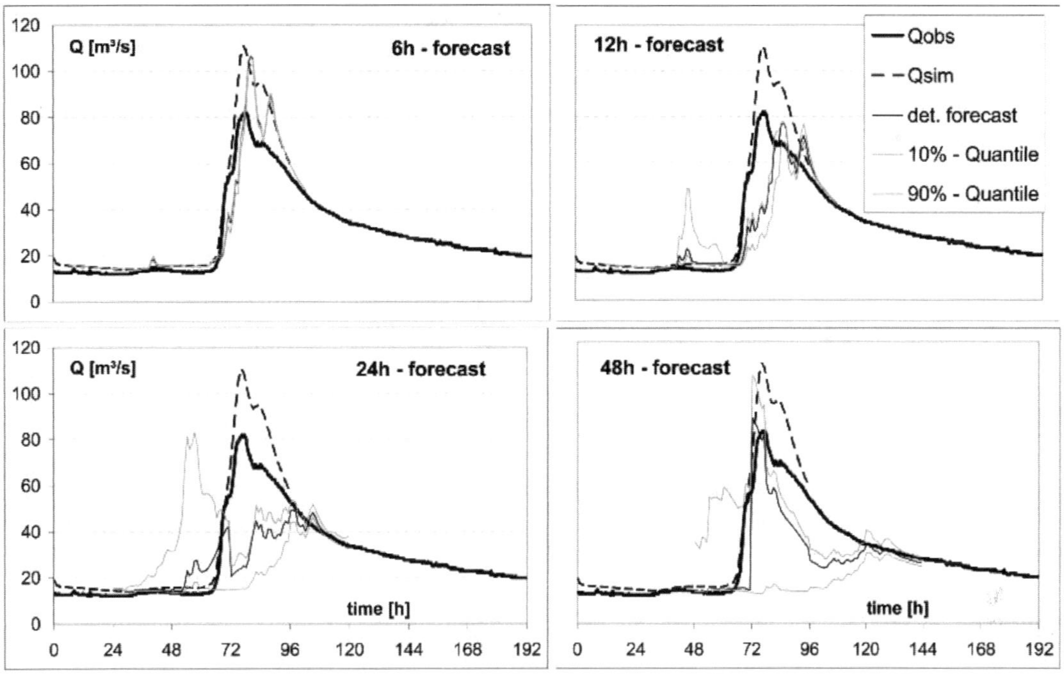

Figure 4. Multi-step-ahead analysis for one flood event for different forecasting time steps (τ=6 h, 12 h, 24 h, 48 h)

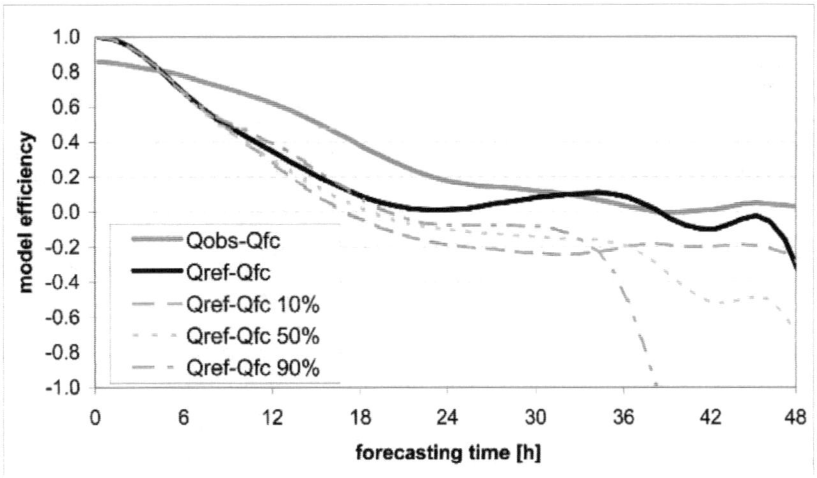

Figure 5. Mean Nash-Sutcliffe model efficiency for forecasting time steps up to 48 h for three flood events in the Traisen basin.

4. Conclusions

The presented modelling frameworks show recent examples of matching complex hydrological modelling with the requirements of operational real time flood forecasting. Models with high spatial resolution are applied and run with high temporal resolution. Operationally, meteorological forecasts with ensembles are used as input. Ensemble spread provides an indicator to assess potential forecast errors.

For lead times smaller than 12 hours, for the Traisen Basin, the ensemble spread is very narrow as forecasted runoff in this range of time is more dependent on catchment response characteristics and observed rainfall. For small lead times updating procedures show good results for improving forecasts.

For longer lead times larger uncertainties remain inherent to hydrological forecasts. Therefore uncertainty must be taken into account within the decision making process.

The practical use of real time early warning systems for decision making also depends upon the stable administration and easy-to-use visualization of the data involved in the hydrological modelling. In all three described applications the data management and visualization is provided by the system Delft-FEWS by Deltares (formerly WL | Delft Hydraulics).

Acknowledgements
The presented work has been carried out in projects funded by the Provincial Governments of Lower Austria, Upper Austria and Salzburg, VERBUND-Austrian Hydro Power AG and Salzburg AG. Data was provided by these institutions, the Provincial Government of Styria, the Federal Ministry of Agriculture, Forestry, Environment and Water Management, the Ennskraftwerke AG and the Central Institute for Meteorology and Geodynamics (ZAMG)

References
[1] Nachtnebel H P, Baumung S and Lettl W 1993 *Abflussprognosemodell für das Einzugsgebiet der Enns und Steyr (in German)* Report, Institute of Water Management, Hydology and Hydraulic Engineering, University of Natural Resources and Applied Life Sciences Vienna, Austria
[2] Fuchs M 1998 *Modeling snow-melt runoff in an alpine watershed* Master thesis, Institute of Water Management, Hydrology and Hydraulic Engineering, University of Natural Resources and Applied Life Sciences, Vienna, Austria
[3] Kling H 2002 *Development of tools for a semi-distributed runoff model* Master thesis, Institute of Water Management, Hydrology and Hydraulic Engineering, University of Natural Resources and Applied Life Sciences, Vienna, Austria
[4] Bergström S 1995 The HBV model *Computer Models of Watershed Hydrology* ed VP Singh (Highlands Ranch, CO: Water Resources Publications) pp 443-476
[5] Herrnegger M, Haberl U, Stanzel Ph and Nachtnebel, H P 2008 Identifying subsurface karstic drainage areas with a continuous hydrological model in the Austrian Kalkalpen *Geophysical Research Abstracts, Volume 10 (EGU General Assembly 2008, 13-18 April, Vienna)* European Geosciences Union EGU2008-A-09587
[6] Habersack H M, Krapesch G, Holzmann H and Koboltschnig G 2006 *Hitze- und Dürrefolgen für die Wasserwirtschaft Österreichs (in German)* Verbund, 72
[7] Reichel G 2008 *FluxDSS und FLORIS2000 – ein leistungsfähiges Paket zur Modellierung der Fließvorgänge in komplexen Fluss-Systemen (in German)* Available for download in March 2008 at: http://www.scietec.at
[8] Wiesenegger H and Kirchlechner P 1998 HYDRIS Hydrologisches Informations-System zur Vorhersage von Hochwässern im Land Salzburg (in German) *Proc. XIXth Conference of the Danube Countrie (15-19 June 1998, Osijek)* ed D Gereš and D Trninić (Zagreb: Hrvatske vode) pp 205-214
[9] Haiden T, Kann A, Stadlbacher K, Steinheimer, M and Wittmann, C 2007 *Integrated Nowcasting through Comprehensive Analysis (INCA) - System overview* ZAMG report, Vienna, Austria
[10] Kahl B and Nachtnebel H P 2008 Online updating procedured for a real-time hydrological forecasting system *Proc. XXIVth Conference of the Danube Countries (2-4 June 2008, Bled)* ed M Brilly and M Šraj (Ljubljana: Slovenian National Committee for the IHP Unesco)

IOP Conf. Series: Earth and Environmental Science **4** (2008) 012006 doi:10.1088/1755-1307/4/1/012006

The necessity of flood risk maps on Timiş River

Geogr. Cătălin Aldescu

National Administration "Apele Române", Banat Water Directorate, Timişoara, Romania

E-mail: catalin.aldescu@dab.rowater.ro

Abstract The paper aims to clarify the necessity of risk reduction in flood prone areas along the Timis River. Different methods to reduce risk in flood prone areas are analyzed as well. According to the EU Flood Directive it is mandatory for the European countries to develop flood maps and flood risk maps. The maps help to assess the vulnerable zones in the floodable (i.e. flood prone) areas. Many European countries have produced maps which identify areas prone to flooding events for specific known return periods. In Romania the flood risk maps have not been yet produced, but the process has been started to be implemented at the national and regional level, therefore the first results will be soon available. Banat Hydrographical Area was affected by severe floods on Timis River in 2000, 2005 and 2006. The 2005 flood was the most devastating one with large economic losses. As a result of these catastrophes the need for generating flood risk maps along the Timiş River was clearly stated. The water management experts can use these maps in order to identify the "hot spots" in Timis catchment, give the people a better understanding of flood risk issues and help reducing flood risk more efficient in the identified vulnerable areas.

1. General presentation

Banat Hydrographical Area is positioned in the South Western part of Romania; the Serbian border limits the South and West of the area. The Northwest limit is the Hungarian border. Banat occupies an area of 18.320 km^2, which represents 7,7% of the Romanian territory.

Banat Hydrographical Area is characterized by the presence of all forms of relief, whose altitude decrease from South-East to North-West.

Banat Hydrographical Area has a moderate continental temperate climate with Mediterranean influences as a result of Atlantic and Mediterranean air circulation. This type of climate generates the moderate character of the thermal regime: warmer periods in wintertime as well as high multi-annual precipitation, between 600-1400 mm/an.

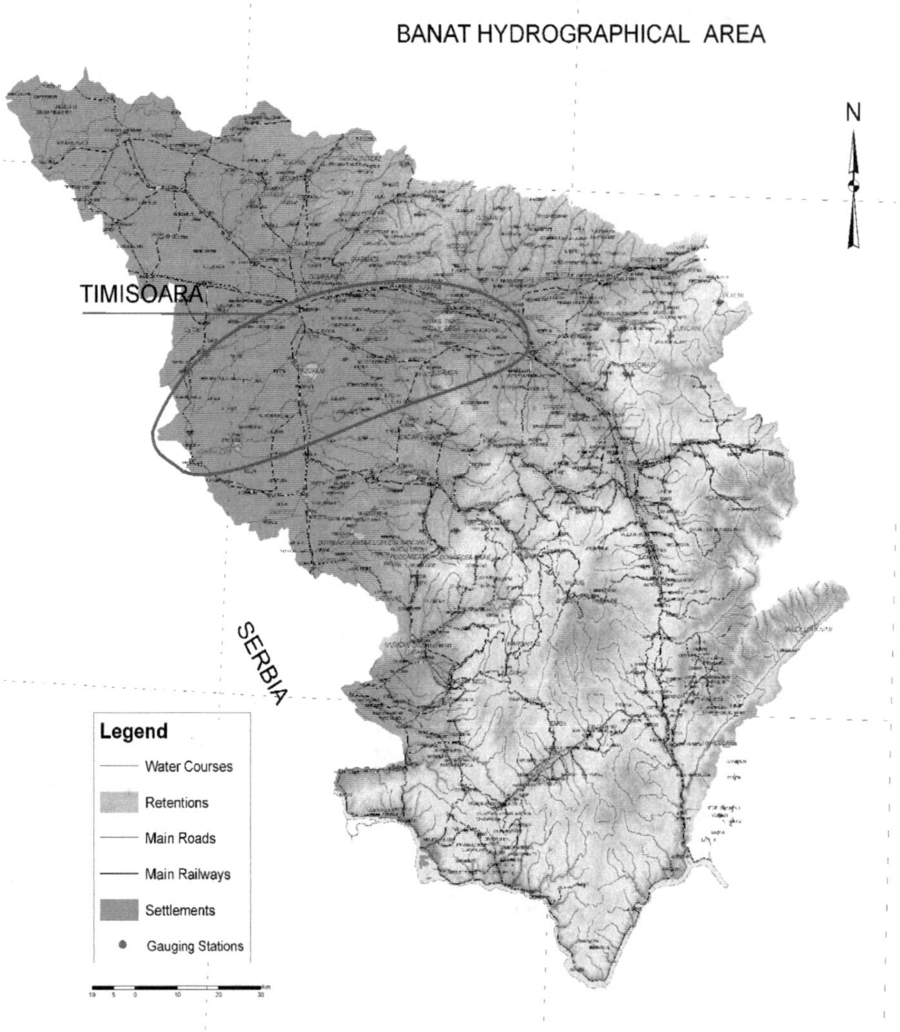

Figure 1. Banat Hydrographical Area (Source: [1])

Banat Hydrographical Area includes the following river basins:

Aranca R. B.	L= 389 km	S= 1201 km^2
Bega R. B.	L= 1418 km	S= 4470 km^2
Timiş R. B.	L= 2434 km	S= 7310 km^2
Caraş R. B.	L= 502 km	S= 1280 km^2
Nera R. B.	L= 574 km	S= 1380 km^2
Cerna R. B.	L= 524 km	S= 1360 km^2
Danube R. B. (partially)	L= 455 km	S= 1319 km^2

R.B. = river basin; L=river network length; S=catchment area

Timiş River with a river basin area of 5.673 km^2 and a main stream length of 241,2 km, has the largest extent of all rivers from the Banat Hydrographical Area. Its stream represents the main drainage artery of the rivers found in the catchment of the Banat-Godeanu-Ţarcu and Poiana Ruscă mountains.

Timiş River springs from the eastern slope of the Semenic Mountains, from the Piatra Goznei peak, on an altitude of 1.135 m. The upper stream has a typically mountainous character, with a slope of 20 m/km. The situation changes radically in the lower course (from Lugoj to Grăniceri) where the slope decreases up to 25 cm/km.

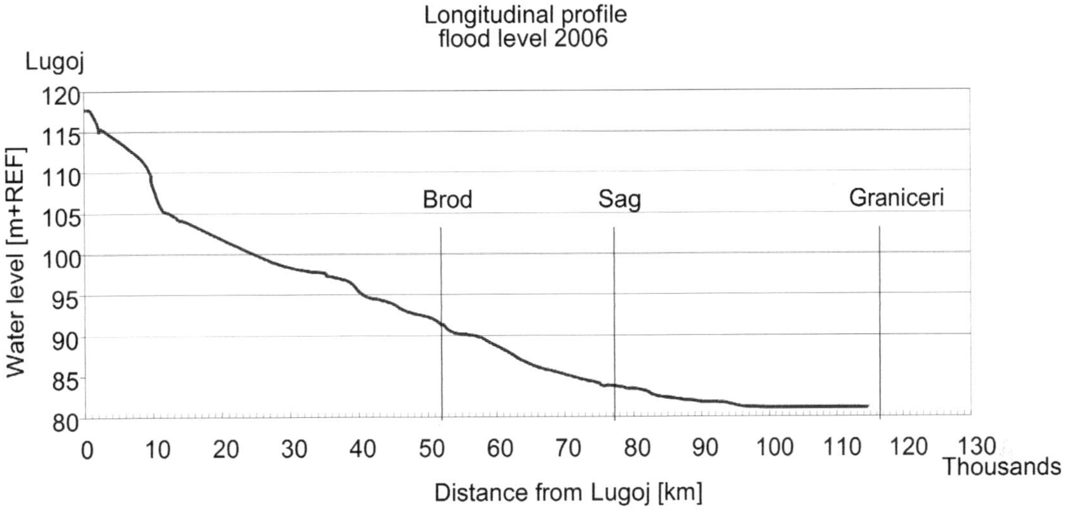

Figure 2. Calculated longitudinal profile during the 2006 flood (Source: [2])

2. Historical floods

In 1966 a catastrophic flood occurred on Timis River when the dyke breached on the left bank of the Romanian part of the river near the Romanian Serbian border. After a 30 years period without important high water events, significant flood waves occurred in February 1999 and April 2000. The 1999 flood wave passed without any inundations. However the 2000 flood wave caused breaching of the left embankment on Romanian territory flooding an area of about 17.500 ha in both Serbia and Romania, similar to the situation in 1966.

In April 2005, about the middle of the month, rapid snowmelt combined with heavy rainfall in the upstream catchment, generated an extraordinary flood wave on Timis River. In Serbia and Montenegro normal flood defence was declared on 16.04.2005, which turned to an emergency flood defence the next day. This resulted in applying the necessary operative flood defence measures. On 19.04.2005 the Romanian side informed about the heavy rainfall with a forecast of very high water levels on Timis River, similar to the ones in the catastrophic year 2000. On 20.04.2005 the Romanian territory was flooded through three breaches in the right embankment (km 6+700, 6+900 and 8+250 upstream of the Romanian – Serb Montenegrin border), flooding 30.000 ha and several villages in both countries.

The Timis River flood defence system is a complex one including several structures with protection purpose: a longitudinal defence line consisting in general of dykes, lateral water retention areas and a network of drainage channels for water divertion.

The dykes are located along the main stream in the lower part of the river. On Serbian territory, at Tomaşevăţ, in the border sector, the dyke on the left bank along Timis River forms a joint defence line for both Timis and Danube-Tisa-Danube Canal. The right bank is also protected; the embankment is continue from the Romanian territory until the junction with Danube-Tisa-Danube Canal at Botos, location where inundation is not possible due to the high grounds.

Water retentions areas and polders are located on Romanian territory. Most of them are located near the river main course in order to retain the high water from Timis River, but there are other lateral

IOP Conf. Series: Earth and Environmental Science 4 (2008) 012006 doi:10.1088/1755-1307/4/1/012006

non-permanent reservoirs which are built on tributaries in order to mitigate the simultaneous flood waves that might occur Timis tributaries.

The diverting channel network that exists on the Serbian territory within the Danube-Tisa-Danube Hydro technical System (D-T-D) improves the flood defence conditions in Banat Hydrographical Area. A unique classical system, formed by inland waters and the D-T-D canal, ensures the possibility of levels redistribution according to discharge on each river stream and according to level coincidence in Danube and Tisa emissaries.

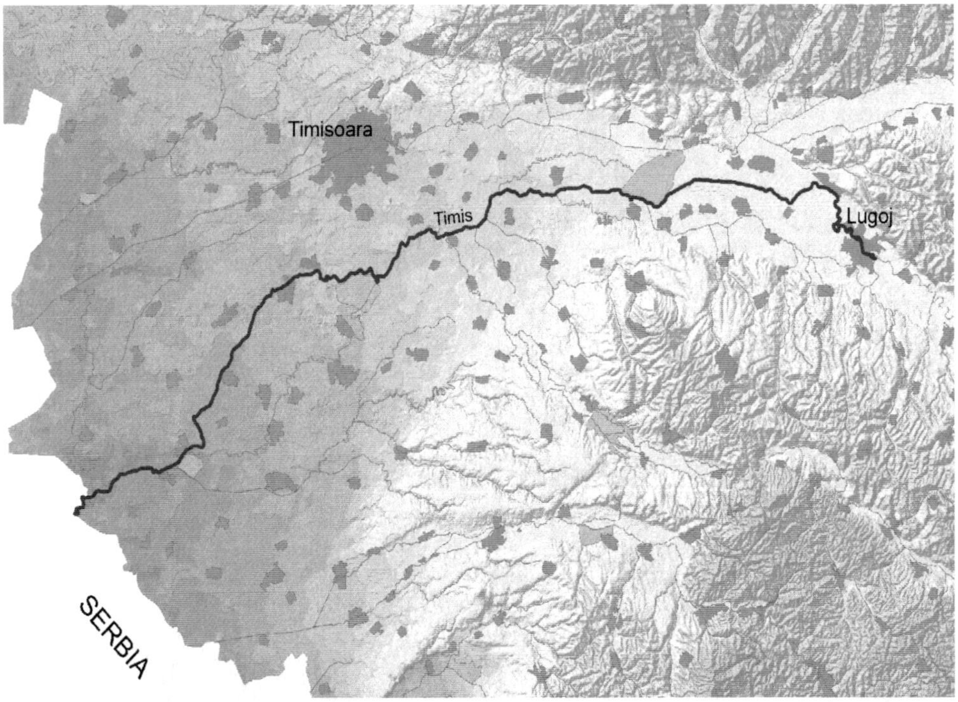

Figure 3. Timiş River – Lugoj-Grăniceri section (Source: [2])

The classical method of reducing flood risk consists mainly of structural measures like linear defence systems (dykes), water retention areas and drainage channels. However raising the dikes do not necessary reduces the flood risk. On the contrary, it can even increase the risk. Therefore vulnerable zones in floodable areas should be identified. Combined with floodplain maps and flood maps, flood hazard maps, flood damage maps and flood risk maps can be produced. These maps and the so called non-structural measures can help to reduce flood risk as well.

On January 18th 2006 the European Parliament and the EU Council issued a proposal for a Directive on the assessment and management of floods that hereafter is referred to as the "draft EU Flood Directive". The draft EU Flood Directive considers all elements of flood risk management (i.e. prevention, protection, preparedness, emergency response and recovery). The draft EU Flood Directive proposes that all EU member states should prepare flood risk maps and flood risk management plans.

The EU Flood Directive states that flood maps and indicative flood damage maps (or flood risk maps) should be prepared by EU member states for river basins, subbasins and stretches of coastline. These flood maps should be prepared for different probabilities of occurrence or return periods (T):

- Frequently occurring flood events (T = 10 year);
- Less frequently occurring flood events (T= 100 year), and;
- Extreme flood events (T >> 100 year).

For the zones assessed as vulnerable in the floodable areas non structural measure to reduce flood risk can be taken into account. The non structural measure that this paper focuses on is the development of floodplain maps, flood maps, flood hazard maps, flood damage maps and flood risk maps.

3. Flood risk maps

Floodplain maps. A floodplain map shows the area that will be flooded during a particular flood event, having a certain probability of occurrence (or return period). Basic information depicted on a flood plain map is the inundation depths. In addition flood plain maps might contain information on maximum flow velocities.

Flood hazard maps. A flood hazard map aims at providing information on the danger level of a certain flood prone area. Parameters of interest that might be depicted on a flood hazard map are for instance:

- maximum water depths,
- maximum flow velocities,
- maximum speed in which water levels can attain certain flood depths, including associated flow velocities.

Flood damage maps. A flood damage map provides information on the potential damage invoked by a particular flood event, having a certain probability of occurrence (or return period). Damage is usually expressed as the total number of casualties and the total amount of Euros expressing the economic losses.

Flood risk maps. A flood risk map provides information on the flood risk, expressed as the probability of flooding multiplied by its potential damage. Flood risk is usually expressed as the number of casualties per year and the amount of euros lost per year.

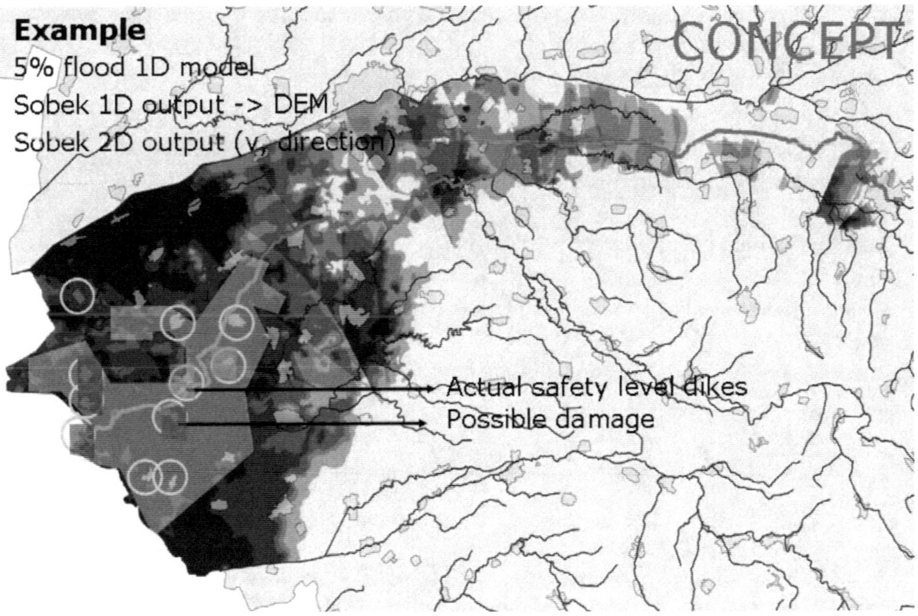

Figure 4. Possible: Flood Risk Map (Source: [2])

In order to draw flood maps a large amount of data is necessary: historical flood data, hydro-meteorological data, geographical data, socio-economic data.

The following data on historical floods have a great importance in drawing up flood maps:

- The probability of occurrence (or return period) of the flood,

- The extend of the flood,
- Observed discharge hydrographs, preferably as function of time,
- Observed flood levels, preferably as function of time,
- Maximum flood depths,
- Observed or estimated local maximum flow velocities,
- Dikes (or other water barriers) that can be overtopped,
- Details on possible dike breaches or collapsed water retaining structures,
- The effectiveness of evacuation procedures,
- The effectiveness of flood mitigation measures,
- Number of casualties,
- The loss of life-stock,
- Economic damage (agriculture, housing, industry, infrastructure, etc),
- Environmental damage (destroyed eco-systems, deposition of toxic agents, etc),
- The loss of historical and cultural values/the damage of historical and cultural assets,
- Indirect economic damage (i.e. loss of economic productivity).

The following hydro-meteorological data is required:

- Location of gauging stations, including the list of observed parameters,
- Location of meteorological stations, including list of observed parameters,
- Observed water levels at river gauging stations,
- Observed discharges at river gauging stations,
- Stage-discharge relationships, including stage-discharge data and the cross-sectional profile of the measuring section (preferably covering the river, including its floodplain areas),
- Hydraulic characteristics of structures (culverts, weirs, reservoirs etc); characteristics such as structural details and structure equations,
- Bathymetric data (i.e. river bed elevations),
- Type of river bed material (silt, sand, gravel etc) and type of vegetation in floodplain areas,
- Information on floods (observed flood levels and maximum flooding extend),
- Precipitation data.

Following geographical data is required:

- Digital elevation models,
- Location and elevation of vertical line elements, that can act as water barriers (i.e. dikes and (rail)roads),
- Location of culverts in vertical line elements,
- Land-use maps.

At last the socio-economical data present a special importance:

- Number of inhabitants,
- Number and value of life-stock,
- Investments in agriculture, housing, industry, etc.
- Investments in infrastructure (dikes, (rail) roads, communication-lines etc.),
- Location of toxic depots as well as the type and quantities of chemicals stored,

The preparation of the flood maps is a complex process, requiring overlaying various fields of expertise. The diversity of the experts that actually should be involved in developing flood maps for particular areas may be influenced by the following issues:

- The type of hydro-meteorological phenomena, that result in floods,
- The complexity of the hydraulic phenomena occurring in the river basin,

- Type of water retaining structures in the flood prone area,
- Degree of influence of flood mitigation measures effected in riparian states,
- The population density in the area (needs for evacuation).
- Type of investments in economic properties and life-stock,
- The presence of toxic depots,
- The presence of environmental, historical, and cultural assets.

The experts that may be involved in preparing these floods maps could be GIS experts, hydraulic engineers, hydrologists, structural engineers, environmental experts, chemical experts, institutional experts, socio-economists, statisticians. The expert list offered here is of course incomplete and may differ according to the actual needs.

4. Conclusions

As a result of the catastrophes mentioned the need for generating flood risk maps along the Timiş River was clearly shown. The water management experts can use these maps to identify the "hot spots" in Timiş catchment, give the people better understanding of flood risk and help to reduce the flood risk especially in vulnerable areas.

All classic or modern tools and means for fighting floods and their effects should be considered while designing or implementing flood protection measures.

Of course, the classical method of reducing flood risk with structural measures like linear defence systems (dykes), water retention areas and drainage channels is still needed. But nowadays experience shows that non structural measures like developing flood maps, flood hazard maps, flood damage maps and flood risk maps are of great importance to reduce flood risk. These measures will not stop floods but will increase public awareness of the danger that can occur during a flood event and will help the water manager to increase flood protection and prevention more effectively and more efficiently.

References

[1] * * * Banat Water Directorate Archive, Timişoara
[2] Joh Udo et. al. 2008 *Development of a strategy for improved protection against flooding and flood risk reduction along Timiş River, Final report* (HKV consultants 2008)
[3] Asselman N E M and Jonkman S N 2003 *Consequences of floods; the development of a method to estimate the loss of life* (Delft Cluster report DC1-233-7)
[4] European Commission {SEC(2006) 66} 2006 *Proposal of the European Parliament and of the Council on the assessment and management of floods* (Brussels, January 18th 2006)
[5] EXCIMAP Questionaire 1 2006 *European exchange circle on flood mapping (EXCIMAP), Questionnaire 1 ¬ flood mapping current practices* Ministére de l'Ecologie et du Développement Durable & Schweizerische Eidgenossenschaft, February 2006
[6] Jonkman S N, van Gelder P H A J M and Vrijling J K 2002 An overview of loss of life models for sea and river flood *Proc. Flood Defence `2002'* ed Wu et al (Science Press)
[7] Van Mierlo M C L M, Vrouwenvelder A C W M, Calle E O F, Vrijling J K, Jonkman S N, De Bruijn K M and Weerts A H 2003 *Effects of River System Behaviour on Flood Risk* Delft Cluster Project nr. DC 02.01.01. (http:/www.library.tudelft.nl/delftcluster ......Risk due to flooding)
[8] Werner M G F 2001 Impact on Grid Size in GIS Based Flood Extent Mapping Utility using a 1D Flow Model *Phys. Chem. Earth* B **26** 517-22
[9] Delft-FEWS Configuration Guide 2005 *National Flow Forecasting System, Delft-FEWS Configuration Guide* (prepared for the Environment Agency, UK)

IOP Conf. Series: Earth and Environmental Science **4** (2008) 012007 doi:10.1088/1755-1307/4/1/012007

Analyzing temporal changes in maximum runoff volume series of the Danube River

Dana Halmova[1], Pavla Pekarova[1], Jan Pekar[2] and Milan Onderka[1]

[1] Institute of Hydrology SAS, Racianska 75, 831 02 Bratislava, Slovakia
[2] Department of Applied Mathematics and Statistics, FMPI CU Bratislava, Mlynska dolina, 842 48 Bratislava, Slovakia

E-mail: halmova@uh.savba.sk

Abstract. Several hypotheses claim that more extremes in climatic and hydrologic phenomena are anticipated. In order to verify such hypotheses it is inevitable to examine the past periods by thoroughly analyzing historical data. In the present study, the annual maximum runoff volumes with *t*-day durations were calculated for a 130-year series of mean daily discharge of Danube River at Bratislava gauge (Slovakia). Statistical methods were used to clarify how the maximum runoff volumes of the Danube River changed over two historical periods (1876–1940 and 1941–2005). The conclusion is that the runoff volume regime during floods has not changed significantly during the last 130 years.

1. Introduction

Apart from the flood peak discharge another very important streamflow characteristic of a river is its runoff volume. Some issues in water management and engineering hydrology require that peak discharge ($Qmax$) and the shape of a flood wave or a flood runoff volume ($Vmax$) are known. In applied hydrology, it is often difficult to assign exact values of a flood wave volume to a particular probability of exceedance and hence to its corresponding T-year discharges. Such relationships are very irregular in nature, so a flood wave hydrograph of a given exceedance probability must be a priory known. Knowledge of flood-wave volumes – as an important hydrological characteristic – was apparent during a flood on the Danube at Komarno (Slovakia) in 1965 (Zatkalik, 1970; Hladny et al., 1970). During this flood event, the river dikes broke under the pressure exerted by the long duration of high water stages, but not because of the extremely high water stages themselves.

In assessing the climate change impacts on the river's runoff regime, it is expected that the increase in air temperature has caused an increase in the extreme flows and flood volumes Vt. Since a 130-year series of the mean daily discharge of the Danube at Bratislava gauging station [11, 12, 13, 14, 15] is available, we could calculate the 130-year series of the highest (annually) *2-, 5-, 10-, 15-, 20-, 25-, 30-* and *60-* consecutive days wave volumes. These series were subsequently divided into two 65-year sub-sets and changes in their probability distribution functions were analyzed.

The aims of the study are to:

1. assess the maximum annual runoff volumes Vt,max lasting *2-, 5-, 10-, 15-,20-, 25-, 30-,* and *60-* days, of the Danube River at Bratislava for the period 1876–2005;
2. determine theoretical exceedance probability curves for maximum runoff volume series;
3. analyze the changes in the maximum runoff volumes for periods: 1876–1940 and 1941–2005.

2. Methods

In Czechoslovakia, Bratranek [2] was the first who investigated the issue of runoff volumes. He used direct and indirect methods of peak vs. flood volume assessment.

The direct method was based on compiling runoff volumes higher than a chosen discharge threshold with assistance of the probability of exceedance (related to the *T*-year return period). The *T*-year discharges were then determined by extrapolating probability curves to the domain of low exceedance probability. In the latter method the author used generalized results of empirical relationships among several flood hydrograph characteristics. In order to determine the runoff volume $W_{20,Q1}$ (Equation 1), the author used not only the highest discharge Q reduced to $(Q_{20} - Q_1)$, but also values of the flood wave duration t_{20}, related to the catchment area.

$$W_{20,Q_1} \approx \frac{3600.t_{20}.Q}{2.10^6} \approx \frac{t_{20}.Q}{556} \quad (1)$$

where: $W_{20,Q1}$ – runoff wave volume with the extreme discharge Q_{20} above the selected threshold Q_1 [mil. m^3];
Q – reduced extreme cumulative discharge $(Q_{20} - Q_1)$ [m^{3}s^{-1}];
Q_{20} – extreme *20*-year discharge;
Q_1 – extreme *1*-year discharge;
t_{20} – the average flood wave duration of the reduced discharge $(Q_{20} - Q_1)$ [hrs].

The main disadvantage of this method is that it is not applicable for flood wave volumes with smaller exceedance probabilities than once in 20 years.

Cermak [3] studied the flood wave volumes of flood events with peaking above the long-term annual mean discharge Q_a on the regional scale. Beard [1] also used theoretical exceedance probability curves and the parameter t – for flood wave duration, to determine the annual extreme runoff volumes. The values of t- parameter depend on the catchment characteristics of the river being studied [5, 9]. However, the author selected only one value for the extreme runoff volumes in every year of the data sets.

A method described in detail by [10] was used to calculate the maximum runoff volume series for various runoff wave duration $t = 2, 5, 10, 15, 20, 25, 30,$ and *60* days.

3. Results

3.1. Assessment of the maximum t-day runoff volume (1876–2005)

In calculations of the annual maximum runoff volumes of particular durations, a reconstructed 130-year series of mean daily discharge (1876–2005) of the Danube at the Bratislava gauging station was used (Figure 1), Pekarova et al. (2007a). The maximum annual *t*-day flood volume was determined for each year. The runoff volume series are shown in Figure 2. Considering the *2*-day and *5*-day maximum runoff volumes, the flood of 1899 was the highest one within the period 1875–2005. But considering the *10*- to *60*-day runoff volumes, the highest flood was that of 1965.

3.2. Theoretical exceedance probability curves of the maximum Danube runoff wave volumes

First, a theoretical probability distribution function was chosen for the given duration of runoff and then the *T-year* extreme runoff volumes were calculated. The theoretical distribution function was chosen.

IOP Conf. Series: Earth and Environmental Science **4** (2008) 012007 doi:10.1088/1755-1307/4/1/012007

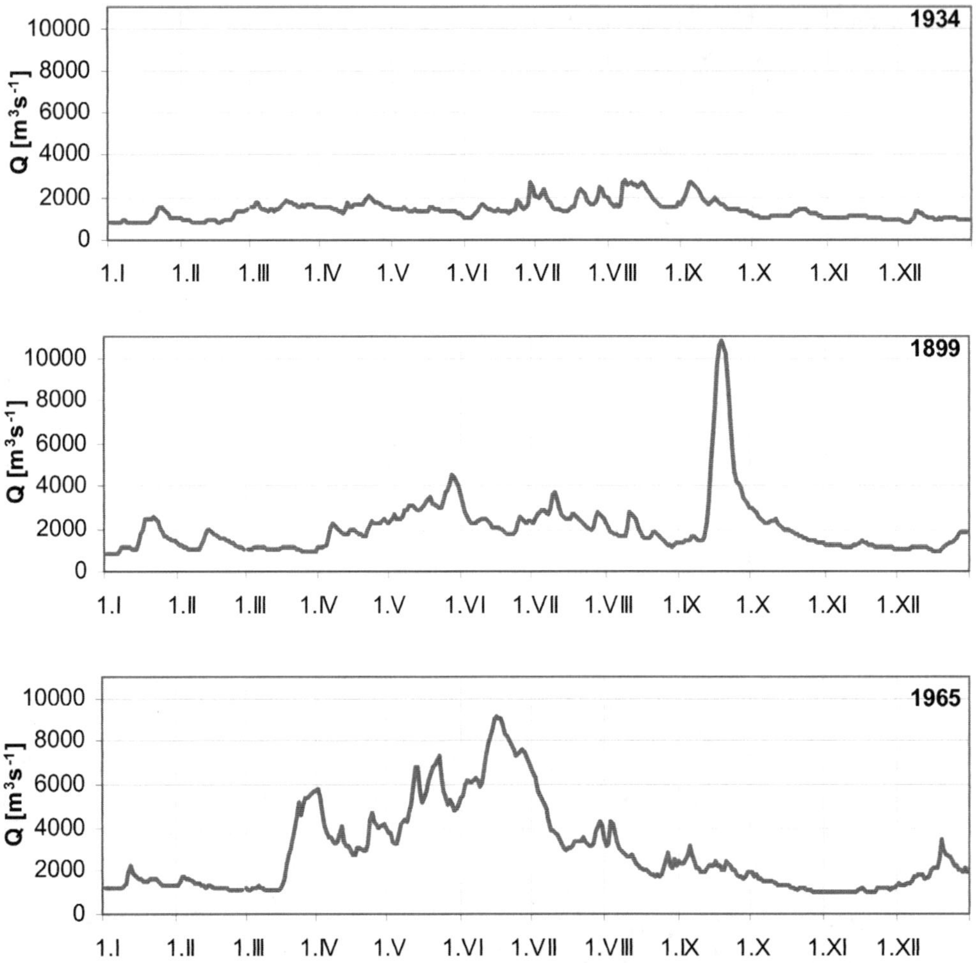

Figure 1. Example of the mean daily discharges (Q).
Year 1934 – the lowest *60*-days runoff volume V_{60};
year 1899 – the highest *2*-days runoff volume V_2; and
year 1965 – the highest *60*-days runoff volume V_{60}.

The relationship between the probability of exceedance p of a certain V_{max} value of the year and the mean return period T is $p = 1/T$. In Slovakia, it is customary that the following formula is used [4, 7, 15]:

$$p = 1 - e^{-1/T}. \tag{2}$$

A log-normal distribution was selected to calculate the *T-year* maximum runoff volume with the given runoff time duration (t). The theoretical (log-normal) exceedance probability curves of the runoff volumes with durations t equal to *2-, 5-, 10-, 15-, 20-, 25-, 30-*, and *60-* days are demonstrated on a logarithmic-probability scale in Figure 3.

The results suggest (Table 1) that the *100*-year maximum of *2*-day runoff volume (V_2) is 1766 mil m^3, 4079 mil. m^3 for *5*-day (V_5), and 7004 mil. m^3 for *10*-day (V_{10}) runoff volumes.

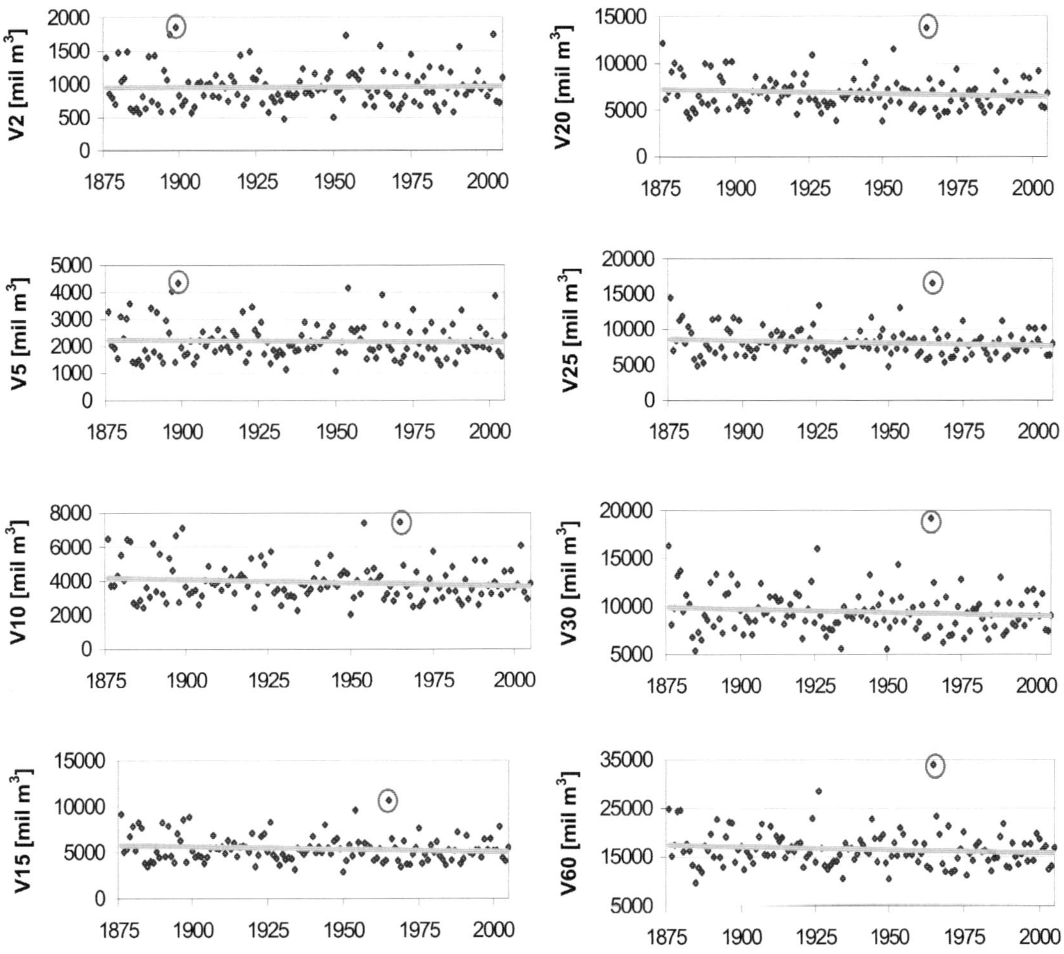

Figure 2. Flood wave volume series of the Danube for various flood durations *t* (e.g. V_{25} means runoff volume in 25 days).

Table 1. The *T*-year peak flow *Qmax* [m^3s^{-1}] and the *T*-year runoff volumes V_t [mil m^3] of the Danube at Bratislava within 1876–2005.

T	1000	500	200	100	50	20	10	5	2	1
P=p.100 %	0.1	0.2	0.5	1	2	4.9	9.5	18	39	63
Qmax	12920	12195	11231	10494	9743	8740	7959	7147	6011	5089
V2	2189	2062	1894	1766	1636	1462	1327	1188	993	836
V5	5070	4773	4379	4079	3775	3370	3055	2730	2278	1915
V10	8566	8100	7480	7004	6519	5868	5359	4829	4084	3477
V15	11281	10706	9936	9343	8735	7916	7271	6596	5638	4848
V20	13874	13189	12271	11562	10834	9849	9073	8256	7093	6130
V25	16172	15402	14368	13567	12742	11625	10741	9807	8473	7362
V30	18350	17504	16364	15479	14567	13328	12345	11305	9811	8563
V60	30658	29352	27586	26209	24782	22833	21276	19618	17217	15187

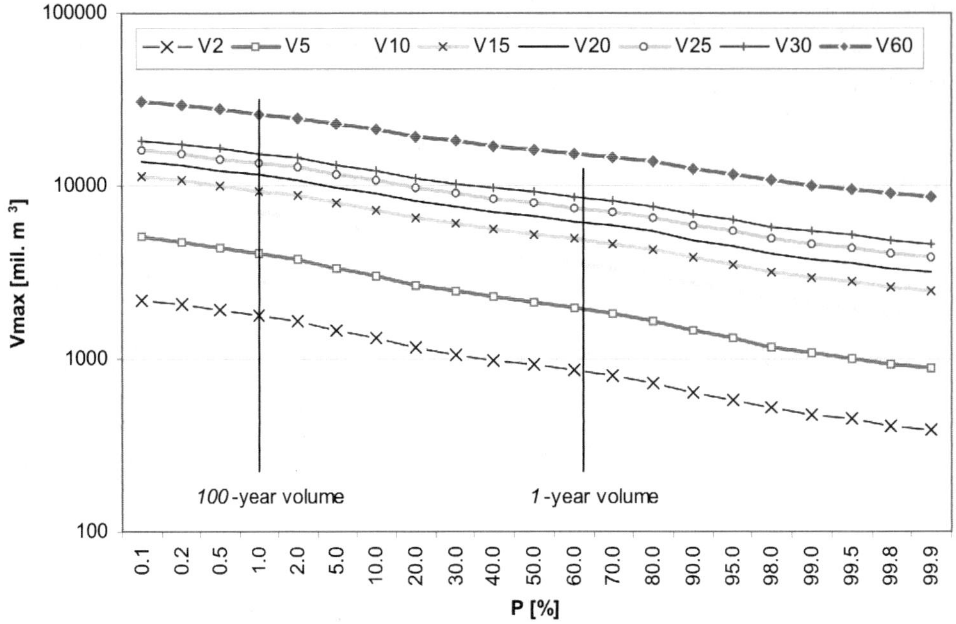

Figure 3. Theoretical (log-normal) frequency curves of the runoff volumes of the Danube at Bratislava, for duration *t* = *2-, 5-, 10-, 15-, 20-, 25-, 30-,* and *60-* days, on logarithmic-probability scale.

3.3. Maximum runoff volumes analysis for two periods: 1876–1940 and 1941–2005

The IPCC report [6] projects that the increase in air temperature may cause (or already has caused) elevation of extreme streamflows, flood volumes and the number of flood events [8]. A hypothetical example illustrating how the probability density curves of precipitation and runoff series could change is illustrated in Figure 4.

The long-term observations of discharge of the Danube (at Bratislava), and the evaluated series of flood volumes are suitable for testing the severity of the climate change impacts. We divided the original series of 130 years into two sub-sets of 65 years. The basic statistical characteristics of both sub-sets are shown in Table 2.

Table 2. Statistical characteristics of the sub-sets with flood volumes *Vt* [mil m^3] (*Vta* = 1876–1940 and *Vtb* = 1941–2005) for duration *t* = *2-, 5-, 10-, 15-, 20-, 25-, 30-*, and *60-* days.

	V2a	*V2b*	*V5a*	*V5b*	*V10a*	*V10b*	*V15a*	*V15b*
Sum	61201	63014	142199	142866	260081	251060	359266	346474
Aver	942	969	2188	2198	4001	3862	5527	5330
Stdev	**299**	**262**	**702**	**619**	**1166**	**1044**	**1467**	**1383**
Min	477	501	1156	1093	2265	2030	3160	2924
Max	1852	1737	4336	4152	7124	7457	9203	10696
	V20a	*V20b*	*V25a*	*V25b*	*V30a*	*V30b*	*V60a*	*V60b*
Sum	452333	435796	539295	521892	623038	606208	1092510	1067070
Aver	6959	6705	8297	8029	9585	9326	16808	16416
Stdev	**1761**	**1699**	**2030**	**1947**	**2257**	**2205**	**3641**	**3653**
Min	3882	3854	4798	4759	5413	5587	9708	10511
Max	12147	13799	14500	16509	16335	19179	28549	33998

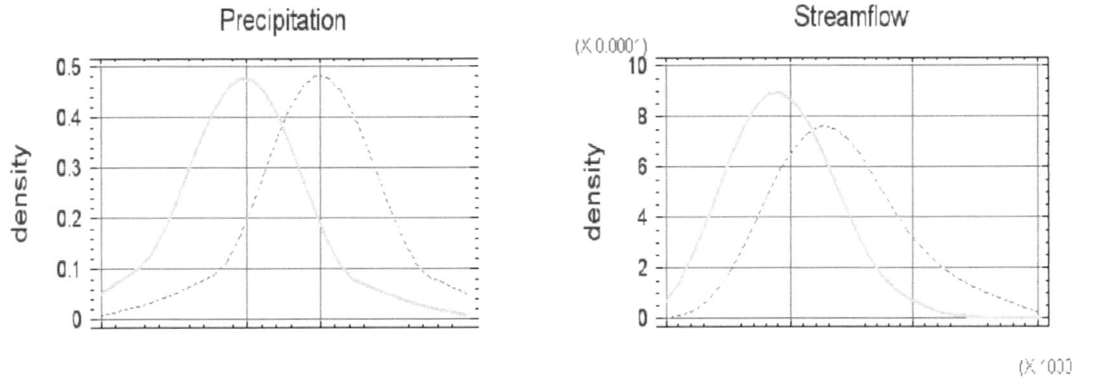

(a)

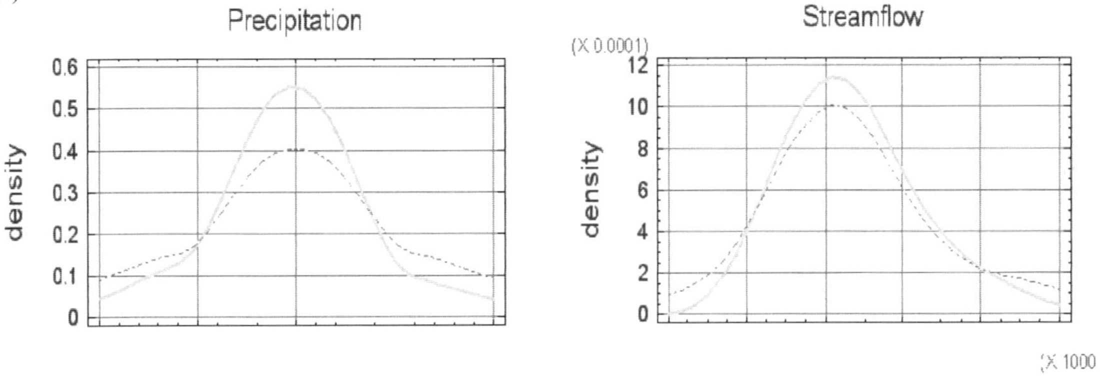

(b)

Figure 4. Hypothetic change of the density curves of the historical (solid line) and recent periods; a) changes in mean; b) changes in variance.

We tested the hypotheses to find whether the investigated subsets (1876–1940 and 1941–2005) are from the same population. Changes in mean values and in standard deviation were subject to an analysis also. For testing of the hypotheses we used well-known statistical tests t-test (for changes in means) and F-test (for changes in standard deviations).

Test results for means (t-test) are listed in Table 3. The *P*-value is in all cases above 0.05, therefore the null hypothesis stating that both periods have the same mean can not be rejected at the 0.05 confidence level. Also, similar results were found in the case of standard deviations (F-test).

The relevant density curves are demonstrated in Figure 5. Figure 6 illustrates how the probability density curves of the flood volumes changed. It is evident that the curves changed in both directions.

Table 3. Results of the tests of means (variable F and P-value).

	F	*P*-value		F	*P*-value		F	*P*-value
V2	0.319	0.5727	V15	0.619	0.4328	V30	0.437	0.5095
V5	0.008	0.9297	V20	0.702	0.4035	V60	0.374	0.5418
V10	0.511	0.4759	V25	0.589	0.4443	Qmax	0.897	0.3452

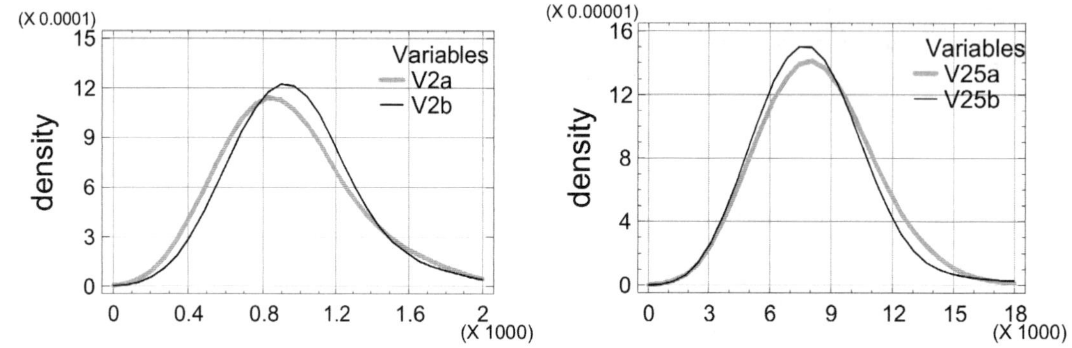

Figure 5. The probability density curves of runoff volume Vt for two periods:
V_{ta} – historical period 1876–1940; V_{tb} - recent period 1941–2005; t =2 and 25 days.

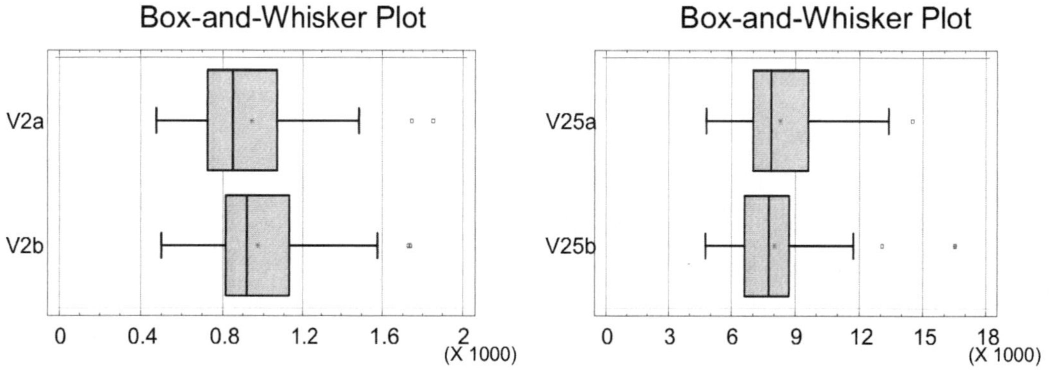

Figure 6. Changes in the statistical characteristics runoff volume Vt
of the historical and recent series for flood durations of: a) 2 days, and b) 25 days.

4. Conclusion

Based on the results presented in Table 2 and Table 3 we can conclude that the hypothesis claiming that both series are from the same distribution can not be rejected at the 0.05 significance level. The changes in mean and variance are insignificant (Figure 5 and Figure 6). These results have been confirmed by statistical tests (test of mean equality and test of standard deviation equality) of historical and recent periods for all runoff volume series *V2–V60*).

The conclusion is that the runoff volume regime during floods (in terms of mean values, variances and consequently the probability distribution function) has not changed substantially during the last 130 years, which is of importance to water management. This conclusion pertains not only to the short-term flood runoff episodes (*V2*), but also the long-term ones (*V60*). There is also a possibility to assess the potential flood extent in the case that the flood bank of the Danube River would break again, like in 1965.

With respect to the fact that the Danube River at Bratislava is, to a great extent, a typical Central-European mountainous river with its characteristic flow regime, it is possible to extrapolate these results (particularly those claiming no climate change impact) also to other rivers in Slovakia originating in the region of the Carpathian Mts. In the future, however, it will be desirable to confirm

IOP Conf. Series: Earth and Environmental Science **4** (2008) 012007 doi:10.1088/1755-1307/4/1/012007

this hypothesis also on other rivers in Slovakia with satisfactory long runoff data series, possibly on the base of reconstructed discharge values by indirect methods (analogy, mathematical runoff modeling etc.).

Acknowledgement
This study was supported by projects MVTS "Low flow and hydrological drought in Danube basin" and VEGA-0096/08, and the EU project GOCE 037063 "GEO-BENE".

References

[1] Beard L R 1956 Statistical Evaluation of Runoff Volume Frequencies Symposium Dorcy IASA Dijon

[2] Bratranek A 1937 Determination of the flood protection storage in reservoirs (in Czech) *Publication No 20* State Hydrological and Hydrotechnical Institute Prague

[3] Cermak M 1956 Flood wave volumes *J. Hydrol. Hydromech.* 4

[4] Dzubak M 1969 Determination of the peak daily precipitation law by means of the global exceedance curves (in Slovak) *J. Hydrol. Hydromech.* **17** 3 pp 209–225

[5] Hladny J Dolezal F Makel M and Sacherova D 1970 Peak runoff volumes of a given duration (in Czech) *Hydrological conditions of the CSSR* Volume III

[6] IPCC 2001 Climate Change 2001 The Scientific Basis. Contribution of Working Group 1 to the Third Assessment Report of the IPCC Cambridge Univ Press UK 944 pp

[7] Kohnova S and Szolgay J 2003 Regional estimation of the index flood and the standard deviation of the summer floods in the Tatry Mountains *J. Hydrol. Hydromech.* **51** 4 pp 241–255

[8] Lapin M and Damborska I 2007 Methods of extreme weather events selection and some results of elaboration Int Conf Bioclimatology and natural hazards Sept 2007

[9] Makel M 1987 Determination of the flood and discharge hydrograph volumes characteristics and their areal variability (in Slovak) Publication 29/I of the SHMU ALFA Bratislava

[10] Mitkova V, Pekarova P and Babiakova G. (2002): Maximum runoff volumes of different duration of the Danube River in dry and wet periods *Acta Hydrologica Slovaca* **3** 2 185–191

[11] Pekarova P, Pacl J, Skoda P and Miklanek P 2007a Supplementation of average daily discharge data of the Danube River in Bratislava with the historical period 1876–1890 *Acta Hydrologica Slovaca* **8** 1 pp 3–12

[12] Pekarova P and Pekar J 2007 Multiannual variability of Danube runoff characteristics in Bratislava gauge *Acta Hydrologica Slovaca* **8** 1 pp 12–21

[13] Pekarova P, Skoda P, Onderka M, Pacl J and Pekar J 2007c Analysis of Discharge Variability of the Danube River in Bratislava for 1876–1940 and 1941–2005 Ed Josip Petraš *10 th International Symposium on Water Management and Hydraulic Engineering* Croatia Zagreb Faculty of Civil Engineering ISBN 978-953-6272-21-1 10 pp

[14] Pekarova P, Miklanek P and Pekar J 2007d Long-term Danube monthly discharge prognosis for the Bratislava station using stochastic models *Meteorological Journal* **10** 4 pp 211–218

[15] Szolgay J Kohnova S Hlavcova K and Minarik M 2007 On the usage of Langbein formula for recurrence period in flood risk assessment (In Slovak) In Cejp Julinek Riha eds. Rizika ve vodnim hospodarstvi *Prace a studie FAST VUT* Brno pp 494–499

[16] Zatkalik G 1970 Calculation of the basic parameters of the discharge hydrographs (in Slovak) PhD Theses p 71

IOP Conf. Series: Earth and Environmental Science **4** (2008) 012008 doi:10.1088/1755-1307/4/1/012008

Hydrological analysis of high waters and flash floods occurred in September 2007 in Slovenia

M Kobold[1], M Sušnik, M Robič, F Ulaga and B Lalić

Environmental Agency of the Republic of Slovenia, Vojkova 1b, SI-1000 Ljubljana, Slovenia

E-mail: mira.kobold@gov.si

Abstract. Heavy and intense precipitation which fell in just a few hours across the western, north-western and northern Slovenia on 18 September 2007, caused quick rise of river discharges especially in the region of Baška grapa, Davča, the Cerkljansko and Škofja Loka hills. In that area the streams caused huge destruction on infrastructure, homes, business buildings and other property. More than 300 mm of rain was recorded on some precipitation measurement stations. The return period of the highest precipitation was more than 100 years. The amount of precipitation decreased from the west to the east of the country where above 100 mm of precipitation was recorded and torrential streams and rivers flooded in the region of Karavanke and foothills of the Kamnik-Savinja Alps, Kranj and Domžale fields, the Tuhinj valley and extensive Celje region. Observed discharges of streams and rivers on the most affected area exceeded periodical maximum discharges. The simulation of flood hydrograph for Železniki was done by HEC-1 model. The return period of floods was more than 100 years. Besides flooding many landslides were triggered. The result of this catastrophe was enormous economic damage and loss of six people's lives.

1. Introduction

Almost every year local thunderstorms and flash floods appear in different regions of Slovenia. Some small catchments have experienced flash floods with return periods in excess of 100 years in recent years. On 18 September 2007, flash floods which occurred in western, north-western and northern Slovenia were focused on small watersheds. Those floods caused, besides enormous material damage, loss of six people's lives. Meteorological forecast predicted precipitation for Tuesday, September 18, but not in the quantity and intensity as than happened. Weather situation led to the catastrophic hydrological situation and raging power of nature. The greatest damage was done on the Selška Sora catchment.

During the event national hydrological forecasting service issued hydrological warnings and data, and predictions regarding to meteorological forecasts. Prediction of precipitation shows great uncertainty in the amount of precipitation in an alpine region with distinctive topography and climate variability [1], so it is almost impossible to predict such events correct and reliable in advance. Discharge measurements using ADCP measurement equipment [2] were tried to perform during the flood on the Sora, Tržiška Bistrica and Lipnica. It was turned out that the discharge measurements

[1] To whom any correspondence should be addressed.

were impossible to carry out in such extreme hydrological circumstances combined with high velocity of water and debris flow.

2. Weather situation

There was a region of low air pressure over the Northern Europe on 18 September 2007. Cold front moved over western and middle Europe towards the Alps. At the same time a high valley of cold air moved over western Europe towards east. South-western wind was getting stronger and stronger over Slovenia. Main reasons for strong precipitation were the diversity of terrain, constant inflow of moist air from southeast, very unstable atmosphere and wind shearing in higher atmospheric layer.

In the morning of September 18, an extensive convective system was formed in the north-western part of Slovenia and that system remained there for a few hours. The first precipitation zone moved over western Slovenia towards east on 18 September 2007 between 5 and 7 a.m. in the morning. A short brake has followed and after 8 a.m. some thunderstorms appeared in the hilly area of western Slovenia. Very heavy rain showers arisen after 9 a.m. (figure 1). A strong thunderstorm line was established from Posočje over Idrija and Cerkljansko hills and Škofja Loka hills to northern part of Ljubljana basin. The thunderstorm was located there nearly two hours.

Next distinctive stationary thunderstorm line was built on 18 September at 13:30 in the direction from Tolmin to Radovljica. Precipitation weakened at 5 p.m. in Bohinj, but has not stopped yet. New thunderstorm cells were formed over northern Slovenia over and over during the afternoon and precipitation strengthened in northeast part of Slovenia (figure 1). Wind had started to blow in lower atmosphere from northeast to northwest. Thunderstorms were formed when the cold front was passing, and moved towards south. Precipitation stopped in western part of Slovenia around 9 p.m., in northwest at midnight, and in southeast during 2 and 3 a.m. next morning.

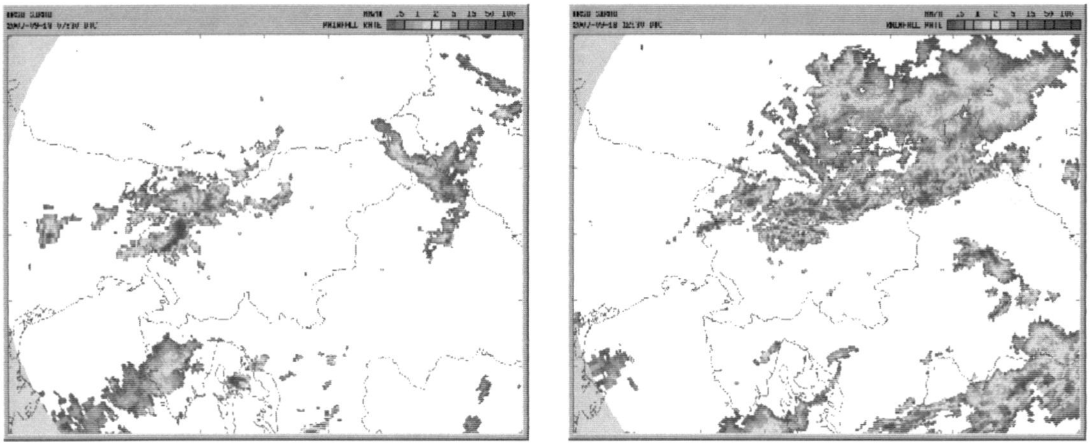

Figure 1. Radar images of precipitation intensity over Slovenia at 9:30 a.m. and 2:30 p.m. local time.

In the region of the highest precipitation unexpectedly heavy rainfall fell in a few hours, mostly within six hours. On some precipitation measurement stations more than 300 mm of rain was recorded. The highest precipitation was recorded in the region from Bohinj and hills of Cerkljansko to Celje valley (figure 2). The spatial distribution of precipitation was very different. High differences in the amount of precipitation were even at small distances. In the region with the highest precipitation more than average amount of precipitation for September was fallen in 24 hours (figure 3). The return period of the highest precipitation was more than 100 years [3]. Shorter time intervals had even greater return periods. The amount of precipitation decreased to the east of the country. Above 100 mm of precipitation fell in north part of Ljubljana valley and in some parts of Štajerska region, where the highest precipitation fell in the surroundings of Celje.

IOP Conf. Series: Earth and Environmental Science 4 (2008) 012008 doi:10.1088/1755-1307/4/1/012008

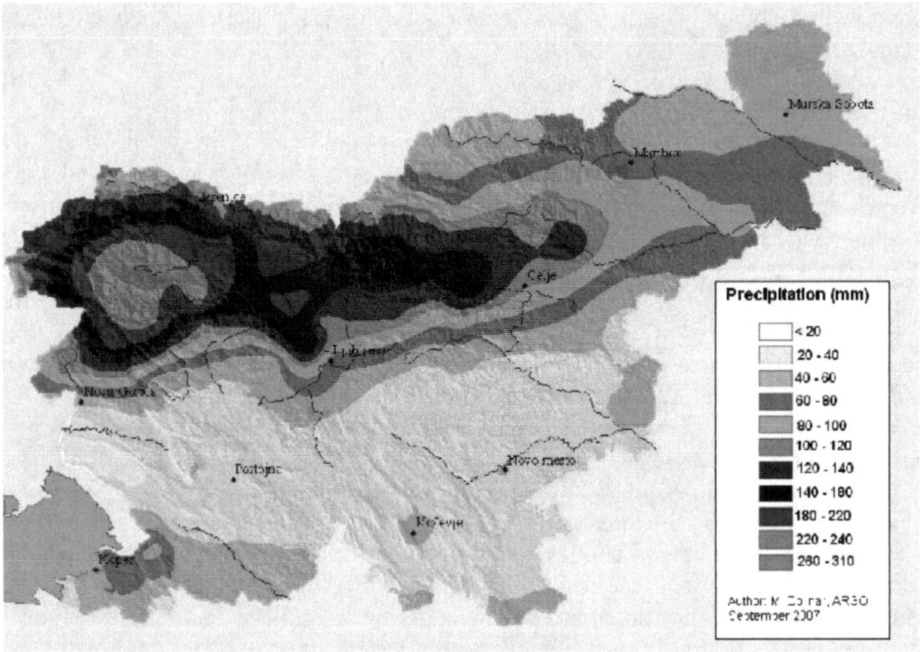

Figure 2. Daily amount of precipitation from 18 September at 8 a.m. to 19 September 2007 at 8 a.m.

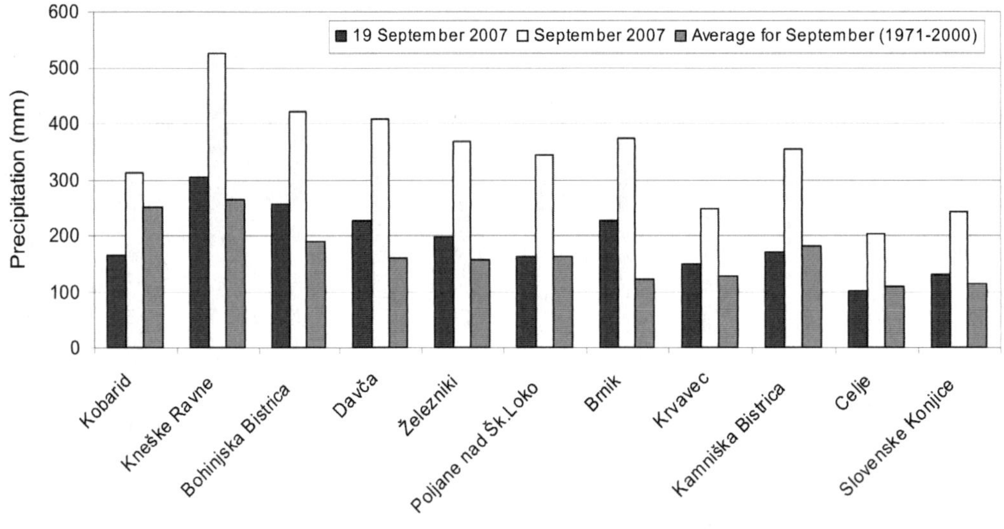

Figure 3. Amount of precipitation in the region with the highest precipitation.

3. Hydrological response and river flows

There was not a lot of precipitation in Slovenia from the beginning of September 2007 (figure 4) and discharges of the most rivers were low early in the morning of 18 September. The soil was mainly dry. Only some rivers in eastern and western part of the country had mean yearly discharges. Following to the weather situation on 18 September (figure 2), the northwestern part of Slovenia got the highest and intense precipitation which caused the quick rise of discharges. High and rapid water caused the destruction of infrastructure, homes, business buildings and other property. The greatest damage was done in urbanized densely populated river valleys like Selška Sora, Davča and Kroparica. Six people lost their lives in those floods. The hydrological situation was the worst on small and medium sized catchments in north-western and northern Slovenia: the Selška Sora, Cerknica, Bistrica, Lipnica, Tržiška Bistrica, Pšata, Kamniška Bistrica, Dravinja and some tributaries of the Savinja (figure 5). Observed discharges of some streams on that region exceeded periodical maximum discharges and the return period of floods was more than 100 years.

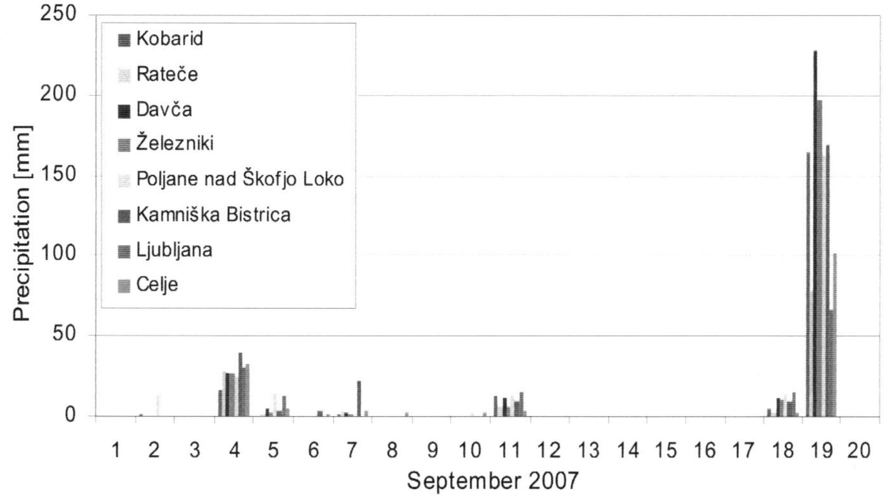

Figure 4. Daily amount of precipitation from 1st to 20th September 2007.

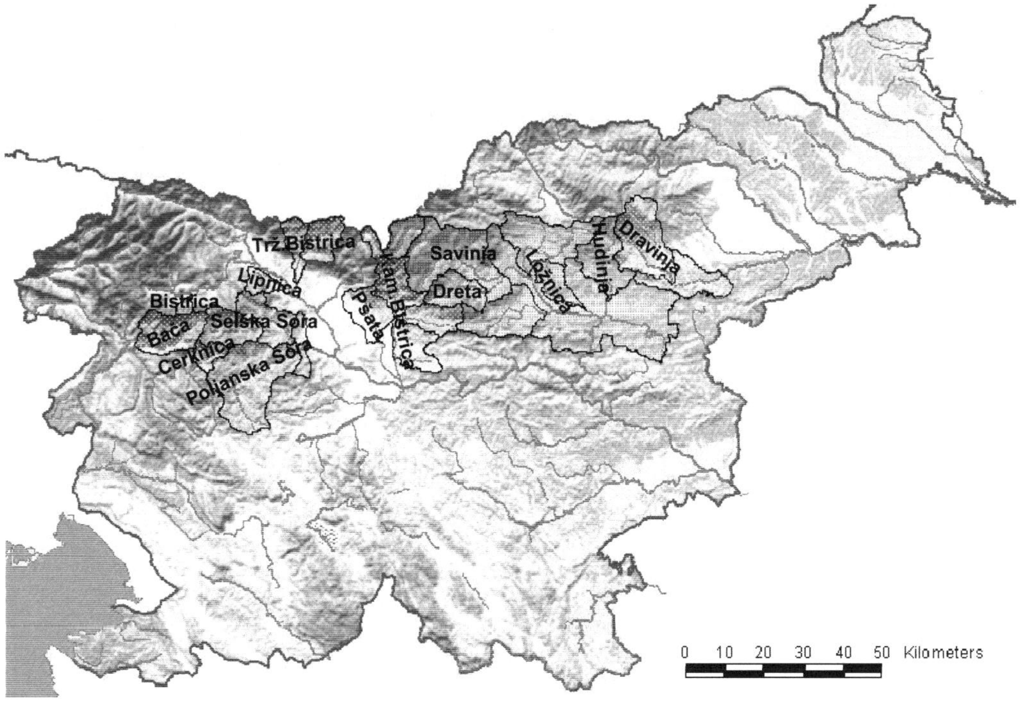

Figure 5. Catchments affected the most by the floods on 18 September 2007.

3.1. The Sora catchment

Under the influence of very intense precipitation in northwestern part of Slovenia, streams and rivers from the catchments of Davča, Cerkljansko hills and Bohinj ridge started to rise. The Bača and Cerknica rised very quickly in only a few hours between 10 to 12 a.m. onward and they started to overflow and flood. At the same time the Selška Sora and its tributary Davča started to rise. There was registered 80 mm of rain on precipitation measurement station in Davča fallen in only 50 minutes. The Davča torrent in Davča and the Selška Sora in Železniki rose extremely quickly and caused destruction and enormous material damage, the most in Davča and Železniki (figure 6). Three people lost their lives during the flood in Železniki. Recorded precipitation for 18 September 2007 in Železniki was the highest in the period of observation (figure 7).

Figure 6. Destruction in Železniki after the flood on 18 September 2007 and traces of flood.

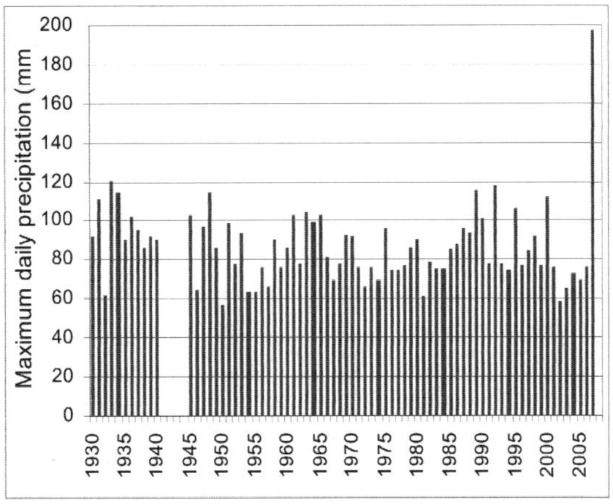

Figure 7. Maximum daily precipitation in Železniki.

The equipment on the water measurement station at Železniki stopped to work during the flood and the flood wave was not recorded entirely. The highest water level 551 cm was determined after the flood according to flood trace and it happened around 13:30. In that event the Selška Sora exceeded the highest water level from the period of observation 1991-2006. The peak discharge was estimated to 300 m^3/s by extrapolation of rating curve, and it exceeded 100-year return period of maximal floods. The highest discharge from the observation period 1991-2006 is 148 m^3/s occurred in 1995 (figure 8).

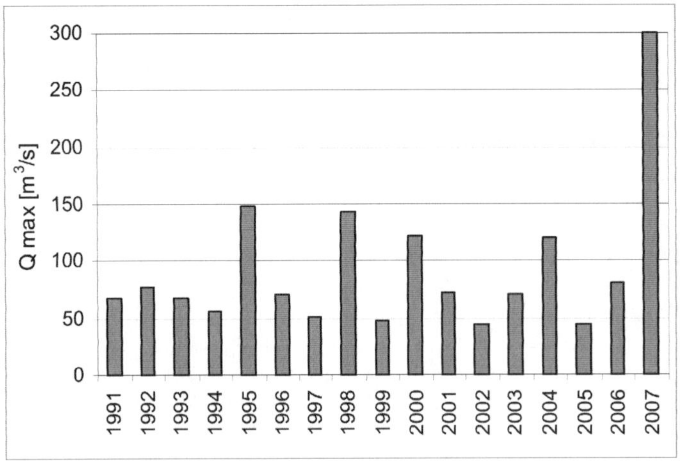

Figure 8. Annual maximum discharges (Q max) of the Selška Sora at Železniki.

3.1.1. *Simulation of flood hydrograph.* The flood of the Selška Sora in Železniki was an extreme hydrological event. The drainage area of the catchment to Železniki is 104 km^2 and downstream to Vešter is 213.8 km^2. The simulation of flood hydrograph for Železniki was done using the HEC-1 model and Watershed Modelling System [4] for setting up the model for the Sora catchment. The input data into the model was precipitation from the precipitation measurement stations on the catchment (figure 9).

Figure 9. The Sora catchment.

The simulated hydrograph for Železniki is shown in figure 10. The peak of flood wave is estimated to 278 m^3/s, what means 2670 l/s/km^2 of specific runoff. This is the extreme value for the Sora catchment and upper Sava watershed [5]. The cumulative areal precipitation for the catchment to Selška Sora amounts to 219 mm, while the effective precipitation which caused direct runoff was only 57 mm.

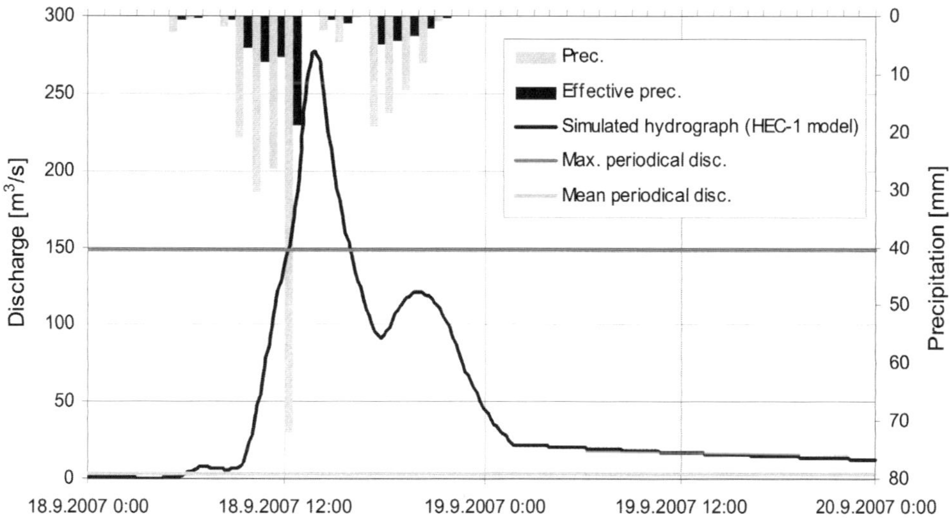

Figure 10. Simulated hydrograph of the Selška Sora at Železniki, periodical mean and maximal discharges, and areal hourly intensity of precipitation to Železniki.

In next few hours flood wave of the Selška Sora moved downstream towards Škofja Loka. At the water measurement station Vešter, situated 15 km downstream from Železniki, the maximum discharge of 335 m^3/s was registered at 16:15 (figure 11), what is 25-year return period of maximal floods considering the period of observations from 1988 onwards. The amount of precipitation decreased downstream on the Selška Sora catchment. On the catchment from Železniki to Vešter the average areal precipitation amounts to 187 mm. That is 32 mm less precipitation than in the upper part to Železniki. In late afternoon the flood water was decreasing (figure 12).

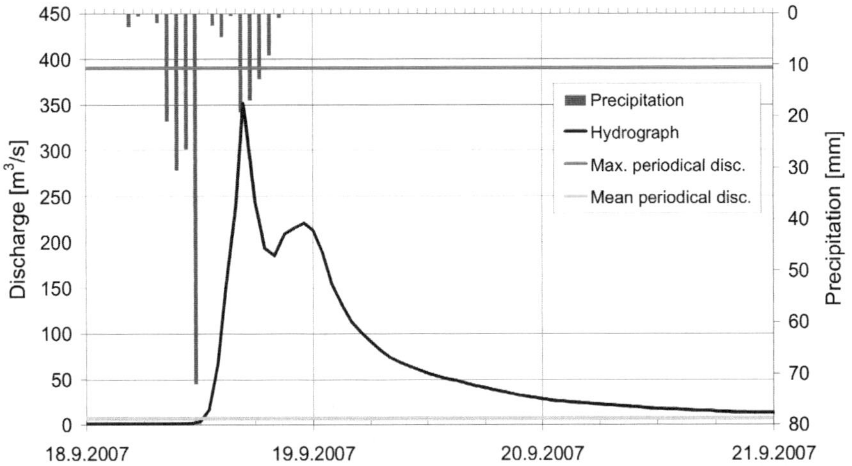

Figure 11. Hydrograph of the Selška Sora at Vešter, periodical mean and maximal discharges, and hourly intensity of rainfall in Davča.

Further analysis of flood waves (figure 13) and calculation of runoff coefficients showed that the coefficients were very small in the Sora catchment. The coefficients of direct runoff were about 0.27 and the coefficients of total runoff about 0.31. Regarding to small antecedent soil moisture and consequently low flows before flood event, the infiltration into the soil was very high and that was a reason for triggering of numerous landslides. Precipitation caught and held by vegetation presented the remaining losses.

Figure 12. The Selška Sora at Vešter on 18 September 2007 at 17:30.

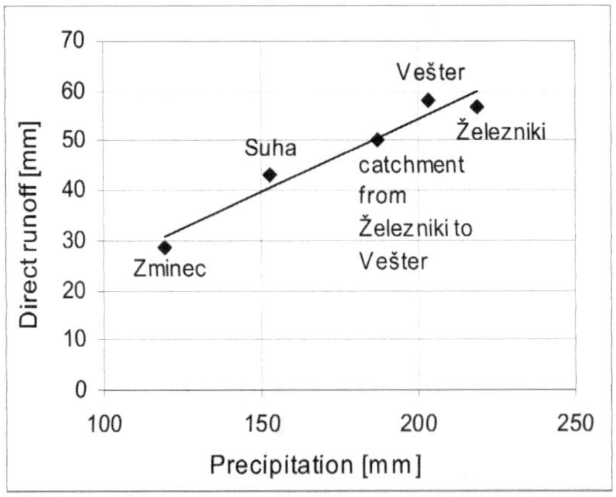

Figure 13. The dependence of direct runoff from precipitation on the Sora catchment.

The Poljanska Sora was not so high. The peak discharge at Zminec was 122 m^3/s what is less than 2-year return period of flood. The maximal discharge of Sora at Suha was 440 m^3/s, what is 10-year return period of flood.

3.2. The catchments of the Lipnica and Tržiška Bistrica

Discharge with a return period over 100 years was recorded on the Lipnica at Ovsiše (figure 14) and Tržiška Bistrica at Preska as well (figure 15). The maximum discharge in Tržiška Bistrica was 155 m^3/s at 16:30 (figure 16).

Figure 14. The Lipnica at Ovsiše on 18 September 2007 at 17:13.

Figure 15. High water of the Tržiška Bistrica at Preska on 18 September 2007 at 16:39.

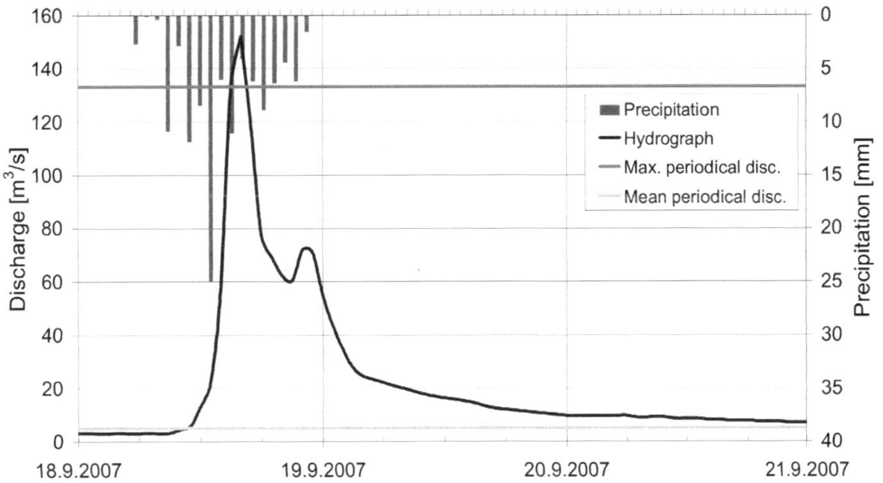

Figure 16. Hydrograph of the Tržiška Bistrica at Preska, periodical mean and maximal discharges, and hourly intensity of rainfall on Zgornje Jezersko.

3.3. The Bohinjska Bistrica catchment

The Bistrica at Bohinjska Bistrica gathers water from the southern side of the Bohinj ridge. On 18 September the extreme amount of precipitation fell in that area which caused huge rise of the Bistrica and flooding of the surrounding buildings. The peak of flood wave reached 108 m^3/s at 17:30 and exceeded the periodical maximum discharge at this location. The return period of flood was more than 100 years (figure 17). Other rivers and streams in Bohinj and its surrounding rose significantly. The roads were closed and the railway traffic was unfeasible due to the number of landslides which were triggered in that area.

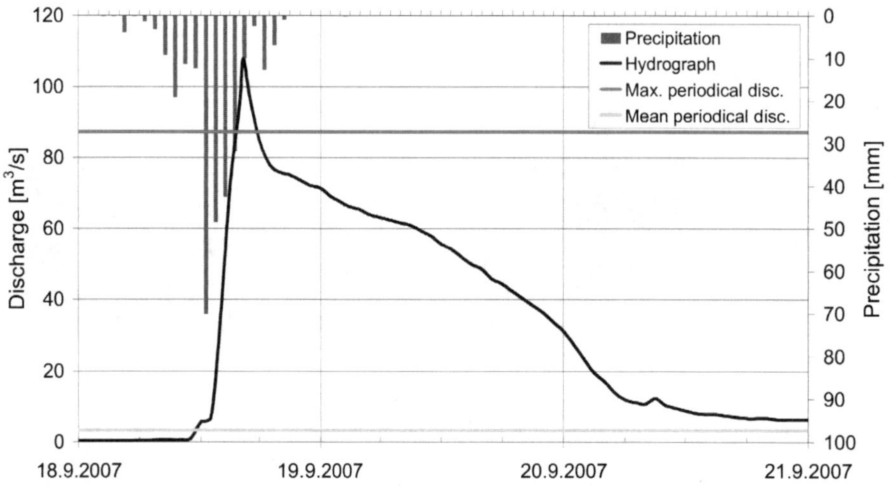

Figure 17. Hydrograph of the Bistrica at Bohinjska Bistrica, periodical mean and maximal discharges, and hourly intensity of rainfall in Bohinjska Češnjica.

3.4. The catchment of the Pšata and Kamniška Bistrica

In the afternoon on 18 September the precipitation moved towards northeast of Slovenia. The streams increased in the foothill of Kamniške Alps, on the field of Domžale and in the Tuhinj valley. The Nevljica at Nevlje had a maximal discharge of 68 m^3/s late in the evening with a return period of 100 years. The Pšata flooded in Komenda and its neighborhood. At Topole the runoff reached 52 m^3/s, what is 100-year return period (figure 18).

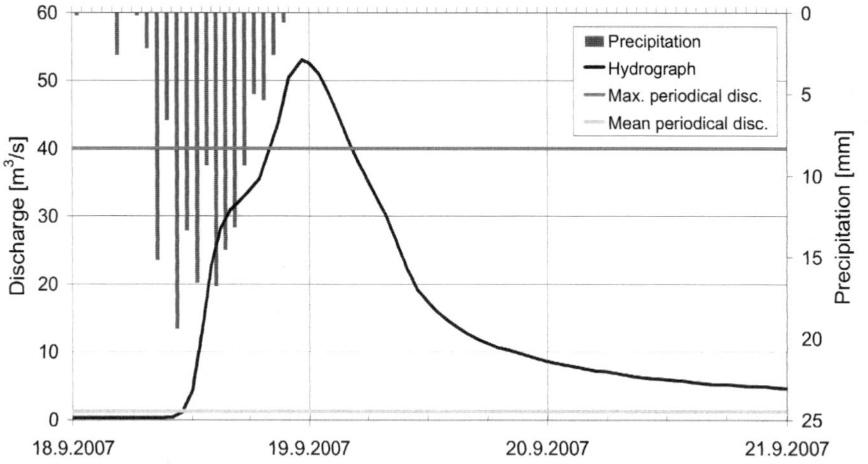

Figure 18. Hydrograph of the Pšata at Topole, periodical mean and maximal discharges, and hourly intensity of rainfall on Krvavec.

The Kamniška Bistrica at Kamnik had 10-year return period with maximum discharge of 140 m^3/s, but downstream the discharge increased to 50 to 100-year return period due to the high Pšata inflow. The Rača with 5 to 10-year return period additionally contributed to the increase of discharge of the Kamniška Bistrica on the confluence with the Sava river.

3.5. The Savinja catchment

In the late afternoon when rivers in western part of Slovenia started to decrease, the majority of precipitation moved to central and eastern part of Slovenia, to the Savinja and Dravinja catchments. Several smaller streams and torrents flooded on the Savinja catchment, as well as the Savinja river in lower stream. The discharges of the tributaries Hudinja and Ložnica exceeded 100-year return period. The maximal discharge of the Hudinja at Škofja vas was 173 m^3/s and Ložnica at Levec 120 m^3/s (figure 19). The Dreta, Paka and Bolska had peak discharges with lower return period and they did not exceed the periodical maxima.

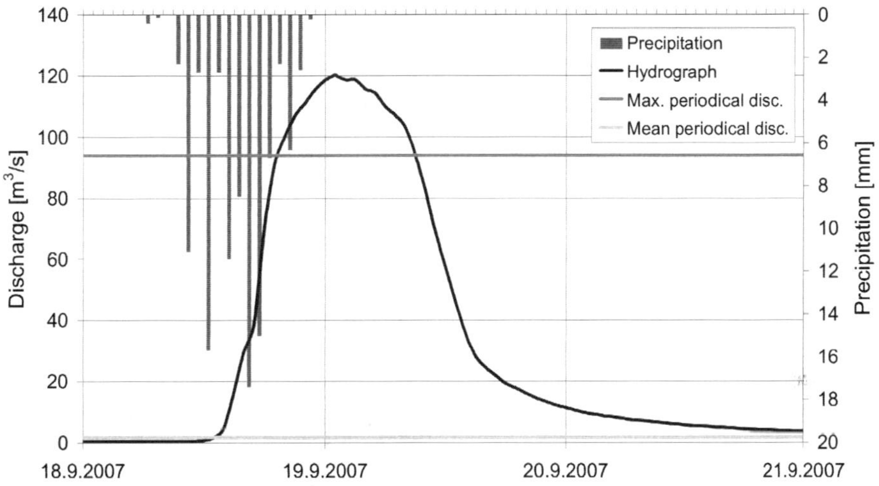

Figure 19. Hydrograph of the Ložnica at Levec, periodical mean and maximal discharges, and hourly intensity of rainfall in Levec.

The Savinja at Solčava had a maximal discharge of 29 m^3/s on 18 September at 9 p.m. This is only mean high discharge in a period. The flood wave was increasing downstream due to the high tributaries. During the night the Savinja in Celje reached the top of the embankment, but it did not flood the city. The maximal discharge at Laško was 1254 m^3/s reached on 19 September at 3 a.m. The return period of flood was 50 to 100 years. The maximal periodical discharge at Laško is 1406 m^3/s occurred in November 1990.

3.6. The Dravinja catchment

The Dravinja in its upper part at Loče exceeded the periodical maximum discharge on 18 September at around 10 p.m. The peak discharge was 78 m^3/s and it is the highest recorded discharge in 25-years observing period on this location. The return period was 20 to 50 years (figure 20). In Makole, which lies 18 km downstream, a flood wave reached a peak discharge of 118 m^3/s early in the morning on 19 September, what is 50 to 100-year return period. The same return period of flood was at Videm with peak discharge of 272 m^3/s and was reached on 19 September at 14:30.

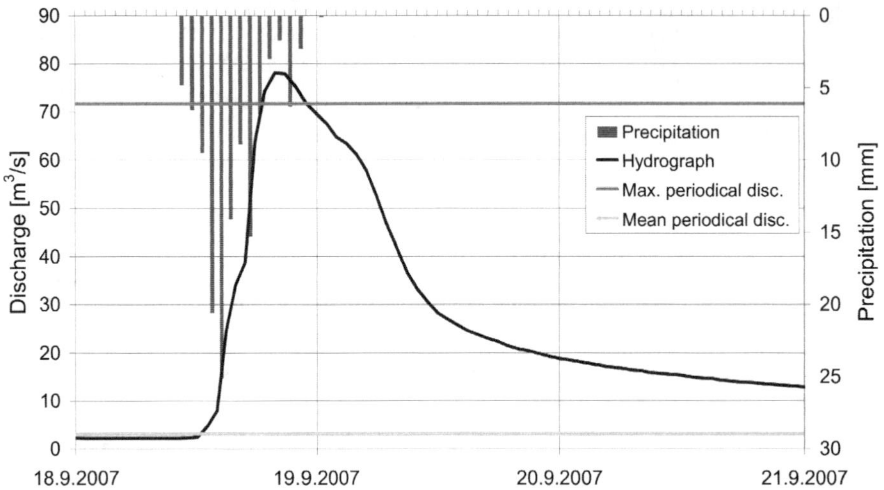

Figure 20. Hydrograph of the Dravinja at Loče, periodical mean and maximal discharges, and hourly intensity of rainfall in Slovenske Konjice.

3.7. The Sava River

Due to the high discharges of the Sora, Bohinjska Bistrica, Lipnica, Tržiška Bistrica, Kamniška Bistrica, Savinja and other rivers, Sava was raising too. The maximal discharge at Šentjakob was 1157 m^3/s, what is around 10-year return period (figure 21). The return periods of the flood wave in the middle and lower Sava were about 20 years. The peak discharge at Hrastnik was 1669 m^3/s.

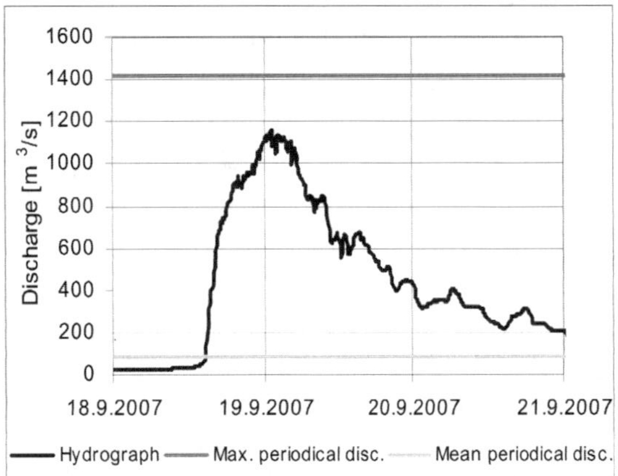

Figure 21. Hydrograph of the Sava at Šentjakob and periodical mean and maximal discharges.

Figure 22. Discharge measurement of the Sava in Hrastnik on 19 September 2007 at noon.

4. Conclusions

Heavy rainfall with high intensity caused huge floods on 18 September 2007 in north-western part of Slovenia. Beside the mentioned rivers the others streams (Bača, Cerknica) and torrents in northwest Slovenia were flooding too. The time of concentration was very short and discharges of the rivers and streams rose very quickly on the most affected area and caused destruction of infrastructure, homes and business. The situation was the worst in the catchment of Selška Sora in Železniki where floods caused enormous economic damage and loss of human lives. Observed discharges of some rivers like Selška Sora, Bistrica at Bohinjska Bistrica, Tržiška Bistrica, Pšata, Dravinja, Ložnica and some others exceeded periodical maximum discharges. That was an event with more than 100-year return period on that area. The transport of sediment load was also very high during that event and the measured values of sediment load at some locations exceeded periodical values.

Due to the large amount of precipitation and strong intensity, several landslides were triggered. One of them destroyed the hospital Franja, a historical monument from the Second World War. The estimated damage caused on the streams is around 91.5 million EUR and the whole damage was estimated over 200 million EUR.

The floods had not been predicted and people had not been warned against floods to protect their property. Meteorological forecasting service predicted precipitation for the September 18, but not in the amount and intensity of precipitation which than happened. It is difficult to predict flash floods for the alpine and mountainous region due to the complex topography and high variability of precipitation. On the other side, the resolution of meteorological models is still not high enough for mountainous terrain.

References

[1] Kobold M, Zgonc A and Sušnik M 2005 Uncertainty of precipitation measurements and predictions in flash flood modelling *Acta hydrotechnica* **23/39** 79–98

[2] Simpson M R 2001 *Discharge Measurements Using a Broad-Band Acoustic Doppler Current Profiler* (Sacramento: U.S. Geological Survey) p 123

[3] ARSO 2007 Exceptional precipitation event on 18 September 2007 (In Slovenian): *http://www.arso.gov.si/vreme/poro%c4%8dila%20in%20projekti/padavine_18sep07.pdf*

[4] Kobold M and Sušelj K 2005 Precipitation forecasts and their uncertainty as input into hydrological models *Hydrol.Earth Sys. Sci.* **9(4)** 322–332

[5] Kolbezen M and Pristov J 1998 Surface streams and water balance of Slovenia (Ljubljana: Hidrometeorološki zavod RS) p 98

IOP Conf. Series: Earth and Environmental Science **4** (2008) 012009 doi:10.1088/1755-1307/4/1/012009

Precipitation intensity probability distribution modelling for hydrological and construction design purposes

Georgy Koshinchanov [1], **Dobri Dimitrov** [2]

[1] National Institute of Meteorology and Hydrology, Hydrological Department, 66 "Tsarigradsko shosse" blvd., 1784 Sofia, Bulgaria
[2] PhD, Deputy Director of the National Institute of Meteorology and Hydrology, 66 Tzarigradsko Shose blvd, 1784 Sofia, Bulgaria

E-mail: georgy.koshinchanov@meteo.bg

Abstract. The characteristics of rainfall intensity are important for many purposes, including design of sewage and drainage systems, tuning flood warning procedures, etc. Those estimates are usually statistical estimates of the intensity of precipitation realized for certain period of time (e.g. 5, 10 min., etc) with different return period (e.g. 20, 100 years, etc). The traditional approach in evaluating the mentioned precipitation intensities is to process the pluviometer's records and fit probability distribution to samples of intensities valid for certain locations ore regions. Those estimates further become part of the state regulations to be used for various economic activities. Two problems occur using the mentioned approach: 1. Due to various factors the climate conditions are changed and the precipitation intensity estimates need regular update; 2. As far as the extremes of the probability distribution are of particular importance for the practice, the methodology of the distribution fitting needs specific attention to those parts of the distribution. The aim of this paper is to make review of the existing methodologies for processing the intensive rainfalls and to refresh some of the statistical estimates for the studied areas. The methodologies used in Bulgaria for analyzing the intensive rainfalls and produce relevant statistical estimates:
- The method of the maximum intensity, used in the National Institute of Meteorology and Hydrology to process and decode the pluviometer's records, followed by distribution fitting for each precipitation duration period;
- As the above, but with separate modeling of probability distribution for the middle and high probability quantiles.
- Method is similar to the first one, but with a threshold of 0,36 mm/min of intensity;
- Another method proposed by the Russian hydrologist G. A. Aleksiev for regionalization of estimates over some territory, improved and adapted by S. Gerasimov for Bulgaria;
- Next method is considering only the intensive rainfalls (if any) during the day with the maximal annual daily precipitation total for a given year;
Conclusions are drown on the relevance and adequacy of the applied methods.

1. Introduction

Bulgaria is situated in the moderate-continental climatic zone, with high influence of the Mediterranean Sea. Its territory could be roughly divided into three major climatic zones: north one belonging to the Danube river basin with predominantly continental climate; south one influenced significantly by the Mediterranean climate; east one being narrow strip close to the Black sea sore. In the recent years this zone is characterized by the flash floods that occur (Figure 1). This became very evident in the last 10-15 years. That's why it is very important to study the characteristics of the intensive rainfalls. The most important of them are the intensity for a certain period and the whole duration of the rainfalls. When the intensity is over above threshold for the certain type of soil there will be induced surface runoff.

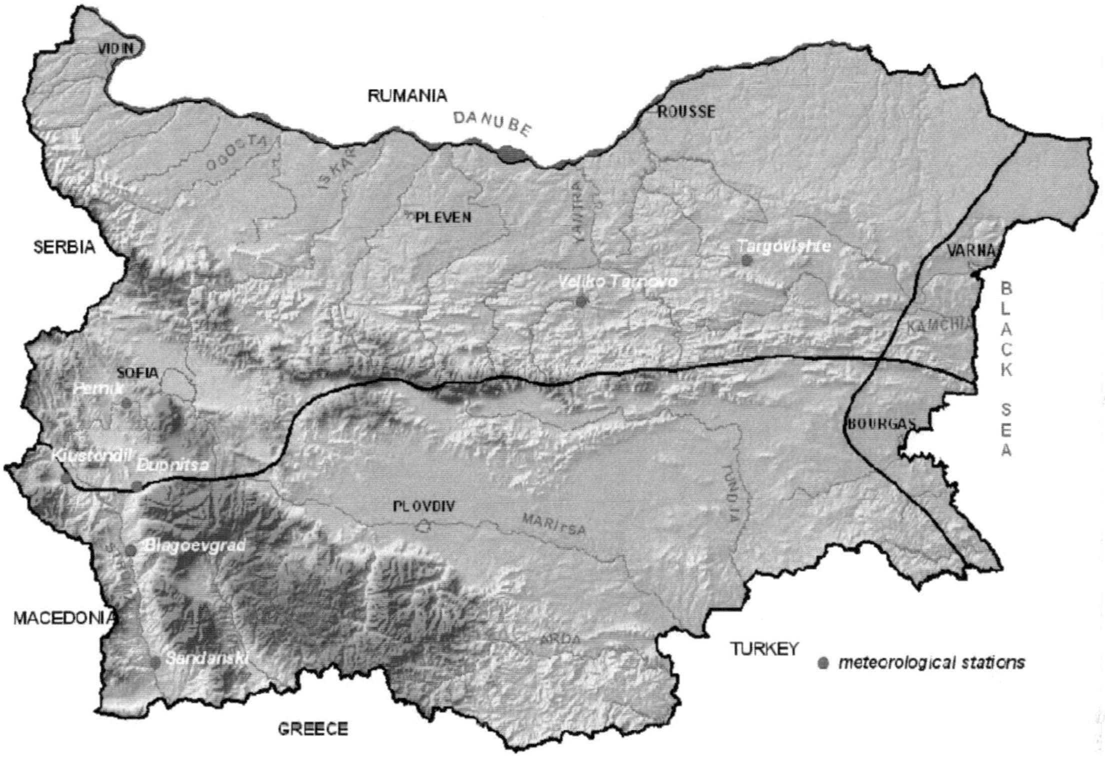

Figure 1. Delineation of the 3 climatic zones and location of the studied stations

1.1. Data availability

There are several definitions for "intensive rainfall" used in Bulgaria. They differ mainly by the threshold values for the minimal intensity and the duration used to estimate the intensity. In Bulgaria this threshold is 0,18 mm/min despite the duration of the rainfall. There are also different type of precipitation gauges which data might be used to determine intensive rainfall characteristics. One possibility, which is used for the present paper is the pluviometer recording on paper tapes. Typical result of the pre-processing of those tapes is given in the table below. As one can see, each rainfall event is pre-processed separately. For the particular event the highest intensity pattern is determined first having duration at least 5 minutes. The rainfall with lower intensity having a duration at least 10 minutes and including the first one is determined next and so on. It should be mentioned that data from automatic telemetric stations recording precipitation intensities of equal intervals are not used in this study, because they are few and their series are short.

Table 1. Example of input data – Turgovishte

№ station	Year	Month	Day	5	10	15	20	25	30	40	50	60
22010	1962	4	5	1.4								
22010	1962	4	16	1.2	2.2							
22010	1962	6	3	1.8	2.5							
22010	1962	6	12	1.0	1.9							
22010	1962	6	12	2.6	3.5	4.3	5.0	5.7	6.3			
22010	1962	6	28	1.0								
22010	1962	7	2	2.8	4.8	6.8	7.9	8.5	8.9			
22010	1962	7	3	0.9	1.8							
22010	1962	7	12	1.4	2.2							
22010	1962	7	24	1.2								
22010	1962	7	29	9.4	18.8	24.7	29.8	33.2	36.5	40.4		
22010	1962	7	31	1.2	1.9							
22010	1962	8	12	0.9								
22010	1962	8	12	2.3	4.1	5.4	6.7	7.9	9.0	11.0	11.7	12.1
22010	1962	8	18	2.8	4.9	6.7	8.1					
22010	1962	8	18	1.3	1.8							

Data from 7 rainfall gauges are used for this study. Some of them are situated in the Struma river valley at southwest Bulgaria, while the others are located in the central part of north Bulgaria, part of the Danube river basin. The stations are situated in general in three different climate zones: Sandanski, having quite low elevation, is in a region influenced quite significantly by the Mediterranean climate; the next 4 stations having elevation between 400 and 700m asl, are mostly in the mountain zone; while the rest two stations belong to the continental climate zone. The temperature conditions of the stations are relatively similar. The annual amount of precipitation also doesn't differ significantly being coherent with the elevation increase. The maximal intensity for 10 minutes interval correlate quite well with the annual precipitation totals average, but differs quite significantly being double in the high mountain zone. Some general type climate statistics are given in table 2 below.[3]

Table 2. Basic climatic characteristics for the used locations

Parameters	Sandanski	Blagoevgrad	Dupnica	Kustendil	Pernik	Veliko Turnovo	Turgovishte
Elevation	206	416	555	520	726	195	226
Absolute min T (till 1976)	-21	-25.2	-30	-27	-26.8	-28.1	-22.9
Absolute max T (till 1976)	42.1	41.6	39.3	39	38.1	41.1	42.4
Average Multi Annual T (till 2006)	13.8	11.8	10.4	10.7	9.3	11.3	11
Average Multi Annual Precipitation	533	560	627	625	606	680	646
Max Int (10min) (till 2006)	1.85	1.63	3.09	2.02	1.95	3.65	2.03

2. Methods

There were several studies on the intensive rainfalls in the recent years. One of them was made during the realization of the PHARE-CBC project "Early warning system in the Struma river". The data from this project are from 5 locations – Pernik, Kustendil, Dupnica, Blagoevgrad and Sandanski. In this paper are included also data from 2 locations in North Bulgaria – Veliko Turnovo and Turgovishte. The analysed period is as follows: for the stations in the Struma river basin from the opening of the station till 1998 and for the other two stations – till 2006.

The following methodologies were used for analyzing the intensive rainfalls:

- The maximum intensity method, used in the National Institute of Meteorology and Hydrology to process and decode the pluviometer's records, followed by distribution fitting for each precipitation duration period;
- Method is similar to the first one, but with a threshold of 0,36 mm/min of intensity;
- Another method proposed by the Russian hydrologist G. A. Aleksiev for regionalization of estimates over some territory, improved and adapted by S. Gerasimov for Bulgaria;[1]
- Method using the maximum annual intensity for different durations, followed by distribution fitting for each precipitation duration period;
- Next method is considering only the intensive rainfalls (if any) during the day with the maximal annual daily precipitation total for a given year.

All the locations were analyzed with the first 3 methods in this paper. The 4th method was used only for Veliko Turnovo and Turgovishte and the 5th method was used only for Veliko Turnovo.

3. Analyses of the distributions

After several experiments it became clear that the following 3 theoretical distributions fit best to the empirical curves of the intensive rainfalls:

- - Pearson III distribution
- - Polynomial regression
- - Lognormal distribution of the type ALOG(Y-COR)=ALN*X+BLN with optimization of the parameters ALN and BLN.

After assessment of each of the 3 distributions for a single location and method a distribution was chosen which is accepted to be valid for a given location for all the duration periods. It should be noted that Pearson III and lognormal distributions are left bounded, while the polynomial is not. The latest is visible especially at probabilities close to 100% (intensities close to the threshold value or zero). As far as for practical purposes we are using that part of the distribution, which is at the low probabilities and high intensities, that particularity is not considerable for the study.

To characterize generally the results for the various distributions related to the set of duration intervals, some basic statistics of those distributions are calculated. The averages do not differ too much with the duration, but always decrease (0.5 0.35) with the increase of the duration. North of Bulgaria and especially Veliko Tarnovo gives higher averages than the other stations. The variation generally also decreases with the increase of the duration. The asymmetry is always positive and varies between 1 and 3.

To analyze the results of the probability distributions established by different methods, range of graphs and tables were prepared. First the probability versus rainfall intensity of rainfall events having duration of 10 min. is plotted for all the available stations in Figure 2. Generally the type of the relations is quite similar for all the stations. The 10 min. duration is selected as typical one, because the results for all the duration intervals can not be presented due to their large volume. Considering Figure 2 becomes visible that the intensity is rapidly decreasing with the probability following the exponential law. Only for Veliko Tarnovo the intensity is little higher for probabilities 0.05 to 0.35 than the ensemble of exponential curves received for the other stations. For all the stations the curves asymptotically tends to intensities 0.18 – 0.2 mm/min for probabilities above 40%. Relatively similar are the results received for all the stations using the reduction curves method, given in Figure 3. Comparing Figure 2 and Figure 3 it could be mentioned that in the second one the intensities for the lowest probabilities are almost twice higher; the differences between the locations are more visible; station Veliko Tarnovo is still the highest; the curves as a whole asymptotically tend to 0.5 – 0.6 mm/min for the highest probabilities with values 0.7-1 mm/min for probabilities above 40%.

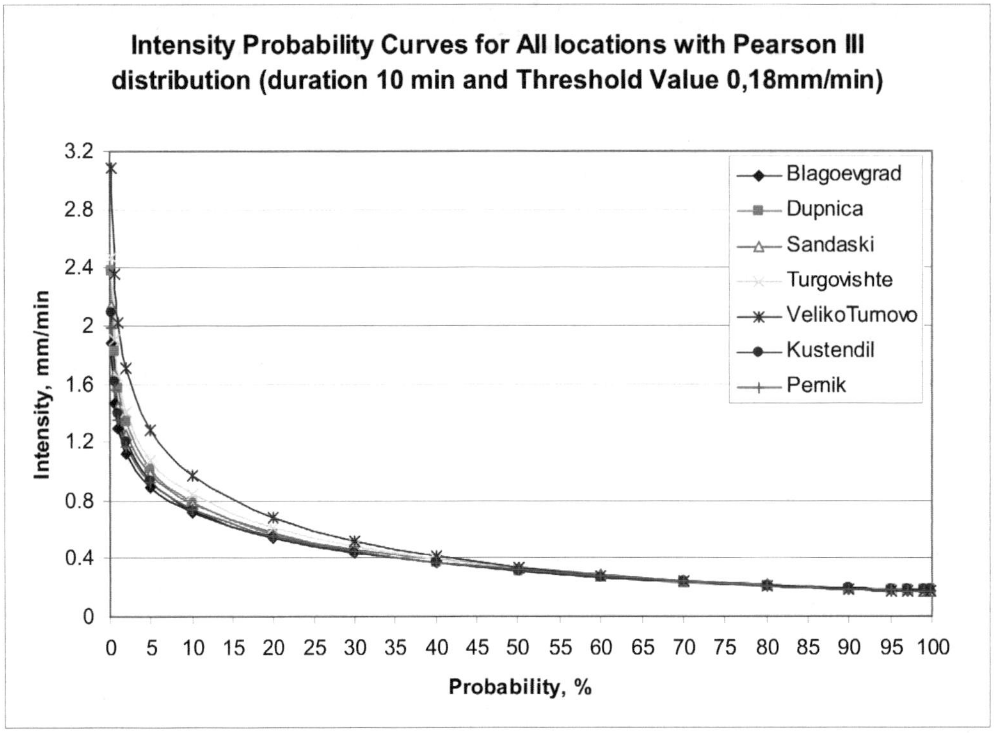

Figure 2

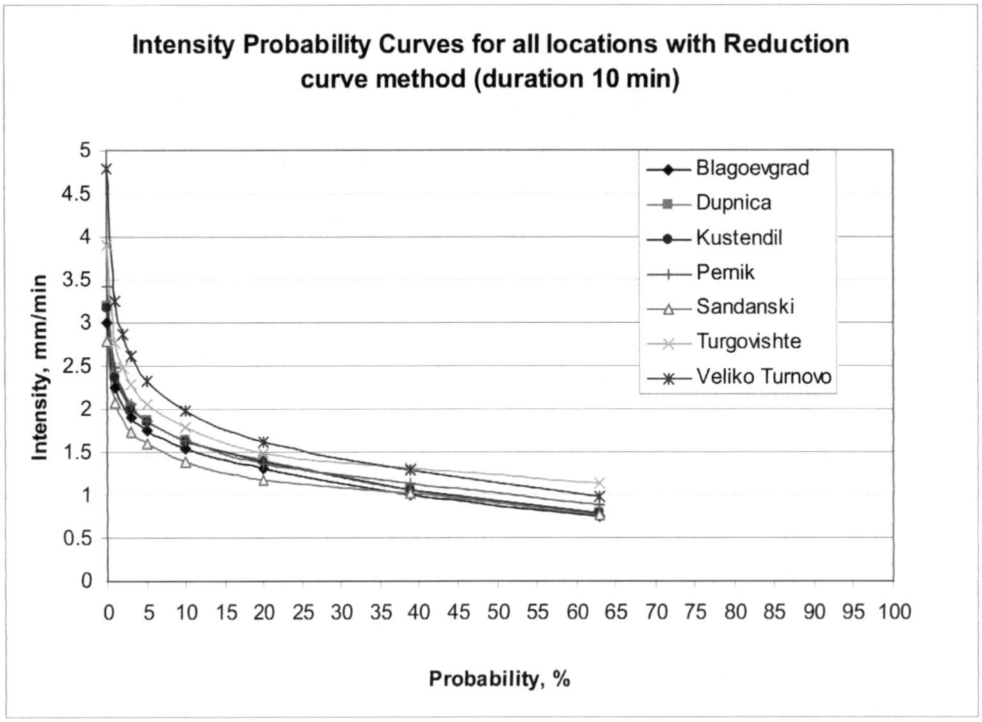

Figure 3

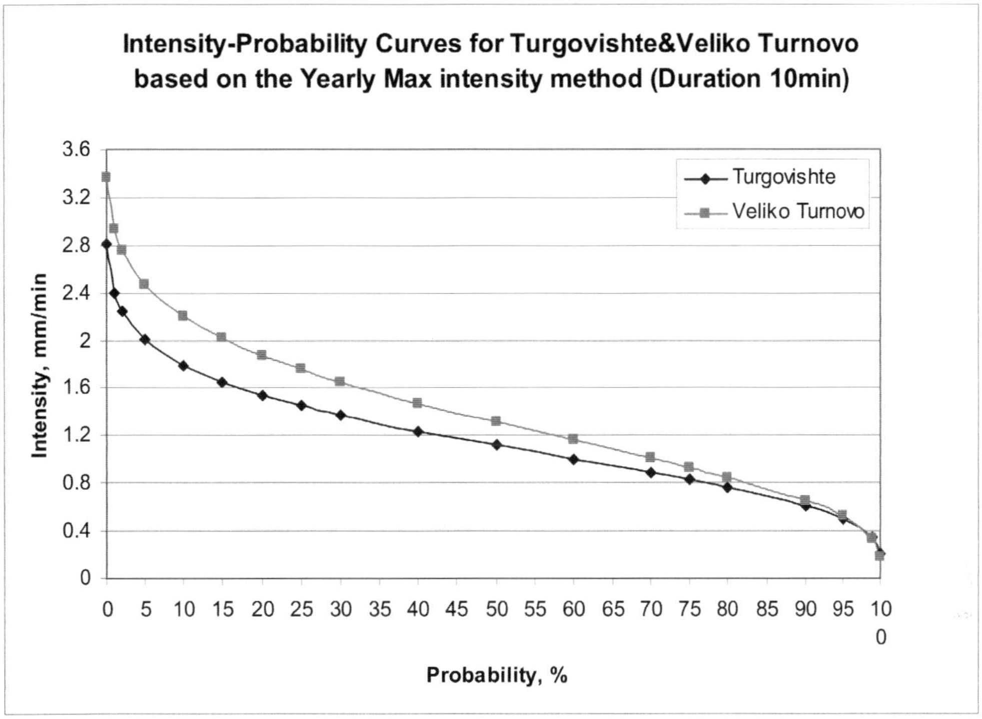

Figure 4

Similar to the above, probability distributions were established for the annual maximal intensities of the rainfall events having duration of 10 minutes for two stations – Targovishte and Veliko Tarnovo (Figure 4). The left part of the distribution - probabilities lower than 30% - shows intensities between those in Figure 2 and the reduction curves given in Figure 3. The most important is the fact that the reduction curves gives estimates higher than even the distribution of the annual maximums. The latest is a proof that the reduction curves overestimate the distribution with at least 50%.

Plots of the distributions received by different methods but for the same location are given below to help the comparison of the different methods. As could be expected highest estimations of the quantiles always gives the reduction curves method. The estimates based on real observations of rainfall intensities are always lower.

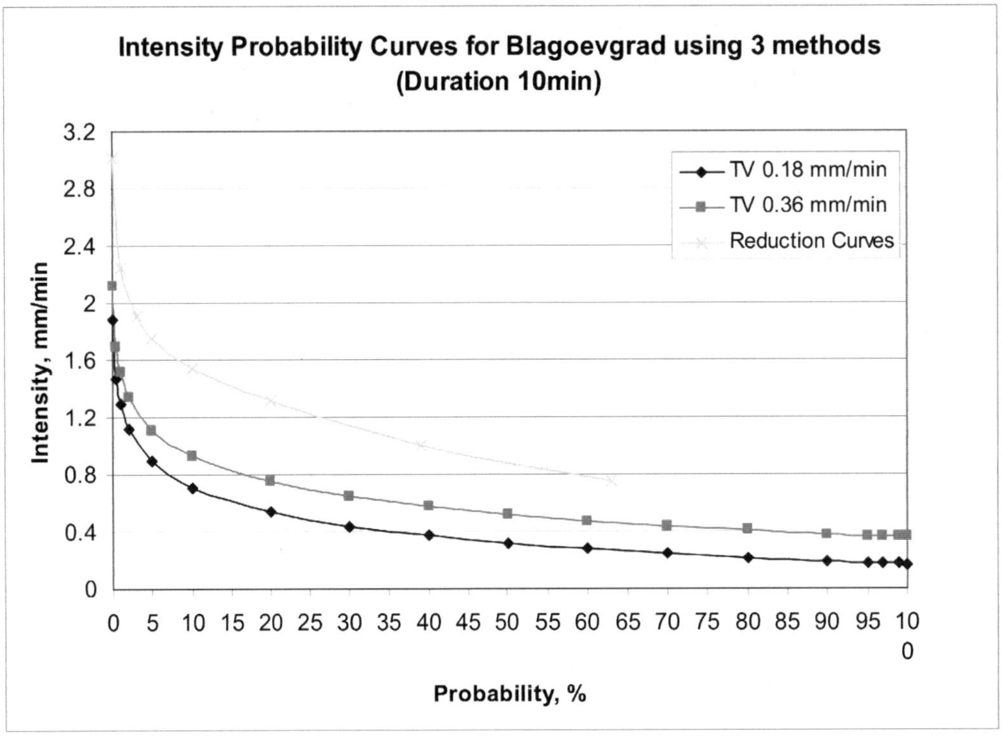

Figure 5

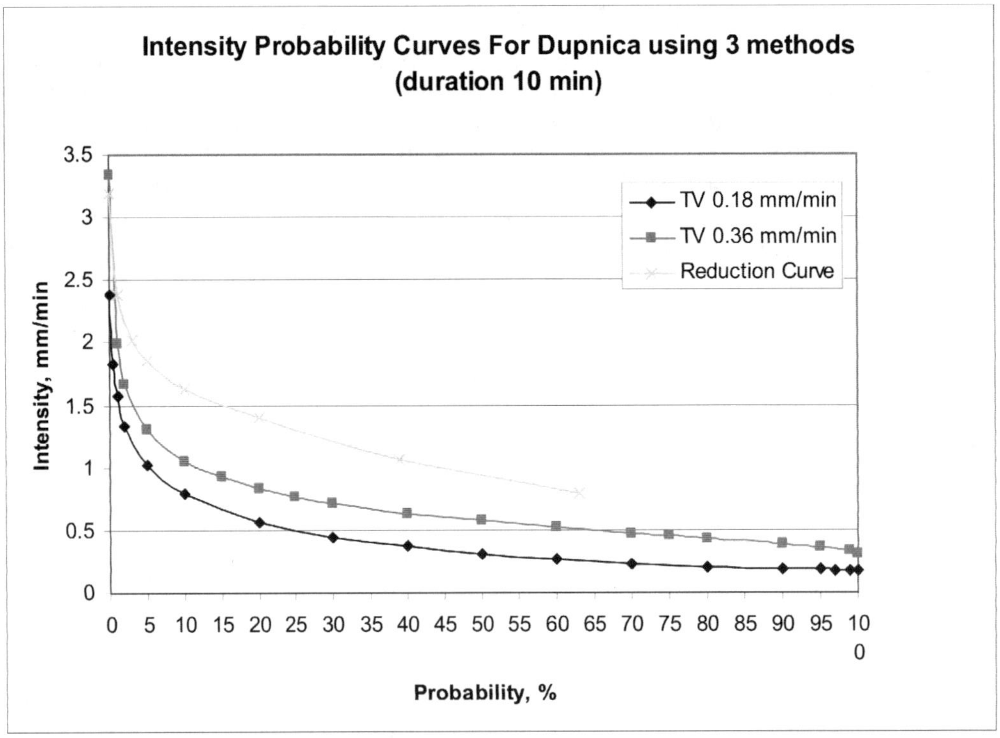

Figure 6

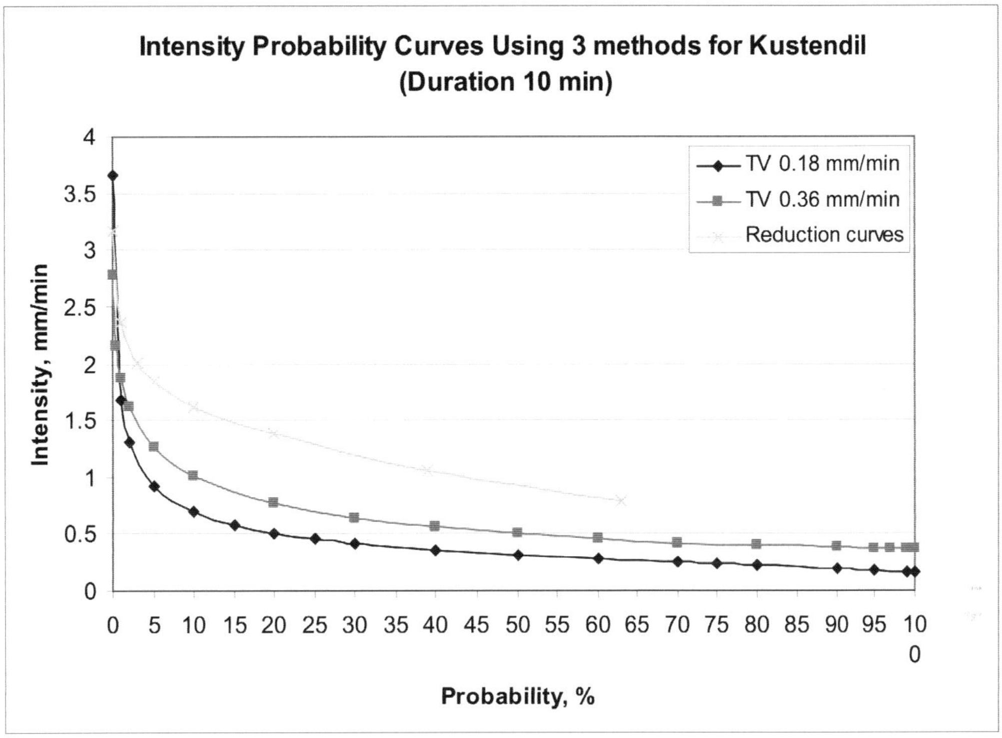

Figure 7

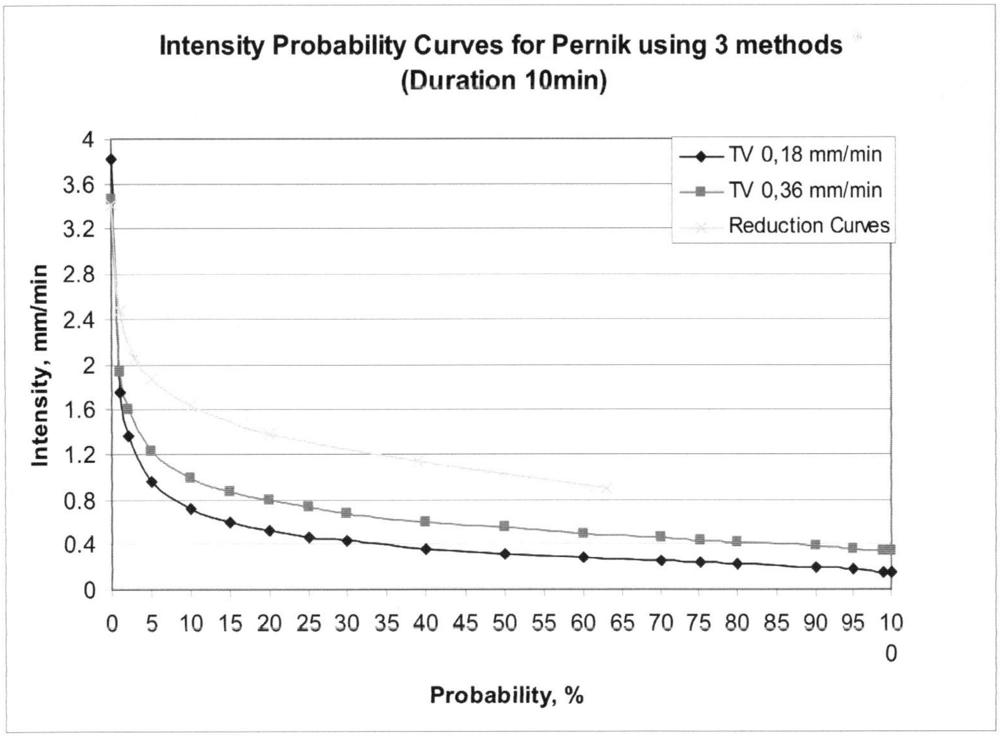

Figure 8

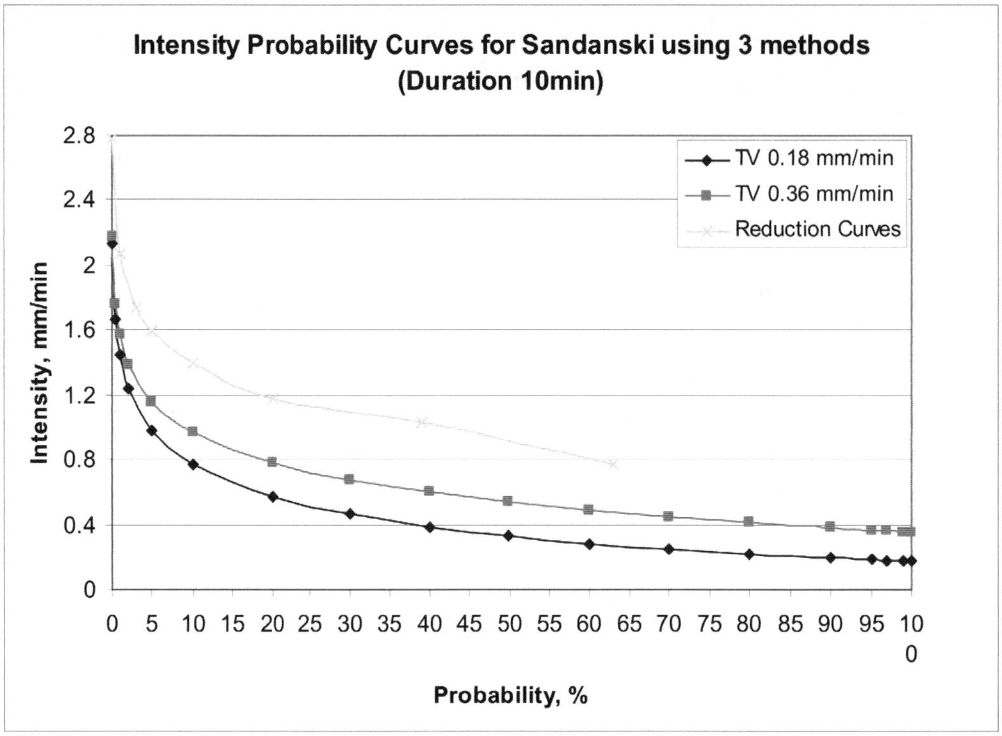

Figure 9

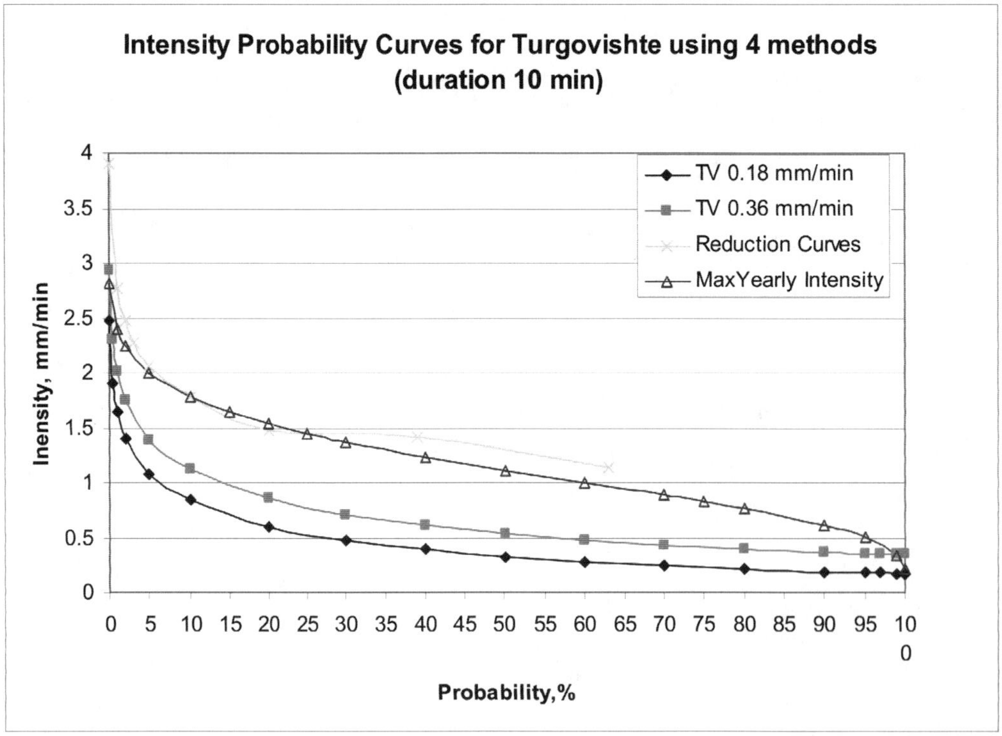

Figure 10

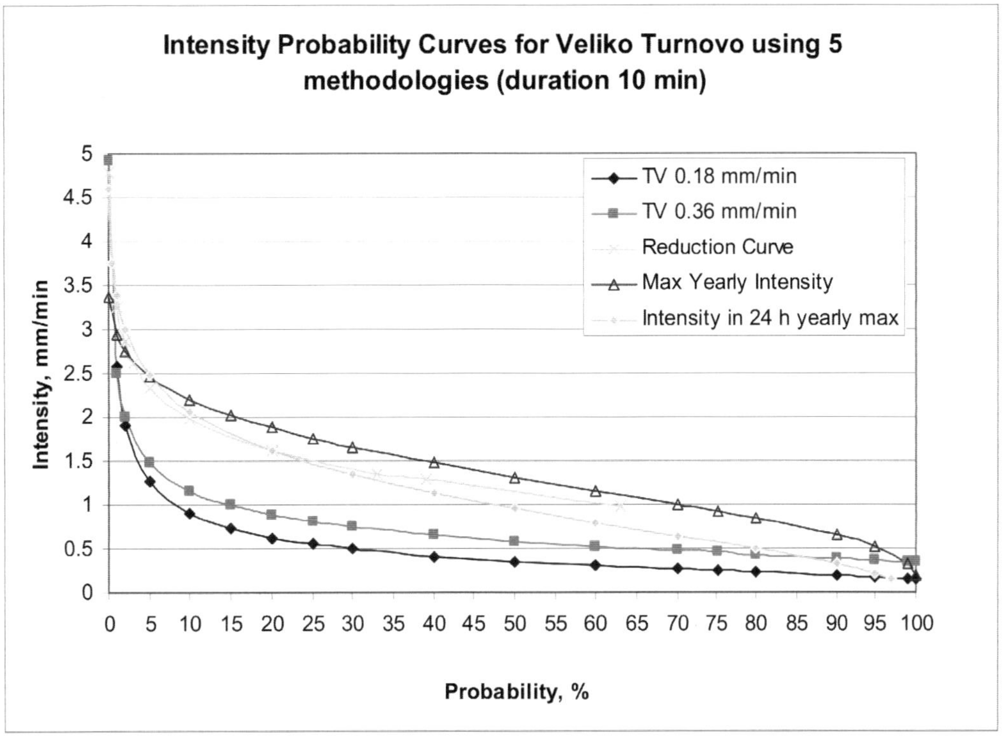

Figure 11

4. Data comparison for two periods – till 1976 and till 2006

It is quite interesting to see how the various characteristics of the intensive rainfalls are changing during the years, e.g. for the recent 50 – 60 years. To analyze those fluctuations, different parameters are calculated for the range of frequencies and all stations for two different periods of time: past period - from the start of observations (most of them early 50s) to 1976; whole period - till nowadays.[4]

Table 3. Annual number of intensive rainfall events averaged for the two periods with different duration despite their intensity

Duration	Sandanski		Blagoevgrad		Dupnica		Kustendil		Pernik		Veliko Turnovo		Turgovishte	
Min	till 1976	till 1998	till 1976	till 1998	till 1976	till 1997	till 1976	till 1998	till 1976	till 1998	till 1976	till 2006	till 1976	till 2006
5	15.75	19.32	15.39	20.72	23.12	29.45	16.68	22.84	22.42	25.78	21.71	19.34	20.88	26.32
10	10.75	12.22	10.07	13.30	16.69	19.39	11.29	14.28	14.05	15.71	14.74	12.00	14.41	17.96
15	7.12	8.20	7.04	9.00	12.50	13.71	7.25	9.32	9.47	10.24	10.00	8.83	8.94	12.32
20	5.38	5.96	4.68	6.36	8.12	9.79	4.82	6.10	6.21	6.76	7.23	6.55	6.47	9.38
25	3.50	4.48	3.29	4.58	5.38	6.82	3.04	4.12	4.10	4.66	4.87	4.79	4.82	7.23
30	2.75	3.30	2.32	3.14	3.75	5.05	2.21	3.08	2.84	3.34	3.32	3.43	3.41	5.26
40	1.38	1.68	1.82	2.16	2.06	3.26	1.57	1.78	1.89	2.00	1.84	2.21	2.65	3.70
50	0.75	0.80	1.07	1.16	1.56	2.00	0.75	0.98	1.00	0.93	0.90	1.51	1.53	2.34
60	0.38	0.42	0.64	0.76	1.06	1.26	0.46	0.58	0.68	0.54	0.52	1.02	0.94	1.62
>60	0.25	0.16	0.57	0.46	0.31	0.82	0.29	0.22	0.26	0.15	0.23	0.96	0.47	1.32

The average of the annual number of rainfall events (in absolute numbers) has increased for the short and average durations for the second period. There is almost no change in the events with high duration. The latest is in correspondences with the observed general decrease of precipitation totals in the southwest part of Bulgaria. For the stations in north Bulgaria it could be mentioned the following:

- Veliko Turnovo– for the duration up to 25 minutes there is a decrease in the number of events. With increasing the duration the absolute number is also increased.
- Turgovishte – for all the durations there is an increase in the absolute number of events for the second period.

It can be summarized as a conclusion that for both stations, situated in North Bulgaria there is significant increase in the events (in absolute figures and percent from all the events) with high duration.

Table 4. Average intensity (mm/min) of the rainfall events with different duration

Duration	Sandanski		Blagoevgrad		Dupnica		Kustendil		Pernik		Veliko Turnovo		Turgovishte	
Min	till 1976	till 1998	till 1976	till 1998	till 1976	till 1997	till 1976	till 1998	till 1976	till 1998	till 1976	till 2006	till 1976	till 2006
5	0.497	0.433	0.465	0.422	0.471	0.442	0.504	0.437	0.487	0.431	0.516	0.485	0.485	0.455
10	0.445	0.420	0.427	0.399	0.429	0.424	0.426	0.402	0.440	0.410	0.446	0.486	0.470	0.435
15	0.419	0.398	0.392	0.382	0.405	0.411	0.409	0.378	0.404	0.387	0.438	0.458	0.486	0.432
20	0.397	0.371	0.368	0.358	0.386	0.394	0.399	0.370	0.382	0.365	0.410	0.429	0.479	0.415
25	0.387	0.346	0.357	0.343	0.379	0.382	0.413	0.368	0.373	0.346	0.375	0.425	0.465	0.397
30	0.339	0.317	0.348	0.334	0.371	0.374	0.397	0.346	0.360	0.334	0.367	0.417	0.467	0.393
40	0.345	0.313	0.311	0.306	0.379	0.351	0.377	0.340	0.324	0.307	0.328	0.393	0.424	0.374
50	0.282	0.282	0.315	0.314	0.339	0.341	0.336	0.307	0.315	0.289	0.301	0.384	0.395	0.365
60	0.254	0.269	0.308	0.283	0.302	0.315	0.328	0.297	0.280	0.257	0.302	0.376	0.366	0.347
>60	0.208	0.224	0.252	0.260	0.241	0.263	0.271	0.267	0.241	0.236	0.257	0.341	0.354	0.292

Table 4 shows that there is a general decrease in the intensity for the short and middle durations. There is slight increase in the intensity of the events with long duration. Only for Veliko Turnovo there is increase in the figures after the 10th minutes. The above general tendency is in accordance with the observed overall decrease of the annual precipitation totals, due to climate changes.

Similarly the maximal intensity of the rainfall events was checked for the two periods referenced above. It should be noted that most of the stations situated at the southwest zone influenced by the Mediterranean climate do not show significant differences of those maximums. The stations influenced by the continental climate show significant increase of the maximal intensities, which especially for the shortest intervals is more than 80 – 100%. The latest is in accordance with the recently observed in other countries increase of flash floods, being obviously due to the increase of the torrential rainfalls.

5. Conclusion

The following general conclusions could be outlined:

- The results of the establishment of probability distribution of the rainfall intensity for different duration periods received by different methods applied are similar.
- The major difference between the methods consists of the fact that the "reduction curves" method overestimates the quantiles by 30 – 60%. The latest might increase the security of the eventual construction design, but obviously increases the costs.
- The "reduction curves" method is very simple and does not need additional data from observations. It should be applied, when no such data are available for the respective region.

IOP Conf. Series: Earth and Environmental Science **4** (2008) 012009 doi:10.1088/1755-1307/4/1/012009

References

[1] Ivanov I etal 1980 *Hydrological handbook* vol 2 (Sofia: Tehnika) pp 24-37

[2] Koshinchanov G and Dimitrov D 2002 Struma River Basin Data Acquisition And Processing System *ERB and Northern European FRIEND Project 5 conference* (Demanoska Dolina, Slovakia, September 2002)

[3] Kuchukova M et al 1983 *Climatic handbook of Bulgaria* vol 3 (Sofia: Science and Art) pp 221-277

[4] Kuchukova M et al 1986 *Climatic handbook (intensive rainfalls in Bulgaria)* (Sofia: Bulgarian Academy of Science) pp 9-11, pp 40-90

IOP Conf. Series: Earth and Environmental Science **4** (2008) 012010 doi:10.1088/1755-1307/4/1/012010

Groundwater drought in different geological conditions

A Machlica and M Stojkovova

Department of Hydrogeology, Faculty of Natural Sciences, Comenius University in Bratislava, Slovak republic

E-mail: machlica@fns.uniba.sk, stojkovova@fns.uniba.sk

Abstract. The identification of hydrological extremes (drought) is very actual at present. The knowledge of the mechanism of hydrological extremes evolution could be useful at many levels of human society, such as scientific, agricultural, local governmental, political and others. The research was performed in the Upper part of the Nitra River catchment (central part of Slovakia) and in the Topla and Ondava River catchments (eastern part of Slovakia). Lumped hydrological model BILAN was used to identify relationships among compounds of the water balance. Presented results are focused on drought in groundwater storage, soil moisture, base flow and discharges. BFI model for baseflow estimation was used and results were compared with those gained by BILAN model. Another item of the research was to compare results of hydrological balance model application on catchments with different geological conditions.

1. Introduction

The problem of drought identification is considered to be one of the most important hydrological research problems. The problems related to the definition, identification and prediction of drought have not been solved yet [1].

The primary cause of drought is the lack of precipitation over a large area and for an extensive time period. Such a drought is called a meteorological drought. Water deficit in the atmosphere propagates through the whole hydrological cycle and gives rise to different types of droughts. Combined with high evaporation rates a soil water deficiency might cause a soil moisture drought to be developed. The term agricultural drought is used when soil moisture is insufficient to support crops. Subsequently, groundwater recharge and streamflow are reduced and a hydrological drought may develop. A reduced recharge leads to lower groundwater heads and storage [2].

Within the hydrological cycle, groundwater is normally the last to react to a drought situation, unless surface water is mainly fed by groundwater. In deep aquifers the slow response of groundwater implies that only major meteorological droughts will finally show up as groundwater droughts. The lag between a meteorological and a groundwater drought may amount to months or even years, whereas the lag between a meteorological and a streamflow drought varies from days in a flashy catchment to months in a groundwater-fed catchment [2].

The aim of our work was to evaluate the occurrence of hydrological drought and its influences to hydrological balance components in the upper part of the Nitra catchment (central part of Slovakia) and in the Topla and Ondava catchments (eastern part of Slovakia) with the focus on groundwater compounds of the hydrological cycle. Primary attention was paid to individual water balance components in the zone of aeration and possible influence on the aquifers. We were also looking into decrease or increase in groundwater runoff.

2. Methodology

Database creation necessary for compiling of the hydrological models was the first step of our work. The data were acquired from the Slovak Hydrometeorological Institute (SHMU) in Bratislava. All the data were verified and edited according to requirements of BILAN and BFI models.

The BILAN model has been developed for assessing the water balance components of a catchment using a monthly step. The structure of the model is formed by a system of relationships describing basic principles of the water balance on the land surface, in the zone of aeration, including the effect of vegetation cover, and in aquifers bearing the groundwater. The input data of the model are monthly series of catchment precipitation and air temperatures. Furthermore, relative air humidity or potential evapotranspiration series are required. Monthly runoff series at the outlet of the catchment are used to calibrate eight model parameters. Model uses an optimization algorithm for calibration of these parameters. Several optimization criteria are available to attain the best fit between the observed and simulated runoff series.

The model simulates monthly series of potential evapotranspiration, actual evaporation, and infiltration into the zone of aeration, percolation of water towards the groundwater aquifer and water storage in the snow cover, zone of aeration (soil) and groundwater aquifer for a catchment. All these variables are used for a catchment as a unit (lumped physically based model). The total runoff consists of three components which are direct runoff, through flow (interflow) and base flow.

The output from the model can be expressed in two types of series – monthly value series and monthly averages (monthly characteristic). Monthly characteristics include average monthly values, monthly minimum and maximum. These were derived from monthly value series for each month in the year and for each variable. Series of outputs can be saved in different forms (numeric and graphics data series) [2].

Model BFI performs a separation of the base flow from the total streamflow and calculates the Base Flow Index (BFI). It is written as a Visual Basic Application within Excel. The index gives the ratio of base flow to total flow calculated from a hydrograph smoothing and separation procedure using daily discharges. Thus the BFI is considered as a measure of the river's runoff that derives from stored sources and as a general catchment descriptor it has found many ways of application, including low flow estimation and groundwater recharge assessment [2].

Input data are time series of daily streamflow data. The time series must be continuous and should not be interrupted. The output is a graph of the base flow separation, showing the total streamflow (input data) and the calculated base flow values [2].

One of the hydrological balance components is also groundwater runoff. This component is not measured directly and the most frequently is separated from hydrograph of discharges using separation methods or mathematic-statistic procedures. The Castany's method [3] was used for the groundwater runoff estimation. The method of Castany was used for comparison of groundwater runoff estimation results gained by BFI and BILAN models. The method was chosen because it gives the lowest estimates of groundwater runoff among the most often used methods. In hydrogeology, groundwater runoff values estimated by the method of Castany are considered for the minimum groundwater runoff values from a catchment and under certain conditions (low precipitation totals) they can be considered as a manifestation of lack of the groundwater in the aquifer. The method is based on a selection of 30 minimum daily discharges following one after another in each year and on calculation of the average value of them. Average groundwater runoff for certain time period (usually for a decade of 10 years) is calculated as a median value from the average values of all years of the assessed period. Groundwater runoff values were calculated for the period of hydrological years 1983-2003.

Some of the model input data were modified according to morphology of research areas. These corrections were done with ArcView 3.2 application in GIS system. Using ArcView application, the mean catchment altitude was estimated and according to this mean value, data on air temperature were recalculated.

The model BILAN was run after modification and verification of all model inputs. It was necessary to verify whether the model is working properly and whether the optimization criterion does not have a very high value. In the case of optimization criterion closest to zero the model setting of parameters is very good and the model conditions are close to real conditions. In the case that the optimization criterion is far from the zero value, the model conditions don't correspond to reality. After finishing the calibration process, creating of hypothesis, analysis of different curves of water balance compounds in more detail has started. The observed time series were divided into specific time periods in order to investigate some relationships describing changes in water balance compounds in extremes conditions. It means that the whole observed time period was classified according to yearly precipitation amounts; very dry, dry, normal, wet and very wet years were defined. The basic period for classification of the wetness character of single years was period from 1950 to 1980.

We were looking for periods of very dry and dry years. Values of the baseflow and total runoff decrease were estimated for dry years and they were expressed in a percentage to the average value. The groundwater storage amount was analyzed in detail considering the influence of the soil water storage.

Time series (1983-2003) of discharges in six discharge gauging stations were processed and analyzed. Four gauging stations were selected in the upper part of the Nitra River catchment, namely profiles: Tuzina on Tuzina brook (catchment area 35.60 km^2), Chvojnica on Chvojnica brook (catchment area 17.82 km^2), Handlova on Handlovka River (catchment area 40.18 km^2), and Chalmova on Nitra River (catchment area 601.11 km^2). Two gauging stations were chosen in the eastern part of Slovakia, namely Bardejov profile on Topla River (catchment area 325.8 km^2) and profile Stropkov on Ondava River (578.4 km^2).

The Tuzina and Chvojnica brooks are the right-side tributaries of the Nitra River in the mountainous part of the catchment. Handlovka is a left-side tributary. Topla is a right-side tributary of the Bodrog River and Ondava is its left-side tributary.

The upper part of the Nitra River catchment belongs to the moderately warm, humid climatic region (Chvojnica, Tuzina, Handlovka streams) up to warm, moderately humid (Nitra at Chalmova). Mean annual precipitation varies from 600 up to 700 mm for Nitra at Chalmova up to 800-900 mm for Chvojnica, Tuzina and Handlovka streams (observation period 1961-199). Mean annual air temperature varies in the interval from 7-8° C for Chvojnica, Tuzina and Handlovka streams up to 8-9°C for Nitra at Chalmova (the same observation period).

The Topla River catchment at Bardejov profile belongs to the moderately warm, moderately humid climatic region, the Ondava River catchment at Stropkov belongs to the warm, moderately humid climatic region with a cold winter. Mean annual precipitation ranged from 700 up to 800 mm for both catchments (observation period 1961-1990). Mean annual air temperature ranged from the 7° up to 8° C in the same observation period.

The period of 21 hydrological years (1983-2003) was evaluated. The data on the mean relative air humidity and mean air temperatures from the Prievidza meteorological station were available for all four profiles in the upper part of the Nitra River catchment. Mean precipitation totals from Nitrianske Pravno precipitation gauging station were used for the Tuzina and Chvojnica catchments. Mean precipitation totals from the Prievidza station were used for the Handlovka and Nitra at Chalmova profiles.

Meteorological data for the Topla River catchment were used from the Bardejov meteorological station and data for the Ondava River catchment from the Stropkov meteorological station.

The Chvojnica brook catchment has very similar geological, hydrogeological and climatic conditions to the Tuzina brook catchment. The upper part of the Nitra River catchment is typical of combination of different geological conditions and human influences. It is characteristic by the lowest terrain altitude and a flashy character. Also Topla and Ondava River catchment have very similar geological, hydrogeological and climatic conditions to each other. decrease or increase in groundwater runoff.

3. Results and discussion

In our work we tried to get a complex view on hydrological drought problems assessing localities in the upper part of the Nitra River catchment (central part of Slovakia) and the Topla and Ondava River catchment in the eastern part of Slovakia. Individual subsurface water balance components were analyzed for the zone of aeration looking for possible influences on the aquifer. The decrease or increase of groundwater runoff values was studied. Some of the results follow.

3.1. Analysis of the discharges in the upper part of the Nitra River catchment

Similar courses of the discharges hydrographs were obtained for the Tuzina and Chvojnica streams. Both streams have expressive maximum discharges in the spring months March and April (from 1.5 up to 1.9 $m^3.s^{-1}$ on the Tuzina brook and from 0.6 up to 1.9 $m^3.s^{-1}$ on the Chvojnica brook). Both streams have their minimum discharges from July up to December (from 0.1 up to 0.2 $m^3.s^{-1}$ on the Tuzina brook and from 0.03 up to 0.08 $m^3.s^{-1}$ on the Chvojnica brook). Shape of the discharge time series graph for the Chvojnica brook is almost identical with that one of the Tuzina brook, differing only in generally lower values of discharges (figure 1).

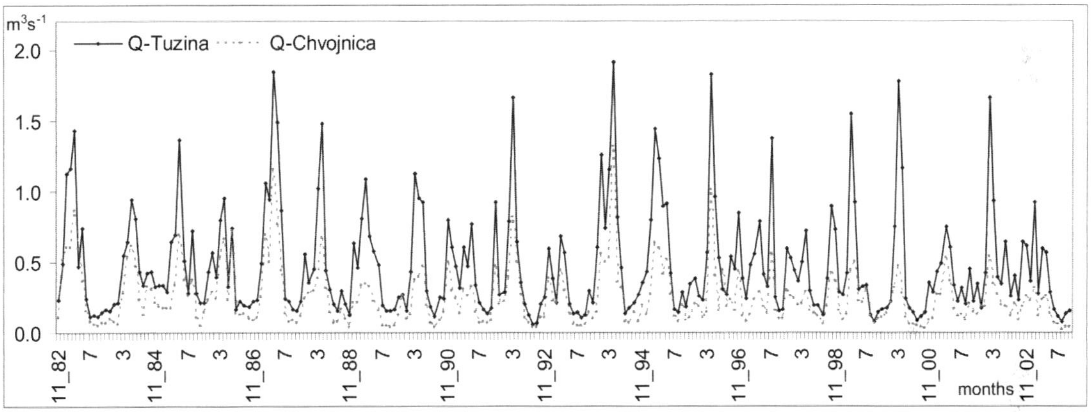

Figure 1. Time series of discharges (Q) on the Tuzina and Chvojnica brooks

Maximum discharge values on the Handlovka River vary from 1 up to 1.5 $m^3.s^{-1}$ (March and April). Minimum discharge values are in summer and autumn months (from 0.2 up to 0.3 $m^3.s^{-1}$). A similar shape of the flow hydrograph was gained for discharges at the Chalmova profile on the Nitra River. Maximum values are typical for spring months and minimum values for the autumn months (about 2 $m^3.s^{-1}$). Value of the discharge grows downstream. Minimum values at Chalmova are ten-times higher than at Tuzina and Handlova. The Nitra River is largely influenced by Handlovka River (figure 2).

The discharges in the Handlova and Chalmova profiles had a similar course up to year 1997. This situation repeated again in years 2001 to 2003. All the streams had a certain regular shape of discharges up to year 1995 with maximum values in spring and minimum values in summer and autumn periods. In the period of years 1987-1992 and 1997-2000, there was an expressive decrease of maximum discharges on the Handlovka River, with consequence on discharges in the Chalmova profile. The maximum values of discharges increased in years 1994-2002.

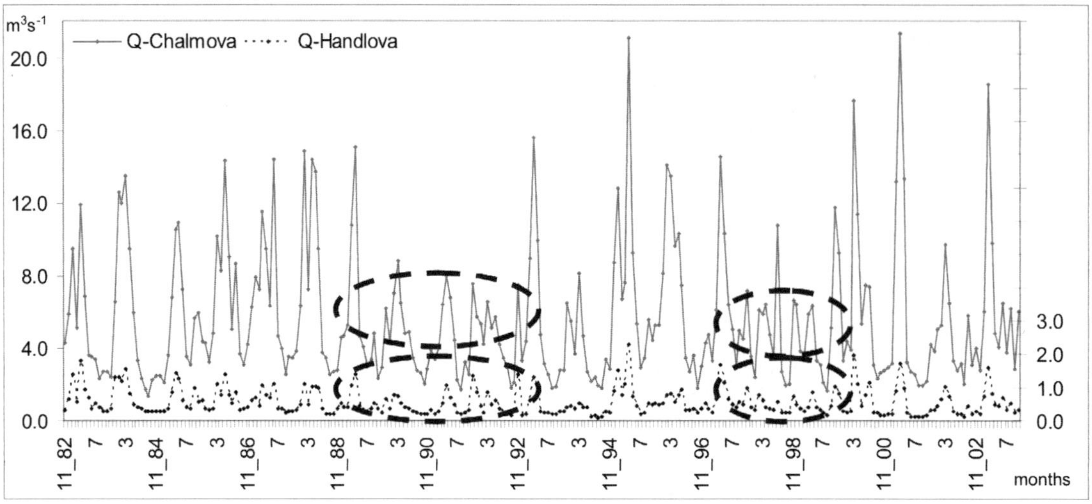

Figure 2. Time series of discharges (Q) on the Handlovka and Nitra Rivers

The maximum monthly discharge values in the Topla River are typical for spring months March and April (influence of snow melting) with the values from 4.55-14.74 $m^3.s^{-1}$. The minimum monthly discharge values occur from October up to December (0.38-1.11 $m^3.s^{-1}$). Periods of minimum discharges on the Topla River catchment are shown in figure 3.

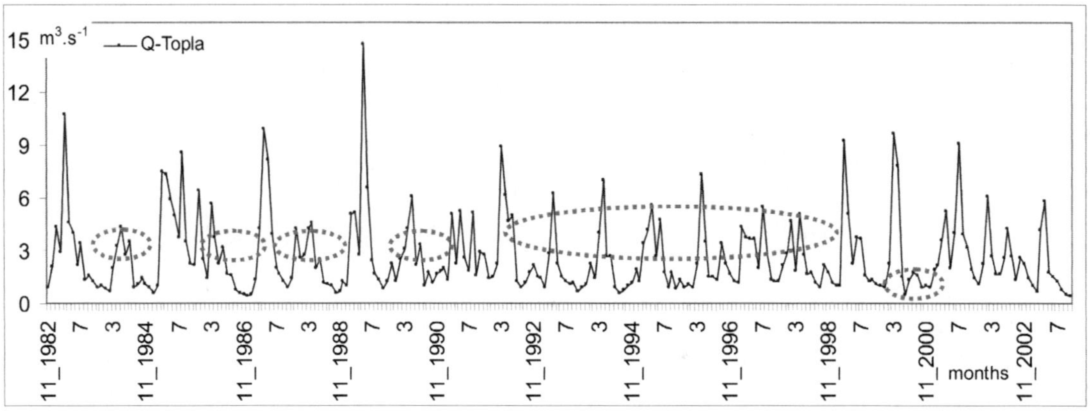

Figure 3. Time series of discharges (Q) on the Topla River

The maximum monthly discharge values on the Ondava River occur in spring months March and April (influence of snow melting) reaching values from 8 up to 25 $m^3.s^{-1}$. The minimum monthly discharge values occur in August and September ranging from 0.6 up to 0.9 $m^3.s^{-1}$.

3.2. Comparison of selected compounds in the time scale

The whole assessed period was classified according to yearly precipitation amounts; very dry, dry and normal years were defined. The Topla River catchment is characterized by occurrence of 33 % of very dry years, which is the maximum value among all evaluated catchments. Frequency of dry years reached 33 % in all gauging stations except of Chvojnica and Tuzina. In the case of normal year occurrence, the highest values were estimated for Chvojnica, Tuzina and Ondava (47 %). For the rest of the gauging stations it was 33%. Our research was focused on changes in baseflow values in very dry years.

3.2.1. *Tuzina Brook.*According to modeling results, the groundwater storage (GS) value reaches about 200 mm. In years 1990, 1996 and 1997, groundwater storage (GS) values ranged from 117 to 130 mm per month (figure 4). This fact points on a deficit in snow cover during the winter period. It was also confirmed by snow water storage (SW) parameter which was so low that there was no possibility to display SW value in the time series course. The parameter SW reached the highest value in 1987 and there was a visible positive relationship among snow cover, SW and groundwater storage.

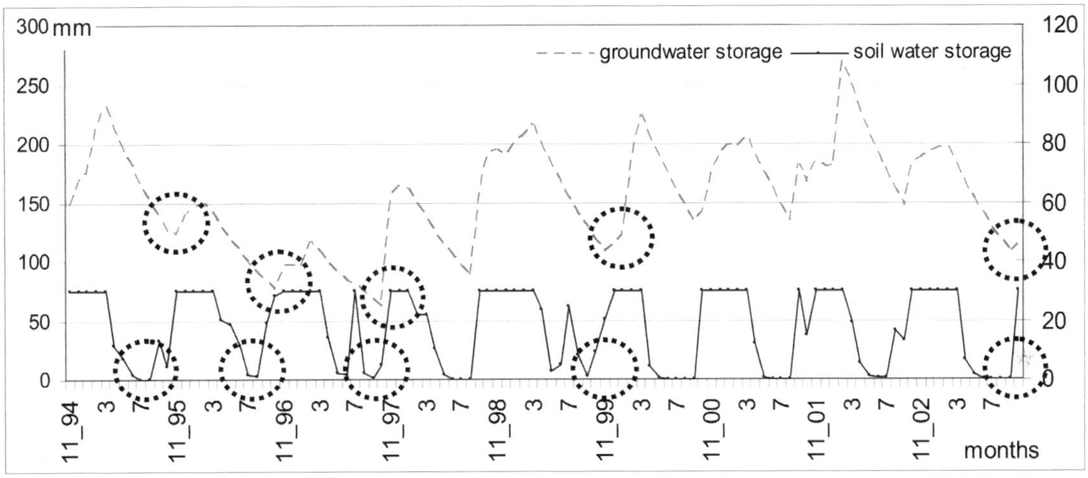

Figure 4. Decrease of soil water storage and ground water storage in summer months

Parameter GS varied in the range from 100 to 130 mm per month in the summer months till 1995. From 1995 to 1997 the GS values were the lowest in the whole observed period. Soil water storage curve (SS) had the similar course within the whole period, with the exception of summer months of 1986, 1988, 1997, 1999 and 2001. In those years the soil water storage reached its maximum values in the summer months, the zone of aeration was fully saturated. Precipitation amounts reached the values from 130 to 150 mm per month. Such values are more than two times higher than the average precipitation amount (61 mm per month). The lowest soil water storage values within the whole observed period were recorded from 1995 to 1997, what was similar to the values of the GS parameter (figure 4).

3.2.2. *Handlovka River.* The Handlovka River is a stream influenced by mining activities. Average GS value is about 283 mm per month. Average GS is varying around 370 mm per month in winter months and around 200 mm per month in summer months. The more significant decrease of GS values occurred in January and February 1984 (252 mm), 1990 (202 mm) and 1998 (294 mm). An approximately regular course of GS curve was disconnected for the whole period from 1993 to 1997. No clear natural course of runoff can be seen in the whole observed period. This can be caused by anthropogenic activities in the area, most probable by mining activities in brown coal mines. Based on analyzes of GS curves courses we assume that the precipitation amount together with water discharged to surface stream from mining works dewatering is the main compound which affects the amount of groundwater in the winter months (figure 5). Therefore the significant decrease of GS in summer months does not occur, as it is typical for Paleogene flysch rock environment (Topla, Ondava Rivers) or for areas built of less permeable Neogene rocks [4].

IOP Conf. Series: Earth and Environmental Science 4 (2008) 012010 doi:10.1088/1755-1307/4/1/012010

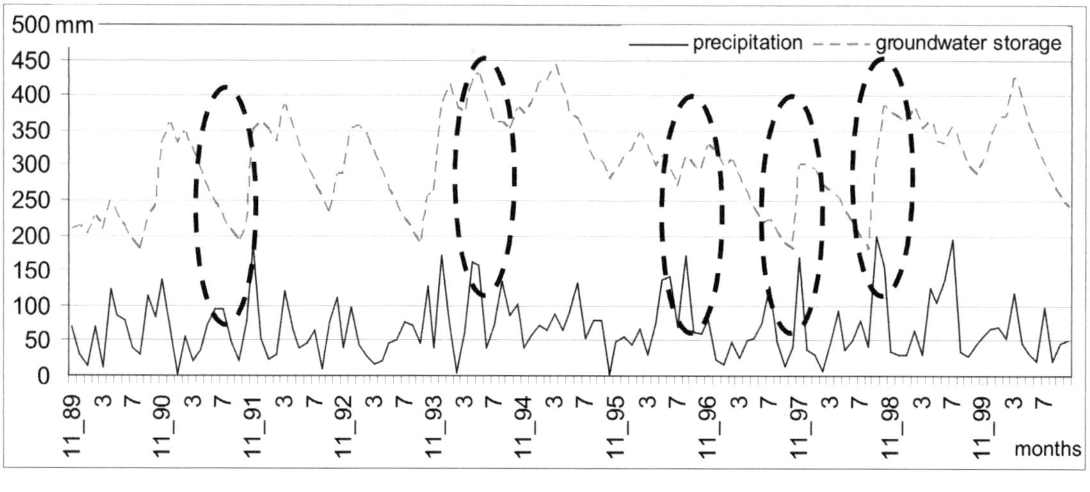

Figure 5. Dependence between precipitation amount and groundwater storage in the Handlovka catchment

3.2.3. *Topla River.* The Topla River is situated in a different geological environment. The area is built by Paleogene flysch sediments (alternation of sandstone and claystone), covered by Quaternary alluvial deposits. Climatic conditions are similar to those in the Upper Nitra region. Modeling results showed different courses of some water balance compounds (GS, SS and BF – base flow). Groundwater storage reaches the value of about 110 mm per month, in the summer months it varies in the range from 40 to 60 mm per month. Dry periods were visible on courses of GS curves only in 1984, 1990, 1998 and 2003. The courses of SS curves reflected the same situation as GS curves, in dry and very dry years there was the biggest decrease of soil water storage (figure 6).

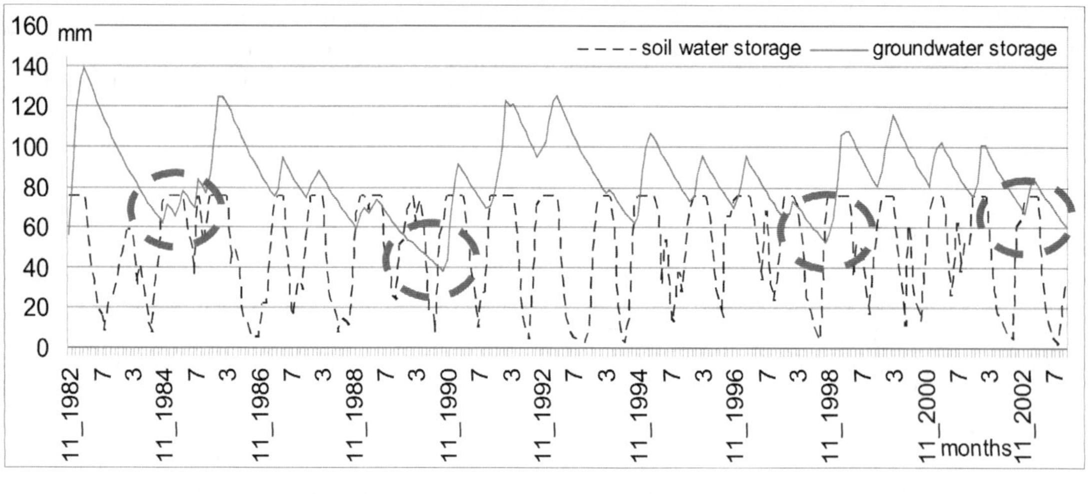

Figure 6. Synchronic decrease of soil water storage and groundwater storage in the Topla river catchment

3.2.4. *Ondava River.* The Ondava River has very similar conditions to those of the Topla River. The range of GS values is from 60 to 145 mm per month with the similar seasonal course as for the Topla River (maximum values in the winter, minimum values in the summer period).

Very dry and dry years have similar effect on GS and SS curves as on the Topla River. The most significant effect was manifested in 1984, 1990 and 2003.

3.3. Base flow estimation and evaluation

Very similar base flow values were estimated using Bilan and BFI models, the lowest values [5] were calculated by the method of Castany (table 1). Because of different catchments areas, it was necessary to re-calculate the base flow values into specific groundwater runoff values. The comparison and analysis of the base flow courses was done at first for very dry years in the whole observed time period and then especially for winter, spring and summer months.

Table1. Base flow values estimated by different methods

Profile	Stream	Baseflow ($m^3.s^{-1}$)			Specific groundwater runoff ($l.s^{-1}km^{-2}$)		
		BFI	BILAN	Castany	BFI	BILAN	Castany
Chvojnica	Chvojnica	0.15	0.12	0.05	8.19	6.45	2.81
Tuzina	Tuzina	0.28	0.16	0.12	7.95	4.49	3.37
Handlova	Handlovka	0.3	0.35	0.19	7.37	8.7	4.73
Chalmova	Nitra	3.71	3.37	2.05	6.16	5.6	3.41
Bardejov	Topla	1.48	1.49	0.71	4.55	4.57	2.18
Stropkov	Ondava	1.80	2.03	0.65	3.12	3.5	1.12

The lowest values of the base flow were estimated for catchments built by flysch sediments (Topla and Ondava catchments). This environment has no capacity to accumulate bigger amounts of groundwater. The worst conditions of water-saturation have the rock formation with dominance of claystones. They have poor fissure permeability, the infiltration is very limited. Better conditions have rock formations with dominance of sandstones with small portion of micro-conglomerates. Estimated values of the base flow correspond to given geological conditions. Influence of external conditions (precipitation amount) is reflected in different intensity of the base flow curves courses.

The Chvojnica and Tuzina streams have their springs in Mesozoic limestones and dolomites; however the main part of these river courses is located in the area made of granitoids and migmatites mainly. Down the Upper Nitra valley, both streams flow through the area built by Quaternary alluvial cones (sand and gravels deposits). Catchments of these streams have the lowest base flow values in the Upper Nitra River catchment. Central part of the area along the Nitra River up to the gauging station at Chalmova is built by fluvial deposits mainly, which are available to store bigger amounts of groundwater; this we can observe on the base flow values. The highest base flow values in the Upper Nitra catchment were estimated on the Handlovka River. The Handlovka River springs in neovolcanic sedimentary rocks along the stream flow through inner Carpathian Paleogene covered by Quaternary sediments. The highest base flow value among the all assessed profiles can be affected by mining activities in this area. Artificial recharging of the stream can also be one of the reasons of quite high groundwater storage and base flow values estimated by modeling and method of Castany [6].

4. Conclusion

Presented results document how the drought in surface and subsurface hydrological cycle components could be manifested. Factors (in monthly time step) leading to groundwater drought were evaluated by interpretation of shape of the time series generated in BILAN model and by groundwater runoff estimates calculated using BFI model. On the other hand, monthly time step is not satisfactory for detailed monitoring of drought development and propagation in the subsurface compounds of the hydrological cycle. Use of hydrological models with daily time step would be more suitable for such specific analysis. At present, works on results re-evaluation using BILAN model in a daily time step are going on.

Acknowledgement
This work was supported by FP6 Project 036946 WATCH and Slovak RTD grant agency APVV project No. APVV-0355-06.

References

[1] Bonacci O 1993 *Hydrological identification of drought, Hydrological Processes* vol. 7; pp 249-262.

[2] Tallaksen L M and van Lanen H A J 2004 *Hydrological drought, Processes and estimation methods for streamflow and groundwater, Development in water science 48*, (Elsevier, Netherlands) p 579

[3] Castany G, Margat J and Albinet M 1970 *Evaluation rapide des resource en eaux d'une region, Simposio internacional sobre aguas Subterraneas de Palermo*, (Asociacion internacional de Hidrogeologos, Ente de Desarrollo Agricola, Palermo) pp 462-682

[4] Machlica A and Fendekova M 2006 *Separation of runoff components using BILAN model in Chvojnica catchment*, Podzemna voda vol 12, No. 2., SAH Bratislava, [in Slovak with English Summary] pp 135-147

[5] Stojkovova M 2007 *Groundwater runoff on the territory of Slovakia*, Podzemna voda, vol. 13, No.2, (SAH Bratislava), [in Slovak with English Summary] pp 146-152

[6] Machlica A, Stojkovova M and Bara M 2007 *Impact of meteorological drought on the base flow formation*. 14th Slovak hydrogeological conference (SAH Bratislava) p 100

IOP Conf. Series: Earth and Environmental Science **4** (2008) 012011 doi:10.1088/1755-1307/4/1/012011

Low flow analysis of the lower Drava River

T Mijuskovic – Svetinovic and S Maricic

Faculty of Civil Engineering, University of Osijek, Drinska 16 a, 31 000Osijek, Croatia

E-mail: tatjanam@gfos.hr and smaricic@gfos.hr

Abstract. Understanding the regime and the characteristics of low streamflows is of vital importance in several aspects. It is essential for the effective planning, designing, constructing, maintaining, using and managing different water management systems and structures. In addition, frequent running and assessing of estimates of low stream-flow statistics are especially important when different aspects of water quality are considered. This paper attempts to present the results of a stochastic analysis of the River Drava low flow from the gauging station, Donji Miholjac [located at rkm 77+700]. Currently, almost all specialists apply the truncation method in low-flows analysis. Taking this into consideration, it is possible to accept the definition of a low streamflow, as a period when the analysed characteristics are either, equal to or lower than the truncation level of drought. The same method has been applied in this analysis. The calculating method applied takes into account all the essential components of the afore-mentioned process. This includes a number of elements, such as the deficit, duration or the time of the occurrence of low flows, the number of times, the maximum deficit and the maximum duration of the low flows in the analysed time period. Moreover, this paper determines computational values for deficits and for the duration of low flow in different return periods.

1. Introduction

The Drava River springs from South Tyrol, near Dobiaca (Italy) and runs for almost 750 km, through five different countries (Italy, Austria, Slovenia, Croatia and Hungary). It flows into Croatia at the Slovenian border, near Ormoz, stretching 325 km away from its source. Seventy-five km downstream, the Mura River flows into Drava and from that point to Donji Miholjac, it forms the Croatian-Hungarian borderline. The Mura River runs through Croatia for 305 km, before it flows into the Danube near Aljmas. Its water level (over 40 000 square kilometres) constitutes about 5% of the entire Danube catchment area.

The Mura River provides one third of the lower Drava water balance. The River Drava has a fluvio-glacial hydrological system (consistent outflow throughout the year, it reaches its maximum level in spring - May, and its minimum in winter - January, February).

The Drava River is a source of water used for many purposes, but it is also a natural wastewater recipient. Furthermore, it has a number of tributaries, which are also used as recipient for communal and industrial wastewaters, stemming from various residential as well as industrial areas situated along the waterways.

Understanding the nature and the characteristics of low streamflows is of vital importance in several aspects. It is essential for the effective planning, designing, constructing, maintaining, using and managing different water systems of varied structures. In addition, frequent running and assessing of estimates of low streamflow statistics are also important, not only when different aspects of water quality systems are considered, but also for the purpose of water quality management. One of the main characteristics of a low streamflow is that a minimum quantity of water is in the stream, added to that is the wastewater received from various sources. Thus there is a minimum capacity for the dilution of the pollution level which was brought in as a result.

Low streamflow regime is conditioned by the precipitation level and the nature of distribution over the year. Winter low fows are associated with the retention of solid precipitation in the seasonal snow pack while summer low flows are related to the relatively large moisture deficits in the lowland catchments during summer. During dry periods low streamflows are supplied and replenished by groundwater. The intensity of supply is mainly dependent on the level of groundwater re-emergence - accumulated river-basin capacity. Hence, the interpretations of these processes are explained in the principles of filtration, or in order words, in the discharge and drainage of these water resources [1].

It is often required to define the functioning of low streamflow of any stream in terms of analysing its hydraulic capacity. This entails its ability to provide the nearby residential or industrial areas with water, ensuring in such a way the supply of sufficient quantities of water needed for irrigation or water power, as well as minimum conditions required for the growth and the survival of various plant- and animal species in the streamflow. Some of the low flow studies are used for determining the parameters, while analysing the low flow. Sometimes they are also used for determining the parameters representing the indicators of the disposable quantities of groundwater, which replenish the river network of the examined river-basin [2].

The stochastic low flow analysis of the River Drava, at Donji Miholjac gauging station [located at rkm 77+700] is presented in this article. The probabilities of a minimum flow phenomenon at different durations of low flow periods are analysed, as well as the stochastic low flow processes, where the used method takes into account all the important components of the above mentioned process: deficit, duration and time of low flow phenomenon, as well as the highest deficit and the highest low flow duration in the given time unit.

2. Applied model and approximations

Currently, almost all specialists apply the truncation method in droughts and low flows analysis. Taking this into consideration, it is possible to accept the definition of a drought or a low streamflow, as a period when the analysed characteristics are either, equal to or lower than the truncation level of drought [3]. The same method has also been applied in this analysis presented in this paper. The stochastic low flow process analysis followed is structured according to the method which, for the chosen locations on the river, provides the approximate solution for the low flow calculation as two-dimensional random variable (D_i, T_i), while only analysing the flow equal to or less than the chosen truncation discharge $Q<(Q_r = Q95)$ (see figure 1). Q95 flow is the river flow that would be exceeded 95% of the time. This low flow characteristic is widely used in Europe and has been chosen because of its relevance for numerous water resources management applications The obtained solution has been reached not by determining the function of distribution of the above mentioned two-dimensional variable, but by testing the approximate equality of phenomenon probability, in terms of its deficit and duration, and also by the detailed stochastic analysis of all of the important components of such process. The low flow phenomenon, duration and deficit, are "estimated" in relation to the pre-accepted truncation discharge Q_r.

The method treats low streamflow as a random number of random variables in the time unit [0,t], where both the number of low flow η (t) appeared under some referential value Q_r in the given time unit [0,t], and the magnitude of their duration are random variables. The method also enables the defining of computational elements i.e. artificial 10-, 50- and 100-year low flows.

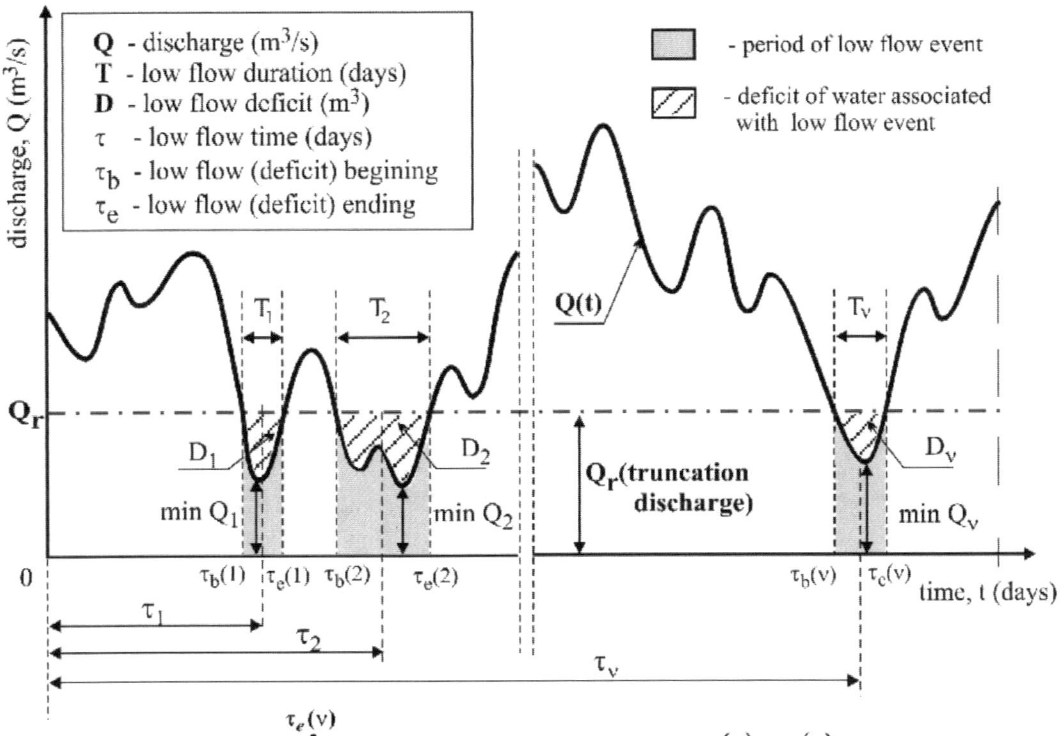

$$D_v = Q_r T_v - \int_{\tau_b(v)}^{\tau_e(v)} Q_t dt \quad \text{za } Q_t < Q_r \qquad \tau(v) = \frac{\tau_b(v) + \tau_e(v)}{2}$$

Figure 1. Main parameters of low streamflow event series - hydrograph of instantaneous river discharge at a certain location in the time unit [0,t]

For the truncation discharge Q_r, through which the original hydrograph is traversed, the flow of 95 % (Q95) of low flow is accepted from the flow duration curve of the mean daily flows. Hence, the lower extremes zone approximately separated from flows which belong to the broader mean values zone and the flood flow zone. The attention is concentrated only to the partial duration series, so-called deficits, developed by the adequate low flow phenomenon.

The theoretical consideration of this method was given by Zelenhasic et al. [4].

However, some simplifications have been introduced to the application of the method:

- All observed very small deficits D_i, i = 1,2,... which satisfy the condition D_i<0,005 max D_{rec} are neglected as they are considered insignificant and are not taken into consideration in this analysis. They are insignificant compared to the deficit of mean value or rare high deficits;
- The other hypothesis refers to those deficits which are kept in the analysis i.e. to those higher than the mentioned limits - when the flow Q_t of the examined river and its location surpasses the truncation Q_r during a very short time unit $\Delta T_{v, v+1,}$ and thus it artificially separates, practically, one low flow into two parts i.e. events. If it is the case of short time unit $\Delta T_{v, v+1}$, in relation to time T_v respectively T_{v+1} and if the volumes $\Delta T_{v, v+1}$ are small in relation to the deficit D_v or in relation to the deficit $D_{v+1,}$ than the events E_v, and E_{v+1} are practically referring to one low flow. $\Delta T_{v, v+1}$ = 7 days is accepted as the time-limit. If the time $\Delta T_{v, v+1}$ is high in relation to the time T_v namely T_{v+1} then the deficit D_v and D_{v+1} are considered mutually independent;
- When the time unit [0,t] is taken as representing the whole calendar year, some low flow

might happen to start in one and finish in the next. If it is the case, the time for this low flow is calculated by using the following formula: $\tau = (\tau_b + \tau_e)/2$. This low flow is not divided into two years as an event but its whole deficit and its whole duration are put into one year, that one of time τ.

3. Low flows stochastic analysis of the Drava River near Donji Miholjac

3.1. Basic data

The low flow stochastic analysis of the Drava River, near Donji Miholjac, was made on the basis of the sample of mean daily flows from 1980 to 2007.

The mean annual flow of sample is Q_{sr} = 500 m^3/s; the maximum observed flow is Q_{max} = 1891 m^3/s; and minimum flow is Q_{min}= 157 m^3/s.

The value of Qr = Q95= 243 m^3/s is taken as the truncation discharge which represents the flow of 95 % of low flow from the flow duration curve of mean daily flows. From these hydrological data, 38 low streamflows have been observed during 28 years; it appears that there has been a period of 10 years without low flows.

The auto-correlation analysis has been performed on 38 observed deficits, to get the total given sample. The auto-correlation coefficient of the first series indicates that the linear dependence of deficits in successive periods (interval, lagging, lag of one deficit) has been negative according to the direction, and the intensity has been of very inferior quality, which is $\rho(1) = -0.1385$. The other auto-correlation coefficients are equal to zero. This leads to the conclusion that the deficit phenomena are independent and non-correlated.

3.2. Low flow number distribution

The monthly variation of the low flow phenomenon at Drava River, at Donji Miholjac gauging station, is presented in figure 2. It can be noticed that only during the months of May and June that the low flow has not been recorded. Low flow most often appears during the winter months, from December to February and hardly ever during the spring months, May, June and July.

According to references [4] and [5], the expected number of low flows in the time unit [0,t] represents the time-dependent of Poisson process (this distribution is assumed as the partial duration series approach, in addition to the peaks-over-a-truncation approach). Poisson's distribution of the low flow number for the interval of 365 days has been calculated, and the test of goodness-of-fit of theoretical distribution, to the observed one, has been performed. Poisson's function distribution parameter has been evaluated through statistics and equal to λ_1 =1.35714. Chi-squared test is close to satisfactory result at the class probability of α=5%, thus the theoretical Poisson's distribution can be accepted for the low flow distribution (see figure 3).

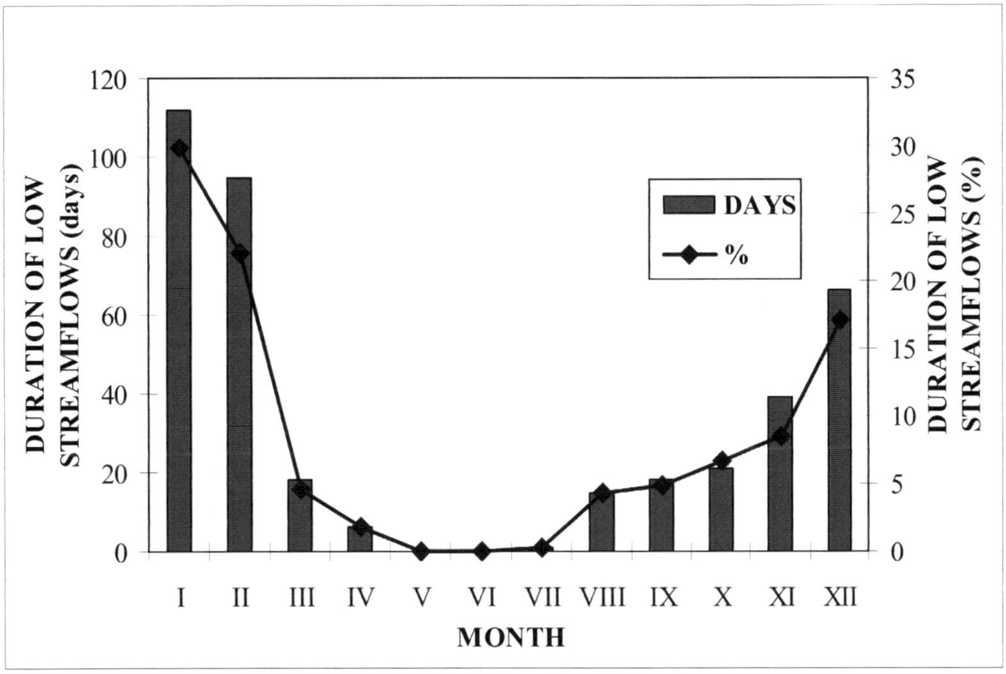

Figure 2. Monthly variations of appearance and duration of Drava low flows close to Donji Miholjac (Q_r = 243 m^3/s; 1980-2007).

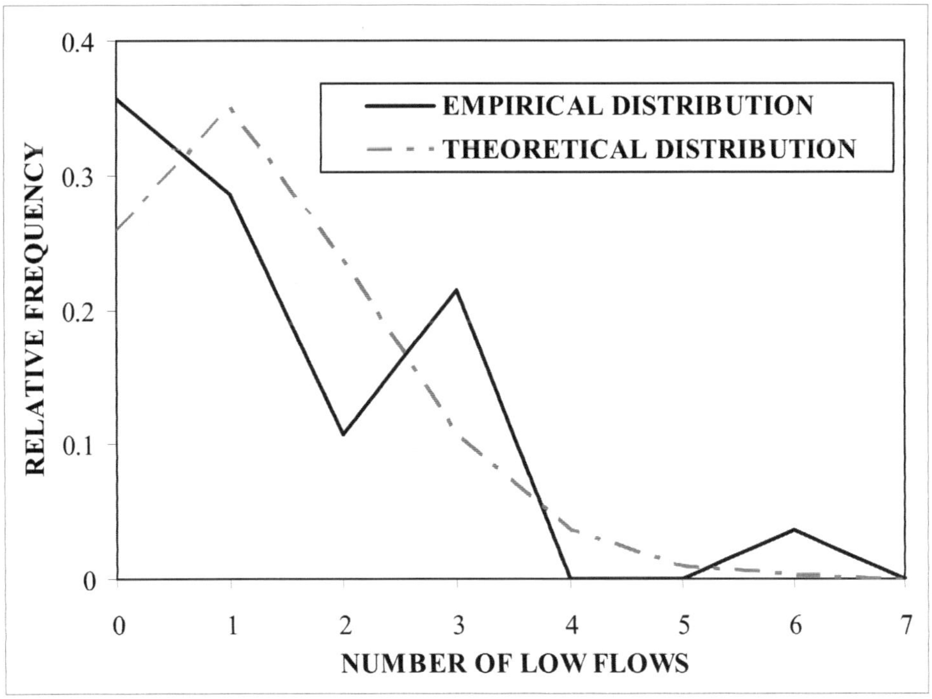

Figure 3. Empirical and appropriate theoretical (Poisson) distribution of the number of low streamflows in the time period of 365 days.

3.3. Deficit distribution

As for the determining of the deficits distribution, i.e. their magnitudes during the time unit [0,t], the whole year has been taken into consideration. The sample has contained 38 event covering the period of 28 years. Basic statistical characteristics of the obtained deficit sample are:

- MAX DEFICIT 177,379,200 m^3
- MIN DEFICIT 950,400 m^3
- AVERAGE DEFICIT 34,561,600 m^3
- STANDARD DEVIATION 38,707,428 m^3

It is known that if one event series occurs in time, according to Poisson's probability distribution, then the phenomenon of the time of k-event from this series follows the Gamma distribution. The observed and appropriate theoretical distribution functions are presented in figure 4, from which it can be seen that the exponential distribution (this distribution is assumed according to references [4] and [5]) is very well adjusted to the observed one i.e. the empirical deficit distribution. The coefficient λ_2 has been obtained through statistics and is equal to $\lambda_2 = 3.11*10^{-8}$. The theoretical function is indicated by the equation:

$$H(x) = 1 - \exp(-3.11*10^{-8} x), \text{ for } x \geq 0, \quad (1)$$

In this equation, x represents the deficit and is measured in m^3. The goodness-of-fit test, according to Kolmogorov-Smirnov, has proved the good coincidence i.e. the closeness of two distributions shown in figure 4.

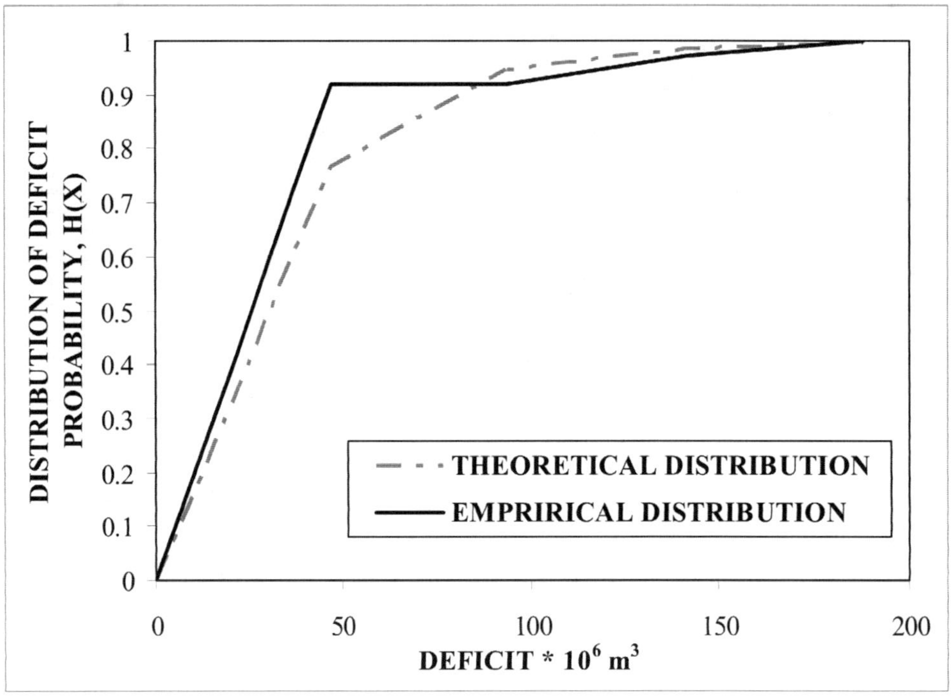

Figure 4. Theoretical and empirical distribution of deficit (low streamflows).

3.4. The highest deficit distribution

The function of the highest, extremely high, deficit distribution, during the time unit [0,t] for the Drava River near Donji Miholjac, derived from the basis of the obtained results:

$$F_t(x) = F_t \exp[-\lambda_1(t)\exp(-\lambda_2 x)], \text{ for } x \geq 0, \qquad (2)$$

In the above mentioned equation, the time unit [0,t] can be any part of the year calculated from the first of January, or be the whole year, depending on what the value of determining function $\lambda_1(t)$ is based on. The summary function of the highest deficit distribution for a time unit of the whole year, which is the most frequent in practice, for the Drava River near Donji Miholjac, is indicated in the equation:

$$F_t(x) = F_t \exp\left[-1.357143 \exp(-3.11 * 10^{-8} x)\right], \text{ for } x \geq 0, \qquad (3)$$

where t is equal to the time unit of one year, and x represents the deficit and is measured in m^3. The theoretical and observed function diagram of the highest deficit (i.e. extremely high) of low flow for the examined example is shown in figure 5. Both tests (Chi-squared test and Kolmogorov-Smirnov test, for the level of accuracy α=5%) have shown good coincidence of theoretical and empirical functions.

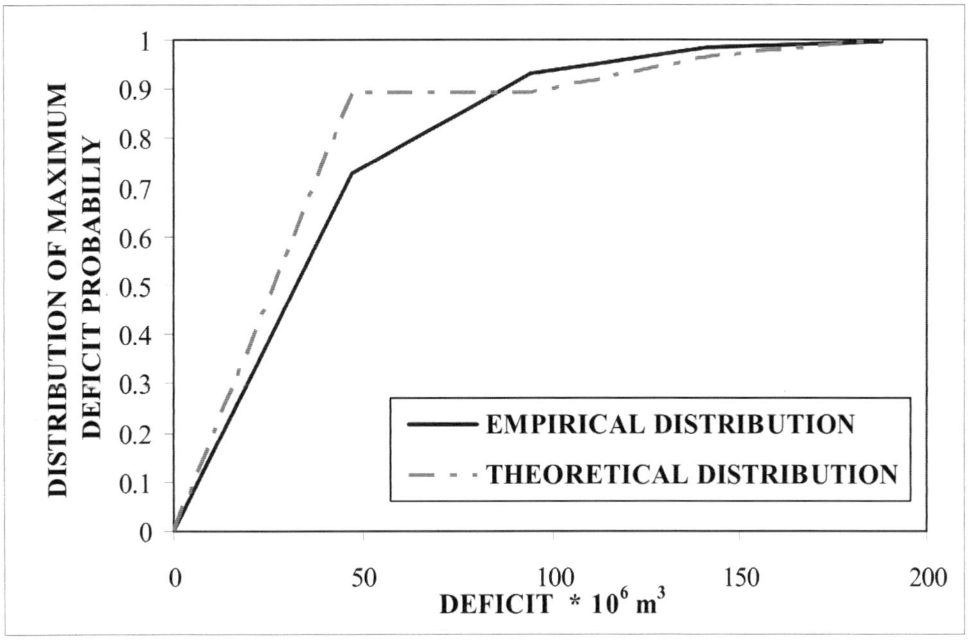

Figure 5. Theoretical and empirical functions of distribution of maximum low streamflow deficits in one year period.

3.5. Low flow duration distribution

A time unit [0,t] of one year is chosen for low flow duration distribution. The sample has contained 38 occurrences during the low flow duration, whose arithmetic mean and standard deviation may be stated as follows:

$\overline{T}$ = 17.58 days,

σ (T) =18.76 days.

The maximum observed low flow duration during 28 years of available data of available data of flow observation is:

max T_{os}= 94 days (1986).

The observed and suitable theoretical functions of low flow duration distribution for the Drava River near Donji Miholjac are shown in figure 6. It can be very clearly seen that the exponential distribution function

$$G(u) = 1 - \exp(-\lambda_3\, u) = 1 - \exp(-0.05688u) \qquad (4)$$

represents the observed, i.e. empirical distribution. In this equation u represents duration and is measured in days. Kolmogorov-Smirnov test for the level of accuracy of α=5% has shown that there is a good coincidence of the two distributions represented by figure 6.

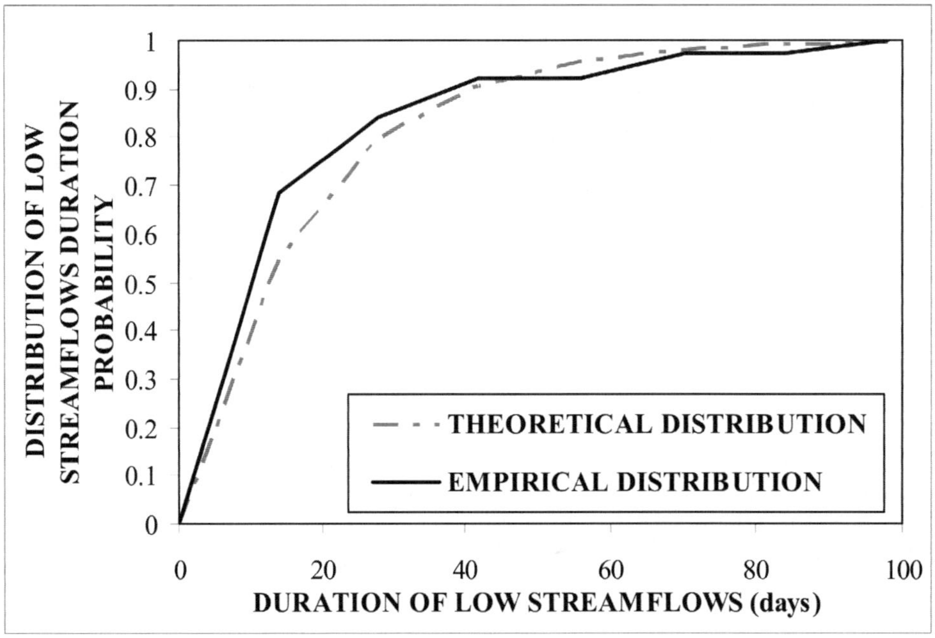

Figure 6. Theoretical and empirical functions of low streamflow duration probabilities.

3.6. Distribution of the highest low flow duration

The period of an entire year is considered for the purpose of determining the distribution of the highest, i.e. rare and extremely high low flow duration T (lar), for [0,t]. This hydrological magnitude does not depend only on duration magnitude T_v(all previously observed) of low flow, but also on their number k. The summary function of the low flow duration distribution for the time unit [0,t] is generally given according to equation [4]:

$$G_t(\mathrm{v}) = P(E_0^t) + \sum_{k=1}^{\infty} P\left[\bigcap_{v=1}^{k} (T_v \le v) \bigcap E_k^t\right] \qquad (5)$$

As to the Drava River near Donji Miholjac for [0,t] = 365 days, the equation is built as follows:

$$G(v/t = 365\,\text{dana}) = \exp\left[-\lambda_t \exp(-\lambda_3 v)\right] = \exp\left[-1.357143 \exp(-0.05688v)\right], \text{for } v \ge 0 \qquad (6)$$

where v represents maximum duration expressed in these days, with

$\overline{T}$ (lar) = 17.5 days, and

$\sigma\left[T(lar)\right]$ = 21.0416 days

The observed and appropriate theoretical functions of the highest duration distribution of the low flow, for the Drava River near Donji Miholjac, are shown in figure 7. The adjustment quality of the theoretical to the empirical function is satisfactory, which can be seen in the figure, of both statistical tests (Kolmogorov-Smirnov and Chi-squared test for the level of significance α=5%), which can clearly imply the good coincidence of two distribution functions.

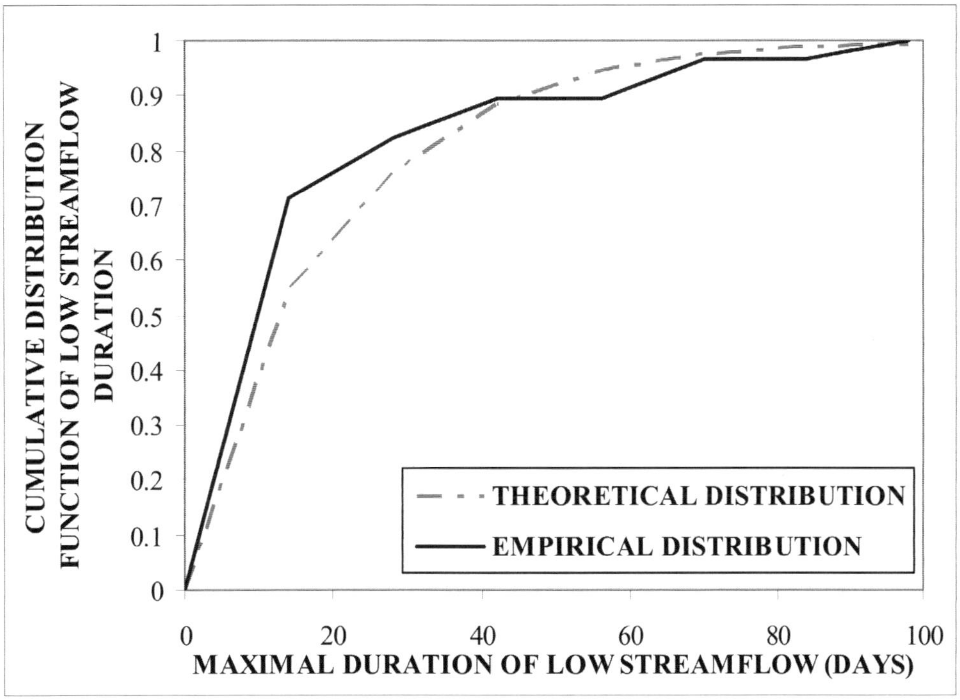

Figure 7. Theoretical and empirical distribution of low streamflow maximum duration T (lar), for [0,t] = 365 days.

3.7. Deficit ranks and low flow duration

In order to examine the possibility of performing the computational 10-, 20-, 50- or 100-year low flow, it has been decided to check the hypothesis about (at least approximate) the equality of the phenomenon probability of rare, i.e. extreme deficits and low flows durations for the Drava River near Donji Miholjac. The twenty observed and most extreme low flows have been ranked in descending order and presented in table 1. Each deficit has been associated to an appropriate deficit rank m_D. For the same river low flows, the durations (T) and accompanying ranks m_T have been allocated equally. The table indicates that there is closeness between river low flow deficit (m_D) and duration (m_T) ranks. Since the low flow, as a hydrological magnitude, is the two-dimensional random variable, which is characterized by its deficit and duration in cases where the condition of equality of phenomenon probability of maximum deficits and low flow duration is at least approximately satisfactory, it is possible to use the deficit magnitudes and duration of the same return period for determining the computational low flow of this return period [6,4]

The computational deficits, as well as low flow durations of 10-, 20-, 50-, 100- and 1000-year return period, have been calculated in equations (3) and (6), on the basis of phenomenon probability of the highest deficit, namely the highest duration, which are presented in table 2.

Table 1. Rank of deficits and durations of Drava River low flow for the 20 maximum observed deficits.

Deficit No.	Date of low streamflow	Deficit D (m^3)	Duration T (days)	Rank of deficit m_D	Rank of duration m_T
1	23-November-01	186,019,200	94	1	1
2	6-December-86	139,449,600	69	2	2
3	1-January-89	100,483,200	66	3	3
4	6-February-90	37,324,800	36	4	4
5	20-November-83	35,164,800	31	5	5
6	14-September-92	23,673,600	21	6	8
7	15-October-85	23,500,800	29	7	6
8	3-January-85	20,995,200	20	8	10
9	16-January-90	19,353,600	12	9	16
10	28-October-95	18,921,600	11	10	17
11	12-August-03	16,156,800	20	11	9
12	26-December-06	14,169,600	23	12	7
13	23-September-03	14,083,200	12	13	15
14	29-August-92	10,022,400	5	14	25
15	2-November-01	9,849,600	7	15	21
16	6-February-91	9,676,800	19	16	11
17	6-December-89	8,553,600	7	17	20
18	9-November-86	8,035,200	16	18	12
19	12-February-06	6,307,200	5	19	24
20	16-September-90	6,134,400	4	20	28

Table 2. Calculated Drava River low flow and duration for different return periods.

Return period (years)	Calculated low flow deficit (10^6 m^3)	Calculated duration of low flow (days)
10	82	45
20	106	58
50	135	74
100	158	86
1000	232	127

IOP Conf. Series: Earth and Environmental Science **4** (2008) 012011 doi:10.1088/1755-1307/4/1/012011

4. Conclusion

Streamflows at Donji Miholjac gauging station, from 1980 to 2007, were analysed, based on available data. The following conclusions can be made:

- Average yearly flow amounts to $Q_{sr} = 500\ m^3/s$, maximum observed flow to $Q_{max} = 1891\ m^3/s$, and minimum to $Q_{min} = 157\ m^3/s$;
- Only the deficits lower than accepted truncation discharge have been stochastically analysed. Hence, the attention has been focused only on the lower extreme zone that is to the partial duration series. As for the truncation discharge from the flow duration curve of the mean daily flows, the flow Q95, the river flow that would be exceeded 95% of the time, has been accepted, $Q_r = 243\ m^3/s$;
- The number of 38 low flows have been observed within 28 years;
- There has been a period of 10 years without low flows;
- Low flows have not been observed only in May and June, whereas in April and July, 6 i.e. 1 day with low flows have been recorded;
- Days with maximum observed low flows were in January - a total of 112, in February – 95, and in December - 66;
- The highest probability for only one (k=1) low flow a year, i.e.
  P(k=1) = 0.349, followed by:
  P(k=0) = 0.257
  P(k=2) = 0.237
  P(k=3) = 0.107, etc.
- There is a probability of 0.5, that the extreme low flow deficit will be $23*10^6\ m^3$, resulting from equation 3. Similarly, the deficit of the required return period can easily be determined. For example, 10-year return period $D_{10} = 82*10^6\ m^3$, i.e. 100-year return period $D_{100} = 158 *10^6\ m^3$, during one year interval;
- There is a probability of 0.5, that the highest low flow duration in a year will be 12 days or longer, which is indicated in the equation (6). Similarly, the extreme low flow duration of 10-year return period $T_{10} = 45$ days, i.e. 100-year return period $T_{100} = 86$ days, for one year interval;
- On the basis of equations (3) and (6), as well as on the basis of table 1 and the discussion in chapter 3.7., the possibility of defining computational low flows of any desired return period can be easily examined, e.g. 10, 20, 50 or 100 years.

References

[1] Srebrenovic D 1986 *Primijenjena hidrologija* (*Applied Hydrology*) (Zagreb: Technical book)

[2] Zelenhasic E and Srdjevic B 1986 Statistički proračun malih voda reka za potrebe analize kvaliteta voda (Statistical analisis of low streamflows for water quality purpose) *Voda i sanitarna tehnika* **4** 41-44

[3] Bonacci O 1993 Hydrological indentification of drought *Hydro.l Proc.* **7** 249-62

[4] Zelenhasic E et al 1986 *Stohasticka analiza malih rečnih voda (Stohastic Analisis of Low River Flow*) (Novi Sad: Faculty of Agriculture University of Novi Sad)

[5] Hall M J 1996 *Statistics and Stochastic Processes in Hydrology* (Delft: International Institute for Infrastructural Hydraulic and Environmental Engineering)

[6] Askew A J, Yeh W W G and Hall W A 1971 A comparative study of critical drought simulation *Wat.Rrcsour.Rcs.* **7(1)** 52-62

IOP Conf. Series: Earth and Environmental Science **4** (2008) 012012 doi:10.1088/1755-1307/4/1/012012

T-year maximum discharges on water courses in Slovakia

J Podolinská and H Šipikalová

Slovak Hydrometeorological Institute, Regional Centre Banská Bystrica, Zelená 5, 974 04 Banská Bystrica, Slovak Republic

E –mail: jana.podolinska@shmu.sk, helena.sipikalova@shmu.sk

Abstract. T-year maximum discharges serve the purpose of background for design, building and operation of water management constructions and facilities, regulation of water courses, flood and environmental protection. The Slovak Hydrometeorological Institute (SHMI) processes and provides these data according technical standards. It is a tradition that the whole territory of Slovakia is processed and the system of river network is considered. This way was also applied for data updating (finished in 2006) with new methods and technologies (GIS). Background materials were time series of hydrological data obtained from surface water stream-gauging stations during the whole observation period (minimum 20 years). T-year maximum discharges in 340 stream-gauging stations were calculated using mathematical-statistic methods. Pooling scheme of 100-year maximum specific discharge ($q_{100.max}$) was used in the profiles without observation. The cluster analysis was applied for regional types and relative estimate deviation $q_{100.max}$ determined using non-linear regression formula expressing dependence of this value on specific physical-geographical characteristics. Input data were time series from 197 selected stations with a catchment area from 20 to 300 km^2 and minimum impact of anthropogenic activities. T-year maximum discharges were determined in stream-gauging stations and significant profiles by using the aforementioned methods and synchronized in the system of river network.

1. Introduction

The T-year maximum discharges ($Q_{max.N}$) estimation is one of the most difficult tasks in hydrology. To choose the method for design discharge calculation is also complicated and depends on the catchment size and data availability. We use statistical methods if data are available, for small catchments and if data are not available a pooling scheme is preferred. Regional methods of estimation belong between traditional and standard hydrological tools. They are used for the interpolation of hydrological information between monitored sites, to sites without monitoring. These methods we used for T-year maximum discharge updating on water courses in Slovakia [1].

2. Data sets and data analysis

Background materials were time series of hydrological data obtained from stage-discharge gauging stations during the whole observation period (minimum 20 years). We used the annual maximum discharges in each hydrological year (1.11.-31.10.) in 340 stage-discharge gauging stations. In every stage-discharge gauging station there was a very important and difficult data analysis made and adjustment of annual maximum discharges (figure 1).

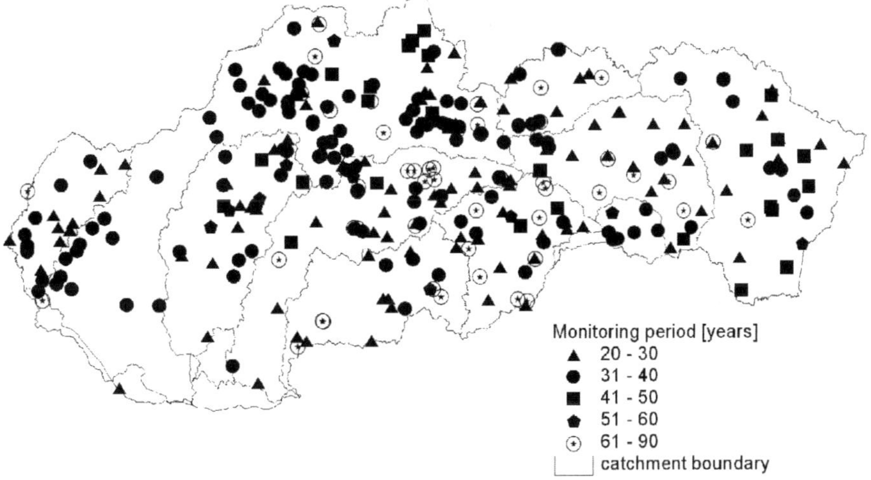

Figure 1. Map of stage-discharge gauging stations with monitoring period in Slovakia.

Course of annual maximum discharges and their trend in Moravský Svätý Ján (Morava River Basin) is shown at the figure 2.

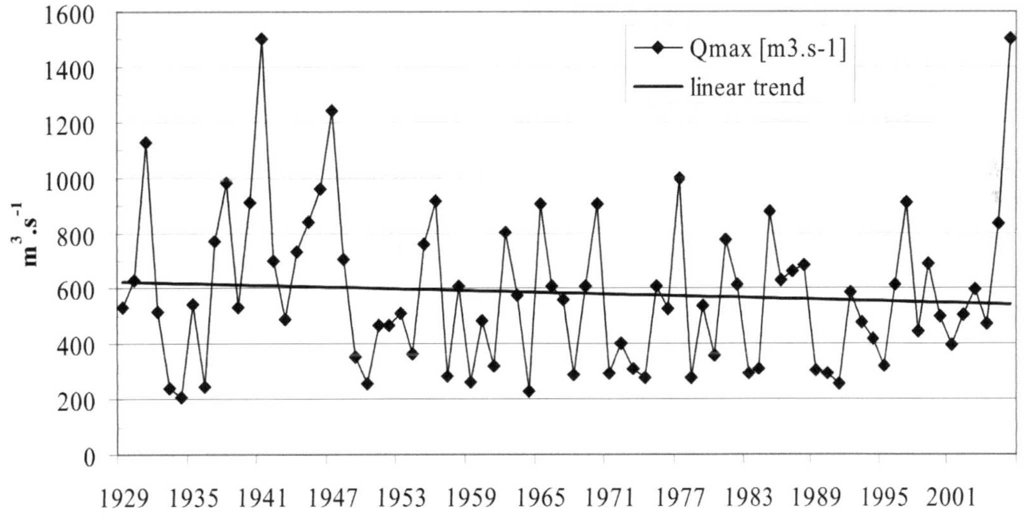

Figure 2. Course of annual maximum discharges Q_{max} and their trend in Moravský Svätý Ján (Morava River Basin) in 1929-2006.

3. T-year maximum discharges in the stage-discharge gauging stations with longer observation period

Statistical calculations of T-year maximum discharges were processed by HQ-EX. software Several types of theoretical distributions and methods of parameter estimations were applied:

- Extremal type1 - Gumbel (E1) - MM, MLM and WGM parameter estimation
- Generalised extremal (GEV) - MM, MLM and WGM parameter estimation
- Rossi distribution (ME) - MLM parameter estimation

- 3-param. Log. Normal (LN3) - MM, MLM and WGM parameter estimation
- Pearson type III (P3) - MM, MLM and WGM parameter estimation
- log. Pearson type III (LP3) - MM and MLM parameter estimation
- 3-param. Weibul's (WB3) - MM, MLM and WGM parameter estimation,

where acronyms stand for the following:
MM – Moments Method
MLM – Maximum Likelihood Method
WGM – Weighted Moments Probability Method

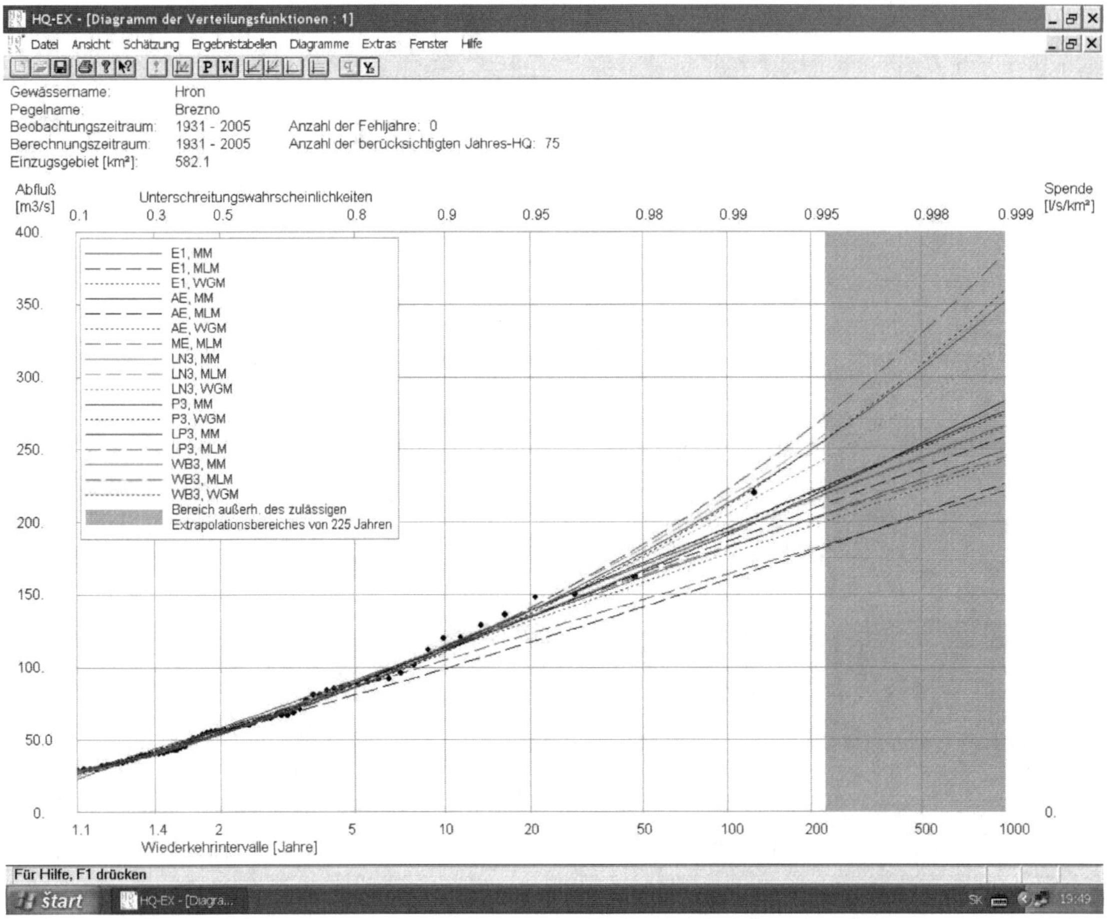

Figure 3. Theoretical distribution lines in Brezno (Hron River Basin).

4. Estimation of T-year maximum discharges in ungauged catchments

The problem of ungauged basins is one of the central problems in hydrology. Estimates can be obtained by a range of methods. Regional estimation methods are traditional and use standard hydrological tools. They are used for interpolation between gauged and ungauged sites. These methods are based on splitting homogenous regional elements (regions or regional types), which are identified according physiographical components, on the set of catchment physiographical characteristics, eventually according similar hydrological regime characteristics.

Homogeneous regional types were selected by:

1. cluster analysis
2. according residual deviations of regression relation 100-year maximum specific discharge ($q_{max.100}$) from physiographical and climatological catchment characteristics for the Slovak Republic. Input data

for both methods were time series from 197 selected stations with an observation period minimum of 25 years, a catchment area from 20 to 300 km^2 and minimum impact of anthropogenic activities.

According to Pearson´s correlational matrix, four independent variables were selected for determination of regression relations:

A – catchment area [km^2],
F – forest coverage [%],
P_{3180} – long-term precipitation on the basin area in the period 1931-1980 [mm],
HG_{char} – hydrogeological index [%].

The HG_{char} index is based on the rock permeability coefficient. There are four categories in Slovakia - low (n), medium (s), high (v), very high (vv). For all selected catchments a percentage proportion of individual areas was specified and index was calculated according to the formulae

$$HG_{char} = 1.n + 2.s + 3.v + 4.vv \tag{1}$$

4.1. Homogeneous regional types were selected by cluster analysis

Cluster analysis, K-means clustering with Euclidian metrix were applied. Regional homogenity was tested by tests of regional heterogenity [2]. 10 homogeneous regional types were selected (figure 4).

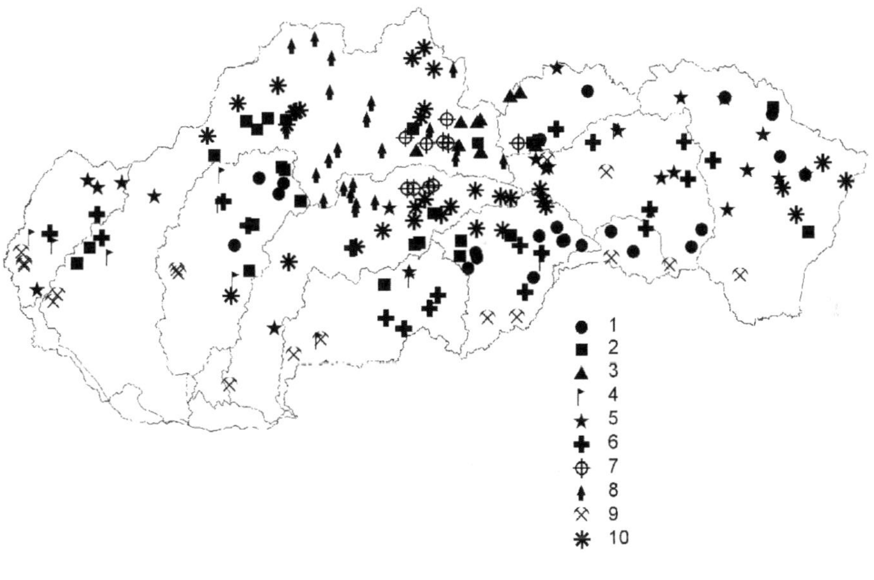

Figure 4. Map of homogeneous regional types selected by cluster analysis.

For $Q_{max.N}$ estimation of this regional pooling scheme estimation of indexed flood according to multiparametric relations by multiple regression method was calculated.

Independent variables were used: soil index, length of the stream, concentration time according to Kirpich, catchment area, HG_{char} index, precipitation, slope of the basin, forest coverage. According to regional distributive function and indexed flood T-year maximum discharges quantities were estimated [3, 4].

4.2. Homogeneous regional types according residual deviations of regression relation $q_{max.100}$ from physiographical and climatological catchment characteristics

According to regression analysis a non-linear regression relation was selected for Slovakia:

$$q_{max.100stat} = b1 \; . \; P_{3180}{}^{b2} \; . \; [\lg(A)]^{b3} \; . \; F^{b4} \; . \; HG_{CHAR}{}^{b5}, \tag{2}$$

where $q_{max.100stat}$ were derived according to DVWK methodology– Regeln 101/1999 [5] and $b1$ - $b5$ are multiplicative parameters.

Map of 8 homogeneous regional types according to estimated $q_{max.100}$ was created (figure 5).

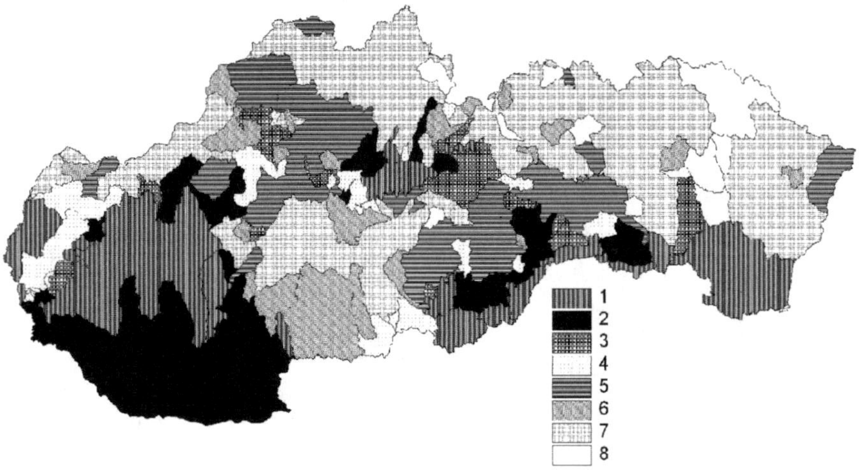

Figure 5. Map of homogeneous regional types according residual deviations of regression relation $q_{max.100}$ from physiographical and climatological catchment characteristics.

5. T-year maximum discharges in the system of stream network

Applied methods of T-year maximum discharges estimation are influenced by uncertainties like: uncertainty of data quality, statistical uncertainty, model uncertainty, regionalization uncertainty.

Logical principles have to follow in the system of stream network (e.g. T-year maximum discharge under the confluence is less or equal to sum of tributaries, and generally the discharge is higher with an increasing catchment area...). In cooperation with the Institute of Hydrology, Slovak Academy of Sciences a new modern method top-kriging was applied in the river Hron basin [6, 7], which reduces subjectivity. The method is based on geostatistical conspectus of top-kriging method, which was adapted by the authors for a solution of interpolation in the stream network. It takes account of the topology of the stream network and the contribution of area size.

A comparison of estimated $Q_{max.100}$ and evaluated by top-kriging method for the chosen profiles in the Hron river basin is shown in table 1.

Table 1. Comparison of estimated $Q_{max.100}$ and evaluated top-kriging method.

River	Profile	Catchment area	Observation lenght	Estimated $Q_{max.100}$	Top-kriging $Q_{max.100}$	Deviation
		(km^2)	(year)	($m^3.s^{-1}$)	($m^3.s^{-1}$)	(%)
Hron	Zlatno	83.670	69	58	58	0.0
Hron	above Havraník	83.701	-	58	58	0.0
Havraník	mouth, Zlatno	16.719	34	20	20	0.0
Hron	under Havraník	100.420	-	65	68	4.6
Hron	under Hronec	272.624	-	140	136	-2.9
Hron	Polomka	329.540	-	155	154	-0.6
Hron	above Bacúšsky c.	361.150	-	165	164	-0.6
Bacúšsky creek	mouth, Bacúch	28.675	43	19	19	0.0
Hron	under Bacúšsky c.	389.825	-	175	173	-1.1
Hron	above Rohozná	475.976	-	195	190	-2.6
Rohozná	Michalová	59.040	27	60	60	0.0

Rohozná	mouth	90.896	-	80	80	0.0
Hron	under Rohozná	566.872	-	225	226	0.4

6. Conclusion

For $Q_{max.100}$ estimation we used data sets from gauged stations. We applied new modern methods and space interpretation of characteristics in the GIS environment. Then we analysed the results obtained by the described methods and synchronized them in the system of stream network.

The $Q_{max.100}$ were estimated in gauged stations and important profiles. In the majority of profiles no significant revision were proposed. A revision of values is in the catchments, where significant maximum discharges occurred in the last period or where the length of the observed data was prolonged. $Q_{max.100}$ in partial catchments and their comparison is shown in table 2.

Table 2. $Q_{max.100}$ and comparison in partial catchments.

River	Profile	Catchment area	$Q_{max.100}$ "old"	$Q_{max.100}$ "new"	Deviation
		(km^2)	(m^3.s^{-1})	(m^3.s^{-1})	(%)
Poprad	state border	1889.21	1180	1200	2
Morava	mouth	26577.77	1320	1330	1
Dunaj	state border	178530.53	9100	9100	0
Váh	mouth	19660.98	2000	2000	0
Nitra	mouth	4501.15	420	420	0
Hron	mouth	5464.56	800	900	13
Ipeľ	mouth	5151.04	670	670	0
Bodrog	under Roňava	11959.7	1400	1400	0
Slaná	state border	3225.1	530	530	0
Hornád	state border	4340.14	1010	1010	0
Bodva	state border	865.52	170	170	0

The created methodology and updated values of T-year maximum discharges are a modern tool for their determination. They are utilized for the evaluation of "standard" values by SHMI. For important water management buildings with money spending investments and impacts, hydrological data could be provided by studies.

References

[1] Podolinská J et al 2006 *Spracovanie hydrologických charakteristík – N-ročné maximálne prietoky Final report* (Bratislava: SHMÚ) p 90
[2] Hosking J R M and Wallis J R 1997 *Regional Frequency Analysis: An Approach Based on L-moments* (Cambridge: Cambridge University Press) p 224
[3] Kohnová S et al. 2003 *Regionalizácia maximálnych prietokov pre územie Slovenska na základe vybraných charakteristík povodí Final report* (Bratislava: SvF STU)
[4] Kohnová S et al 2006 *Metodika odhadu návrhových maximálnych prietokov pomocou regionálnej frekvenčnej analýzy a jej overenie na vybraných povodiach Slovenska Final report* (Bratislava: SvF STU)
[5] DVWK Regeln 101/1999 1999 *Bemessungshochwassers. Empfehlung zur berechnung der Hochwasserwahrscheinlichkei* (Hamburg: Verlag Paul Parey)
[6] Skoien J O, Merz R and Blöschl G 2006 Top-kriging – geostatistics onstream networks *Hydrol. Earth Syst. Sci.* **10** 277-287
[7] Parajka J 2006 *Priestorová interpolácia 100-ročného maximálneho špecifického odtoku v povodí Hrona metódou TOPKRIGING Final report* (Bratislava: Ústav hydrológie SAV Liptovský Mikuláš)

IOP Conf. Series: Earth and Environmental Science **4** (2008) 012013 doi:10.1088/1755-1307/4/1/012013

Multiple-coincidence of flood waves on the main river and its tributaries

S Prohaska, A Ilic and B Majkic

Institute for the Development of Water Resources, „Jarsolav Cerni", Jaroslava Cernog St. No. 80, 11226 Pinosava, Belgrade, Serbia

E-mail: stevan.prohaska@jcerni.co.yu, ilicsaleksandra@gmail.com and brankica.majkic@jcerni.co.yu

Abstract. This paper addresses the definition of multiple coincidences of flood waves on the main river and its tributaries. Contrary to previous studies of partial coincidences of various flood parameters (Prohaska 1999) for the main river and one of its tributaries, this procedure allows for the definition of complex (multiple) coincidences of a single parameter for the main river and several of its tributaries. In particular, coincidence is defined for the major parameter which characterizes a flood (i.e., the flood wave volume). The paper gives a practical example of the analysis of simultaneous flood waves on the Danube and its main tributaries in Serbia: the Tisa and the Sava rivers. The procedure for potential use of the established coincidence functions in applied water management and forecasting is also described in the paper.

1. Introductory remarks

The development of flood protection systems requires comprehensive information on flood characteristics. The design of engineering structures (i.e., determination of their capacities), requires facts on peak discharges. In addition to the maximum discharges, analyses and evaluations of the effects of reservoirs and/or river channels on flood wave transformation involve data on flood wave volume and shape of the hydrograph. Therefore, the safety of the protected area/structure, or the risk to which it is exposed, can be related to the exceedance probability of appropriate flood characteristics.

The conventional approach to the evaluation of flood risk leads to an evaluation of the probability that a pre-defined magnitude of a relevant flood characteristic will be exceeded. This is equivalent to the determination of the return period of the considered event. This task involves statistical analyses of hydrologic time series for adjacent gauging stations. The approach yields satisfactory results in solving many engineering problems, particularly those dealing with flood protection in simple cases, such as flood protection along a river reach without any significant tributaries.

However, when the river receives a significant tributary, the above-mentioned approach does not yield reliable estimates of flood properties. Namely, flood concentration on two merging rivers is usually different, so that the floods on them do not appear simultaneously. Furthermore, a flood wave occurring on one river may significantly affect the flow pattern of the other. Yet, regular processing of hydrologic time series cannot yield an assessment of the impact of one river on another, for the data are commonly collected at gauging stations located beyond the area of influence. Under these circumstances, it is often important to assess the coincidence of flood waves on the main river and its tributary. If there are several significant tributaries, the problem becomes even more complex and there is a need for a comprehensive consideration of hydraulic and hydrological conditions within a broader region. Random variables that describe the hydrological processes should be viewed as a multi-dimensional event in order to fully determine the risk which is the most frequently adopted criterion for the design of flood protection structures and/or systems. The application of a two-dimensional or even a multi-dimensional approach to hydrological random variables, which is proposed in this paper, can help in overcoming any difficulties which may be encountered.

2. Metodology for the evaluation of coincidence of multi-dimensional flood waves

The term coincidence is used to denote the simultaneous occurrence of floods at two (or more) rivers. The degree of coincidence is measured by the probability of the events. The theoretical background is founded upon the practical application of a multivariate probability distribution function, or more precisely, by its conditional probabilities. In this particular case, one considers the quantitative characteristics of two simultaneous flood waves on the main river and its tributary. Usually, subjects of analysis are: maximum discharge of the flood hydrograph, volume of the flood, and duration of the flood, or the time between maximum flows during the analyzed flood.

The procedure for the calculation of multi-dimensional coincidence of flood flows on complex river systems stems from the developed methodology for the evaluation of coincidence of floods on two adjacent streams (Prohaska 1999). The coincidence is equivalent to the probability that two variables, X and Y, will occur simultaneously, where X and Y represent a random event (flood wave characteristics) on the considered river.

According to the theory of statistics, a two-dimensional probability distribution function of a normally distributed bi-variate random process, X and Y, is defined as follows:

$$f(x,y)=\frac{1}{2\pi\cdot\sigma_x\cdot\sigma_y\cdot\sqrt{1-\rho^2}}\cdot\exp\left\{-\frac{1}{2(1-\rho^2)}\left[\frac{(x-\mu_x)^2}{\sigma_x^2}-\frac{2\rho(x-\mu_x)(y-\mu_y)}{\sigma_x\sigma_y}+\frac{(y-\mu_y)}{\sigma_y^2}\right]\right\} \quad (2.1)$$

The symbols in expression (2.1) stand for the following:

X and Y – random variables (flood flow characteristics of the main river and its tributary, or at two adjacent profiles of the main river),
x and y – simultaneous realization of the random variables X and Y, respectively,
μ_x and μ_y – expected values of X and Y variables,
σ_y and σ_x – standard deviations of X and Y variables and
ρ – the coefficient of correlation between X and Y variables.

For a joint probability density function - j.p.d.f., f(x, y), the marginal densities, f(x,•) and f(•,y), are defined by:

$$f(x,\cdot)=\int_{y=-\infty}^{y=\infty} f(x,y)dy \quad (2.2)$$

$$f(\cdot,y)=\int_{x=-\infty}^{x=\infty} f(x,y)dx \quad (2.3)$$

The marginal cumulative probability functions are determined from:

$$F(x,\cdot)=\int_{t=-\infty}^{t=x} f(t,\cdot)dt \quad (2.4)$$

and

$$F(\bullet, y) = \int_{z=-\infty}^{z=y} f(\bullet, z)dz \tag{2.5}$$

The cumulative probability function - c.d.f., F(x,y), is assessed from:

$$\mathrm{F}(\mathrm{x}, \mathrm{y}) = \mathrm{P}[\mathrm{X} \le \mathrm{x} \cap \mathrm{Y} \le \mathrm{y}] = \int_{t=-\infty}^{t=x} \int_{z=-\infty}^{z=y} \mathrm{f}(\mathrm{t}, \mathrm{z})\mathrm{dtdz} \tag{2.6}$$

The exceedance cumulative probability, Φ(x,y), can be obtained from the following relation:

$$\Phi(x, y) = \int_{t=x}^{t=+\infty} \int_{z=y}^{z=+\infty} f(t, z)dtdz = P[X > x \cap Y > y] = 1 - P[X < x \cup Y < y]$$
$$= 1 - F(x, \bullet) - F(\bullet, y) + F(x, y) \tag{2.7}$$

In bi-variate statistical analyses of flood characteristics, hydrologists encounter two basic obstacles which must be resolved in the practical implementation of the proposed model. The first obstacle comes from the fact that most flood characteristics are not normally distributed. It is, however, customarily assumed that the considered variables follow a log-normal distribution. Therefore, their logarithmic transformations:

$$\mathrm{U} = \log \mathrm{X} \tag{2.8}$$

$$\mathrm{W} = \log \mathrm{Y} \tag{2.9}$$

are said to be normally distributed.

The previously described model appears to be the direct method for the evaluation of the cumulative distribution functions. However, this can be lengthy and tedious, for it involves extensive calculations in a three-dimensional space, X, Y, and ρ. For current sophisticated computers it does not necessarily represent a severe obstacle. However, at the time when this project was first conceived, the method of direct computation was almost impossible to apply. For that reason, a more convenient grapho-analytical scheme was implemented. This scheme was described by M Abramowtz and A Stegun (1972), and is briefly discussed in the ensuing text.

The scheme deals with standard normal variables. A transformation of non-standard variables into standardized variables can be achieved using the well-known procedure:

$$\psi = \frac{(u - \overline{U})}{\sigma_u} \tag{2.10}$$

$$\xi = \frac{(w - \overline{W})}{\sigma_w} \tag{2.11}$$

The variables ψ and ξ are, according to the above assumption, normally distributed, with the expected values $\mu_\psi = \mu_\xi = 0$, and the standard deviations $\sigma_\psi = \sigma_\xi = 1$.

With the above transformations, the joint probability density function can be stated as:

$$f(\psi, \xi) = \frac{1}{2\pi \cdot \sqrt{(1-\rho^2)}} \exp\left\{-\frac{1}{2\cdot(1-\rho^2)}\left[\psi^2 - 2\rho\psi\xi + \xi^2\right]\right\} \tag{2.12}$$

The values of the correlation coefficient ρ should be replaced with R, which can be calculated from observed data using the standardized series. With this parameter, and after simplifying the notation in equation (2.12), the following relation can be written:

$$\iint_A f(\psi,\xi)\,d\psi\, d\xi = 1 - \exp\left\{-\frac{\lambda^2}{2(1-\rho^2)}\right\} \tag{2.13}$$

Namely, the integral given by equation (2.13) over the area A (i.e., the integral over the space $\psi, \xi \in A$), represents the probability that events $\psi = h$ and $\xi = k$ will be realized in the area A, which is contoured by an ellipse given by the following equation:

$$\psi^2 - 2\rho\psi\xi + \xi^2 = \lambda^2 \tag{2.14}$$

The newly introduced symbol λ is obviously related to the constant value of the integral (2.13). Consequently, it is related to the variables ψ and ξ, as well as to the correlation coefficient.

Hence, for each value of λ=const, the probability contained within the ellipse of equation (2.14) can be calculated:

$$\xi^2 - 2\rho\psi\xi + (\psi^2 - \lambda^2) = 0 \tag{2.15}$$

As previously stated, any particular value of λ = const corresponds to an ellipse. Furthermore, any given value, $\psi = h$, intersects the ellipse at two different values of ξ, let us say $\xi = kl$ and $\xi = k2$.

Hence, by solving the quadratic equation (2.16) for any particular value of λ = const corresponding to the required level of probability given by equation (2.13), one obtains two particular coordinates (values of ξ, say $\xi = kl$ and $\xi = k2$), representing the intersection of the ellipse and the straight line $\psi = h0$. By repeating the calculation for several selected values of λ while changing the values of $\psi = h0$, a series of the ellipses can be constructed. It should be kept in mind that after each calculation, an appropriate transformation should be performed in correspondence with equation (2.11) and equation (2.10) to obtain non-standardized, natural values of the flood characteristics, instead of standardized logarithmic values.

The described computational scheme is rather direct. However, the results are not of great use, except to give the analyst general insight into the relation of the considered flood characteristics.

When it comes to the evaluation of the cumulative distribution function, the direct method, as previously outlined, is not convenient. To overcome computational difficulties, the Abramowitz and Stegun procedure was implemented in this study. The computational scheme uses a grapho-analytical procedure which defines the cumulative probability, Φ(h,k,ρ), in terms of the probability Φ(h,0,r) and the probability Φ(k,0,r), where instead of the correlation coefficient, ρ, the value r = r(h,k,ρ) is used. The value r is related to h and k, as well as to ρ itself. More specifically, the probability Φ(h,k,ρ) can be assessed from:

$$\Phi(h,k,\rho) = \Phi\left(h, 0, \frac{(\rho h - k)\cdot \mathrm{sgn}h}{\sqrt{h^2 - 2\rho hk + k^2}}\right) + \Phi\left(k, 0, \frac{(\rho k - h)\cdot \mathrm{sgn}k}{\sqrt{h^2 - 2\rho hk + k^2}}\right) - \begin{cases} 0 & \text{if } hk \geq 0 \text{ and } h+k \geq 0 \\ \frac{1}{2} & \text{for all other cases} \end{cases} \tag{2.16}$$

where (sgnh) and (sgnk) are equal to 1 when h or k, respectively, are greater or equal to zero, while they become -1 whenever h and k are smaller than zero.

Further elaboration of the theory of multiple flood coincidence in a complex river system was adjusted to a river reach where the main river receives two significant tributaries and negligible lateral inflow. Flood volume was selected to represent the flood hydrograph. The probability distribution functions of flood volumes represent a starting point in assessing a multi-dimensional probabilistic dependency of a flood wave on the main river (at the downstream profile), as a function of floods at all the considered entry profiles (of the main river and its tributary). It can be stated as:

$$P(W \geq w) = p \qquad (2.17)$$

where:
$P(W \geq w)$ – probability that flood volume at all inflow/outflow profiles will be exceeded,
p – exceedance probability.

In addition, the exceedance probability of a simple coincidence between two inflow profiles, j and k, is required:

$$P((Wj \geq wj) \cap (Wk \geq wk)) = q \qquad (2.18)$$

where:
$P((Wj \geq wj) \cap (Wk \geq wk))$ – exceedance probability of coinciding flood volumes at two considered inflow profiles, j and k,
q – exceedance probability.

The procedure for setting the required relationship consists of a step-by-step balancing of the inflowing flood-volumes which correspond to different probabilities according to the following balance equation:

$$Wiz,p{=}\theta = Wi,p{=}T + Wj,r + Wk,z \qquad (2.19)$$

where:
$Wiz,p{=}\theta$ – flood volume at the exit profile for a fixed probability $p{=}\theta$,
$Wi,p{=}T$ – flood volume at the i-th inflow profile for a selected return period T,
Wj,r – flood volume at the j-th inflow profile for an arbitrarily chosen probability r, in accordance with the exceedance probability $P((Wj \geq wj) \cap (Wk \geq wk))$,
Wk,z – flood volume at the k-th inflow profile for an arbitrarily chosen probability z, in accordance with the exceedance probability $P((Wj \geq wj) \cap (Wk \geq wk))$.

For a fixed exceedance probability of the outflow hydrograph, θ, and a selected return period, T, using equation (2.17), theoretical values of the flood volume, $Wiz,p{=}\theta$ and $Wi,p{=}T$, can be determined. Arbitrarily chosen probability r or z, let us say r, allows for the calculation of theoretical values of Wj,r and Wk,z., on the condition that equation (2.18), i.e., $P((Wj \geq wj) \cap (Wk \geq wk)) = q$, is satisfied. The corresponding probability, Wk,z, can also be obtained from equation (2.17). In that way, all values of corresponding probabilities can be defined: $p(Wiz,p{=}\theta)$, $p(Wi,p{=}T)$, $p(Wj,r)$ and $p(Wk,z)$. From this, a multi-dimensional relationship of the exceedance probabilities of flood-volumes for the main river and its tributaries can be produced in the following form:

$$p(Wiz,p{=}\theta) = P((p(Wi,p{=}T) \cap p(Wj,r) \cap p(Wk,z)) \qquad (2.20)$$

where:
$p(Wiz,p{=}\theta)$ – exceedance probability of flood-wave volume at the outflow profile,
$p(Wi,p{=}T)$ – exceedance probability of flood-wave volume at the i-th inflow profile,
$p(Wj,r)$ – exceedance probability of flood-wave volume at the j-th inflow profile,
$p(Wk,z)$ – exceedance probability of flood-wave volume at the k-th inflow profile.

These relationships can be set up for a set of arbitrarily chosen values of θ at the downstream profile of the considered river reach. Therefore, for known volumes of flood hydrographs at upstream (inflow) profiles, using equation (2.17) and equation (2.20), the flood volume at the outflow profile can be calculated.

3. Practical application

The developed methodology for the evaluation of multiple coincidences of flood waves on the main river and its tributary was applied to a stretch of the Danube River in Serbia, with two tributaries: the Tisa and Sava rivers. In this particular case, the probabilities of multi-dimensional coincidence of flood events on the Danube River was evaluated for the gauging station at Pancevo (downstream from Belgrade), as a function of the probabilities associated with the flood waves at the inflow profiles: GS Bogojevo (the Danube River inflow profile), GS Senta and GS Sremska Mitrovica (on the tributaries – The Tisa and Sava rivers, respectively). A schematic outline of the considered stretch is shown in figure 1.

Figure 1. A schematic representation of the Danube River stretch used for flood coincidence analyses.

Flood volume was selected as a referent variable for analysis. Due to the pronounced complexity of flood origin on the considered rivers, the largest annual flood-volumes above discharge, corresponding to 50% duration of the daily flow, were analyzed. It was assumed that a flood wave on the Danube River at GS Pancevo with the exceedance probability θ, is formed by corresponding floods at Bogojevo having the exceedance probability p (return period T) and arbitrarily chosen combinations of two-dimensional coincidences of floods on the Tisa River near Senta and the Sava River near Sremska Mitrovica. Using the devised procedure, the relationship of the probabilities of coincidence of flood events at the considered stations is:

$$p(Wiz,p=\theta)^{Pan} = P((p(Wi,p=T)^{Bog} \cap p(Wj,r)^{Sen} \cap p(Wk,z)^{Sr..M.}) \qquad (3.1)$$

where:

$p(Wiz,p=\theta)^{Pan}$ – exceedance probability of flood volume of the Danube River at the exit profile near Pancevo,

$p(Wi,p=T)^{Bog}$ – exceedance probability of flood volume at the inflow profile of the Danube River at GS Bogojevo,

$p(Wj,r)^{Sen}$ – probability of a flood wave of the Tisa River near Senta (inflow profile of the considered river section),

$p(Wk,z)^{Sr.M.}$ – probability of a flood wave of the Sava River near Sremska Mitrovica (inflow profile of the considered river section).

The results of an evaluation of multiple coincidences of floods on the Danube, Tisa and Sava rivers for a 100-year flood on the Danube River near Pancevo, are presented in figure 2.

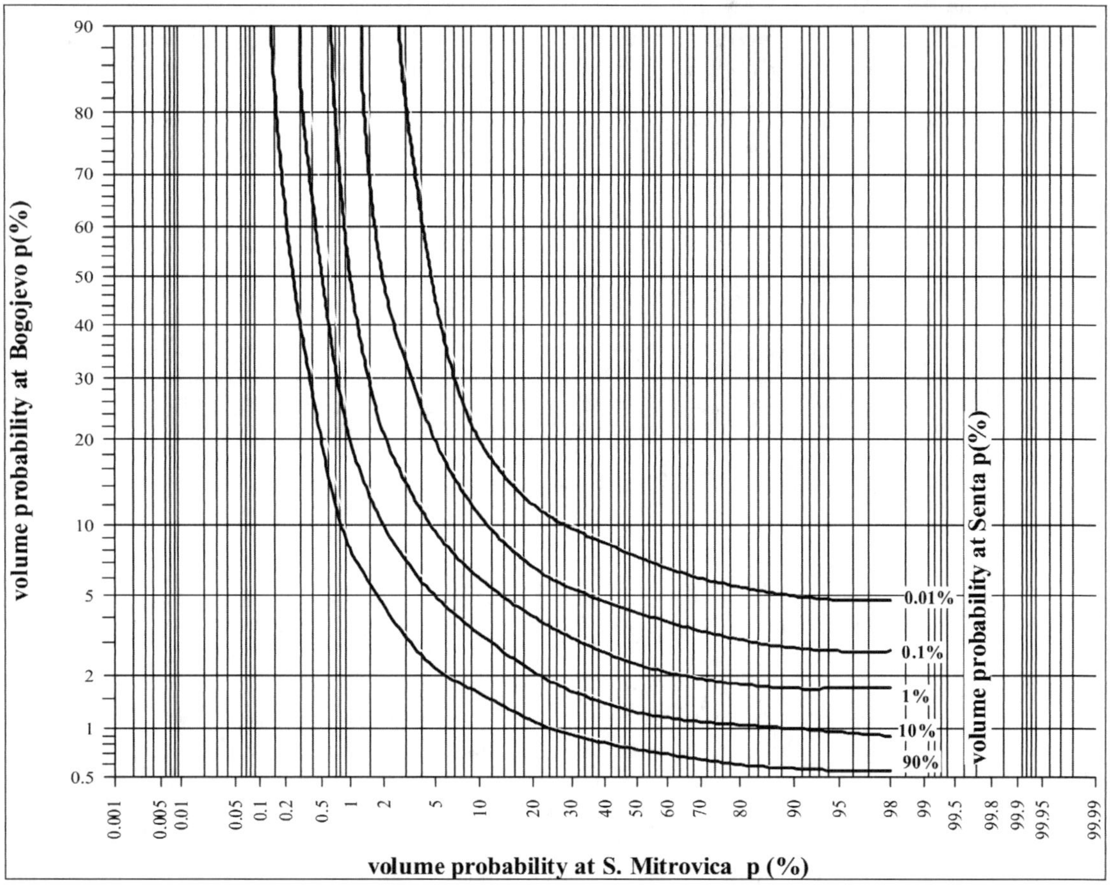

Figure 2. Multiple coincidences of a 100-year flood of the Danube River near Pancevo as a function of floods of the Danube River near Bogojevo and the Tisa and Sava rivers near Senta and Sremska Mitrovica, respectively.

4. Practical importance of results

The significance of concrete results of multiple coincidences of floods on the Danube River and its tributaries is multifold; they:

- allow for an assessment of the flood genesis pattern on the Danube River and its tributaries;
- provide the groundwork for the development of line-systems for flood protection within the zone of interaction of the main river and its tributaries;
- assist in setting up a strategy for the development and improvement of riverside flood protection systems within the zone of confluence of significant river flows;
- constitute a hydrological database for the assessment of the statistical significance of the Danube River floods in Serbia;
- provide support in preparing real-time hydrologic forecasts and flood warnings; and
- offer tools for flood risk assessments.

5. Conclusions

The long list of practical uses of the developed methodology for analyses of flood coincidence on the Danube River and its tributaries, the Tisa and Sava rivers, suggests broad opportunities for application

IOP Conf. Series: Earth and Environmental Science **4** (2008) 012013 doi:10.1088/1755-1307/4/1/012013

in water resources management. Its applicability is generally tied to flood protection and river training. The methodology can also be used in real-time forecasting and for generating flood warnings. This paper contributes to a better practical knowledge in the area of engineering hydrology, particularly with regard to the assessment of flood events and the performance of comprehensive flood-risk analyses.

References

[1] Petkovic T and Prohaska S 2004 Estimation of the Maximum Water Discharge on theUngauged River Basin in the Zapadna Morava Basin *22nd Conf. of the Danubian Countries on the Hydrological Forecasting and Hydrological Bases of Water Management* (Brno, Czech Republic)

[2] Prohaska S, Isailovic D, Srna P and Marcetic I 2000 Coincidence of Flood Flow of the Danube River and its Tributaries *The Danube and its Basin – A Hyd. Monograph follow-up volume 4 Reg. Co-operation of the Danube Countries in the Frame of the Int. Hyd. Prog. of UNESCO* (Bratislava: Water Research Institute) pp 1-187

[3] Prohaska S, Isailovic D and Srna P 2002 Flood Coincidence of the Danube and Tisa rivers in Serbia and Montenegro *Int. Conf. on Research Basins and Hyd. Planning* (Hefei, P.R. China) ed Ru-Ze Xi, Wei-Zu Gu and Klaus-Peter Seiler

[4] Prohaska S 2006. *Hydrology Part 2* (Belgrade: Institute for the Development of Water Resources „Jaroslav Černi“, University of Belgrade-Faculty of Mining and Geology and Hydrometeorolgical Service of Serbia)

[5] Prohaska S, Simic Z, Orlic A and Stanic Z 2007 Mathematical Model for the Definition of Flood Wave Formulation Genesis on Small Non-monitored Watersheds *Int. Conf. Erosion and Torrent Control as a Factor in Sustainable River Basin Management* (Belgrade: University of Belgrade-Faculty of Forestry)

IOP Conf. Series: Earth and Environmental Science **4** (2008) 012014 doi:10.1088/1755-1307/4/1/012014

Quality management for discharge measurements in Bavaria

Carmen Roth

Hydrology of surface Waters, *Bavarian Environment Agency*, Hof, Germany

E-mail: Carmen.Roth@lfu.bayern.de

Abstract. The hydrology of surface waters is based on the knowledge of the two main hydrological parameters, water level and discharge. These data are not only needed for almost every hydrological planning, but are very important for the management of high and low water. Our long term goal is a complete registration of every natural runoff – covering everything from low to high water – at our 600 measuring stations using limited resources but at the same time achieving high reliability and quality. To this point we are currently enforcing the use of modern data acquisition technology for water level and discharge measurements. The new techniques enable a continuous registration of water levels, actual velocity and consequently the discharge. Currently we are using gauging methods like the well known bubble gauge and radar, while the ultrasonic systems become more and more frequent. The moving boat techniques using the ultrasonic-Doppler-systems (e.g. ADCP, Q-Liner) are gradually substituting traditional methods for discharge measurement. In addition to that hydraulic methods (e. g. SIMK®) are being incorporated in the measuring systems as well. Since 2000 almost all gauges have been equipped with data loggers and automatic data transmission systems. One of the outcomes of these investments is the online availability of hydrological data on our website (www.hnd.bayern.de) where high quality data can be easily accessed. A new project has been initiated to improve the present rating curve. By means of hydraulic calculation the curve will be verified and if necessary adjusted to minimise deviations in the near future, which is of high importance especially for flooding forecasts.

1. Introduction

Water bodies provide water for important uses including drinking water, navigation, power generation, irrigation and recreation. Too much water causes problems of public safety as rivers overflow their banks and inundate critical infrastructure like hospitals. But low data also have quite an impact on water quality. The amount of water available for different uses and for a low and high water management is dependent upon accurate measurements of the rate of flow under widely varying conditions. The discharge measurement is one of the most important issues in hydrology. Besides that, discharge data also impact economy and social security.

In the past, data logging and data control was based on time-consuming technologies and required a lot of manual work. But in times of limited resources especially in terms of manpower, new and more efficient technical equipment has to compensate these short comes. Unlike in the past, the demand on hydrological data has changed as well. For our high water management system accurate data are needed right when they are generated. In an ideal scenario the measured water levels should be automatically transmitted from the gauging stations and converted on-line in to discharge values. These data must have already high quality, because they form the basis for flood forecasts and therefore the basis for disaster control. In addition they are an integral part of the highly priced flood protection planning.

These facts make it even more important to define quality standards and to have a high quality management system in place. The approach taken by the Bavarian hydrology department to establish this quality standard will be highlighted subsequently. The major topic is on improving discharge measurements and the defining of the rating curve.

2. Quality improvements in water level control

A first step to reduce the need of resources is to improve the monitoring network. During a substantial revision of the water gauge system the water level registration has been adjusted to achieve state of the technology. Every gauge has been categorized into a quality classification system (A, B or C) based on its hydrological importance. According to this classification, the technical equipment of the gauge, the time needed for the analysis of the data and the data logging are the key criteria applied. At present we have 600 stations for hydrological services. 200 of them belong to category A and 400 to category B. In addition to these, we operate about 200 of category C for other purpose.

Thanks to modern regionalization procedures it is now possible to improve region-wide forecasts. This optimization of the water gauge stations in terms of measurement quality enabled a fine tuning of the network, which is currently ongoing.

The second step towards an improved water level registration is the optimization of the technical equipment both in terms of the measurement itself as well as the data recording and transmission. Therefore in Bavaria all hydrological station are now equipped with an additional transducer and data logger, so that the reliability - even under extreme circumstances like flooding - is guaranteed. With the on-line data processing of the measured data points the responsible hydrology department can easily identify data discrepancies and technical issues, which in turn allows for a fixation of the problem in more or less real time. Overall, the maintenance of the gauging stations became easier and more effectively managed as data control and comparison will be more and more automated.

3. Quality improvements of discharge measurements

A quality management in the water level data is the equivalent to quality management in discharge measurement. Despite the fact that improvements in the water level registration have been made during the revision of the water gauge system, it is still difficult to predict discharge data. Especially in alpine regions considerable sediment transport during high waters negatively impacts the rating curve and therefore potentially causes false predictions.

One of the major reasons for that can be found in the measurement technology used. In Bavaria the basic method for discharge measurement is still the traditional current meter. Direct measurements are taken at regular intervals across the river with mechanical meters from bridges or cableways. While this method under normal weather conditions fulfils its purpose, in case of flooding, the direct measurements of discharge by current meters are unreliable, unsafe or even straight out impossible because of high velocities, drifting logs, stumps and debris, to mention only a few. During such conditions discharge has to be determined indirectly by surveying high-water marks left by the flow and applying indirect discharge calculations. Another important aspect of the direct measurements is that only data from selected stations are captured. To comply with our modern high water management system requirements however, it is necessary to continuously monitor discharge.

In an effort to address these issues a project investigating new and emerging technologies has been initiated. The expectations from this initiative are that a more cost effective, accurate, safe and robust method of stream gauging will emerge eventually.

The project comprises:

- Acoustic Doppler current profilers (ADCPs) and tracer sensors as alternative measurement techniques during flooding
- Reduction of current meter measurements in the river cross section

- Continuous measurements technologies like ultrasonic systems or stationary ultrasonic Doppler systems
- Measurements of point-velocities in combination with hydraulic calculations (e.g. SIMK®)
- Hydraulic calculations of the rating curve

The above mentioned projects have been performed in close collaborations with the state water offices (WWA), the technical university of Munich and the Bavarian Environment Agency. Three of the above mentioned projects will be described in more detail.

3.1. Ultrasonic Doppler technologies

The most efficient technology for improving direct measurement is the use of ultrasonic Doppler systems. One representative of this technology is the moving-boat system which is used in Bavaria since the beginning of 2008.

Figure 1. StreamPro during a discharge measurement (WWA Landshut)

The frequently used systems are the ADCPs (StreamPro or Workhorse RioGrande) - as shown in Fig.1 - and the QLiner. These sensors record water velocity through the water column by measuring the Doppler shift in the frequency of the acoustic signals reflected from materials suspended in and moving with the river flow. The most important improvement of this method is the possibility to measure velocity of the entire river cross section as compared to the single points captured by the mechanical current meter methods. Both ADCP (continuous measurement across the cross section) and QLiner (measurement in single verticals) allow an on-line control of the measurement data in less time and without a loss of accuracy. Due to the direct transmission of the measurement onto a pocket-pc via blue tooth a real time control of the result is enabled. If necessary, additional measurements can be performed quickly. The huge variety of sensors and boats allows discharge measurements in almost every river which contributes to the superior effect of the Ultrasonic Doppler technology.

In addition, the bottom track signal allows the determination and surveillance of the river bed as well. If additional GPS-antennas or similar equipment are used, discharge measurements by instable river bed or sediment transport are also possible. To this purpose a research project was ratified and initiated at the Inn River. A mapping of different river beds during low floods is exemplified in the figure below:

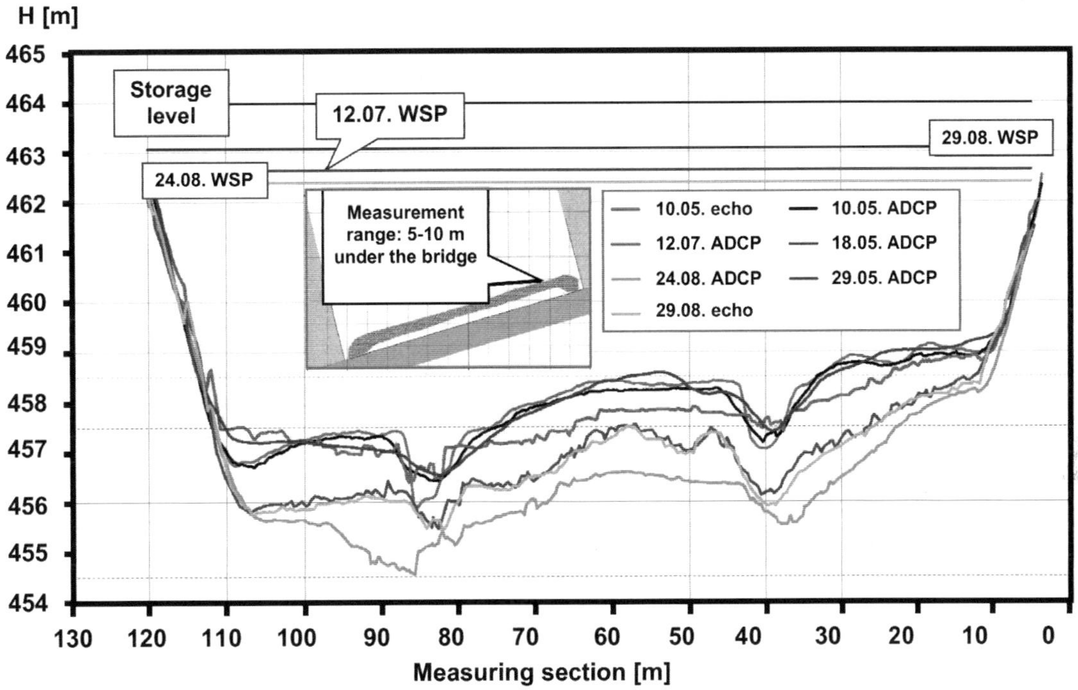

Figure 2. Comparison of ADCP-measurements during flooding at the Inn River 2007 [2]

The third major advantage of the ultrasonic Doppler technology is the measurement of suspended solids. Experiments performed at the *Versuchsanstalt der Technischen Universität München in Obernach (VAO)*, demonstrated an enormous increase of the echo signals and sound levels in a suspended solid cloud. Given an adequate calibration process of the signals a quantitative conclusion of the suspended solid concentration is therefore possible. [2] Unfortunately an on-site experiment in a natural river couldn't be carried out due to the absence of flooding.

3.2. Tracer sensors

A completely different kind of discharge measurement is the use of tracer sensors which are mainly used in mountain torrents to assess the velocity of the water flow. The inserted colour tracer is detected by the tracer sensors after having passed a sufficient drifting distance. The result is visible online on the display. Measurement failures can be detected immediately and the measurement itself repeated. First experiences evidenced that the discharge rating curve of the mountain torrents can be improved substantially by this method while testing is still ongoing.

3.3. Improvements of conventional measurements

Apart from the new technologies improvements of the discharge measurements are also achievable with the current conventional meters. One possibility is the adjustment of measuring programs and the amount of measuring points. In steady river cross-sections the reduction of these is possible. A pilot study carried out at a couple of gauging station demonstrated, that there is a close correlation between the middle velocity and the surface velocity, allowing to establish a functional relation s between these two parameters. For this reason during flooding a measurement of the surface velocity is absolutely sufficient to determine the velocity of flow but it is still necessary, however, to control the relation to full measurement.

4. Hydraulic calculation of the rating cure

Another effort to increase the quality of discharge data is to improve the discharge rating curve itself. Despite all new technologies the measuring of extremes – both high and low water - will still be difficult also in the future due to inherent limitations. Therefore, the Bavarian Environment Agency in cooperation with the state water agencies initiated a project to improve the existing rating curve by incorporating hydraulic calculations. The result of the first 21 gauging stations didn't come as a surprise: Almost every calculated curve showed particular variations if compared to the existing ones.

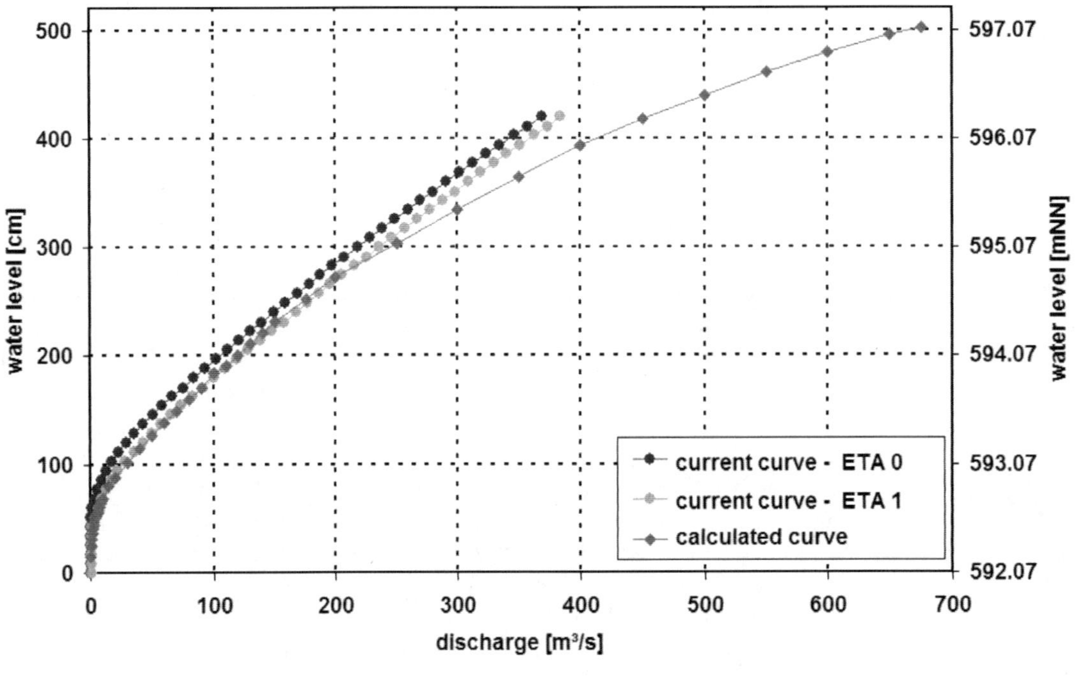

Figure 3. Calculated rating curve (red diamonds) from the gauge Peißenberg / Loisach River [3]

These variations clearly highlight the need for rating curve improvements. At present about 300 gauging stations are being recalculated hydraulically and about 2.5 Mio € will be invested in this project over the next 5 years.

5. Conclusion

There is sufficient evidence at this point that the current problems in discharge measurements can not be solved by addressing technology or hydraulic calculations individually. An improvement of hydrological data in the long run may be achieved by combining new technologies with optimized measuring programs, skilled personal and hydraulic models. The Bavarian water management administration is taking such an approach at present, using quality management as one of the modern tools to enhance the quality of information provided.

IOP Conf. Series: Earth and Environmental Science **4** (2008) 012014 doi:10.1088/1755-1307/4/1/012014

References

[1] Bavarian Environment Agency 2008 *Bericht zum Einsatz von Ultraschall-Doppler-Systemen in der Gewässerkunde in Bayern*, PG Niederschlag und Pegelwesen, unpublished

[2] Technical University Munich – Fiedler, Katharina 2008 *"Operatives Entwicklungsvorhaben – Ermittlung des Abflusses an Pegeln mit bewegter Sohle"*, unpublished

[3] Björnsen Beratende Ingenieure GmbH 2007 Bericht *"Hydraulische Überprüfung von Abflusskurven an Pegeln im Extrembereich"*, unpublished

IOP Conf. Series: Earth and Environmental Science **4** (2008) 012015 doi:10.1088/1755-1307/4/1/012015

The survey of ecologically acceptable flows in Slovenia

Nataša Smolar-Žvanut[1] and Darko Burja[2]

[1] Institute for Water of the Republic of Slovenia, Hajdrihova 28c, 1000 Ljubljana, Slovenia and University of Ljubljana, Faculty of Civil and Geodetic Engineering, Hajdrihova 28, 1000 Ljubljana.

[2] Institute for Water of the Republic of Slovenia, Hajdrihova 28c, 1000 Ljubljana, Slovenia

Email: natasa.smolar@izvrs.si, darko.burja@izvrs.si

Abstract. Excessive water abstractions from watercourses constitute a negative impact on the structure and functioning of aquatic and riparian ecosystems. In order to preserve and improve the aquatic ecosystems it is therefore necessary to maintain adequate quantity and quality of water in watercourses, which can be ensured by providing ecologically acceptable flow (EAF). In Slovenia, a large diversity of watercourses regarding their hydrologic, morphological and ecological characteristics dictates the determination of EAF separately for individual sections of watercourses. Since 1994, the determination of EAF in Slovenia has been carried out primarily for the existing water abstractions such as hydroelectric power plants, fish farms, and to a lesser extent for the abstractions for drinking water, process water, recreation facilities and at the outflows from reservoirs. The results of EAF value analyses showed that the EAF values for individual water abstractions differed widely both with respect to the values of the mean annual minimum flow and the values of the mean daily flow. The results of analyses support the basis for the determination of EAF used in most EU countries, namely that EAF must be determined through interdisciplinary approach where the hydrologic data represent the benchmark values for the determination of EAF.

1. Introduction

Irregular water management practices in the past had a significant impact on the reduction of biodiversity of aquatic and water-related ecosystems. The reduction was mainly due to the interventions for the regulation of watercourses, which changed or hindered the river dynamics and resulted in the disappearance of many habitats and, consequently, the loss of plant and animal species. Impoundments and the related excessive abstractions of water from watercourses adversely affect the structure and functioning of the aquatic ecosystems [1].

The results of previous hydrological analyses in Slovenia pointed to the substantially altered hydrological regime and the dominance of gravel deposits with a considerable share of coarse grains below the barriers. As a result of water abstraction, the physical and chemical composition of water below dams is no longer directed by the conditions in the upper part of the catchment area but determined by the inflow below the impoundment. A constant flow and rare occurrence of high waters downstream of dams bring about a change in the species composition and the abundance of aquatic organisms in comparison with reference sites upstream of the barriers [2].

[1] To whom any correspondence should be addressed

The research results have shown that in the period of constant water flow downstream of dams the periphytic algae generally achieve a high biomass on large pebbles and stones, particularly due to low sediment mobility on the bed of watercourses [2]. Moreover, fragmentation of mesohabitats and the reduction in their diversity, size and hydrological characteristics occur below the dams [3] [4]. The interrupted connection between the aquatic and terrestrial environments, changed hydrology and geomorphology reflect in the loss of the basic functions of aquatic ecosystems such as a primary production, self-cleansing capability and biodiversity.

In the past, simple hydrological formulas were used for the calculation of the water flow values in water abstraction river sections. They were based mainly on the data about the mean minimum flow (MAMF) and the mean daily flow (MDF) [5] [6]. The formulas did not take into account the requirements of ecology, morphology, environmental protection and landscape features; therefore, they are no longer used in most countries, including Slovenia. In the recent years, comprehensive methods have been increasingly used, which are based on interdisciplinary study of the aquatic and riparian ecosystems and the specific features of the watercourse section where water abstraction takes place [7].

The paper presents the survey of the ecologically acceptable flow values (EAF) in relation to hydrological parameters with emphasis on case studies.

2. Determination of the ecologically acceptable flow

2.1. Definition of EAF

The ecologically acceptable flow is defined as the quantity and quality of water ensuring the preservation of the natural balance in and along watercourses. From the environmental protection point of view, EAF denotes the maintenance of the structure and functioning of ecosystems, reflected in the preservation of the species diversity [8]. The EAF value is the quantity of water which warrants the existence and reproduction of organisms in different hydrological environments in watercourses, from rapids to pool habitats. In the Water Act (2002), EAF is defined as the quantity of water that at the legitimate use or acceptable pollution does not deteriorate the ecological status of surface waters or prevent its improvement [9].

2.2. Methods of EAF determination

Using hydrological, morphological and ecological criteria, two different approaches are used for the determination of EAF in Slovenia, a rapid assessment method and a detailed assessment method. Determination of EAF for existing users has prioritized their water requirements so they can remain economically viable. Where new schemes are proposed, there has been much greater scope to prioritise ecosystem requirements [9].

Regarding the rapid assessment method, the basic starting point for the EAF determination are the data on the basic hydrological and hydraulic parameters. If the hydrological data are not available and the method of simultaneous measurements for the determination of flows in not reliable enough, at least a two-year hydrological monitoring is necessary, with particular emphasis on the low flow periods. In addition to the hydrological data, it is necessary to obtain the basic ecological field data, perform the inventory of mesohabitats and identify the morphological characteristics of the watercourse section [9].

The starting points for the definition of an EAF according to both methods are basic hydrological indices, such as MDF, MAMF and minimum average daily flow (MADF). The values of EAF typically vary for different seasons in order to recognize the temporal dynamics of the river system. Normally at least two EAF are prescribed that are minimum flow requirements, one for the winter period (October-March) and one for the summer (April-September) [9].

IOP Conf. Series: Earth and Environmental Science **4** (2008) 012015 doi:10.1088/1755-1307/4/1/012015

3. Survey of EAF determination in Slovenia

The data on the determined EAF values were obtained from the studies carried out within the framework of cooperation between the Institute for Water of the Republic of Slovenia and the LIMNOS Company for Applied Ecology. The data on the EAF values calculated by the equation merely on the basis of individual hydrological data (e.g. EAF values determined by design engineers or an individual fishing society) were not taken into account in the review.

3.1. Hydrological Characteristics of Individual Regions in Slovenia

Water regimes in the regions of Slovenia are very diverse and extremely sensitive to anthropogenic impacts. In Slovenia, the ratio between the MDF and MAMF values is regionally differentiated. In most cases, the ratio value ranges from 2:1 to 6:1; it is particularly higher for the watercourses in the region of Pomurje: Ledava – water-gauging station (WGS) Polana 10:1, Kobiljski potok – WGS Kobilje 24:1, Pesnica – WGS Zamušani 8:1 and in in the littoral region: Dragonja – WGS Podkaštel 36:1. The differences in the ratio values between the MDF and MADF are even bigger. In general, the ratio value ranges from 4:1 to 10:1; however, it is 28:1 for Ledava – WGS Polana and 26:1 for Pesnica – WGS Zamušani 26:1. Dragonja in cross-section – WGS Podkaštel and Kobiljski potok in cross-section – WGS Kobilje dry up, so the MADF = 0.

Most water abstractions take place in small watercourses where there are no data on water quantity. Due to the often specific hydrogeological conditions in those sections, simultaneous hydrometric measurements of flow are needed for the determination of basic hydrological parameters [10].

For the determination of EAF for watercourse cross-sections where there is an active or abandoned water-gauging station nearby with a long operation period, the hydrological parameters such as MAMF, MDF, MADF, Q82, Q95 are available. Water-gauging stations are primarily on larger watercourses, but generally they do not exist on small watercourses of up to some kilometres long. In such cases, the basic hydrological data are determined by the correlation between the flows in cross-sections at water abstractions and one or more water-gauging stations, taking into account the simultaneous hydrometric measurements at the abovementioned cross-sections at different water statuses. For small watercourses where flow data do not exist, often only the MAMF values are determined. In this case, simultaneous hydrometric measurements of flows are carried out during the low-flow periods, since only these values are relevant for the determination of MAMF.

If only one hydrological parameter is considered in the determination of EAF, this is the MAMF value, often representing an important basis for the determination of EAF, in particular for small watercourses.

3.2. Analysis of the EAF Value in Relation to Hydrological Data

From 1994 until the end of 2006, EAF was determined at 187 watercourse sections in Slovenia, for the existing and new water users. The EAF values were determined for water abstractions from watercourses and for outlets from storage basins. 163 determinations of EAF values were carried out for the existing water abstractions and 24 for the planned water abstractions. Of the total of 187 EAF value determinations, 61 were implemented in the sections of watercourses with the catchment area larger than 100 km^2, while 126 EAF value determinations were on smaller watercourses.

Figure 1 shows the results of EAF determinations carried out separately for individual purposes and the existing or planned water abstractions. The term "existing abstractions" means that water abstraction took place during the period of EAF determination and, therefore, the facility had already been constructed, while the term "planned abstractions" means that there were no water abstractions during the EAF determination and the facility did not yet exist.

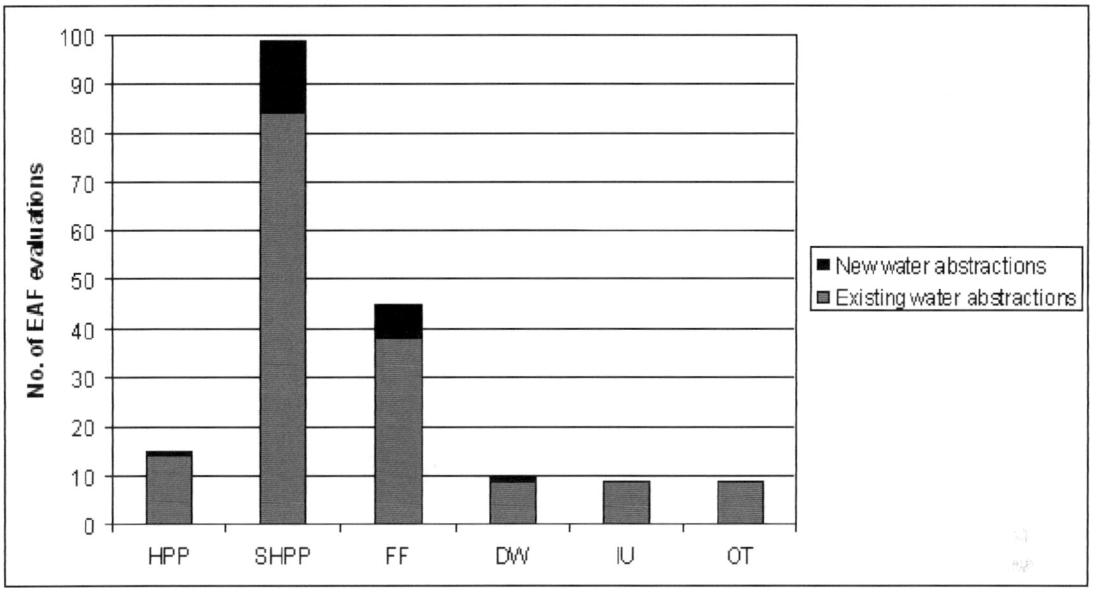

Figure 1. Number of determined EAF values.

According to the review of EAF determination studies, it may be concluded that the determination of EAF values for water abstractions intended for energy purposes prevailed, followed by the EAF determination for fish farms, drinking water and industrial purposes. On the basis of rapid assessment method, the EAF was determined at 145 watercourse sections and 42 on the basis of modified detailed assessment method. It should be noted that, for the needs of the Ministry for the Environment and Spatial Planning, the range of EAF values was given (where the watercourse hydrology was not verified) and the EAF was determined according to the basic principles of the rapid assessment method also in cases where the application of the detailed assessment method should have been necessary. The reasons for this were short time limits given for the determination of EAF and the lack of data. If the determination of EAF was carried out in line with the proposed methodology, it could be expected that for 40% of watercourses the determination would be implemented by the rapid assessment method, and for 60% by the detailed assessment method.

A calculation was done for all determined EAF values, which indicated what proportion of the MAMF value and the MDF value represented the EAF value (Table 1). The determined EAF values ranged from 0.7 % to 353 % of MAMF and from 2 % to 70 % of MDF. In some small watercourses where the monitoring of water quantities did not take place, the MDF value was not determined and, therefore, the comparison of EAF values in relation to the MDF values was not carried out.

Table 1. Values of EAF expressed in percent (%) of MAMF or in percent (%) of MDF for the existing and planned water abstractions (the table gives a range of values from – to).

Purpose of abstraction	Existing abstractions EAF value (in % MAMF)	Existing abstractions EAF value (in % MDF)	Planned abstractions EAF value (in % MAMF)	Planned abstractions EAF value (in % MDF)
Hydroelectric power plants	4 - 90	2 - 18	27[a]	70[a]
Small hydroelectric power plants	23 - 273	3 - 45	51 - 353	3 - 58
Fish farms	13 - 90	10 - 21	40 - 80	6 - 23

Drinking water	0.7 - 150	18 - 37	-	-
Industrial use	66 - 133	3 - 30	-	-
Other[b]	22 - 125	7 - 23	-	-
Range of value	0.7 - 273	2 - 45	27 - 353	3 - 70

[a] the value range is not given because there was only one EAF determination.
[b] outlets from storage basins, abstractions for millstreams, for different water users, fire-extinguishing water.

The determination of EAF for the Kokra River, where the determined EAF value = 0, has not been included into the analysis of EAF values because the abstracted water flows back to the watercourse immediately below the dam.

3.3. Examples of Determined EAF Values in Relation to Selected Hydrological Parameters

The water abstraction for the Možnica SHPP takes place on the Koritnica River, in northwest part of Slovenia. The Možnica SHPP has been in operation since 1911, with the installed flow of 2.7 m^3s^{-1}. The abstraction of water from the riverbed takes place in a relatively short section, at a distance of 380 m, where also very narrow gorge is located. The Koritnica River flows through the Natura 2000 site, an ecologically significant area of Slovenia, the protected area of the Triglav National Park and the protected area of the Soča River and its tributaries. In addition to the marble trout, three other fish species occur in the section of the Možnica SHPP. The site of the dam for SHPP Moznica on the river Koritnica has a catchment area of 37.7 km^2, with the mean minimum annual flow of 1.07 m^3s^{-1}. The hydrological data for the Koritnica River indicate a big gap between the data on the flow downstream of the dam and the EAF determination in relation to the hydrological principles demanding a higher flow below the dam, which would considerably reduce the present production of electric power, but the ecological conditions of the watercourse would not improve substantially. On the basis of hydrological, hydraulic, morphological and ecological characteristics of the river, the EAF of 0.200 m^3s^{-1} (what means 6,6 % of MDF and 18 % of MAMF) should be ensured in the Koritnica River where water abstraction for the Možnica SHPP takes place. The reasons for very low value of EAF are very narrow river bed below the dam and occurrence of high flows during all times of the year. [11].

By the increase in water abstraction from the Krka River from 15 ls^{-1} to 7 m^3s^{-1} for the SHPP on the millstream in the village of Krka, significant changes will occur in the water quantity and the velocity of water flow during the periods of medium and low water flows of the Krka River; namely, the duration of low water flow will be substantially prolonged. Actually, the water abstraction takes place in a short section of the Krka River, but the riverbed is broad in this part. Therefore, the EAF value was determined by month, namely from 2.0 m^3s^{-1} to 4.0 m^3s^{-1}, representing between 123.0% and 245.4% of the MAMFvalue [12].

The water abstraction from the Savinja River for the Prislan SHPP is planned in the section where water abstraction used to take place in the past. The water will be channelled through the old millstream and it will return after 250 m back into the Savinja riverbed. The abstraction foreseen for the SHPP is 6 m^3s^{-1}, which is more than twice the value of MAMF. In the section in question, the Savinja riverbed is narrower in comparison to the upstream section, which further contributes to the greater presence of mesohabitats with high water flow velocity and larger depth of water. Considering the hydrological bases and morphological and ecological characteristics of the Savinja River, the EAF value for the water abstraction was determined in the range from 2.3 to 5.3 m^3s^{-1}, representing between 21.5% and 49.5% of the MDF value [12].

4. Conclusions

The results of the analyses have shown that the EAF values vary substantially both with regard to the MAMF values and the MDF values. This implies that it is not possible to generalize as to which proportion of the MAMF value or the MDF value represents the EAF value. This is proven also by the findings of numerous experts throughout the world [13]. The analysis results support the basis for EAF determination used in most countries of the European Union, namely, that an interdisciplinary approach is needed in the determination of EAF and for each watercourse section separately, whereas the hydrological data represent only a certain base value for the EAF determination. The determination of EAF moves in the direction of the determination of dynamic EAF values, e.g. the determination of acceptable water flow by individual months. It would be unreasonable to set a uniform EAF value for all watercourses (f.e. % of MAMF), as it does not take into account the hydrological dynamics and is not specific to a particular watercourse section. The EAF value must ensure the water regime of watercourses preserving the aquatic and riparian ecosystems while, where ecologically and economically feasible, still enabling water abstraction.

Due to growing use of water, ensuring the EAF in watercourses is becoming increasingly important in particular for watercourses with a large number of water users. Countries are more and more faced with the need for a solution suitable to the requirements of different water users and they will have to set up the priorities for water abstraction, especially within the context of climatic changes.

With the aim of preserving the structure and functioning of the aquatic ecosystem, it is essential to maintain the ecologically acceptable flow in compliance with the Water Directive and the objective to achieve a sound ecological status or good ecological potential by 2015. Determination and provision of EAF is one of the best protection measures for watercourses in Slovenia, as also a principal measure leading to the improvement within the framework of the river basin management plan.

References

[1] Petts G E and Maddock I 1995 Flow Allocation for In-river Needs *The Rivers Handbook, Hydrological and Ecological Principles* vol 1 ed Calow P and Petts G E (Oxford: Blackwell Scientific Publications) pp 289-307

[2] Smolar-Žvanut N 2000 Vloga perifitonskih alg pri določanju ekološko sprejemljivega pretoka v tekočih vodah (The role of periphytic algae in determination of ecologically acceptable flow in running waters) (PhD Dissertation, unpublished) *(Ljubljana: Univerza v Ljubljani, BTF)* p 172

[3] Maddock I P, Smolar-Žvanut N and Hill G 2007 The effect of flow regulation on the spatial distribution and dynamice of channel geomorphic units (CGU's) in the Soča river *Bridging the gap between hydraulics and biology: extended abstracts, posters of the 6th international symposium on ecohydraulic* (Christchurch, New Zealnad, 18-23 february 2007) pp 1-4

[4] Smolar-Žvanut N, Mikoš M and Breznik B 2007 The impact of the dam in the Bistrica river on aquatic ecosystem *Acta Hydrotechnica* **23** 99-115

[5] Schälchli U 1991 Morphologie und Strömungsverhältnisse in Gebirgsbächen: ein Verfahren zur Festlegung von Restwasserabflüssen *Mitteilungen der Versuchsanstalt für Wasserbau, Hydrologie und Glaziologie ETH* **113** 113

[6] Mader H 1992 Festelgung einer Dotierwassermenge über Dotationsversuche (dissertation) *Wiener Mitteilungen, Wasser Abwasser Gewässer* **106** 375

[7] Dyson M, Bergkamp G and Scanlon J 2003 *Flow, the essentials of environmental flows* (Gland: IUCN)

[8] Vrhovšek D, Martinčič A, Krušnik C, Kosi G, Burja D, Muck P, Smolar N and Pogačnik Z 1994 Kriteriji za zagotovitev dopustnih najnižjih pretokov v Sloveniji (The criteria for acceptable minimum flows in Slovenia), Final report *(Ljubljana: Water management Institute)* p 131

[9] Smolar-Žvanut N, Maddock I, Vrhovšek D 2008 Evaluation and Application of Environmental Flows for Running Waters in Slovenia. *Water Resources Development:* **24** 609-619

[10] Burja D, Smolar N and Vrhovšek D 1995 *Osnove določanja ekološko sprejemljivega pretoka (Qes) - The basis for ecologically acceptable flow determination (EAF) (Maribor, 8. december 1995, Mišičev Vodarski dan 95)* pp 67-71

[11] Smolar-Žvanut N and Burja D 2007 *Analiza določenih vrednosti ekološko sprejemljivih pretokov v Sloveniji* vol 18 – *The analysis of ecologically acceptable flows in Slovenia* vol 18 *(Maribor, 10. december 2007, Mišičev vodarski dan) (Zbornik referatov: Vodnogospodarski*

IOP Conf. Series: Earth and Environmental Science **4** (2008) 012015 doi:10.1088/1755-1307/4/1/012015

biro) pp 109-115

[12] Smolar-Žvanut N and Burja D 2007 *Pregled študij določitve ekološko sprejemljivega pretoka Qes (v izdelavi) - The review of studies of ecologically acceptable flow determination EAF (in preparation)* (Ljubljana: IzVRS)

[13] Acreeman M and Dunbar M J 2004 Defining environmental flow requirements – a review *Hydrology and Earth System Sciences* **8(5)** 861-876

IOP Conf. Series: Earth and Environmental Science **4** (2008) 012016 doi:10.1088/1755-1307/4/1/012016

Modelling snow accumulation and snow melt in a continuous hydrological model for real-time flood forecasting

Ph Stanzel, U Haberl and H P Nachtnebel

Institute of Water Management, Hydrology and Hydraulic Engineering, University of Natural Resources and Applied Life Sciences, Muthgasse 18, 1190 Vienna, Austria

philipp.stanzel@boku.ac.at

Abstract. Hydrological models for flood forecasting in Alpine basins need accurate representation of snow accumulation and snow melt processes. A continuous, semi-distributed rainfall-runoff model with snow modelling procedures using only precipitation and temperature as input is presented. Simulation results from an application in an Alpine Danube tributary watershed are shown and evaluated with snow depth measurements and MODIS remote sensing snow cover information. Seasonal variations of runoff due to snow melt were simulated accurately. Evaluation of simulated snow depth and snow covered area showed strengths and limitations of the model and allowed an assessment of input data quality. MODIS snow cover images were found to be valuable sources of information for hydrological modelling in alpine areas, where ground observations are scarce.

1. Introduction

In alpine basins snow melt can contribute severely to flood runoff. Thus modelling snow melt accurately is of central importance to flood forecasting. In order to provide adequate system states for all flood events the processes of snow accumulation and snow melt need to be simulated continuously.

Snow modelling generally faces problems of snow data scarcity. Snow depth is measured at many stations, but mostly in the valleys, and snow density and snow water equivalent are not measured regularly. Available snow measurements typically are not transmitted in real-time and therefore cannot be used as input for real-time flood forecasting. The same applies to measurements of radiation, which could be used in snow modelling algorithms. Though precipitation measurements are known to have considerable errors especially in winter, they are, together with temperature measurements, commonly the only available input for the simulation of snow system states in the context of real-time flood forecasting.

The presented contribution shows results of snow modelling with a continuous hydrological model in the context of a real-time flood forecasting system in an alpine Danube tributary basin. Simulation results were evaluated using discharge measurements and snow depth measurements. In addition to ground measurements MODIS remote sensing snow cover data was tested as a means for hydrological model validation.

2. Snow model

The applied snow model is part of the continuous, semi-distributed rainfall-runoff model COSERO, developed at the Institute of Water Management, Hydrology and Hydraulic Engineering. Processes of snow accumulation and snow melt are calculated for hydrological response units (HRUs) derived from intersection of maps of subbasin boundaries, soil types, land cover and 200 m elevation bands.

Only precipitation and temperature are needed as input. Separation of precipitation into rainfall and snowfall is effected with two threshold temperatures. Snowfall is distributed to five internal snow classes with a log-normal distribution to consider variable snow-depths within a HRU. Snow melt is simulated with a temperature-index approach with spatially distributed, seasonally varying melt factors. Higher reflection of fresh snow is considered by a reduction of the melt-factor after the snowfall event. The model also accounts for evaporation from snow, retention of liquid water in the snow pack, refreezing of retained liquid water, and settling of the snow layer [1]. Glacier areas are considered as deep snow layers.

Subsequently, the transformation of rainfall and melt water to runoff is effected in a conceptual model accounting for processes of infiltration, soil storage, runoff generation and routing. Separation of runoff into fast surface runoff, inter flow and base flow is calculated by means of a cascade of linear and non-linear reservoirs [2].

The model is calibrated against measured catchment runoff with historic time series of precipitation and temperature as input data. Initial model parameters are based partly on physical meaning and partly on previous experiences. Only few of the snow model parameters are adjusted in a manual calibration procedure. Values of minimum and maximum melt factors which determine the sine shaped seasonal course of the snow melt factor are altered. Their spatial patterns, based on radiation patterns [3], are kept constant. Threshold temperatures for snowfall and snow melt and a parameter governing snow depth variability are less influential and modified within close constraints.

3. Application

The model was implemented in a flood forecasting application in the Salzach basin, an Alpine Danube tributary watershed. The presented results were obtained in its most upstream part, where the influence of snow melt on runoff is highest. This catchment from the source to the gauge in Mittersill has an area of 600 km², with altitudes ranging from 800 to 3600 m a.s.l. (figure 1). It was divided into 10 subcatchments and 251 HRUs with sizes ranging from 0,05 km² to 14 km².

The model was pre-calibrated with precipitation and temperature data regionalized from daily ground measurements from 1999 to 2005. Runoff simulation results on a daily basis were evaluated for the period from 1995 to 1998. For the period from 2003 to 2005 hourly precipitation analyses incorporating ground and radar measurements of the Central Institute for Meteorology and Geodynamics (ZAMG) were available. They are based on comparable algorithms as the input for operational flood forecasting and were used as definitive input for calibration. Simulation results presented were derived with these input data. The meteorological analyses have a spatial resolution of a 1x1 km grid and a temporal resolution of one hour. The algorithms for this product [4] were revised within this project in joint hydrological and meteorological research. A general increase of the amount of snowfall of 30% was applied.

4. Snow data

4.1. Snow depth measurements

Snow depth measurements from 11 meteorological stations in and around the catchment area were obtained (figure 1). Most stations are located at elevations between 800 and 1500 m a.s.l., with only 2 stations at approximately 2300 m a.s.l. In the later ablation period, when only higher areas are snow covered, only these two stations provided ground based information.

4.2. MODIS snow cover data

MODIS (Moderate Resolution Imaging Spectroradiometer) is an instrument aboard the two satellites of NASA Earth Observation System, Terra and Aqua. MODIS observations are processed at the U.S. National Snow and Ice Data Center (NSIDC) to obtain daily snow cover data grids with a resolution of

500x500 m. The general applicability of MODIS snow cover images for the evaluation of snow covered area in Austria was shown by [5].

For two snow melt periods in 2004 and 2005 MODIS snow cover products (MOD10A1, Version 4 and 5) were downloaded from the NSIDC Distributed Active Archive Center (DAAC). For each year three snow cover maps with the least cloud cover interference were selected [6].

Figure 1 shows the snow cover map of 02.05.2005. It can be seen that frequently the edges of snow-covered area are mapped as clouds, a general deficiency of the actual MODIS algorithm. For analysis the area classified as snow was interpreted as minimal snow cover extent, this area plus the area classified as cloud as maximal snow cover extent. As only maps of days with very little cloud cover were analysed, this approach lead to rather narrow ranges of snow covered area.

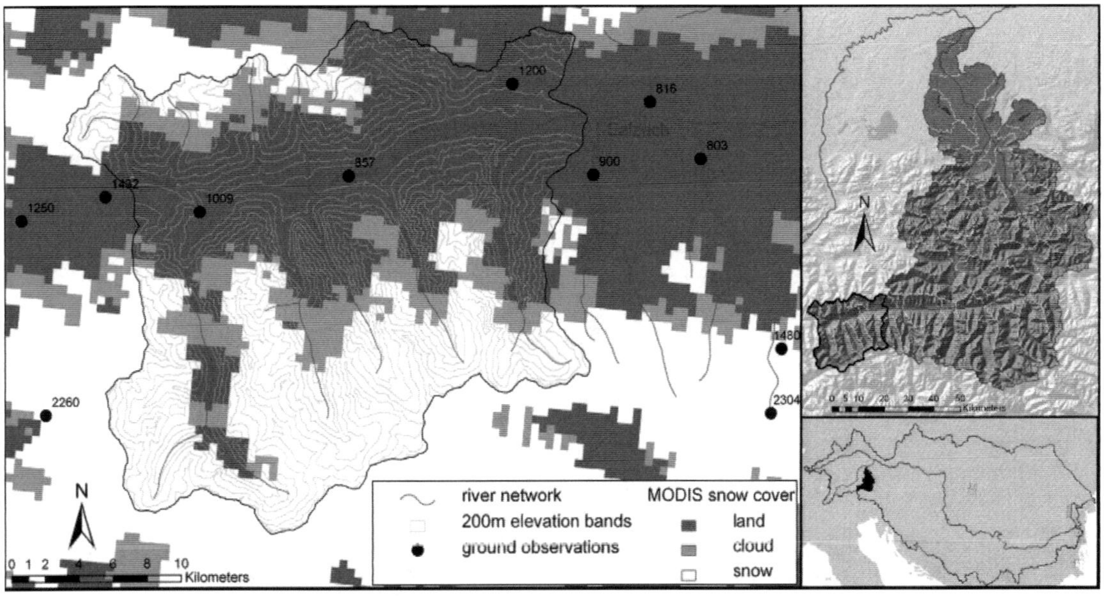

Figure 1. Right side: The Mittersill subbasin as part of the Salzach and eventually Danube basins Left side: MODIS snow cover map (02.05.2005) and ground observation stations with elevations

4.3. Comparison of ground observations and MODIS images

Table 1 shows a comparison of ground observations of snow cover (snow depth larger than 1 cm) and MODIS snow classification for the corresponding pixel. The agreement for both evaluation periods was generally good. While in 2004 the decrease of snow covered area as measured by MODIS was faster than measured on the ground (with one station pixel misclassified as not snow covered), the decrease in 2005 was slower as measured by MODIS than measured on the ground (with two and then one station pixel misclassified as snow covered).

5. Results and discussion

5.1. Evaluation of simulation results

Snow modelling results – after a calibration based on runoff only – were evaluated in three ways. Simulated runoff in snow melt periods was compared with measured runoff in Mittersill. Simulated snow depths were averaged for elevation bands and compared with measurements of stations located within the corresponding elevation range. Simulated snow covered area was averaged for elevation bands and compared with snow covered area shown by MODIS snow cover maps.

Table 1. Snow cover measurements: ground observations and MODIS classification for the pixel corresponding with the location of the ground observation

Total nr. of ground observations: 11 Date	Nr. of ground observations snow	MODIS classification snow
30.04.2004	2	2
08.06.2004	2	1
17.06.2004	2	1 (Cloud)
15.03.2005	11	11
22.04.2005	3	5
02.05.2005	2	3

Results of runoff simulations are shown in figure 2. The seasonal runoff variations due to snow melt processes are well represented. For the simulation from October 2003 to October 2005 a Nash-Sutcliffe coefficient of 0,88 was achieved. The water balance indicated precipitation deficits of 4 % for 2004 and 10 % for 2005 (which is also clearly visible in the hydrograph of summer 2005).

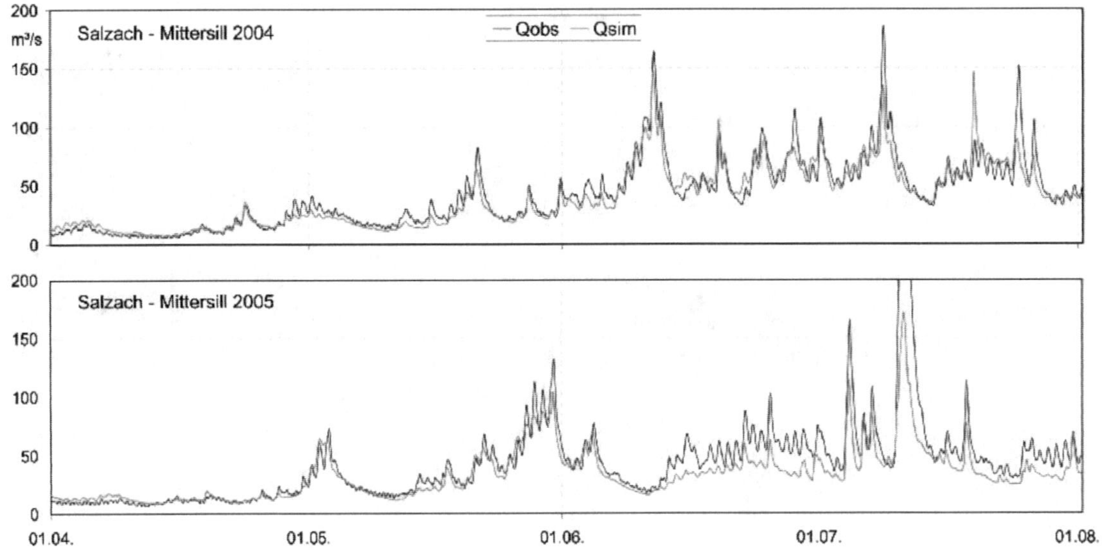

Figure 2. Runoff simulations and observations (April to July 2004 and 2005)

Simulated snow depths were generally lower than measurements. Figure 3 shows results for the elevation band 900-1100 m a.s.l. in comparison with two stations at corresponding elevations. Similar results were obtained for other elevations. Differences between measured and simulated snow depths can stem from differences between assumed and real snow densities. At the end of the accumulation period, when snow has settled, these discrepancies should be low, however. These results suggest that there is still some underestimation of precipitation in the input data in the snowfall period, despite the included increase of 30 %.

The course in time of snow accumulation matches very well. The start of the melting period and the sharp initial decrease of snow depth were reproduced correctly. In the later phase of ablation, however, simulated snow depth decreased slower than measured snow depth, resulting in a later end of snow cover.

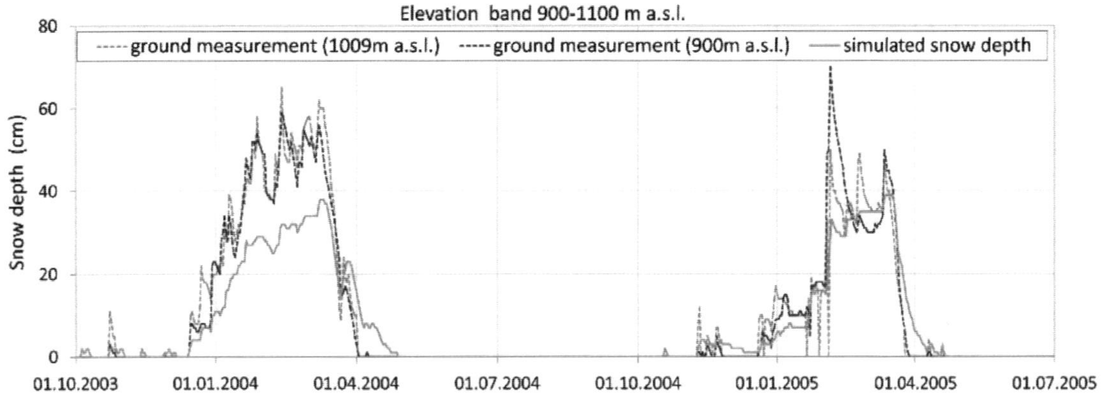

Figure 3. Snow depth simulations (elevation band 900-1100 m a.s.l.) and observations

Figure 4 shows a comparison of simulated snow covered area and results of analysing MODIS snow cover images for all elevation bands. The temporal characteristics of decreasing snow cover in the ablation period were generally reproduced. For the period from 30.04.2004 to 17.06.2004 simulated snow covered area was overestimated. For the period of 15.03.2005 to 02.05.2005 simulations and MODIS observations matched much better (the low values for higher elevations on 15.03.2005 are due to cloud interference). Although results of two different years cannot directly be connected this corresponds with the finding that the earlier ablation phase, evaluated in March and April 2005, is reproduced better than the later ablation phase, evaluated in May and June 2004. This agreement with the results of the evaluation with ground measurements confirms that MODIS snow cover maps can be used in similar evaluations when little or no observations in higher altitudes are available.

5.2. Multi-objective recalibration
First tests of manually recalibrating the snow model taking both runoff and snow information into account were performed. The objective with regard to snow modelling was to achieve a better representation of the duration of snow cover. With minor decreases of model performance with regard to runoff, faster melt in later ablation phases could be obtained when model parameters were altered beyond previous constraints. An evaluation whether this leads to more robust simulations will be carried out with results of the operational test during 2008. Effects of such recalibration on simulated runoff in the periods of largest contributions of snow melt to floods (May and June) were very low.

IOP Conf. Series: Earth and Environmental Science **4** (2008) 012016 doi:10.1088/1755-1307/4/1/012016

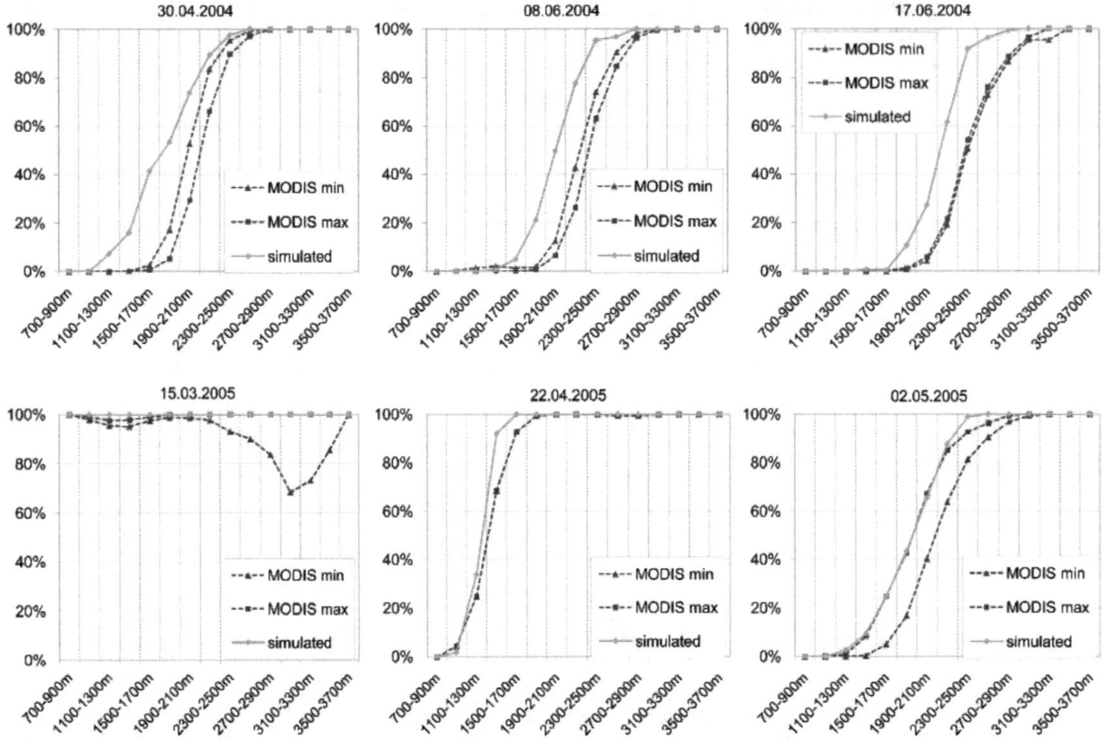

Figure 4. Snow covered area simulations and observations in ablation periods in 2004 and 2005

6. Conclusions

Snow modelling results of a hydrological model using only precipitation and temperature as input data were evaluated. In addition to discharge measurements snow depth measurements and MODIS remote sensing snow cover information were compared to simulation results.

The model was able to simulate the seasonal runoff variations due to snow melt very well. The analyses of simulated snow depths and snow covered area showed strengths and limitations of the model and supported assessment of input data quality. The effect of improving snow cover modelling results on the simulation of flood runoff with high snow melt contributions was very low.

MODIS snow cover maps allowed an evaluation of snow modelling for the whole range of elevations and for the whole ablation period. Good agreement with results of model evaluation with snow depth ground measurements showed that MODIS snow cover images are a valuable source of information in alpine areas, where ground observations are scarce.

Acknowledgements

The ongoing research has been carried out in the framework of a project funded by the Provincial Government of Salzburg, VERBUND-Austrian Hydro Power AG and Salzburg AG. Data was provided by these institutions, the Federal Ministry of Agriculture, Forestry, Environment and Water Management and the Central Institute for Meteorology and Geodynamics (ZAMG).

References

[1] Fuchs M 1998 *Modeling snow-melt runoff in an alpine watershed* Master thesis, Institute of Water Management, Hydrology and Hydraulic Engineering, University of Natural Resources and Applied Life Sciences, Vienna, Austria

[2] Nachtnebel H P, Baumung S and Lettl W 1993 *Abflussprognosemodell für das Einzugsgebiet der Enns und Steyr (in German)* Report, Institute of Water Management, Hydology and Hydraulic Engineering, University of Natural Resources and Applied Life Sciences Vienna, Austria

[3] Kling H, Fürst J and Nachtnebel H P 2006 Seasonal, spatially distributed modelling of accumulation and melting of snow for computing runoff in a long-term, large-basin water balance model *Hydrol. Processes* **20**, 2141-56

[4] Haiden T, Kann A, Stadlbacher K, Steinheimer, M and Wittmann, C 2007 *Integrated Nowcasting through Comprehensive Analysis (INCA) - System overview* ZAMG report, Vienna, Austria

[5] Parajka J and Blöschl G 2006 Validation of MODIS snow cover images over Austria. *Hydrol. Earth Syst. Sci.* **10** 679-89

[6] Hall D K, Riggs G A and Salomonson V V 2006 MODIS/Terra snow cover daily L3 global 500m grid V005, 30.04., 08.06, 17.06.2004; 15.03., 22.04., 02.05.2005 Digital media Boulder, Colorado USA: National Snow and Ice Data Center

The Main Characteristics of the High Water Registered in the River Basin Bega in February 1999

Niculae Iulian Teodorescu

Ph. D. hydrologist, National Administration "Apele Române", Banat Water Directorate Timisoara

e-mail: niculae.teodorescu@gmail.com

Abstract. The year 1999 was characterized by high water events with a rare rate of occurrence. The events were produced in different months of the year, in different river basins in the Banat Region (area situated in the south - western part of Romania). In the River Basin Bega the most important high water appeared in mid February and was generated by both rain and the melting of the snow layer. The cold period of 1999 was characterized by relatively high quantities of precipitation (module coefficients from 1.17 to 2.15), the most part of those being liquid (rainfall). At altitudes higher than 1,000 m, the snow layer was relatively continuous. Weather warming appeared in the middle of February, in the same year, generating the melting of an important layer of snow (10 - 15 cm in the plain and around 20 cm in the hilly area), and the water layer generated by this had an important effect, overlaid on the water layer generated by rainfall. The high water (flood) event, which is the topic of this study, occurred between 20th – 27th February 1999 and after analyzing the data we initiated a comparative analysis between the main element of the event (specific discharge, high water duration, increasing time, shape coefficient) and the mean element of high waters from 10 gauging stations from the river basin. After the analysis we observed that some elements, like high water duration and the increasing time, are bigger than the mean values - which is a characteristic of high waters generated both by rainfall and snow melting, while other elements are smaller (α, overrun layer) - this as a consequence of river basin response to the generating elements. This analysis has enriched our database for hydrologic prognosis, as it is known that in this part of the country warm periods have a frequent occurrence in winter. They produce snow melting and, sometimes generate important high water.
Keywords: Bega River Basin, high water, February 1999

1. General Presentation

The hydrographical area Banat is situated in the south - western part of Romania and has a series of river basins with tributaries from the mountains Semenic, Tarcu-Godeanu and Poiana Ruscai.
The River Basin Bega, subject of this study (Figure 1) spreads towards east – west as a general direction. The main stream has its springs in the Poiana Ruscai Mountains, at an altitude of 890 m.

In the first sector, the river flows along a south-northern direction; after the confluence with first major tributary (Sasa river, F=123 sq km L=28.0 km), the River Bega changes its course and flows from east to west. The most part of the right tributaries are small - the biggest being Chizdia F=233 kmp L=34.0 km; the river basin is more developed on the left part, here it receives the most important tributary in this area, Gladna River F=173sq km L=34.0 km.

Bega River Basin is one of the most important river basins in this part of Romania, it flows along an east - west general direction and has a total length of 256 km, from which in Romania – 170 km. Bega River has in the middle sector a double connection with the Timiş River - the most important river in Banat.

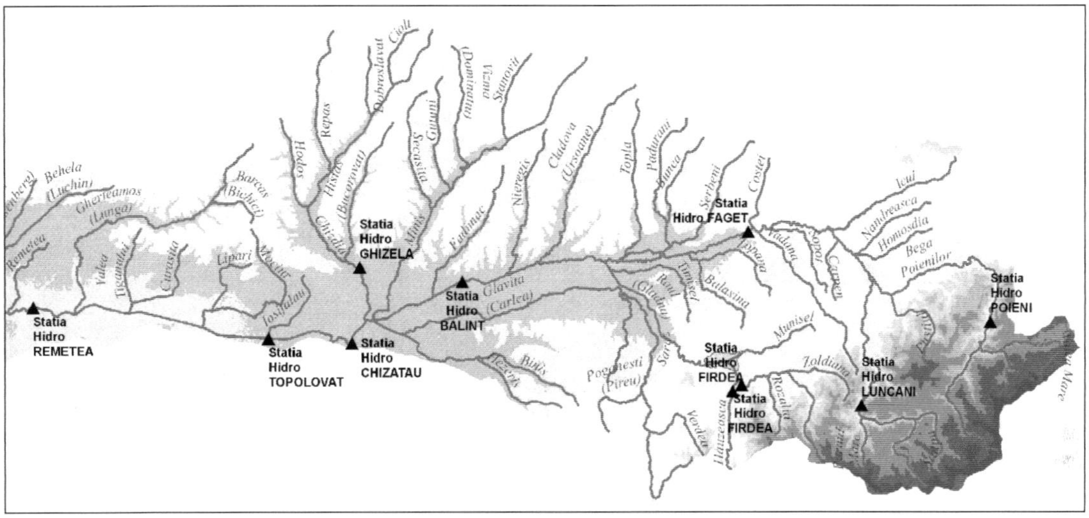

Figure 1. Bega River Basin

These connections are:

Timiş - Bega Supply Canal – with a role in supplementation of discharge during low water;

Bega - Timis Discharge Canal - together with some hydro technical works, is situated on the right bank from the middle sector of the River Basin, downstream Chizatau - with role in satisfaction of enforced condition established by the Hydrotechnical Agreement between Romania and Serbia.

There are 10 gauging stations in the River Basin Bega. The main characteristics of the rivers are controlled by these stations and are presented in Table 1.

The studied area has a climate influenced by the air masses transported from west and north - west, but also Mediterranean air masses (especially in the south part of the region.)

2. Generating factors of high water

As a result of these conditions, in the cold period of the year, between November - February, some warm moist air mass penetrate and generate conditions for the appearance of rainfalls and even the melting of snow- conditions that generate high waters. Such elements appeared in the third decade of February 1999 - a month with the greatest quantity of precipitation registered in the entire period of observations and measurements. February 1999 was a relatively cold month; the first part of it comprising air temperatures below zero degree (Table 2). Instead the precipitation regime recorded an exceeding of the multi annual average in most parts of the river basin – the module coefficients having an oscillation between 2.14 in the lower part of the river basin and 3.41 at Balint gauging station; while in the upper part of the river basin normal values were registered (Figure 2). The precipitations were in this interval mostly of solid nature. This phenomenon led to the accumulation of a snow layer, with its biggest thickness of 20 cm at Surduc on the 15th of February 1999 (altitude of 220 m) (Table 3).

Starting with the 20th of February 1999, we observed that the weather warmed up, the temperatures in the centre of the basin exceeding 7°C, situation, which gradually led to the melting of the snow layer, especially in the plains and hilly areas.

Table 1. The gauging stations in the River Basin Bega

Crt No.	River	Gauging station	Length (Km)	Catchment area		Foundation Year
				Surface (Km^2)	Altitude (m)	
1	Bega	Luncani	14.0	76.0	775	1952
2	Bega	Faget	42.0	483	470	1875
3	Bega	Balint	79.0	1002	335	1875
4	Bega	Chizatau	86.1	1740	278	1919
5	Bega	Topolovatu Mic	89.0	1755	270	1930
6	Bega	Remetea	114	2057	253	1968
7	Sasa	Poieni	17.0	80.0	763	1952
8	Gladna	Fardea	20.5	62.0	456	1980
9	Hauzeasca	Fardea	9.00	30.0	364	1980
10	Chizdia	Ghizela	33.5	210	194	1961

Table 2. The evolution of air maximal temperatures in the period 20-23 February 1999 at some meteorological stations in the region

Meteorological Station	Date			
	20	21	22	23
Tarcu	-5.4	-5.8	-3.6	-4.0
Cuntu	-2.2	-1.0	0.9	0.3
Semenic	-2.2	-0.8	1.1	-0.4
Lugoj	2.0	4.2	6.6	5.6
Timisoara	3.8	7.2	9.0	6.0

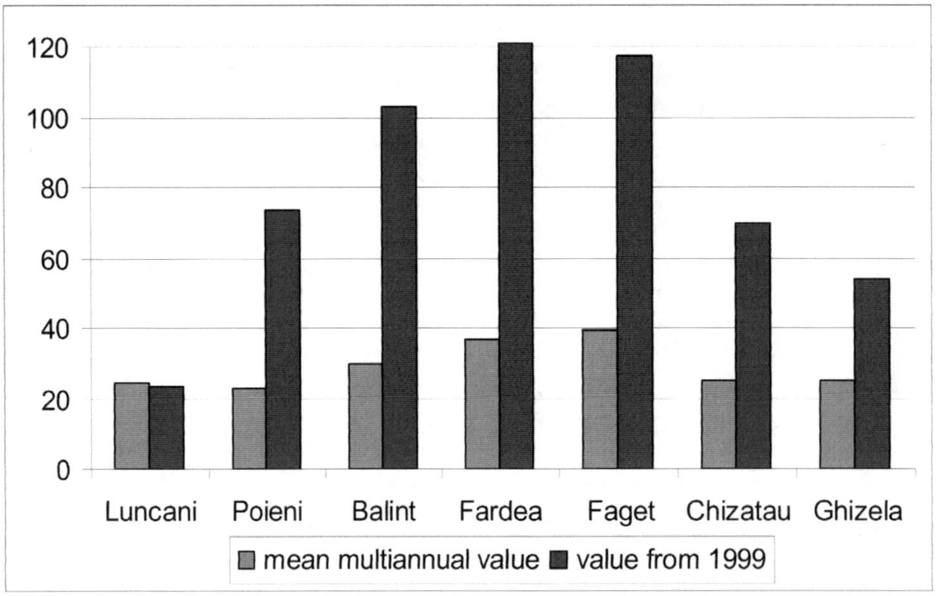

Figure 2. The comparative analysis of multiannual average precipitations and those produced in February 1999

Table 3. The snow layer thickness (cm) in Bega river basin at some neighbouring meteorological stations registered between 15th and 23rd of February 1999

Crt. No.	Meteorological/Gauging Station	Date								
		15	16	17	18	19	20	21	22	23
1	Faget	18	17	16	14	8	6	4	4	4
2	Balint	11	10	10	5	p	P	p	p	p
3	Chizatau	15	15	15	11	8	5	3		
4	Surduc	20	20	20	17	14	11	7	4	8
5	Costei			11	9	7	5	2	1	1
6	Lugoj	19	17	17	9	7	7	4	2	3

p – Patchy (discontinuous)

Starting with the 21st of February, some liquid precipitations were observed; these continued until 24th of February and had a nucleus on 22nd and 23rd February (Table 4).

It must be noticed that in the mentioned interval – 39.6% to 59.7% of the total amount of monthly precipitations were produced.

3. The high water analysis

Given these meteorological conditions, we witnessed the recording of high water that originated from precipitation and snow melting with maximal discharge values, which were, most of them, the highest from all the observation string. Thereby, in the middle and lower section of the river basin – where the amount of precipitation was bigger and where there was a snow layer that melted almost completely – were registered values with a return period probability of 1-2%. In the upper part of the river basin, where the precipitations and temperatures values were:

Table 4. The amount of precipitations (mm) registered in Bega river basin between 14 and 17 February 1999

Crt. No	Observation point	Day February 1999								Amount
		17	18	19	20	21	22	23	24	
1	Faget					8.0	20.8	28.9	1.6	59.3
2	Balint				4.8	5.8	17.4	28.0		51.2
3	Chizatau	0.4					20.8	12.2		33.0
4	Topolovat	0.2	1.3		0.8	4.7	11.2	22.2		38.1
5	Margina				1.3	3.4	9.4	1.4		14.2
6	Bethausen	1.0			3.5	5.6	22.4	20.0		48.0
7	Bunea Mare	0.6			4.2	7.2	2.4	2.5	0.4	12.5
8	Fardea					8.5	16.2	23.2		47.9
9	Hauzesti	1.2			2.6	4.1	23.0	48.2	0.3	75.6
10	Ghizela	0.4			11.0		15.8	2.5		18.3
11	Recas	0.5	1.8		1.9	6.7	24.3	16.4		47.4
12	Timisoara	0.1	1.7	0.3	1.5	6.1	22.8	11.4	0.1	40.4
13	Lugoj	3.4	2.9	0.7	3.5	5.9	18.6	24.3	1.7	50.5

lower, the melting of the snow layer was not caused in totality and the high waters were smaller (gauging station Luncani).

On the other hand, the analysis of the relation between the specific maximal discharge values and the surface of the catchment area (Figure. 3) reflects very well this way of manifestation in the whole river basin level, their distribution on a relation for most of the river basin can be observed very well, except the two gauging stations from the upper part of the river basin (Luncani and Poieni).

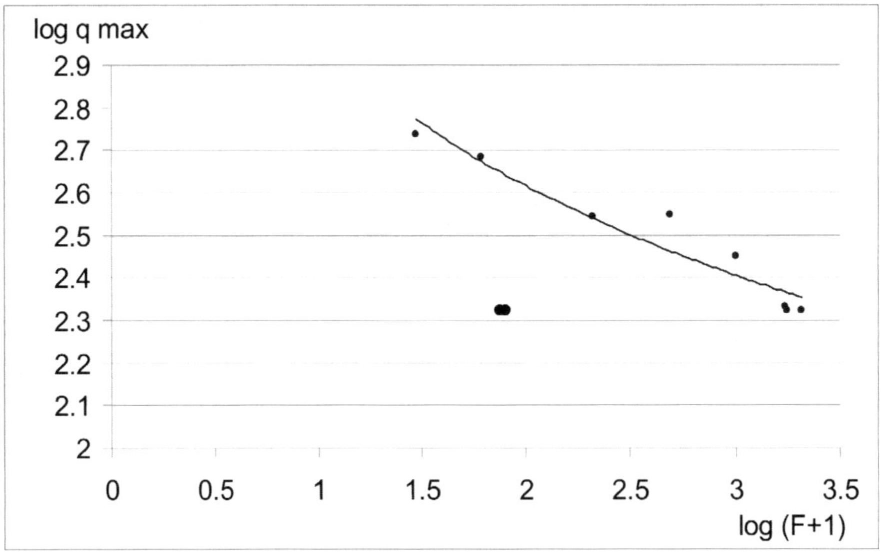

Figure 3. The connection between the specific maximal discharges (l/s.km^2) and the catchment area (F km^2)

Analyzing the relation between the specific maximal discharge (l/s.km^2) and the flowing layer (mm) (Figure 4), we can see that at this moment in some way this difference is fading, but it's still obvious, the upper part of the river basin previously mentioned recording lower values of the flowing layer than the rest of the river basin especially due to previously mentioned conditions.

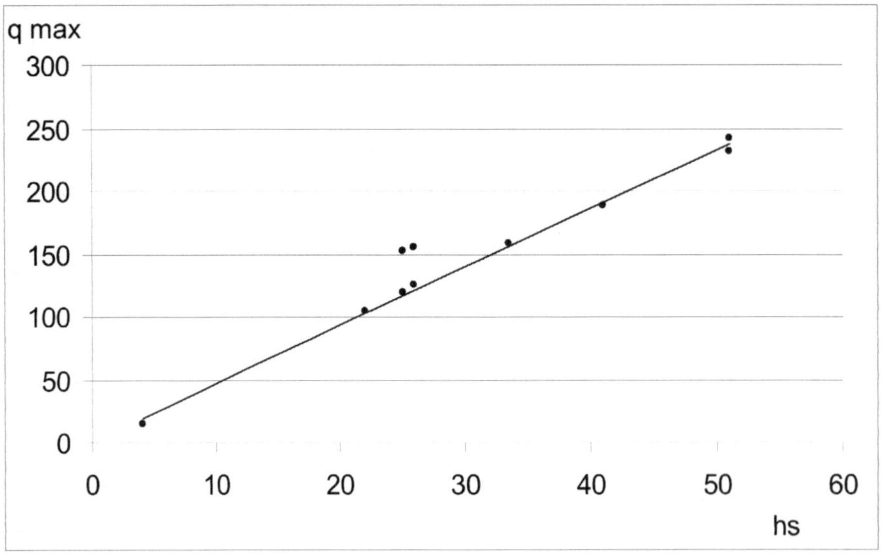

Figure 4. The connection between the specific maximal discharge (l/s.km^2) and flowing layer (mm)

4. Reconstruction of natural flow

In the area of the river basin Bega there are some hydrotechnical constructions that produce major modifications of the maxim flow, consequently we will offer a short presentation of these facilities:

- The reservoir Surduc is situated on the upper flow of the river Gladna. This controls a basin area of 133 km^2 and has the normal level of retention 192.00 moMN and a volume of 28,375 mil m^3 and a total surface of water 37.0ha;
- The alimentation canal Timis-Bega transfers the water from the hydrographical basin of the Timis river in the Bega river with the purpose of supplementing the discharge in the low water period; thus in the flood period the canal is likely to bring a discharge input;
- The discharge canal from Topolovat has the role to discharge a part of the high water from the river Bega in the Timis river- with the purpose of respecting the border regulations that must be observed by the Romanian part, as stated by the Common regulation of defence against floods and freezing. The capacity is around 350 m^3/s and when there is high water we record the decrease of discharge measured downstream on this Canal.

The polder Hitias – has the role to reduce the maximum discharge at high water, it presents a spillway that allows temporary storage of a water volume; the total surface of the polder Hitias is 130.7 ha and has a volume of 20 million m^3. It is worth mentioning that the polder presents a spillway that allows alimentation of the polder during high water from the Timis river; - admission spillway from Bega, 160 m length, situated on the Bega dam, left bank, km 1 + 070 ÷ 1 + 230 upstream Topolovat, with 104.26 moMB level.

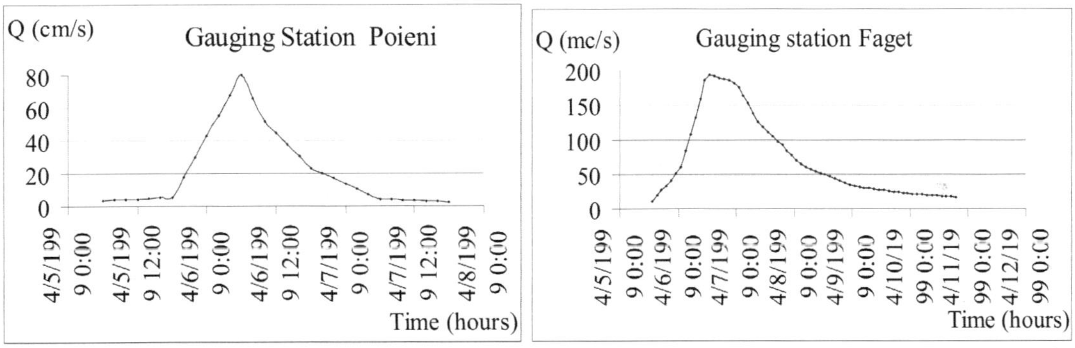

Figure 5. Recorded flows on the gauging stations Poieni and Faget (natural regime)

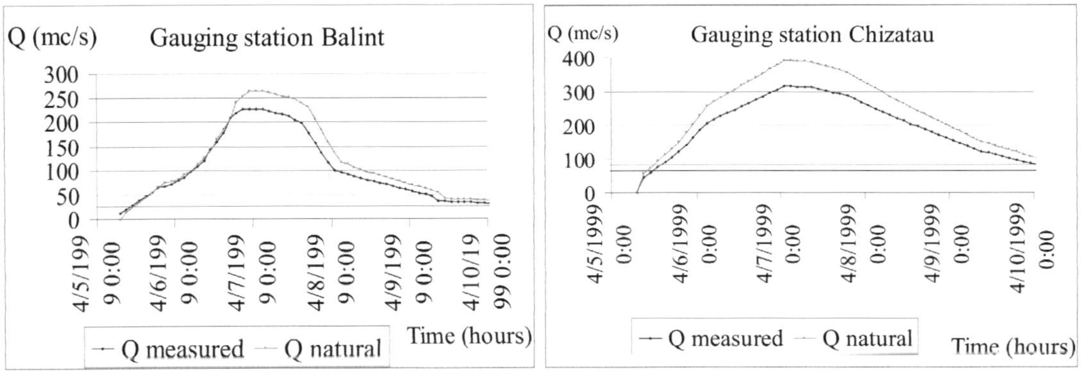

Figure 6. Recorded flows (Q measured) and reconstruction (Q natural) at gauging station Balint and Chizatau

The temporary reservoirs on the right tributaries from the middle and inferior course of the Bega River have the purpose of reducing the value of the maxim discharge previously mentioned to meet the border requirements.

All these facilities lead to the registration of some values of the maxim discharge, which are more reduced than the natural ones. Consequently, we proceeded to the calculation of the natural flow reconstruction.

5. Conclusions

The flood in February 1999 was generated by rain and the melting of snow, in which both the quantity of water from rain and the warming period, that produced the melting of a considerably quantity of snow, played an important role.

The different distribution on the altitude was triggered by the penetration of warm Mediterranean air that came as an additional element to an obvious frontal activity and produced an important quantity of precipitation that, together with the water obtained from the melting of the snow layer, increased the humidity of the soil and resulted in a significant flow. Consequently, some extremely high discharge values with rare rate of occurrence were recorded.

The global discharge coefficient presented high values that were varied from 0.25 to 0.37; of course these values have to be treated with reserve, taking into consideration that the upper stream of the basin represents a "white spot"- we do not have concrete data from this area, this thing is determined by the absence of some observation points. The analysis was made by comparing data with those obtained from the meteorological stations from the neighbouring mountains (Semenic, Tarcu-Godeanu).The information thus obtained enriches the database that will be used in the future to forecast similar events.

It is important to show that the role of the hydrotechnical works in the retention and redistribution of water into the maximum flow has led to the registration of some discharge values that were reduced with 81.2%, their effect was noticed in the down stream of Remetea Mare and led to the observance of the terms imposed by the article of the Common regulation of defence against floods and freezing, negotiated with the Serbian part being at the same time part of the Hydrotechnical Romanian Serbian Agreement.

References

[1] Diaconu C 1988 *Rivers from floods to drought* (Tehnica) p 188

[2] Diaconu C and Serban P 1994 *Hydrological synthesis and regionalisations* (Bucharest: Tehnica) pp 11-277

[3] Macridin Z and Harabagiu C 1999 *Consideration meaning maximum flow in the Bega Veche River basin* (manuscript)

[4] Stanescu V Al 1972 Mathematic models for critical flood wave's determination *Hydrological studies XXXIV* (IMH Bucharest) p 245

[5] Serban P, Stanescu V Al and Roman P 1989 *Dynamic Hydrology* (Bucurest: Tehnica) p 9-55; 145-251

IOP Conf. Series: Earth and Environmental Science **4** (2008) 012018 doi:10.1088/1755-1307/4/1/012018

The minimum flow in the upstream basin of the Timis river

Niculae Iulian Teodorescu

Ph.d. hydrologist, National Administration "Apele Române", Banat Water Directorate Timisoara

E-mail: niculae.teodorescu@gmail.com

Abstract. The minimum discharge represents a very important stage in the water flow, especially during the long periods lacking rainfall, both in the cold season, but actually mainly during the dry season. Due to this fact, the present paper aims at an analysis of this stage of the water flow in the upper basin of the Timis River, upstream the Lugoj gauging station. This part of the hydrographical basin presents a major importance due to the fact that downstream this gauging station at about 7 km there is The Costei Hydro technical Knot, through which at low water levels, the entire quantity of water in the Timis river bed is directed towards the Bega river in order to supplement the quantity of water upstream Timisoara, the location of the water supply plant. The water taken from the river is treated there in order to be distributed for use in Timisoara, the capital city of the county. The hot period of the year 2007 was characterized in Romania by the presence of some very long periods with high temperatures (the maximum values being of over 39^0 C in many regions) and lacking rainfall. Under these circumstances very low values were recorded on rivers, drastic drops of underground water reserves – having as an effect the appearance of drought in certain hydrographical basins. In the analysed hydrographical basin the values of the minimum water quantities during the summer were situated around the assurance values of 75% - that means they were about 75%. The conclusion we have reached is that this summer wasn't among the driest, lower values being recorded in the years 1950-1955, 1957, 1962, 1965,and so on.
Keywords: Timis Upstream Basin, dry period, 2007.

1. General presentation

The Timis River – Danube's tributary on Serbian territory, springs under the Piatra Goznei Peak in the Semenic Mountains – "Banat's water castle" – at an altitude of 1,280 m. Banat territory is crossed by the river on 244 km, developing an area of 7,310 km^2 (Figure1). Its course is divided into 3 parts: the first part in the mountain area, where it gathers several short streams and reduced area water streams; the second part gets through the Timis-Cerna gorge, a tectonic scallop that separates the Banat Mountains from the Southern Carpathians Range (Carpati Meridionali). Here it receives several tributaries from the mountain ranges Tarcu-Godeanu and Semenic – the first ones more developed and presenting a consisting flow. Finally, the last part – after receiving its most important tributary – Bistra (F=919km^2, L=60 km) spreads over the Timis dejection cone and Banat's flooding plain.

Within the Timis River basin there are several important reservoirs – Trei Ape Reservoir (S=52.6 ha, 4.4 mil m^3 at NNR – Normal Level of Retention), Rusca Reservoir (S=120.05 ha., volume 18.6 mil m^3 at NNR).As a result, important modifications of the flowing regime are produced, which have to be considered when analysing the minimum flow.

The upper part of the Timis River is the area controlled by the Lugoj gauging station, due to the fact that the flow practically forms in this area (morph metrical elements are presented in Table 1). At the same time, downstream of this station there is the Timis-Bega supply canal (Figure 2) where during dry periods waters are diverted to supply the low discharge on the Bega river.

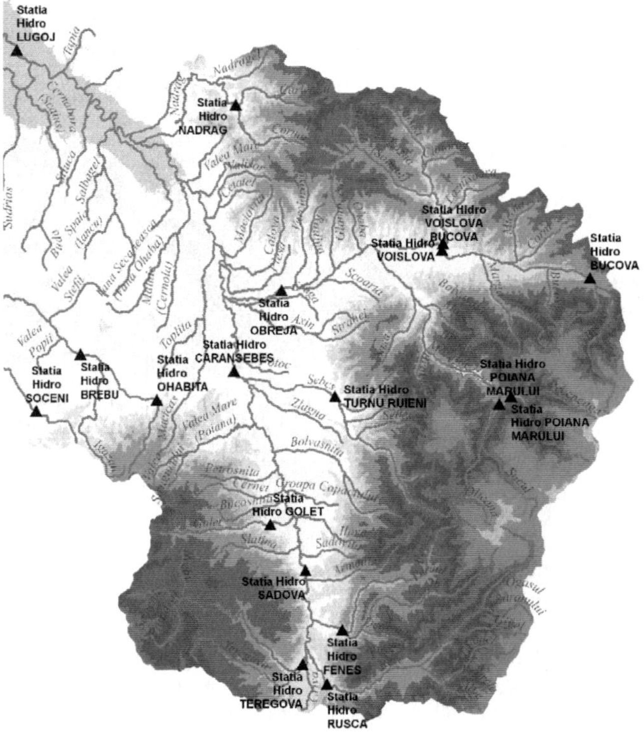

Figure 1. Bega River Basin

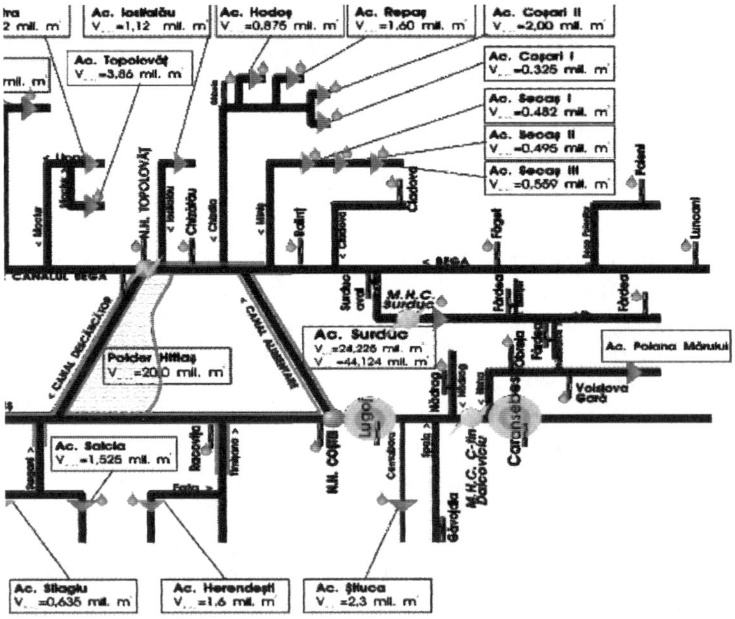

Figure 2. Timis-Bega double connection

2. Annual minimum average monthly flow analysis

Analysing the minimum average monthly flows registered in 2007, it is noticeable that significantly long periods with high temperatures and lack of rainfall were registered in our country during the summer months. This was reflected in the minimum discharge values registered in several areas. Table 1 offers a comparative presentation of the values at different ensurances of this parameter and those registered in 2007.

Within the analysed river basin minimum flow values higher than 80% assurance values – the limit for hydrologic drought were recorded. If we analyse Figure3 the relationship between the specific flow values calculated for the 2 pairs of values is obvious and it shows 2 different distinctive relations. Thus it is necessary to notice the fact that for watercourses that have their springs in the eastern part of the river basin in the Tarcu-Godeanu mountain range, the minimum flow values are higher. The same tendency can be observed when we monitor the 2007 value evolution – the distribution is the same on 2 relations – the values being higher for the river basin with higher altitudes – especially in the mountain areas.

This is due to the fact that in the analysed area the 2007 rainfalls were higher than the multiannual average values (Figure 4), except for the river basin in its lower part, reflected in the registered values at the Obreja and Lugoj gauging stations.

Table 1. Morphometrical and typical discharge elements for the upper part of the Timis river basin

Nr	River	Hydrometric station	F km^2	Hm m	Multiannual Average flow	Annual minimum monthly average discharge Ensurance of	Minimum Monthly Average flow 2007
						80%	
1	Timis	Teregova	167	901	2,28	0,295	0,378
2	Timis	Sadova	560	936	10,9	1,48	2,20
3	Timis	Caransebes	1072	765	17,5	2,25	2,29
4	Timis	Lugoj	2827	666	38,4	7,00	7,58
5	Raul Rece	Rusca	163	1184	4,64	0,810	1,62
6	Fenes	Fenes	125	973	2,82	0,495	0,668
7	Golet	Golet	41	751	0,686	0,105	0,240
8	Sebes	Turnu Ruieni	124	815	2.94	0,675	0,650
9	Bistra	Bucova	64,0	1236	1.65	0,345	0,550
10	Bistra	Voislova Bucova	232	892	3,71	0,545	1,38
11	Bistra	Obreja	863	880	17,1	4,34	3,30
12	Rusca	Voislova	166	1184	2,70	0,540	1,00
13	Bistra Marului	Poiana Marului	79	1442	2,71	0,570	1,10
14	Sucu	Poana Marului	77	1430	3,00	0,590	1,20
15	Nadrag	Nadrag	35	742	0,374	0,140	0,192

There are no significant differences when analysing the seasonal distribution of these values; thus it can be noticed that across the river basin's greatest part that belongs to the main water course – Timis river – values that presented a module coefficient no higher than 1.37 have been registered (at Teregova hydrometric station) and they were lower within the main tributary river basin – Bistra; here the rainfall module coefficients registered values of up to 0.85 in the lower part of the river basin (Obreja hydrometric station) and slowly exceeded in the mountain area (up to 1.15 at Poiana Marului hydrometric station).

IOP Conf. Series: Earth and Environmental Science **4** (2008) 012018 doi:10.1088/1755-1307/4/1/012018

A similar tendency was preserved during the warm period of the year – the time period between June and September, when the module coefficients have exceeded with 7%-16% the normal values on the largest part of the river basin. It can be noted that these values were more reduced in the lower part – at altitudes under 250 m, the 2007 registered values were 4% (Lugoj)-14% (Obreja hydrometric station) were lower.

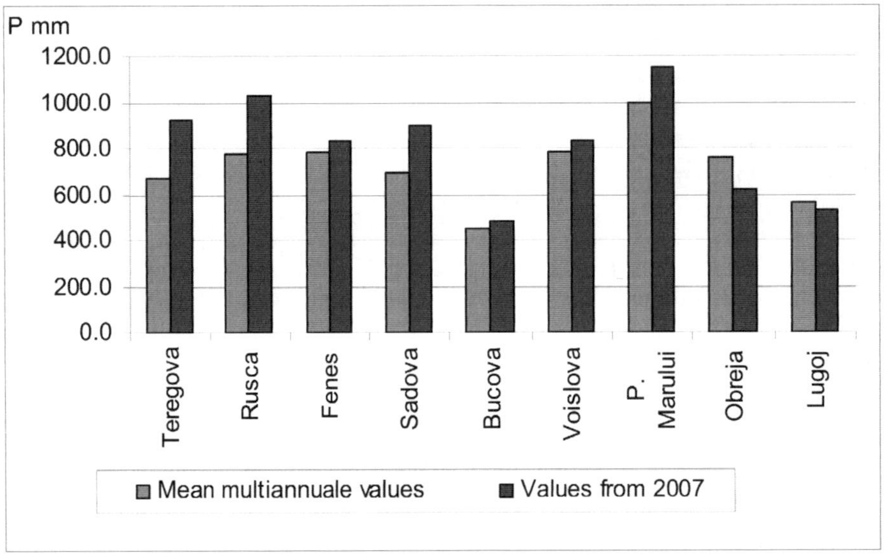

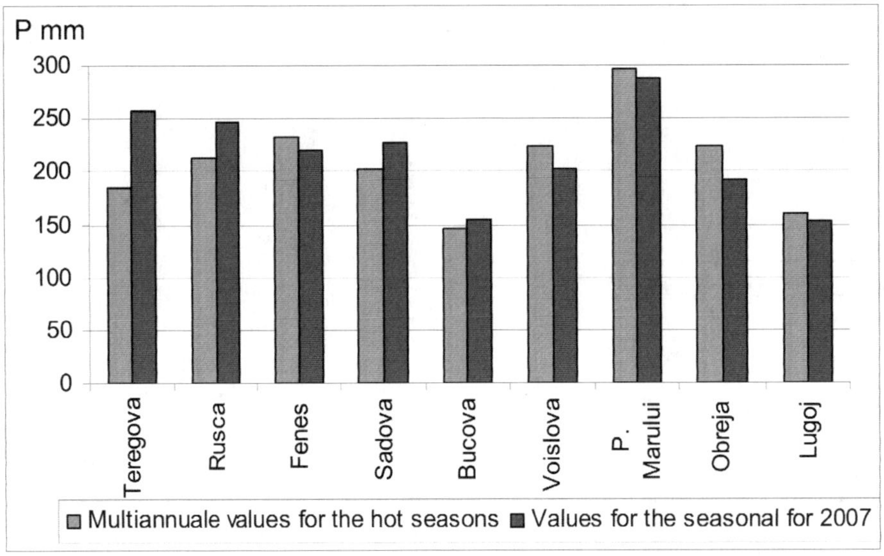

Figure 3. Comparative analysis of rainfall multiannual average values

The analysis of the distribution of the 2007 monthly minimum discharge values shows that in the upper part of the river basin the values were much higher compared to the discharge values with the ensurance of 80% and decreased towards the river basin lower part. Thus within the upper part the module coefficient registered values between 5,39 (at Bucova gauging station) and 3,17 (at Golet gauging station), values that decreased till 1,91 at Lugoj gauging station.

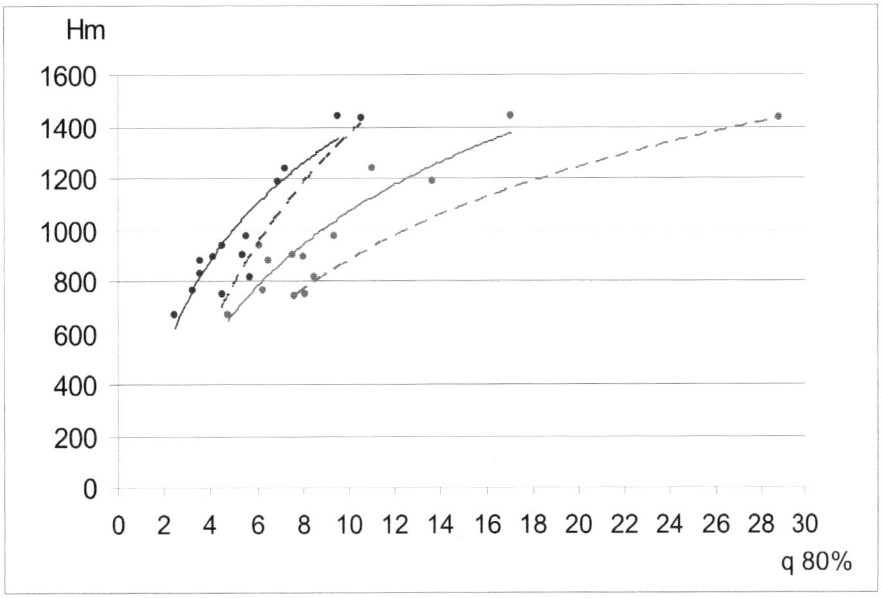

Figure 4. Relation between the multiannual minimum monthly average minimum flow with 80 % ensurance (__--) and the one in 2007 (__--)

3. Daily minimum average flow analysis

In order to be successful in our attempt, annual minimum daily flows were processed, which allowed us to establish the values on different ensurances for this element of the minimum flow. Values are presented in Table 2.

The role of the rainfall upon the minimum flow in this river basin area is considered not to be relevant, mainly due to the fact that the upper part has received during this very time spell a relatively permanent supply of the flow, but also due to the distribution of the water volumes from the above-mentioned reservoirs.

This is very well reflected if we consider the daily average flow graphic for the time period June 1-September 30 2007 at Lugoj gauging station (Figure 5), where no permanent decrease can be noticed due to exclusive supply from the underground water. This happened because of the frequent turbinations upstream and then evacuations from the Zervesti reservoirs – the buffer lake – that sloop the flows into the Sebes and then Timis River. We appreciate that in the absence of these conditions the flow would have registered more drastic decreases (marked with red line in Figure 5) than those represented in the figure, which could have led to lower flow values. This fact allows us to appreciate the positive role of the hydro technical works that redistribute the flow in time, category where the existing reservoirs in this part of the river basin can be included.

Otherwise this fact is more obvious if we notice the yearly minimum flow evolution registered between 1950-2007 at Lugoj gauging station – the station that controls the entire river basin area. As it can be noticed (Figure 6) the 2007 registered value, of 7.58 m^3/s (8.3% higher than the value for the 80% assurances) is the 15th value in a 58-year range. This proves that in this part of Banat area we cannot speak about a hydrological drought, even if for agriculture it has been such.

The not very low minimum flows value has also reflected into the ground waters level, which at fall end registered values close to multiannual average values.

If we consider the number of days with permanent decreasing or equal values this was of 4 to 6 and was registered several times in August and only once in September and October.

Table 2. Morphological elements and characteristic flows for the upper part of the Timis River

Nr.	River.	Gauging station	F km^2	Hm m	Multiannual average flow	Annual minimum average daily flow With ensurances 80%	Minimum daily average flow 2007
1	Timis	Teregova	167	901	2,28	0,298	0,378
2	Timis	Sadova	560	936	10,9	1,49	0,880
3	Timis	Caransebes	1072	765	17,5	2,20	2.29
4	Timis	Lugoj	2827	666	38,4	4,70	7,58
5	Raul Rece	Rusca	163	1184	4,64	0,785	0,668
6	Fenes	Fenes	125	973	2,82	0,491	2,20
7	Golet	Golet	41	751	0,686	0,202	2,29
8	Sebes	Tr. Ruieni	124	815	2.94		0,650
9	Bistra	Bucova	64,0	1236	1.65	0,344	0.550
10	Bistra	Voislova Bucova	232	892	3,71	0,510	1,38
11	Bistra	Obreja	863	880	17,1	2,28	3.30
12	Rusca	Voislova	166	1184	2,70	0.521	1.42
13	Bistra Marului	Poiana Marului	79	1442	2,71	0,605	1.10
14	Sucu	Poiana Marului	77	1430	3,00	0,593	1.20
15	Nadrag	Nadrag	35	742	0,374	0,092	0.240

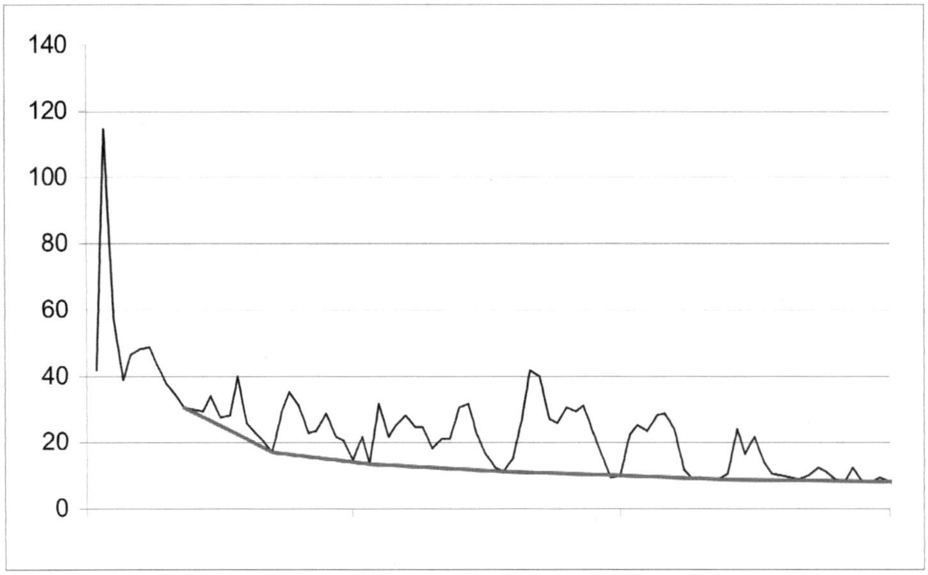

Figure 5. The evolution of the daily average flow values at Lugoj gauging station on the Timis River between June-September 2007

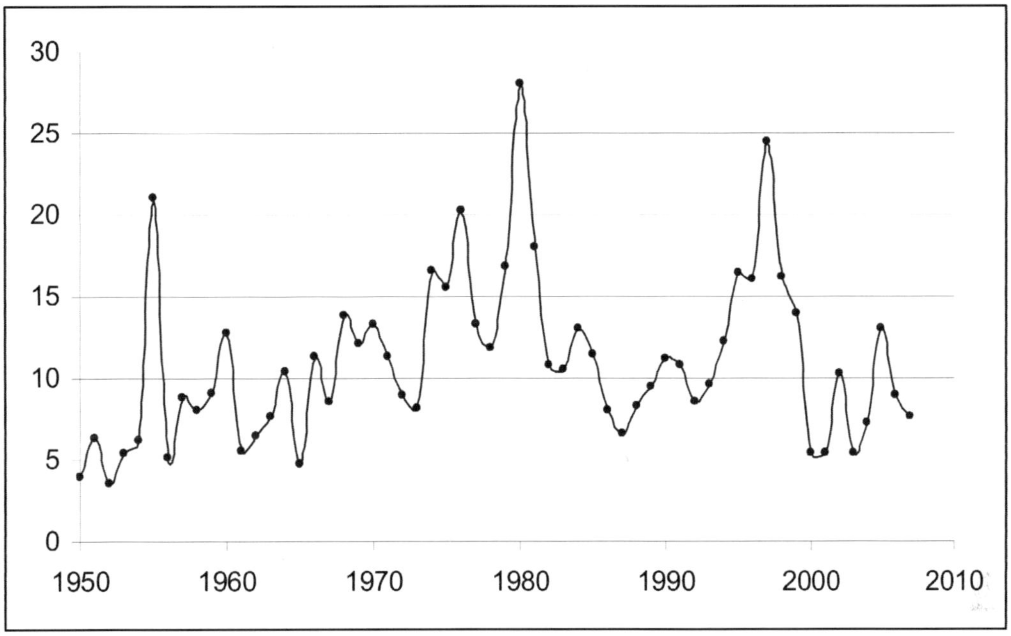

Figure 6. The annual minimum flow evolution between 1950-2007

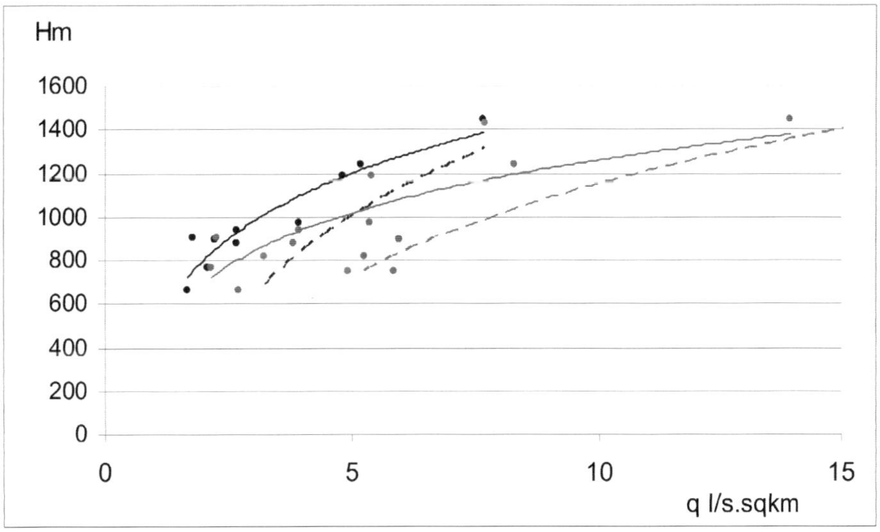

Figure 7. The relation between the multiannual minimum daily average specific flow with ensurance of 80% (__--) and that in 2007 (__--)

The interruptions were caused by some increases either due to sloops into the reservoirs, or of some flows increasing due to rainfall, the most important being the period 4 to 10 August.

If we analyse these values, considering the registered multiannual daily average values, we notice this was approx. 12% higher.

In what concerns their repartition over a time period, we can notice the fact that the lowest values occurred at the beginning of the 50s and mid 60s and as never before 4 years in a row with decreasing values were registered. (e.g. 1971-1974 and 1984-1987).

If we consider the last 10 years, we can notice that we have to deal in 4 cases with low values for the daily average values, 2 of them being lower than the values with the assurance of 80%.

4. Conclusions

Within the Timis river upper river basin we cannot speak about a hydrological drought, although there were periods with low water. These were interrupted by periods with rainfall (the number of rainy days during summer-fall was 37), which determined a good supply of the underground waters as well as the river- beds.

In the mean time we consider it led to a certain supply of the ground levels with positive effects on minimum flow values.

In this paper we did not present the reconstituted flow values due to the fact that these would have been lower (errors from 1.2% to 5.4%).

In order to register drought periods within the Timis River basin upper part it is necessary to register a major rainfall deficit (several months one after another – as it was the case in 2000, when the deficit of rainfalls was present during May-November); otherwise the minimum flow values will not fit into very high value assurances.

References

[1] Diaconu C and Serban P 1994 *Sinteze si regionalizari hidrologice* E.T. Bucuresti
[2] Stanciu P 2007 Caracteristicile secetelor hidrologice *Hidrotehnica* Bucharest **52** 22-24
[3] Topor N 1964 *Ani ploiosi si secetosi in Republica Populara Romana* (Bucharest: Meteorological Institute) p 304
[4] * * * 1970 *Raurile Romaniei* (IMH Bucharest)
[5] * * * Banat Water Branch Archive

Proving the ecosystem value through hydrological modelling

W Dorner[1], K Spachinger[1], M Porter[2] and R Metzka[1]

[1] University of Applied Sciences Deggendorf, Edlmairstr. 6+8, D-94469 Deggendorf, Germany, [2] University of Southern Queensland, Australia

E-mail: wolfgang.dorner@fhd.edu

Abstract. Ecosystems provide valuable functions. Also natural floodplains and river structures offer different types of ecosystem functions such as habitat function, recreational area and natural detention. From an economic stand point the loss (or rehabilitation) of these natural systems and their provided natural services can be valued as a damage (or benefit). Consequently these natural goods and services must be economically valued in project assessments e.g. cost-benefit-analysis or cost comparison. Especially in smaller catchments and river systems exists significant evidence that natural flood detention reduces flood risk and contributes to flood protection. Several research projects evaluated the mitigating effect of land use, river training and the loss of natural flood plains on development, peak and volume of floods. The presented project analysis the hypothesis that ignoring natural detention and hydrological ecosystem services could result in economically inefficient solutions for flood protection and mitigation. In test areas, subcatchments of the Danube in Germany, a combination of hydrological and hydrodynamic models with economic evaluation techniques was applied. Different forms of land use, river structure and flood protection measures were assed and compared from a hydrological and economic point of view. A hydrodynamic model was used to simulate flows to assess the extent of flood affected areas and damages to buildings and infrastructure as well as to investigate the impacts of levees and river structure on a local scale. These model results provided the basis for an economic assessment. Different economic valuation techniques, such as flood damage functions, cost comparison method and substation-approach were used to compare the outcomes of different hydrological scenarios from an economic point of view and value the ecosystem service. The results give significant evidence that natural detention must be evaluated as part of flood mitigation projects. In addition can be stated that the loss of detention due to land use and dikes can be called an externality and results in economic inefficiencies.

1. Introduction

Our environment provides different services and functions we are using on a daily basis. But what happens if we destroy these functions or restrict their ability to provide the services? Since Kyoto it is not new that also environmental problems are threats to our economy and destruction of environmental functions result in economic damages. Most of these services can be measured in economic terms as benefits or their loss as economic damages. To describe these effects requires a deep understanding of environmental dependencies and the knowledge how to value the results in economic terms. In different fields these approaches have been tested. The Kyoto Protocol is one of the applications where the interaction of environmental economics and scientific modeling resulted in a global approach to protect the environment.

Especially with regard to water it is a major concern that natural measures and the protection of the environment is confronted with economic considerations. This includes always the thought that nature stands besides economic interests. This paper does not propose a methodology to "economize" our environment, but provides a methodology and delivers evidence why it is economically necessary to protect riverine landscapes and catchments. It is an approach to couple hydrological and hydrodynamic models and economic analysis to quantify the amount of ecosystem services provided by natural river systems and flood plains and express it in monetary terms. As a form of ex ante analysis the proposed methodology can help to derive economic arguments to protect riverine landscapes. From an ex post point of view and as part of flood protection studies it provides economic arguments for river renaturalisation projects and dike re-allocations to increase the flood plain. Results of the chosen approach can be used in cost-benefit analysis or cost comparison analysis to provide a more complete economic picture of planned activities.

2. Economic theory

Different forms of water use like fresh water or sewage fees already show that in some cases water or water related services are valued. But in some cases these instruments seem to miss or fail. Water use or the use of water bodies can be subdivided into: direct and indirect use. Direct use includes all types of use where water becomes directly part of the product or is used in the process of production. Indirect use sums up all effects, where water is influenced as a side effect of the production, where the water "use" has no direct impact on the success or failure of the production process and the resulting product. Examples of direct uses are hydropower, farming in connection with irrigation or washing water for vegetables. A typical form of indirect use is erosion on agricultural sites. Excessive rainfall does not play a role for the success of agriculture. During such events most rain water is lost, because of low infiltration and fast surface run-off. But through erosive processes because of agriculture and the transport of sediments, nutrients and pesticides the farming process influences the water bodies.

Water can be used physically and chemically. A typical form of physical use of water is hydropower. The elevation energy of water is transformed into motion energy in a turbine, which is transformed to electricity. In sewer systems the transport capacity is used to remove sewage and waste. In power stations water is used to cool the system or transfer heat energy via steam to drive turbines. Classical chemical uses are the ability of water to serve as a solvent, either for other chemical substances (coloring, nutrition,...) or for detergents to remove dirt.

A discussion about the use of water should evaluate also two different forms of use:

1. Consumptive use like for irrigation or thermal power station cooling
2. modifying uses such as discharge of waste water

Dyck [7] names the main functions of water bodies as:

- self-purification,
- biological potential yield,
- ecologic potential,
- transport of material (solids or solved materials),
- potential energy (kinetic energy, heat capacity),
- recreational use and
- flood routing.

We can use a scientific classification scheme for the use of water, water bodies and ecosystems:

- Physical use, like gravitational energy, flow energy, transport capacity for subspended
- and solid substances and temperature.
- Chemical conditions like solvent for substances and fluids.
- Biological conditions, like self cleaning capacity.

This means that the direct use of water e.g. as potable water or for irrigation is only one of several use types or use function aquatic ecosystems provide. Emmerton [8] distinguishes between direct values (production and consumption goods), indirect value (ecosystems functions and services), option value (potential for future use), non-use value (like heritage value, cultural value, ...).

It is not only of importance to look at the use of water itself, but also the functions water bodies provide. Technical intervention and human actions in general do not only use the resource water, but lead to severe damages and irreversible changes in our environment. ``Wasser wird heute zumeist als eine unter vielen Ressourcen angesehen, die für die wirtschaftliche Entwicklung eines Landes notwendig ist und die sich für diese Zwecke nach Belieben aufstauen, umleiten und vermehren läßt. [Today water is seen as one of many resources, necessary for the economic development of a country and that can be detained, redirected and increased for this purpose.]'' [22, p. 165]

"Natural ecosystems yield a wide range of goods and services, many of which are related to either their demand for water or their role in water supplies, and most of which have an extremely high economic value" [8, p. 3]. Daily [3] points out that "Ecosystem services are the conditions and processes through which natural ecosystems, and the species that make them up, sustain and fulfill human life." Ecosystem valuation can be seen as a method to make peoples preferences to pay for ecosystem services visible. Using this valuation technique allows to make ecosystems or their services comparable to other goods and sectors [8]. Emerton [8] gives an overview of the benefits of wetlands. As part of the study "Integrating wetland economic values into river basin management" 12 projects from different continents are presented and evaluated [8]. Different functions of wetlands were identified and valued with regard to their economic value e.g. water purification, habitat and food reserve or flood detention.

From an economic point of view the relevant factor in decision making can be reduced to the well being of individuals or society. Cost-benefit analysis as an evaluation technique to compare different situations is based on this concept. The situation where the welfare, well-being or utility of an economic actor in a situation A is higher than in situation B would result in the preference for situation A. Preferences are measured as the willingness to pay for a certain benefit or willingness to accept compensation for a cost. [19]

For classical markets this fact is expressed in the market price. Classical market theory describes the distribution of goods which are dealt in a perfect market. Offer and demand are responsible for the market price as an equilibrium price. Actors are willing to pay or willing to sell a good if they can achieve a certain price. For a perfect market with numerous actors on both sides the equilibrium price describes the situation were all actors achieve an optimum outcome (see e.g. [17]).

This covers all goods that can be called private. But what if goods are not private and not traded at markets. Do they have no value? And what about the environment and environmental 'goods' and 'services'. Emerton [8, p. 3] states that "the problem is not that water ecosystems have no economic value, but rather that the value is poorly understood, rarely articulated and as a result is frequently omitted from decision –making".

For decision making in water resources management different economic instruments and methods are used like cost-comparison method or cost-benefit analysis (e.g. [15], [20]). Main problem is how to integrated the value of an ecosystem or its services into these calculations to provide a sound basis for decision making for distinct projects or for policy making and strategies e.g. on a catchment scale.

3. Hydrological background

The Kyoto Protocol is a first application of an interaction of economic instruments and sciences. But "Watersheds differ from air sheds, however, in such key aspects as confinement to a channel, non-uniform mixing, and downstream accumulation" [21, p. 410]. For an economic classification and valuation a lot of information about the regional situation, but also in general the hydrologic system interrelation is necessary: How much water is necessary for human and natural use? How much water can be expected for an area and a specific period? What quality will the available water have? What is the most efficient way to use land and regard the impacts on run-off? Its regional and temporal varying availability, the problem of ownership, often referred to as water being a public good, and its allocation make it difficult to evaluate water and aquatic ecosystems in economic terms. These are in most cases scientific questions that need to be evaluated before an economic assessment can take place.

Environmental structures and landscape provide several services that can be identified in technical terms. From the point of view of flood risk management the detention capacity is an important component of each flood protection project. Natural detention is either regarded as given for status quo scenario, if evapotranspiration, infiltration, base flow, flood routing and other types of effects of natural structures are considered. As an extension of these classical hydrological modeling principles used e.g. in engineering hydrology, also the consequences of a loss of these effects or impacts of a redevelopment of these structures can be calculated.

Two ways can be distinguished to identify the benefit of natural detention:

- The substitution approach
- Damage analysis

Natural detention can be substituted by technical infrastructure like levees or reservoirs. Costs for building and maintaining these structures can be seen as the costs to substitute lost natural detention and therefore an equivalent for the value of natural detention. If no counter measures are taken against floods increased flood due to lost natural structures will result in increased flood damages. Also this increase of damages can used as an expression of the value of natural detention or vice versa the avoided damage would be part of the ecosystem service expressed in monetary terms.

We know that human action in the catchment and along the river system affects flood development and consequently the peak, volume and duration of a flood. Maniak [16, p. 10] states ``Beim Ausbau oberirdischer Gewässer wurden vielfach die natürlichen Rückhalteräume in der Talaue verkleinert, um die Landwirtschaft gegen Sommerhochwasser und die Siedlungsgebiete gegen noch größere Hochwasser zu schützen. Dies führt zu Abflussverschärfungen mit größeren Hochwasserspitzen in den unterliegenden Gebieten." [As the river systems were developed the natural detention storage was often reduced in the flood plain. Measures were required in order to protect agriculture from summer floods and the settlement areas against increased floods. There has been an increase in flood water levels and peaks in the lower catchment areas.].

Especially in small scale catchments these quantitative relations between land use and water in the literature have been proven [1, 13]. But linkages between land use, river training and flood development can be shown for larger catchments as well ([14], [5], 18). Dyck [7] describes the enormous losses in flood capacity along the Elbe in Germany over the past 800 years and refers to the inadequate or token attempts to construct flood detention works to compensate these losses within the past 100 years. He states "Infolge Flußregelung und Deichbau haben sich die Retentiosflächen vieler Flüsse verringert. Dies konnte auch durch Rückhaltebecken und HW-Schutzräume in Talsperren meist nicht kompensiert werden. [In-stream works and levee construction have reduced the flood capacities of many river systems. It was not possible to balance this by detention reservoirs or flood storage in dams.]" [7, p. 433].

4. Methodology

This paper will concentrate on the detention function of landscape and river systems with regard to flood mitigation. Other aspects and services provided by the described environmental systems are ignored and would go beyond the focus of this paper. Some of them can be named as reduced erosion and protection of water quality, habitat function, visual quality of landscape and have been partially covered by other authors (e.g. [8]).

Different methods are described in literature that can be used to value public goods and ecosystem services. For goods where real values are difficult to estimate because no comparative goods are available or markets are missing instruments like contingent valuation or choice modeling can be used. In the case of flood protection or flood mitigation substitutes like artificial flood protection works are available, which can be calculated and used to quantify the effect and resulting costs e.g. for building and maintenance. Another alternative are increased damages resulting from floods.

IOP Conf. Series: Earth and Environmental Science **4** (2008) 012019 doi:10.1088/1755-1307/4/1/012019

4.1. Herzogbach catchment study

The study was performed for a small rural catchment the Herzogbach rivulet (70 km²) in Lower Bavaria, a small tributary of the Danube. The catchment is partially located in the Tertiary Hilly Landscape and the fertile plain of the Gäuboden. During the last 20 years landscape structures in the catchment were heavily modified as a result of rural development and land consolidation. Over 90% of all rivulet reaches are heavily modified and lost their natural flood plain.

In this case the detaining effect of land cover and the natural flood plain was compared with technical detention structures and flood damages in settled areas resulting from the loss of natural detention. A combination of hydrological and hydrodynamic modeling was used to assess:

1. Run-off, volumes and peaks of flood waves for different scenarios of land use and river structure (current use vs. pristine catchment)
2. Necessary technical detention volume to mitigate floods
3. Extent and intensity of floods in settled areas

Four hydrological scenarios were evaluated

a. Status quo of land use and river structure
b. Pristine landscape, but river structures of the status quo
c. Pristine landscape and river structures
d. Agricultural land use sustainable techniques like intermediate crops or direct cropping

In a next step costs for building reservoirs were calculated and the flood damages in settled areas were estimated using a simplified water-level-damage function for residential buildings. Only direct damages to buildings and furniture were taken into account, because of easy evacuation possibilities, the rural structure, missing businesses and major infrastructure causing significant indirect damages or other forms of direct damages.

4.2. Flood plains of the Lower Vils

In a second case the effect of natural flood plains was assessed on a local level using hydrodynamic models. The Lower Vils is a lower section of the Vils river with a catchment of about 1000 km² and is located in an intensively used hilly landscape. Especially the floodplains provide fertile and flat areas for farming. During the last 50 years main sections of the Vils were modified. Dikes protect now settlements as well as agricultural areas. A small 100 meter wide flood plain remained of the former 500 to 1000 meter wide natural flood plain. At the Lower Vils the main question was how to improve the flood protection for settled areas. Therefore the status quo of land use and dikes was compared to different scenarios.

Scenario 1 was an increase of dike level for settled areas or a new construction with a higher elevation. From an ecological viewpoint the situation in the flood plain would have remained unsatisfying, resulting also in major annual maintenance works for the heavily modified river bed due to erosion and sedimentation.

Scenario 2 was a minimum increase of the flood plain, where old agricultural levees reduced the detention capacity and restricted the river bed (deconstruction of agricultural levees). This development would result in a wider flood plain and sufficient space for renaturalisation of most sections of the river.

In Scenario 3 a total removal of flood protection works for agricultural sites was assessed. Due to high costs for the purchase of land to establish a new flood plain and necessary for a renaturalisation of the river a major question was weather and after which period the decrease of building costs for protection and maintenance would compensate purchase of expensive agricultural land. The effectiveness of the three scenarios was compared using 2D hydrodynamic models to quantify water levels, flow velocities and bed share for two given hydrological input scenarios as design floods with a probability of 5 and 100 years. Cost comparison method was used the economically compare the three scenarios.

5. Results

5.1. Herzogbach catchment

Bachling is a typical small village in the steeper upper part of the Herzogbach catchment (Tertiary Hilly Landscape). In the hydrologic model only a small sub-catchment is of relevance for the run-off and flood development in the settlement. It has a size of 1.2 km^2. 72% of the catchment consists of fields, mainly used for root crops and vegetables without intermediate crops (result of inspections in summer and early autumn 2004 and 2005). 15% of the catchment is covered by forest, 12% by grassland and less than 1% is sealed and impervious area. The rivulet is straight and with a deep river bed between 1 and 1.4 m below the natural surface.

Within the village, the river bed lies between 1.2 and 1.4 m below the level of the streets or the courts. This section is only represented by the regionalisation approach and not a flood routing approach because of the short length of the rivulet in this sub-catchment. This partially limits the precision of results for river elongation and renaturalisation. The peak flow of 1.5 m^3/s for a 100 year flood event is mainly influenced by land use in the status quo scenario (Scen. A). The catchment with natural land cover (Scen. B) results in a simulated peak flow of 1.2 m^3/s, while a natural river structure (Scen. C) only causes an additional decrease of 0.1 m^3/s to 1.1 m^3/s. The insertion of intermediate crops (Scen. D) causes an decrease of only 0.1 m^3/s to 1.4 m^3/s in comparison to Scenario A.

The changes in river structure and land use have a significant impact on the shape of the flood wave. Land use increased the volume of the 100 year flood wave by 0.005 M m^3 to 0.019 M m^3 (Scenario A). The use of intermediate crops decreases the volume by 0.002 M m^3 to 0.019 M m^3 (Scenario D).

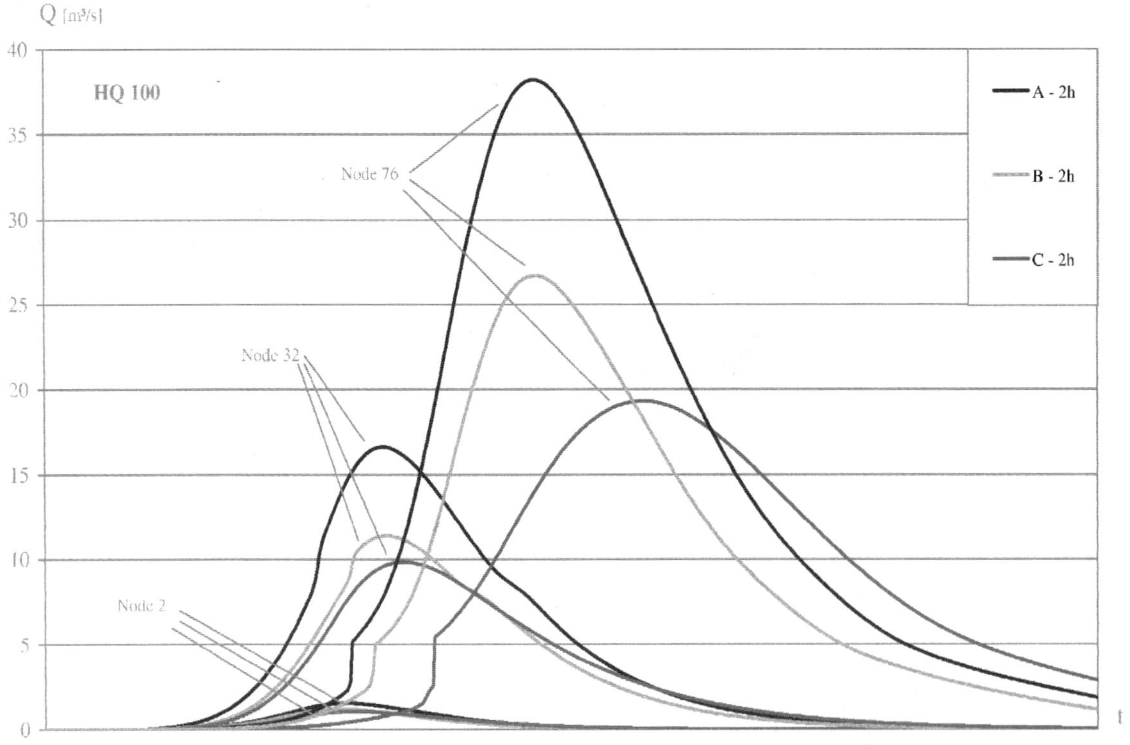

Figure 1. Development of flood waves at three different sections (nodes) of the catchment (2 upper catchment, 32 middle section, 76 close to the outlet) for three hydrological scenarios

In addition a flattened wave would result from a renaturalisation (Scenario C) and decrease the volume necessary for flood detention. A potential pond has a maximum available volume of 6,600 m³. Damages in the village start with a discharge of 0.6 m³/s. For scenario A a volume of 3,900 m³ would be necessary to mitigate a flood wave of 1.5 m³/s to 0.6 m³/s. For scenario B the same reduction could be achieved already with a storage capacity of 2,000 m³. In scenario C a volume of 1,800 m³ would be necessary. Assuming average building costs for detention volume of 30 Euro/m³ this will increase building costs from 54,000 Euro for scenario C to 60,000 Euro for scenario B and 117,000 Euro for scenario A. Splitting up these costs per hectare of farmland means in this catchment 580 Euro/ha for scenario B for 86 ha of farmland in this subcatchment causing this extra runoff. For scenario A the extra costs of detention in contrast to scenario C are 730 Euro/ha. The difference of buildings costs of 6,000 Euro between scenario B and C would represent the extra costs to compensate the effects of river training. For a channel length of 1.1 km in this particular subcatchment this means extra costs of 5.45 Euro/m of channel.

Buchhofen is a settlement in the small Herzogbach valley in the centre of the catchment. It developed firstly along the valley axis but in recent years also towards the flood plain. At this point of the catchment (14 km²) agriculture (69%) is the predominant form of land use. followed by forestry (17%) and grassland (13 %). Sealed and impervious areas are below 3 % of the size of the relevant catchment. Land use as well as river training show a significant impact on the shape and peak of the flood wave. River training increased run off by 3 m³/s (scenario C) to 13 m³/s (scenario B) while land use increased the peak by an additional 4 m³/s to 16 m³/s (scenario A) (Fig. 1).

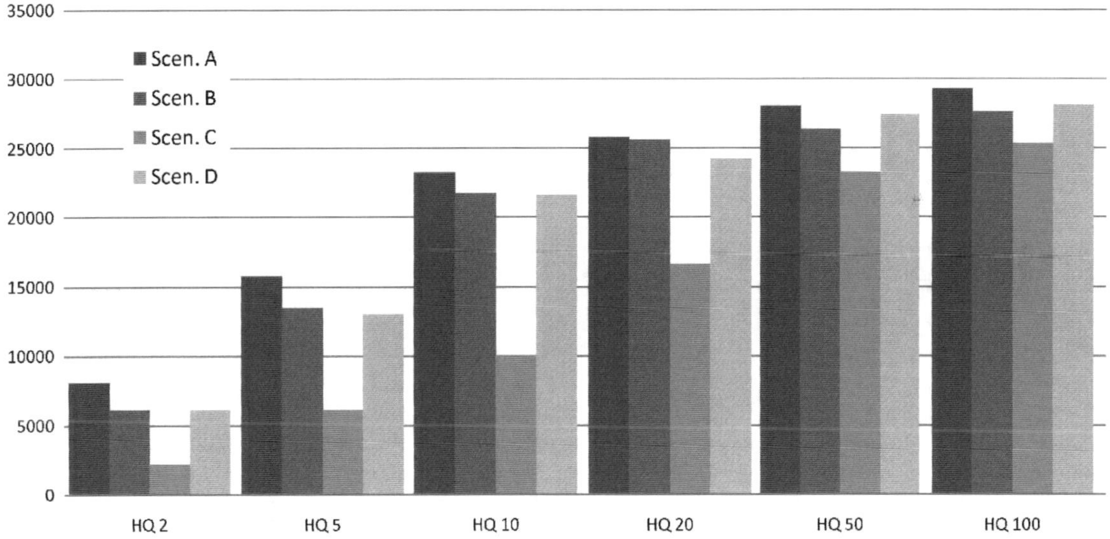

Figure 2. Affected square meter of buildings for the different hydrological scenarios and floods of different probabilities.

Buchhofen is a rural village, but also the centre of an independent municipality. It provides services like city hall, restaurants and shops. Fortunately main businesses and municipal infrastructure are located in the older parts of the village along the valley hillsides. The flood plain is mainly occupied by residential areas and farm estates. Unfortunately the whole village is oriented along the Herzogbach valley. The number of bridges and other intersecting buildings increase the danger from backwater effects and the number of buildings located in the river valley increases the total damage costs.

The distribution of affected building base areas (Fig. 2) shows that land use and river training affects mainly small and common floods events. But the distribution of affected buildings is also a result of the geomorphology of the flood plain, the distribution of buildings in the flood plain and intersecting buildings. In Buchhofen mainly backwater processes are responsible for flooding. Once the discharge limit of an intersecting building is reached it results in a filling of the valley respective the flood plain. The wide valley situation results a flooding. Due to the high number of bridges and pipes in Buchhofen and their restricted diameter this level reached early. The form of the valley restricts also the amount of buildings affected to those at the bottom of the valley.

Buchhofen shows very well the increasing effect of high probability low loss events. Small flood event cause the majority of damages, calculated over a 100 year period (Fig. 3). Although natural detention has only a minimal effect on larger flood events and resulting damages, it can reduce the total accumulated damages.

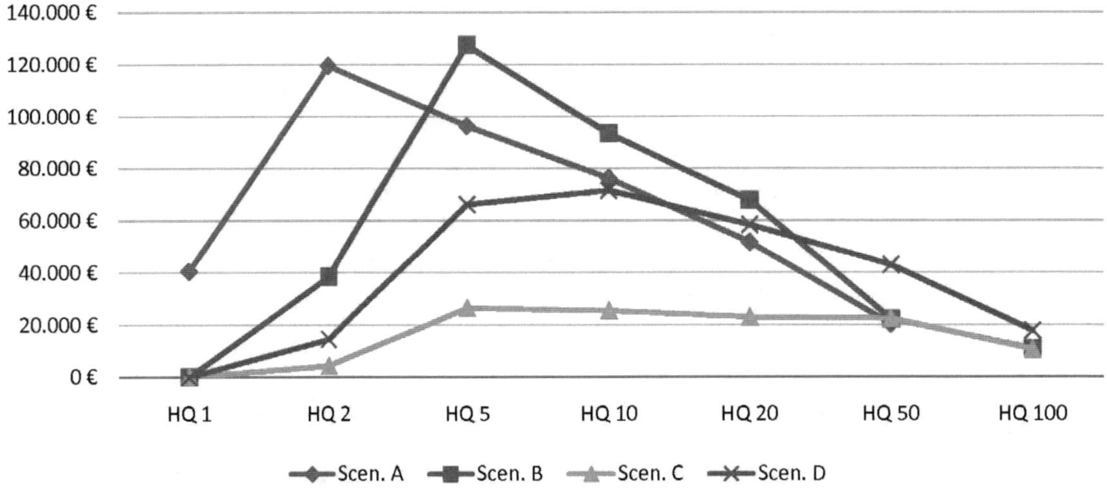

Figure 3. Damage function of the village Buchhofen in the central section of the catchment as cumulated damages per year over a 100 year period.

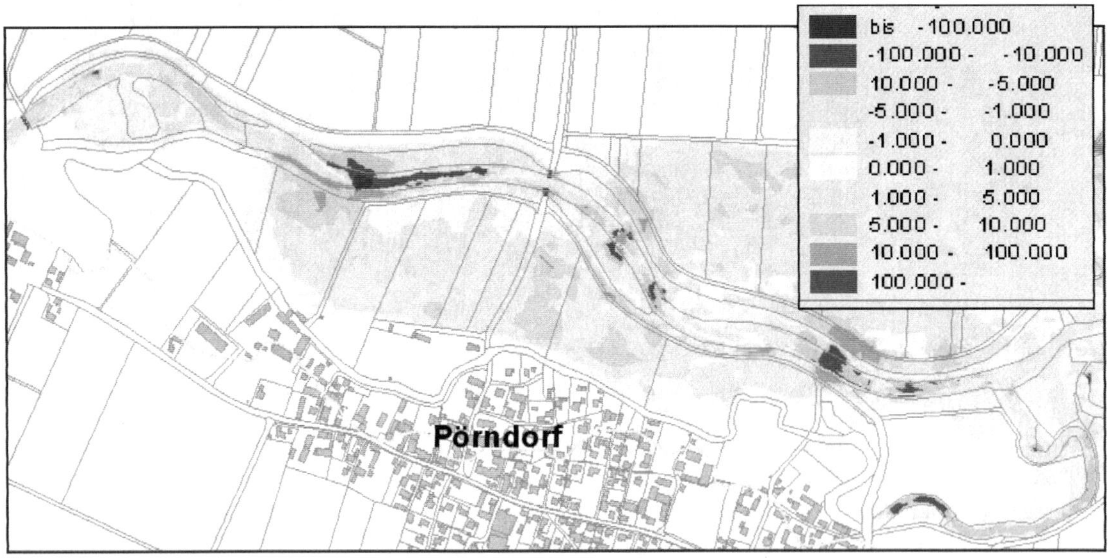

Figure 4. Change of the shear stress for the scenarios 1 and 2 in N/m².

5.2. Lower Vils

As expected the opening of the former flood plain would result in a decrease of water levels. Already the medium scenario would be sufficient so old levees for settled areas could be used again and would provide protection against a 100 year design flood. As an additional benefit a significant decrease of shear stress (Fig. 4) was identified resulting in lower maintenance costs in comparison to the status quo. This was included in the cost-comparison analysis.

From an economic view point the high costs for the purchase of land in scenario 2 (middle) and 3 (maximum) (Fig. 5) would be compensated by the reduced building and maintenance costs over 40 years (Fig. 5). This shows that renaturalisation can be an efficient solution for flood protection and as a substitute for hydro engineering works from an economic and technical point of view if project costs are considered.

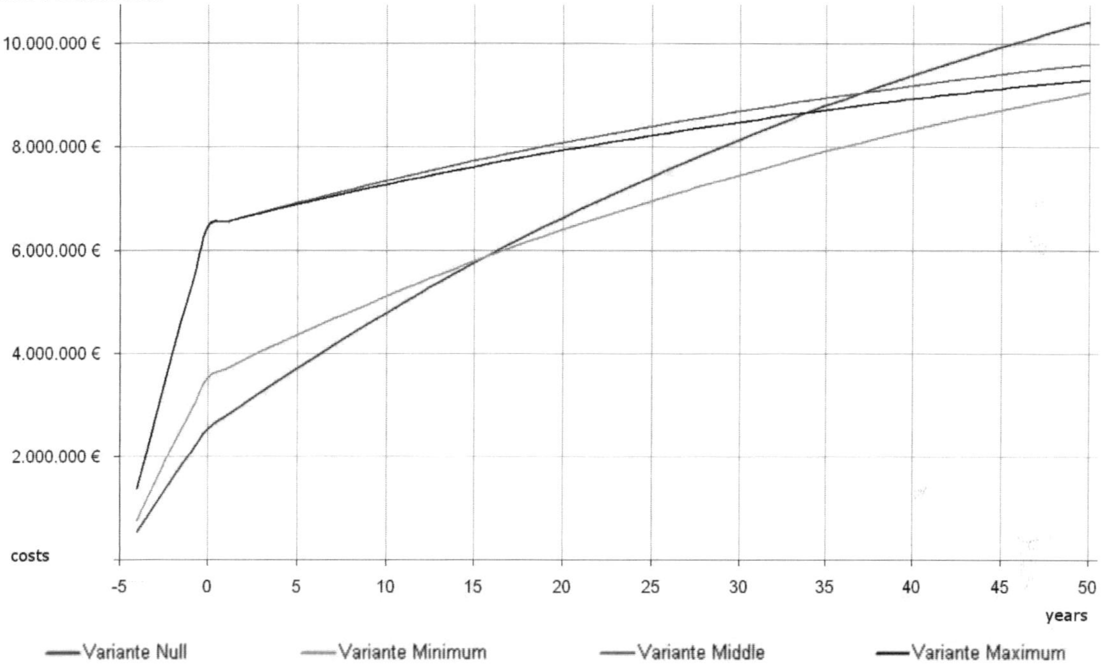

Figure 5. Comparison of the project costs (building, purchase of land and maintenance) for the three scenarios (Minimum, Middle and Maximum) as well as the Null-Scenario

6. Conclusion

In a next step it would be necessary to compare full economic costs and benefits including also other ecological and societal aspects into the equation as part of a cost-benefit equation. But it can be stated in general that different methods can be used to economically quantify the benefit of natural river structures with regard to flood protection. These methods can be used to make river renaturalisations technically and economically comparable with technical flood protection structures. Outcomes can be used in flood protection studies and as input for cost comparison method [15, 20] and cost-benefit analysis [11]. With regard to the economic targets and tasks of the European Flood Directive [10] and Water Framework Directive [9]must be stated that also the loss of natural riverine landscapes and other run-off relevant landscape structures in the past must be critically reviewed and described approaches and concepts are important means to develop the most efficient combination of measures to fulfil the requirements of the directives. Because of the unidirectional dependencies in river basins impacts in the upper catchments impose an economic threat to downstream riparian land owners and users [6] which can result in additional economic externalities that can be evaluated similar approaches.

For the two cases can be concluded that natural river structures provide a significant benefit. It needs to be taken into account in decision making. A major problem is that human interventions over the last 20 to 50 years changed the hydrological characteristic of a lot of river systems. These impacts along long river sections and of large parts of the catchment have from a hydrological view point a cumulating effect. Today's projects for renaturalisation and flood protection only affect small sections of the river system. The cumulating effect of renaturalisations and natural measures is often ignored, because they go beyond the scope of punctually oriented flood protection studies. Therefore natural detention measures and concepts are ignored on a catchment scale.

The complex environmental interactions within river basins and their scientific basis e.g. for modelling and simulation in combination with the growing economic pressures and legal requirements are asking for a new interdisciplinary approaches. Hydroeconomics could be seen as an emerging new scientific branch combining scientific understanding and description of environmental and anthropogenic processes regarding hydrology, hydraulics and aquatic ecosystem sciences and combining it with economic theory and instruments for better decision making.

7. Acknowledgements

The paper presents aspects of a study that was submitted as a Doctoral Thesis at the University of Southern Queensland (Australia). Hydrological and hydraulic data of the study Herzogbach catchment was derived from results of research project "Herzogbach-Angerbach-Lindenbach River basin management in small rural catchments" which was co-financed by the Bavarian Government represented by the Distric government of Lower Bavaria, the city Osterhofen and the municipalities Buchhofen, Wallerfing and Künzing. The results of the project Lower Vils are part of the Interreg project ILUP – Integrated Land Use Planning and River Basin Management which was co-financed by the European Union Interreg IIIB CADSES and the Bavarian Ministery for the Environment, Health and Consumer protection.

References

[1] Bormann H, Diekkrüger B and Hauschild M 1999 Impacts of landscape management on the hydrological behaviour of small agricultural catchments *Phys. Chem. Earth* B **24(4)** 291 – 6

[2] Croke B F W, Merritt, W S and Jakeman A J 2003 A dynamic model for prediction hydrological response to land cover changes in gauged and ungauged catchments *Journal of Hydrology in press.*

[3] Daily V G C 1997 *Nature's Services: Societal Dependence On Natural Ecosystems* (Island Press)

[4] Daily V G C 1997 *Introduction: What are ecosystem services* in Daily V G C 1997 *Nature's Services: Societal Dependence On Natural Ecosystems* (Island Press)

[5] de Roo A, Odijk M, Schmuck G, Koster E and Lucier A 2001 Assessing the effects of land use changes on floods in the Meuse and Oder catchment *Phys. Chem. Earth* B **26(7-8)** 593–9

[6] Dorner W, Porter M and Metzka R 2008 Are floods in part a form of land use externality *Natural Hazards and Earth System Sciences* **8(3)**

[7] Dyck, S. 1995, Grundlagen der Hydrologie, 3 edn, Verlag für Bauwesen.

[8] ed Emerton L 2005 *Values and Rewards: Counting and Capturing Ecosystem Water Services for Sustainable Development* (IUCN)

[9] European Parliament and Council of the European Union 2000 Directive 2000/60/ec of the European parliament and of the council of 23 October 2000 establishing a framework for community action in the field of water policy, Technical report *Official Journal of the European Communities*

[10] European Commission 2006 *Proposal for a directive of the European parliament and of the council on the assessment and management of floods* SEC(2006) 66 Technical report,

IOP Conf. Series: Earth and Environmental Science **4** (2008) 012019 doi:10.1088/1755-1307/4/1/012019

European Commission.
[11] Hanusch H 1994 *Nutzen-Kosten-Analyse Verlage Vahlen*
[12] Kalof V L and Satterfield T 2005 *The Earthscan Reader in Environmental Values* Earthscan
[13] Koehler G 2005 *Vorbeugender Hochwasserschutz durch Maßnahmen in der Fläche, in 'Wasserwirtschaft in der Fläche'* (DWA Landesverband Bayern)
[14] Lammersen R, Engel H, van de Langemheen W and Buitveld H 2002 Impact of river training and retention measures on flood peaks along the rhine *Journal of Hydrology* **267** 115 – 24
[15] LAWA – Länderarbeitsgemeinschaft Wasser 1998 *Leitlinien zur Durchführung dynamischer Kostenvergleichsrechnungen (KVR-Leitlinie)* (Länderarbeitsgemeinschaft Wasser)
[16] Maniak U 1993 *Hydrologie und Wasserwirtschaft* 3 edn (Springer Verlag)
[17] Mankiw G N 2003 *Principles of economics* (Dryden press)
[18] Niehoff D, Fritsch U. and Bronstert A 2002 Land-use impacts on storm-runoff generation: scenarios of land-use change and simulation of hydrological response in a mesoscale catchment in SW-Germany *Journal of Hydrology* **267** 80–93
[19] Pearce D, Atkinson G and Mourato S 2006 *Cost-Benefit Analysis and the Environment* (OECD)
[20] Schmidtke R 1981 *Monetäre Bewertung wasserwirtschaftlicher Maßnahmen – Systematik der volkswirtschaftlichen Nutzenermittlung: Ergebnisse eines interdisziplinären Forschungsvorhabens* (Bayerisches Landesamt für Wasserwirtschaft)
[21] Thurston H W, Goddard H C, Szlag D and Lemberg B 2003 Controlling storm-water runoff with tradable allowances for impervious surfaces *Journal of Water Resources Planning and Management* **129(5)** 409–18
[22] Wallacher J 1999 *Lebensgrundlage Wasser* (München)

IOP Conf. Series: Earth and Environmental Science **4** (2008) 012020 doi:10.1088/1755-1307/4/1/012020

Water Balance of Slovenia 1971 – 2000

P Frantar[1] , M Dolinar[1] and B Kurnik[2]

[1] Environmental Agency of the Republic of Slovenia, Ljubljana, Slovenia, [2] Joint Research Center, Institute for Environment and Sustainability, ISPRA, Italy

E-mail: Peter.Frantar@gov.si, Mojca.Dolinar@gov.si, Blaz.Kurnik@jrc.it

Abstract. The water is becoming more and more valuable natural resource. The increasing water demand and climate changes are making water a precious and not always available valuable. The water balance is the most appropriate way to make a full overview of water cycle in Slovenia, to find general information about hydrological characteristics of drainage basins, precipitation, evaporation and runoff. The article presents the methodology and the results of the Water balance project of Slovenia. Slovenia has the geographical position at the juncture of 4 main European georegions: The Alps, the Panonian Basin, the Mediterranean and the Dinaric Mountains. This makes the territory very diverse also from a hydrological point of view. Our major watershed divides the precipitation runoff into two watershed areas – the Adriatic Sea and the Black Sea. Due to this watershed almost all the Slovenia's rivers have headwaters in our territory. Water balance is calculation of water inputs and outputs over the defined area. The basic elements of the water balance include all the inflows and outflows for a given basin and serve for the computation of the water regime of a catchment area. It is defined by the parameters precipitation (P), evaporation (E), discharge (Q) and the change of the water reserves (dS). Main results of the water balance elements for the 1971 – 2000 period for Slovenia are: Average annual precipitation in Slovenia is 1579 mm, average annual evapotranspiration is 717 mm and calculated runoff is 862 mm. Compared to water amounts in the World, where the average precipitation is 750 mm, evapotranspiration is 480 mm and runoff is 270 mm, Slovenia shows an abundance of water quantities. Also the runoff coefficient with 55 % is much higher as 36 % of the world. The major questions remain if we are capable to live with this water amounts within the limits of sustainable development and what will be the effects of climate change to water balance.

1. Water balance calculation methodology

The water balance calculation is based on the circulation of water. It assesses the quantities of water in a certain area over a certain time period. It must consider all the inflows and outflows as well as changes in storage. For simple systems, such as a container or water reservoir with measurable inflow and outflow, the balance is straightforward and easily understood. The small water cycle schematic, where the main "inflow" is precipitation and the main "outflow" is evaporation, is also straightforward [8]. The water balance of a selected area, such as a country, region, etc., is much more complicated. Such water balance is invariably a simplified depiction of the actual conditions that covers the essential water balance elements and accurately portrays the relations between them.

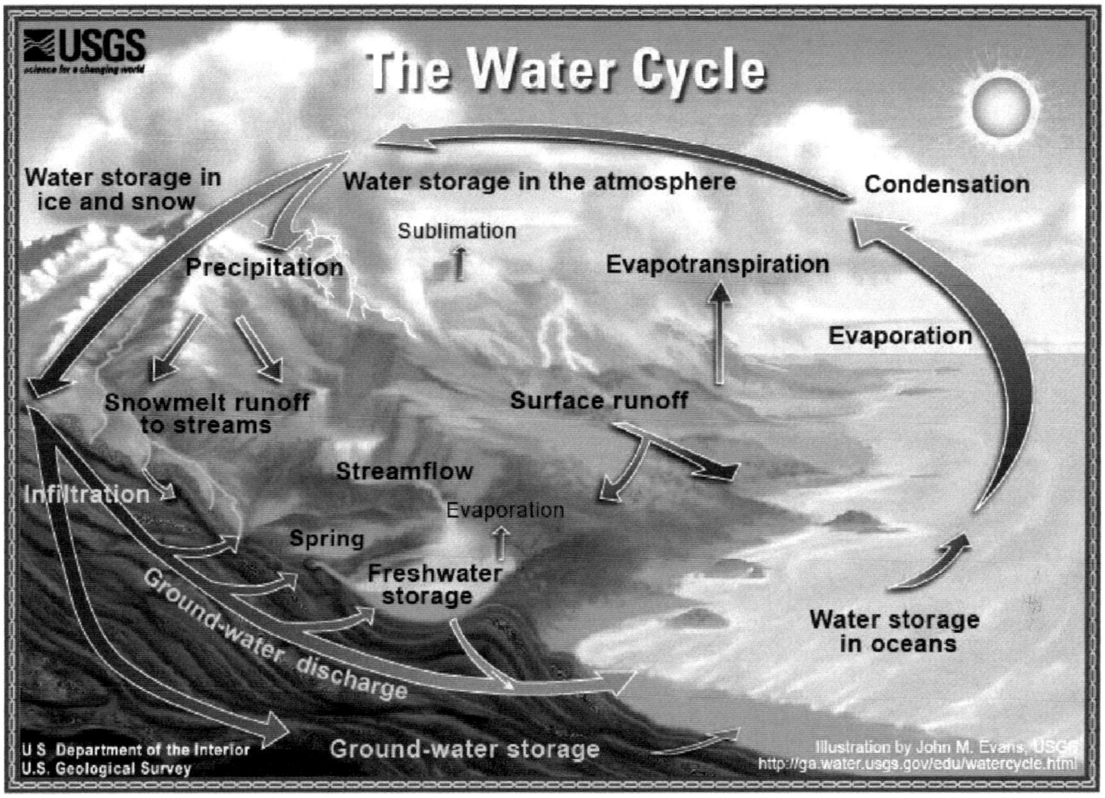

Figure 1. The circulation of water on Earth [5]

The basic water balance equation is based on the circulation of water between the atmosphere and the surface of the Earth [10] [6]:

Precipitation (P) = Runoff (Q) + Evaporation (E) + Changes in the storage (dS)

According to the definition, precipitation is atmospheric water that exits from the air either by condensation or sublimation and falls to or towards the ground because of gravity [9]. The term evaporation signifies the transfer of water into vapor from areas of open water, while transpiration means the transfer through vegetation. The term evaporation here encompasses both forms, which are together called evapotranspiration. When analyzing the water balance, we must limit ourselves to the analysis of the areas – to hydrometric catchment areas – whose runoff can be measured as the water flow rate (Q) of the water gauging station.

In addition to precipitation, it is necessary to consider other inflows of water (Qi) when dealing with the water balance of a selected area. The equation is thus as follows:

P (precipitation) + Qi (inflow)
=
Qo (runoff) + E (evaporation) + dS (changes in the storage)

In the balance for the 1971–2000 period we have not taken into account changes in water storage (dS) as we assume that these can be neglected over a long-time period [4].

For the main building block of the water balance the hydrometric catchment area was used. This area is delimited by the water divide of the balance cross-sections. Aside from precipitation, the

IOP Conf. Series: Earth and Environmental Science 4 (2008) 012020 doi:10.1088/1755-1307/4/1/012020

headwater hydrometric catchment areas do not have other inflows. The precipitation surplus (part of precipitation that does not evaporate) simply flows out of them and is measured as a discharge (Q). Intermediate hydrometric catchment areas receive water from both precipitation and inflow from the upstream hydrometric catchment area. Therefore the correct spatial delimitation of the hydrometric catchment area is very important.

We analyzed and reconciled the water balance elements of precipitation, evaporation and runoff for the selected areas. The runoffs derived from the discharges at water gauging stations (Q) were compared with the runoffs calculated using the water balance equation. The correctness of the relationships between the elements of the water balance was reviewed with the help of balance error analysis.

2. Precipitation

In the 1971-2000 reference period, we measured the quantities of precipitation at 394 stations. We used data from stations with at least 25 years of operation within the period treated for the analysis. To achieve a greater spatial coverage with the stations, we included 8 additional meteorological stations with shorter time series. Finally, there were 201 precipitation stations included in the analysis. For the calculation of the spatial distribution we also used measurements from totalizer rain gauges in 18 locations in the Julian Alps, the Karavanke Mountains and the Dinaric Alps available for the period treated. In addition to data from the stations in the territory of Slovenia, also the data on the daily precipitation from 29 stations located in Austria, Croatia and Italy were included.

The quantities of precipitation are measured in Slovenia using the Hellmann rain gauge. These measurements are partially underestimated the amount of precipitation due to various effects: evaporation, the wetting of the walls of the Hellmann rain gauge and the effect of the wind blowing the precipitation away from the rain gauge [11]. For the purpose of calculating the water balance, we corrected the measured precipitation, considering the effect of the wind, the precipitation intensity and the wetting of the Hellmann rain gauge [7][3]. The missing data was interpolated on a daily and monthly level based on the values from neighboring stations. The average correction factors for the precipitation stations are similar to those calculated for the 1961-1990 period [6]. Differences occurred at some of the individual precipitation stations, but no systematic deviations could be observed.

The differences between the corrected and measured precipitation values are highest in the mountainous regions and the smallest in western and south-western Slovenia.

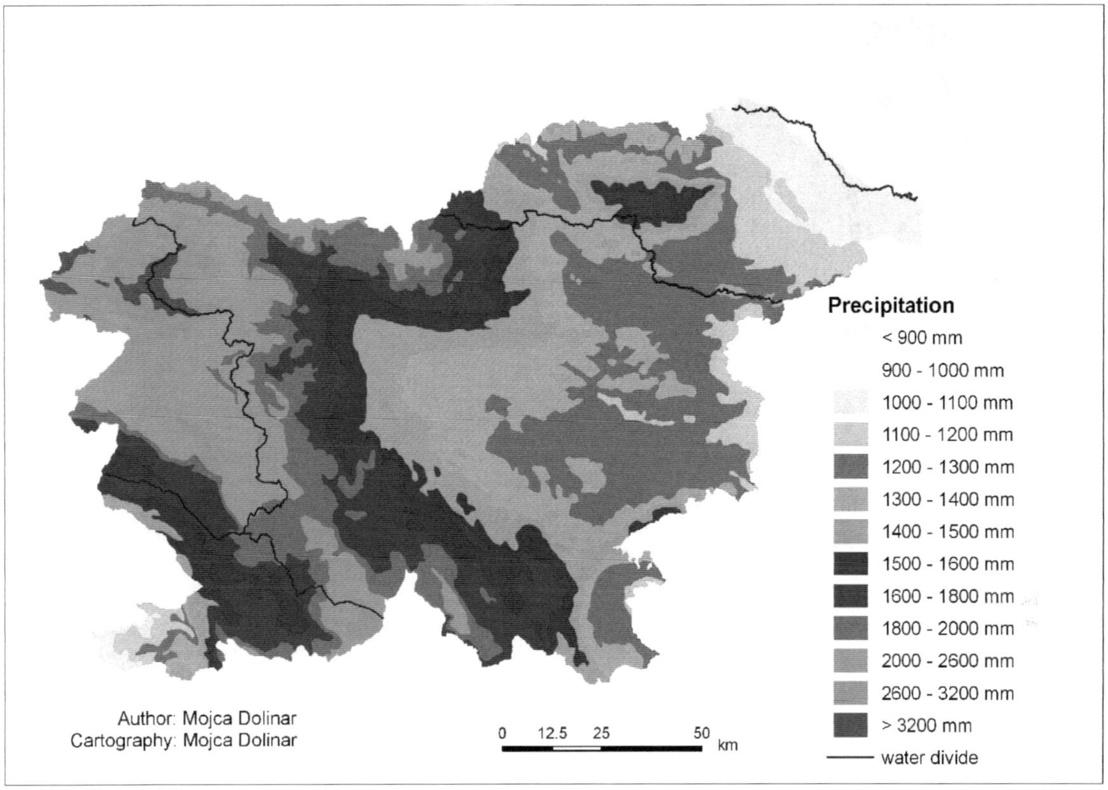

Figure 2. Precipitation in period 1971-2000

3. Evaporation

Evaporation or evapotranspiration (ETP) is the transfer of water in the form of water vapour from the water surface, the ground and through plant stomata into the atmosphere (Allen, 1998). Evapotranspiration is a process that combines evaporation from un-vegetated ground and water surfaces with the transpiration from plants. We have calculated the actual, real evaporation for the 1971-2000 period, which is given in the average height of the water column in mm.

The real evapotranspiration depends on the variety of plant, the phenological phase of the plant, the ground moisture available to the plants and the meteorological conditions, among which the air temperature affects evaporation the most, followed by relative air humidity, wind speed and solar radiation.

The calculation of the real evaporation was performed for 37 climatological stations using the modified Hargreave's method balanced for Slovenia with linear regression coefficients concerning their daily value of potential evaporation according to the Penman-Monteith method [1]. This calculation applies to the warm part of the year and to well-wetted ground covered with grass. Because of the difference in evapotranspiration from different types of land cover (forest, farmland, etc.), the values obtained for potential evaporation were corrected by using standard correction coefficients for individual layers of land cover with respect to the potential evaporation.

The calculated average annual values of the real evaporation for individual meteorological stations vary from 355 mm at the station on Kredarica meteorological station to 845 mm measured at the Bilje pri Novi Gorici meteorological station. Evaporation is strongly affected by elevation, so it also depends on the geographical position of the station. Using interpolation, we produced a map of the spatial distribution of the average evaporation.

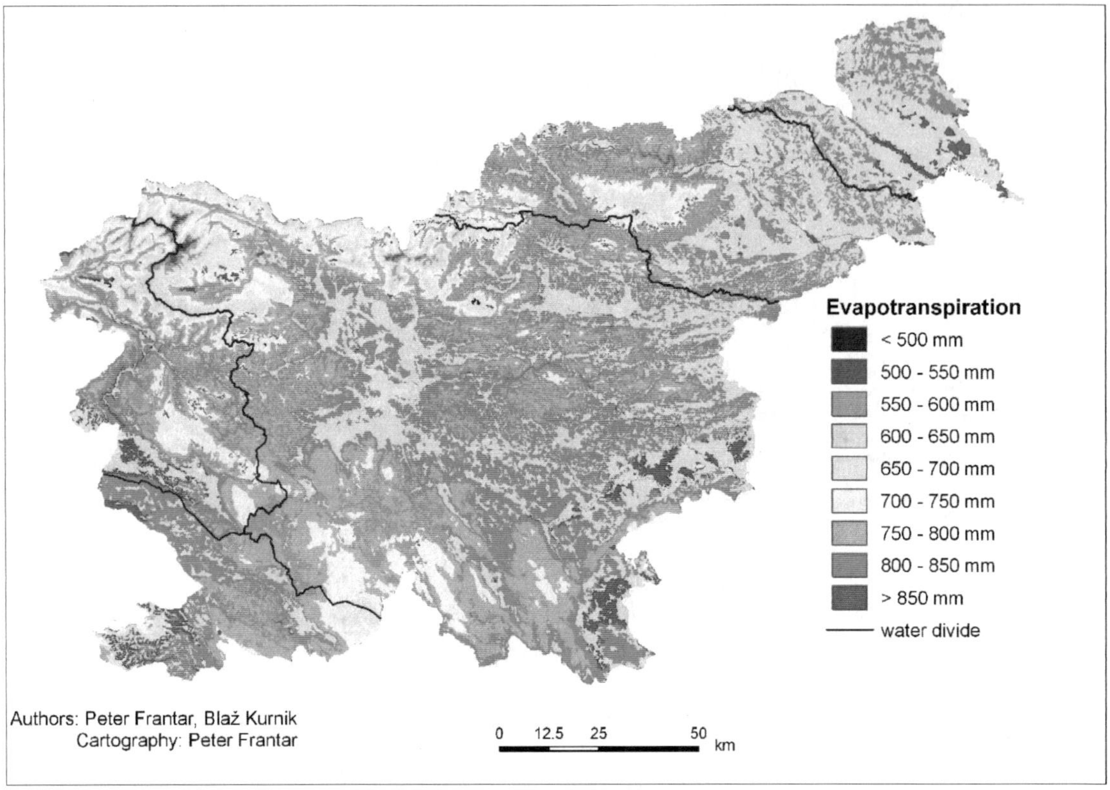

Figure 3. Evapotranspiration in period 1971-2000

4. Runoff

Runoff describes or represents the movement of a certain part of precipitated water into channelled streams or the discharge of water within it [2][10]. It is the surplus of precipitation that does not evaporate and is not used for transpiration. This surplus flows away as runoff. When this surplus is large enough, the runoff is collected into streams, which represent the majority of the runoff from a certain catchment area. In places where the majority of the runoff is collected the runoff can be measured as the discharge. In general, measured discharges are the most reliably measured elements of the water cycle. In suitably located water gauging stations, great majority of water from a certain catchment area runs through the cross-section of the water gauging station.

The characteristics of a discharge at a certain point are a reflection of the entire catchment area [11]. Because of this, knowledge of the physical-geographical space, especially of the gauging profiles and divides is of key importance, as it is only in this way that we can obtain comparable data and analyse smaller river basin units. Discharge data are linked to the space via the hydrometric catchment areas.

Data gaps cause difficulties in the analysis of hydrological data. When reviewing the data on discharges for the 1971-2000 period, it was found that 65 water gauging stations had complete data sets. On other stations, data gaps were supplemented with the use of a statistical method – with the Pearson linear correlation coefficient based on the mean monthly averages of the 1961-2001 period.

4.1. Specific Runoff in the 1971-2000 Period

The specific runoff shows how many mm of water runs off per year on average. The average runoff can be estimated from the measured discharge values at individual water gauging stations or by using the water balance equation – precipitation minus evaporation (Q = P – E). In Slovenia the runoffs measured are very similar to those calculated using the water balance formula, which indicates the correctness of the calculations of precipitation and evapotranspiration. The prevailing characteristic of the specific discharge in Slovenia is that it is highest in the upstream part and gradually decreases downstream [6].

The most important effect on the specific runoff has the climate or, more precisely, the precipitation, which is also reflected in the geographical distribution. The quantity of the specific runoff decreases as you move from the Alps and the Dinaric belt toward the north-east and south-west, which is also shown by the chart of specific runoffs that was produced based on the water balance equation using precipitation and evaporation data.

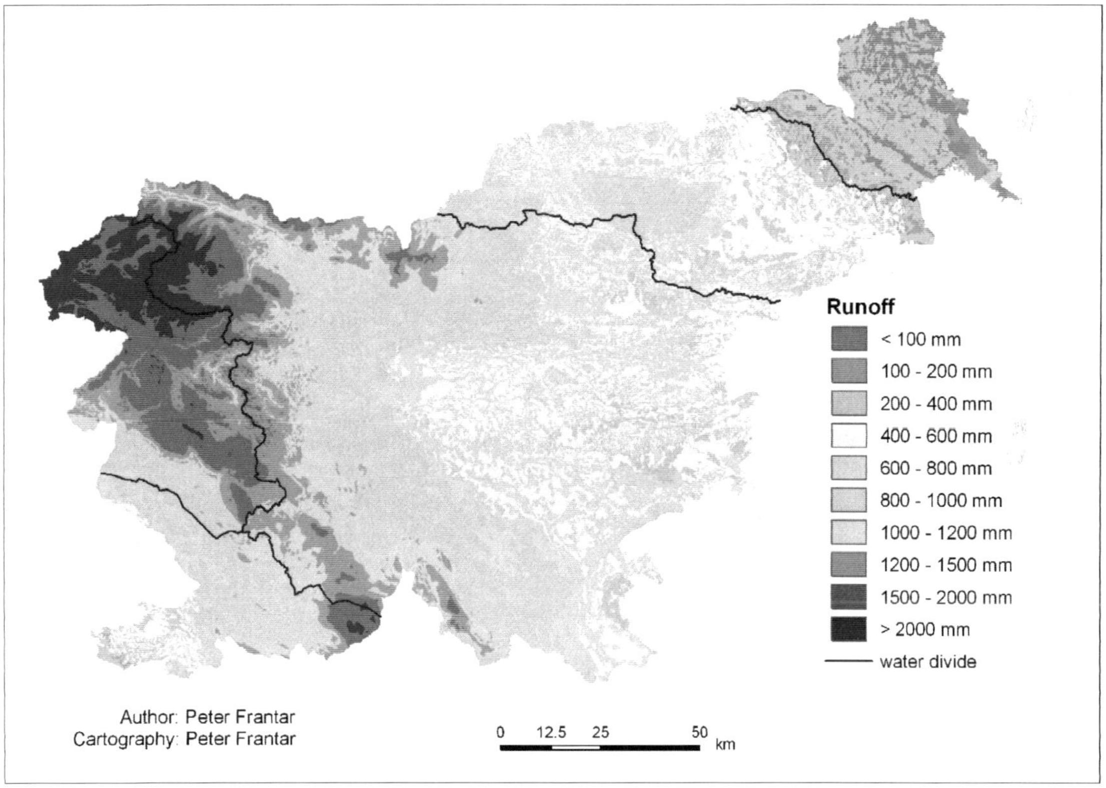

Figure 4. Runoff in period 1971-2000

5. Overview of the water balance by river basins

The results of the water balance elements indicate a great variety of climatic-geographical and hydrogeographical conditions in individual areas of Slovenia.

5.1. Pomurje (The River Basin of the Mura River)

Pomurje has the smallest amount of available water in Slovenia. On average, there is 897 mm of precipitation, 693 mm of evaporation and 204 mm of runoff annually in the entire Pomurje. On the left bank of the Mura, there is less than 900 mm of precipitation, while in the eastern-most part there is less than 850 mm. The most precipitation in the river basin occurs in the Slovenske gorice area. The annual evaporation has only slight geographical deviation – the situation is being similar throughout Pomurje. The result of this evaporation and precipitation are the low runoff values – with the lowest values in the country. The average annual runoff of around 300 mm is on the right bank of the Mura river and decreases to the east of the Pomurje.

5.2. Podravje (The Drava River Basin)
The Slovenian part of the Drava river basin encompasses a more varied landscape than Pomurje. On average, there is 1244 mm of precipitation, where 697 mm evaporate and 547 mm of water run off annually. Throughout Podravje, the measured annual quantities of precipitation are in excess of 1000 mm. The minimum amount of precipitation is in Slovenske gorice. The quantity of precipitation increases from here towards the higher-lying parts of Pohorje and the western Karavanke Alps, where there is more than 1600 mm of precipitation. Evaporation is lowest in the area of Slovenske gorice and the Dravsko polje, as well as in the highest reaches of the Drava river basin – on Pohorje and in the Karavanke Alps. In the remaining area evaporation exceeds 700 mm annually, with the highest occurring in individual areas of Haloze area – more than 800 mm. The runoffs show a similar picture – they are lowest (around 300 mm per year) in the eastern part, and the highest of the higher reaches of Pohorje and the Karavanke Alps, where they reach more than 1100 mm per year.

5.3. Posavje (The Sava River Basin)
Posavje is the largest river basin in Slovenia. In the entire Posavje area, without the Kolpa River basin, there was an average of 1594 mm of precipitation, 716 mm of evaporation and 878 mm of runoff per year. In the wettest areas of the Julian Alps, in the western parts of the catchment areas of the rivers Savica, Bohinjska Bistrica and Mostnica, there is more than 3000 mm of precipitation. The quantity decreases from here towards the east, so that the central area of Posavje exhibits between 1400 and 1800 mm of precipitation. The least precipitation in Posavje occurs in Posotelje region. Evaporation in this river basin is the lowest in the high-mountain Alpine area and in Posotelje region. In general, between 650 and 850 mm of water evaporate from Posavje annually, with less evaporation observed in the north and more in the south. The runoff in Posavje is distributed similar to the precipitation. The most water runs off from the Alpine and pre-Alpine areas and the quantity is decreasing towards the east.

5.4. Pokolpje (The Kolpa River Basin)
Pokolpje is a part of Posavje as the Kolpa river discharges into the Sava River in Croatia. It is a karstic river basin. In the Slovenian part of Pokolpje, there is an average of 1534 mm of precipitation, 757 mm of evaporation and 777 mm of water run off annually. The most precipitation is in the west. Evaporation is very high – from 650 to up to over 850 mm per year, and is strongly dependent on the vegetation and the surface exposure. The runoff is the highest in the upper part of the Kolpa river basin, with more than 1600 mm, and decreasing rapidly from here towards the east. The least runoff occurs in central Bela krajina region, where only around 400 mm of water run off annually.

5.5. Posočje (Soča River Basin)
Posočje is our most water-abundant river basin. In the entire Posočje, there is an average of 2386 mm of precipitation, 726 mm of evaporation and 1660 mm of runoff annually. In some areas the precipitation amount exceeds 3000 mm per year. The amount is decreasing towards the south. The lowest precipitation is measured in the Lower Vipava Valley and in the Vipavska brda area - around

1500 mm annually. Evaporation in Posočje is highest in the southern areas and the amount decreases as you move towards the north in accordance to the elevation. Great differences occur in the terms of runoff. In the majority of the Alps in Posočje the runoff is over 2000 mm (with the exception of the valley ground areas), but in the Vipavska brda the quantity decreases to a mere 650 mm per year.

5.6. The Catchment Area of the Adriatic Rivers without Posočje

The smallest hydrogeographical area of Slovenia is represented by the Reka River - with its catchment area that flows into the Adriatic via the Karst - and the area of small rivers and streams that drain directly into the Slovenian sea. In the entire Primorje area with the Karst region, there is an average of 1619 mm of precipitation, with 748 mm evaporation and 871 mm of water run off annually. The most precipitation falls in the surroundings of Mount Snežnik. The quantity is decreasing towards the west. The coastal areas receive a good 1000 mm of precipitation and the western Karst around 1500 mm of precipitation annually. Evaporation is highest in individual parts of the Koper coastal area and in the south-western part of the Karst. Towards the north and east, the quantity of evaporation decreases, being the lowest in the Snežnik mountain chain – below 700 mm annually. The runoffs are the highest in the eastern-most part (surroundings of Mount Snežnik), where more than 2000 mm of water runs off. Runoffs decrease from there towards the west. The groundwater runoff in the Karst is between 500 and 750 mm, with the lowest being in the Koper coastal area – between 200 and 500 mm annually.

6. Overview

On the territory of Slovenia in the 1971-2000 period, there was an average of 1579 mm of annual precipitation, 717 mm of evaporation and the 862 mm of runoff.

Table 1. The water balance – a comparison of periods (1961-90 source: [6]) for the territory of the Republic of Slovenia

In mm	1961-1990	1971-2000
Precipitation	1567	1579
Evaporation	650	717
Runoff (Q = P – E)	917	862
Runoff coefficient	58.5 %	54.5 %

A direct comparison with the 1961-1990 reference period water balance [6] shows that the quantities of precipitation were almost the same in the 1971-2000 period, but that the quantity of evaporation has increased and runoff has decreased. It shows a significant change in the evaporation and runoffs. Evaporation is higher by 11%, and the runoff is lower by 6%.

References

[1] Allen R G, Perreira L S, Reas D and Smith M 1998 Crop evapotranspiration – guidelines for computing crop water requirements FAO irrig. and drain. p.
[2] Davie T 2004 Fundamentals of hydrology Routledge Fundamentals of Physical Geography
[3] Dolinar M, Ovsenik-Jeglič T, Bertalanič R 2006 Izračun korigiranih padavin v obdobju 1971–2000

IOP Conf. Series: Earth and Environmental Science **4** (2008) 012020 doi:10.1088/1755-1307/4/1/012020

[4] Frantar P, Hrvatin M 2005 Pretočni režimi v Sloveniji med letoma 1971 – 2000 Geog. vest. **77-2** 115–27

[5] Internet: http://ga.water.usgs.gov/edu/watercyclehi.html (7/1/2008)

[6] Kolbezen M, Pristov J 1998 Površinski vodotoki in vodna bilanca Slovenije HMZ RS

[7] Nespor V, Sevruk B 1999 Estimation of wind-enduced error of rainfall gauge measurements using a numerical simulation J. of Atm. and O. Tech. **16** 450-64

[8] Ritter M 2006 The Water Balance

[9] Schöniger M, Dietrich J 2003 Hydrologie Hydroskript

[10] Van Abs D J, Stanuikynas T J 2000 Water Budget in the Raritan River Basin

[11] World Meteorological Organisation 1994 Guide to hydrological practicies WMO – No. 168

IOP Conf. Series: Earth and Environmental Science **4** (2008) 012021 doi:10.1088/1755-1307/4/1/012021

Analysis of Interrelation between Water Quality and Hydrologic Conditions on a Small Karst Catchment Area of Sinking Watercourse Trbuhovica

V Hinić[1], J Rubinić[2], I Vučković[3], I Ružić[2], A Gržetić[1], G Volf[2], M Ljubotina[1] and N Kvas [4]

[1]RiEKO-LAB, Janka Polić Kamova 19., 51 000 Rijeka, Croatia
[2]Faculty of Civil Engineering Rijeka, V.C. Emina 5, 51 000 Rijeka, Croatia,
[3]HRVATSKE VODE, Department of Water Protection, Central Water Management Laboratory Ul. grad Vukovara 220, 10000 Zagreb, Croatia
[4]Grad Čabar, Narodnog oslobođenja 2, 51 306 Čabar, Croatia

E-mail: vhinić@ticri.hr

Abstract: Sinking watercourse Trbuhovica is located at the topping karst of Gorski Kotar in Croatia, near the Slovenian border. About 900 inhabitants live in Trbuhovica catchment area. Sewage system had not been built. The project KEEP WATERS CLEAN (INTERREG III A project) was approved by EU commission and has a purpose of investigating water resources of that area, their appropriate protection and improving management of those resources. This paper presents project's 1st phase investigation results: hydrologic conditions and water quality at several locations on stream and at the springs of Trbuhovica, Mlake and Obrh. Climatologic (precipitation, air temperature and snow cover), basic hydrologic characteristics (flow and water temperature), water quality parameters (pH, electric conductivity, alkalinity, oxygen regime, nutrients and mineral oils) and microbiology indicators have been monitored. Samples of micro invertebrates and samples of periphyton have been collected in the field. Biological results have been elaborated via Saprobial Index according to Pantle-Buck. Analyses results showed a strong connection between hydrologic condition and selected water quality parameters. The groundwater quality changes are very quick. Maximum pollutions occur during the period of intensive rain. Water at the spring of Mlaka is very clean and is classified in the first to second water category, while Trbuhovica shows higher organic pollution..
Key words: transboundary aquifer, karst water resources, hydrology, water quality, international projects.

1. Introduction

This paper elaborates monitoring of hydrologic conditions and water quality of a small karst aquifer whose pollution has transboundary influence on a karst underground. As an example, small natural watercourse Trbuhovica has been chosen. It is located in Croatia, Gorski Kotar region, in a karstic depression (Figure 1) where village Prezid with ca 900 inhabitants is situated. The village does not have a sewage system and most of the wastewater ends in the stream Trbuhovica which sinks underground by the border with neighbouring Slovenia.

Catchment of Trbuhovica forms the highest horizon of Ljubljanica river catchment and it is connected with springs located in Loško polje [1], [2] Because of that, Trbuhovica water pollution influences the water quality of those springs in Loško polje. Some of them are used for water supply [3]

The project Keep Waters Clean was approved during the year 2005. Its purpose is to investigate water resources of that area and their appropriate protection. This paper describes basic results of the 1st investigation– monitoring phase of hydrologic conditions and water quality.

Trbuhovica and Obrh springs compose the majority flow of the stream Trbuhovica. Nearby, Mlaka spring waters form an individual surface stream which ends in near site sinking underground area (Fig. 2.). Springs Trbuhovica and Mlaka are used for water supply system of village Prezid.

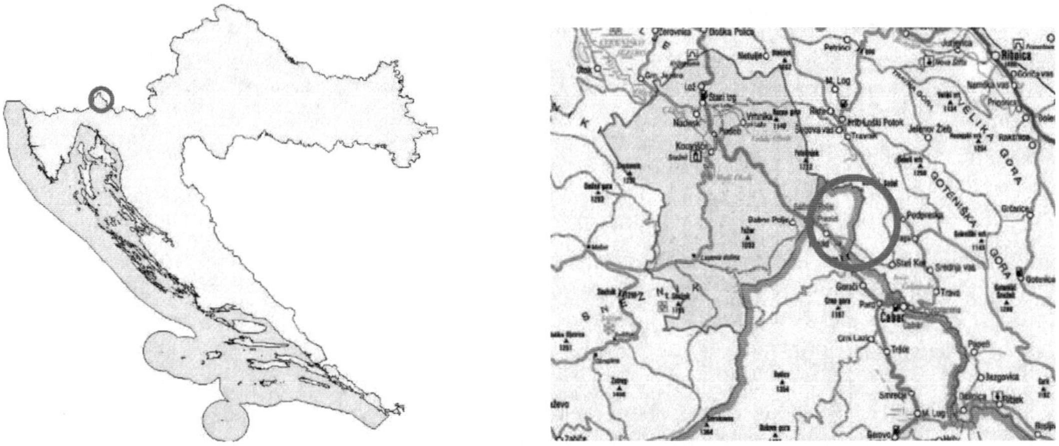

Figure 1. Site of Prezid investigation area

A yearlong hydrological monitoring (from November 2006.-September 2007.) has been established on two locations (stream Trbuhovica and spring Mlaka) with periodical flow measurements on other stations alongside the stream Trbuhovica [4] Data from the State Weather Bureau - climatologically station Parg (about 4 km away from the middle of catchment Trbuhovica) and rain gauge station Prezid (in a catchment of Trbuhovica) have been used. Water quality parameters (physical-chemical and microbiological) were measured monthly at five stations, two located on stream Trbuhovica (at the beginning of permanent flow and before the sinkhole) and stations on springs: Trbuhovica, Obrh and Mlake (Fig. 2). Biological researches have been conducted (structure analysis and structure of biocenosis) only on one station with continual water appearance – the upper part of the Trbuhovica stream [5].

Hydrologic monitoring was not established in Loška dolina area. Water quality (physical-chemical and microbiology parameters) was monitored monthly on ten stations during five months – from February to June 2007 [3]. All characteristic states during the year were not included.

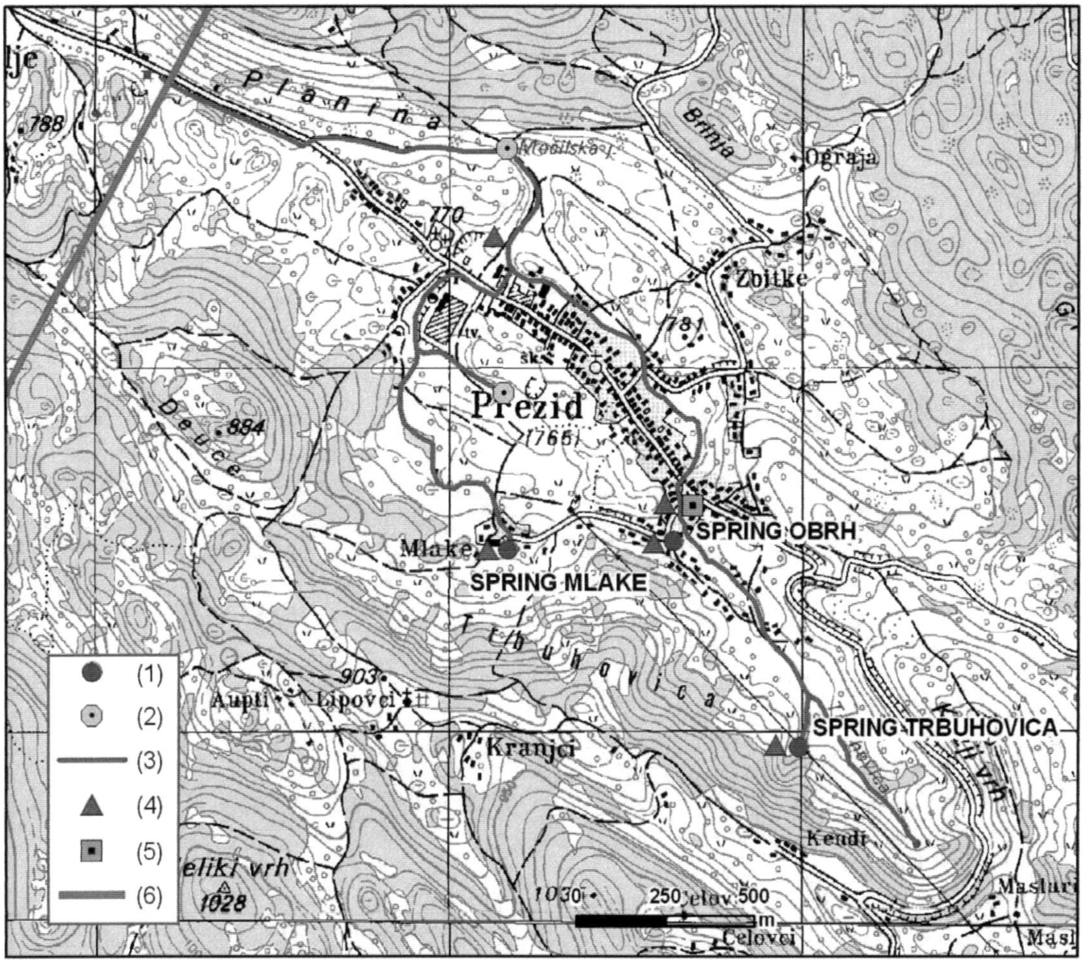

Figure 2. Map of Prezid with locations for monitoring hydrological and water quality characteristics

1-Spring; 2-Swallow; 3-Water course; 4-Water quality monitoring; 5-Hydrological monitoring 6-State border

2. Purpose And Methodology Of The Study

Physical-chemical, microbiological and biological analysis of surface water and groundwater within project Keep Waters Clean has been performed with regards to:

- assessing the quality of underground water intended for human consumption and specifying an appropriate water treatment,
- identifying range and possible consequences of water pollution: influence of polluted surface water on sources of water supply in Loška dolina and hygienic requirements in Prezid village,
- Identifying the source of water pollution: settlement's sewage system, run off from the roadside and other
- defining measurements and water protection plans such as construction of sewage system, individual management of waste water treatment, illegal landfills sanitation, controlled roadside drain construction and
- evaluating the established catchment protection measures.

Those analyses were monitored together with hydrological monitoring whose purpose was to quantify the basic hydrological features of water in the catchment of Trbuhovica and spring Mlaka and to provide interpretation of the results of water quality considering dynamic hydrological conditions in the field.

2.1. Examination Methods

Groundwater and surface water analysis included physical-chemical (temperature, turbidity, conductivity, pH, dissolved oxygen, BOD_5, oxidisability, alkalinity, hardness, ammonium, nitrite, nitrate, chloride, oil and grease and mineral oils) and microbiological parameters (total coliforms, fecal coliforms, Escherichia coli, fecal streptococcus and clostridium). Determination of these parameters has been made according to standard methods: EN ISO methods and Standard Methods (APHA, AWWA and WEF).

Examination of surface water included saprobity analysis of bio fouling and makrozoobentosa and Saprobial Index has been provided according to Pantle-Buck.

Biological analysis has been made only at the monitoring station- Vatrogasni dom, because in the lower section of the Trbuhovica stream, before Mrzla jama sinkhole, the water occurs occasionally (duration is less than 15 days). Sampling of macrozoobentos has been made according to AQEM method.

2.2. Groundwater and Surface Water Quality Assessment

Ground water and surface water quality assessment has been made according to Regulations of water classification in Croatia. Classification for individual parameters and a group of parameters is shown in table according to the group of parameters (Figure 5 and 6).

Croatian drinking water regulations regulates maximum permissible values for drinking water safety parameters. In regard to their use, quality assessment of springs Mlaka and Trbuhovica has been made also according to this regulation.

3. Results

Monitoring has been conducted during 2006/2007 under unusual climate conditions. That hydrological year was unusually warm, with temperature mean deviation almost 1^0C higher than average. The period from October to February showed a monthly temperature mean deviation which was almost 3^0C higher than average. According to precipitation amount regarding average hydrological condition, April 2007 was extremely dry as well as period of time from June to September 2007. During winter period of research in Prezid (2006/2007), the snow- cover lasted for an unusually short time – only 38 days in 5 shorter snow periods, with maximum snow height of 37 cm (registered on March 20. 2007).

These hydrological conditions influenced the conditions of spring and surface runoff. Runoff on stream Trbuhovica was ranged from dry (during 8 sequential days in August 2007) to 3.9 m^3s^{-1} (in September 2007). Monthly runoff mean deviation was between 0.1 (in August 2007) and 0.11 m^3s^{-1} (in September 2007). During research time (January - November 2007), the runoff on spring Mlaka was between 0.001 (in August 2007) and 0.033 (in February 2007) m^3s^{-1}. Monthly runoff mean deviation was between 0.0015 (in August 2007) and 0.018 (in February 2007) m^3s^{-1}. The spring Mlaka has average amount of water for water supply ranging between 0.0013 and 0.002 m^3s^{-1}. Those spring and stream have a small runoff under the regular hydrological conditions. They can achieve multiple maximal runoffs and in that case, due to a limited capacity of sinking zone, they can cause flood in lower parts of Prezid area. During the research, in the period (2006/2007), no such incident occurred.

Recorded precipitation and hydrological conditions were used for interpretation of the results. Figure 3 present daily precipitation and bacteriological water pollution of spring Trbuhovica. It shows that the intensity of bacteriological pollution strongly varies according to hydrological conditions and pollution spike appears during or immediately after the downfall.

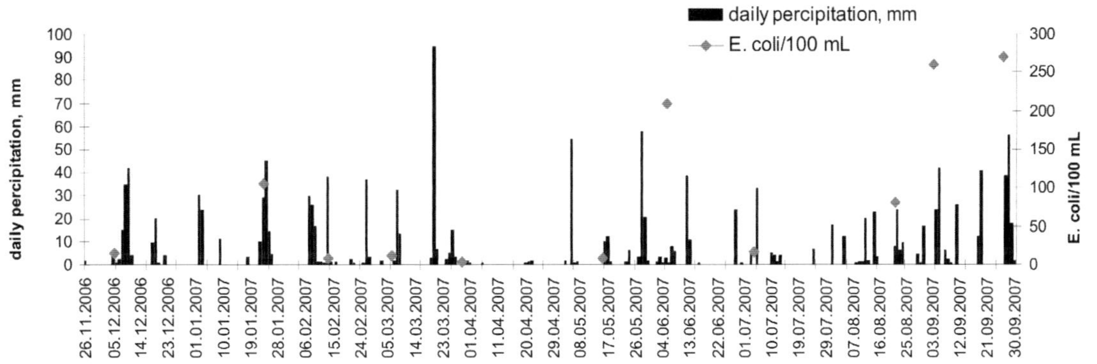

Figure 3. Daily precipitation on hydrological station Parg and bacteriological water pollution of spring Trbuhovica

Figure 4 shows the interrelation between runoff on Trbuhovica (monitoring station Vatrogasni dom) and pollution with ammonium. In period of low water the pollution is very high and under such conditions this water shows characteristics of weakly polluted waste water (ammonium has been chosen only to demonstrate these situations). In period of high water the pollution is lower because of dilution effect.

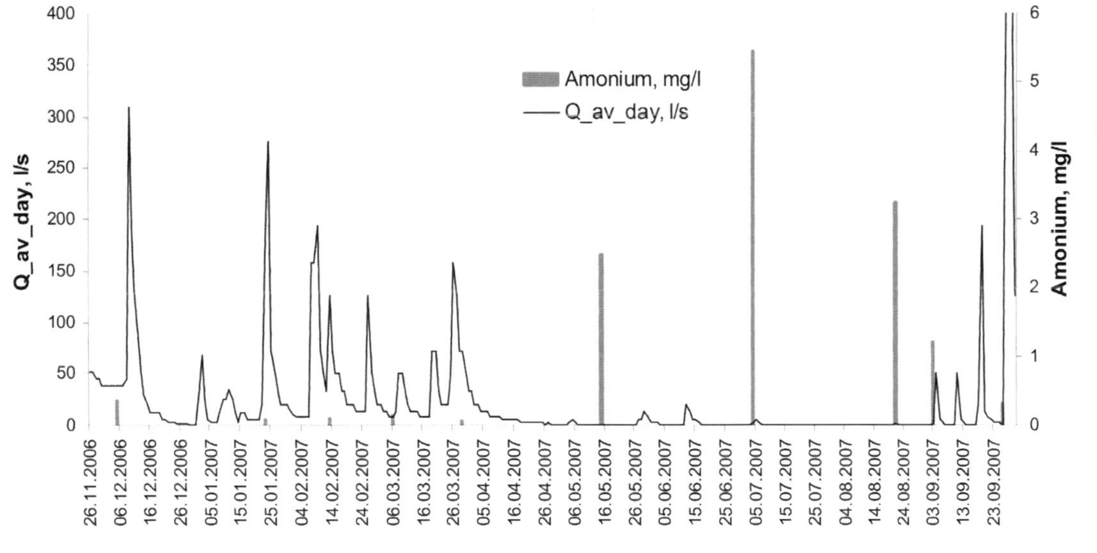

Figure 4. Interrelation between runoff on Trbuhovica and pollution with ammonium (monitoring station Vatrogasni dom)

Based on water quality examination in the analysed area, classification of spring water (Fig. 5) and surface water (Fig. 6) have been made.

Spring Mlaka is groundwater of high quality. Physical and chemical characteristics are very good, without a visible anthropogenic pollution. Low fecal pollution is present only during drain fall.

IOP Conf. Series: Earth and Environmental Science **4** (2008) 012021 doi:10.1088/1755-1307/4/1/012021

Determined water quality is in compliance with catchment area situation - without a significant pollution source (only two small villages without economic facility and with 21 inhabitants).

Spring Trbuhovica has good physical - chemical water quality characteristics. During rainfall the water is turbid but not significantly. Fecal pollution is always present and is very high during heavy rains due to the rinsing pollution from the unsaturated zone in karst aquifer and its quick transport to spring. The pollution is a consequence of sewage system lack in few small villages around the spring. Water quality requires a complex water treatment and the present disinfection with chlorine is not adequate.

Water from spring Obrh is strongly fecally polluted. The spring is situated in the centre of Prezid and few houses are situated just above it!. Present odour on fecal materials, high concentration of organic matter, oxygen deficiency and constant pollution with nitrite indicate an extra high contamination. Fecal pollution is constant and intensive. The water is not suitable for human consumption even after water treatment.

Classification of ground water in Prezid in period from November 2006. to September 2007.			Spring Mlaka			Spring Trbuhovica			Spring Obrh		
Group of indicators	Parameter	Measurement unity	Measurement value	Type P	Type G	Measurement value	Type P	Type G	Measurement value	Type P	Type G
A- Physical-chemical	pH		7,61-7,97	1	1	7,80-8,10	1	1	7,44-7,55	1	1
	Conductivity	µS/cm	414	1		481,8	1		422	1	
	Alkalinity m- value	mg $CaCO_3$/L	230	-		244	-		239	-	
B- Oxygen regime	Dissolved oxygen	mg O_2/L	10,55-10,96	1	1	10,77-11,42	1	1	6,8-8,02	2	3
	Saturated oxygen	%	88-93	1		91-97	1		58-74	3	
	COD- Mn	mg O_2/L	0,7	1		1,5	1		1,2	1	
	BOD_5	mg O_2/L	1,19	1		1,37	1		1,00	1	
C- Nutrients	Ammonium	mg/L	0,001	1	1	0,072	1	1	<0,001	1	1
	Nitrite	mg/L	0,001	1		0,012	1		0,020	1	
	Nitrate	mg/L	1,79	-		4,2	-		4,13	-	
D- Microbiological	Total coliforms	UK/100 mL	60	1	1	536	3	3	650	3	
	Fecal coliforms	FK/100 mL	6	1		339	3		480	3	
	Fecal streptococcus	FS/100 mL	8			223			320		
F- Organic compounds	Mineral oils	µg/L	8,5	1	1	11,7	1		-	-	-

Figure 5. Classification of ground water an analysed area – springs Trbuhovica, Mlaka and Obrh

Classification of watercourse Trbuhovica in period from November 2006. to September 2007.			Watercourse Trbuhovica- Hydrological station			Watercourse Trbuhovica - Before sinkhole		
Group of indicators	Parameter	Measurement unity	Measurement value	Type P	Type G	Measurement value	Type P	Type G
A- Physical-chemical	pH		7,53-8,03	1	1	7,57-8,19	1	1
	Conductivity	µS/cm	400	1		414	1	
	Alkalinity m- value	mg $CaCO_3$/L	225	-		210	-	
B- Oxygen regime	Dissolved oxygen	mg O_2/L	2,02-11,50	5	5	5,4-13,8	3	3
	Saturated oxygen	%	21-94	4		56-113	3	
	COD- Mn	mg O_2/L	3,94	[illegible]		4,90	2	
	BOD_5	mg O_2/L	2,89	2		3,88	2	
C- Nutrients	Ammonium	mg/L	0,310	2	2	0,198	2	3
	Nitrite	mg/L	0,059	2		0,278	3	
	Nitrate	mg/L	2,67	2		3,29	2	
	Total phosphorus	mgP/L	0,14	2		0,12	2	
D- Microbiological	Total coliforms	UK/100 mL	13000	4	4	110000	5	5
	Fecal coliforms	FK/100 mL	1800	4		15000	5	
	Fecal streptococcus	FS/100 mL	450			4800		

Figure 6. Classification of surface water Trbuhovica on hydrological station Prezid and before swallow hole

Type I = type according to parameter
Type G = type according to group of parameters

Water quality of Trbuhovica stream is degraded by disposal of untreated wastewater in its watercourse. Appearance of the first water wave is the most critical period for the pollution of groundwater in the area of Loška dolina. Accumulated pollution in catchment area and stream quickly rinses and concentration penetrates the underground via sinkholes. During research time in July 2007 we encountered that kind of situation. After a long time of dry period the watercourse (before sinkhole) was dry but a two hours long downfall caused a short-time flood of the roads in the lower area of Prezid and great runoff on the watercourse Trbuhovica. The water was very turbid with big flowing sewage and with fecal odour. Chemical and bacteriological examinations indicate great water pollution (ammonium 8.12 mg/L, fecal coliforms 10^4/100 mL). Lower alkalinity and water hardness indicate higher runoff into the watercourse Trbuhovica.

According to saprobic characteristics the monitoring station Trbuhovica (Vatrogasni dom) belongs to an alfa- mezosaprobic area. The water was polluted and anthropogenic influence was evident. The sampling was made in May and September. Current velocity was low. In that part of the stream, mud is in a reduction process which is visible after rising a gas bubble with odour on H_2S. Stream vegetation was strongly developed which indicates an extra organic input. This area is characterized by water species typical of high contents of organic matter such as: algae - *Oscillatoria limosa*, *Oscillatoria putrida*, *Navicula cuspidata*, and animal organisms- *Chironomus thumi* i *Nais sp.* In September macrozoobentos species contain more *Odonata* and *Metgaloptera* which are more likely to occur in autumn in streams with smaller amount of runoff.

4. Conclusion

The result of all conducted analyses has indicated a strong correlation between hydrological conditions and selected water quality parameters. Influence of monitoring period on interpretation of the results has also been shown.

Because of the hydrological characteristic of karst groundwater quality changes are very quick and the maximum pollutions occur in the period of intensive rains. The main problem of spring Trbuhovica and spring Obrh is the microbiological pollution as a result of disposal of untreated wastewater from small villages in their catchment. The spring Mlaka is a clean karst groundwater.

This research indicates that water quality of watercourse Trbuhovica is very disturbed by untreated waste waters disposal directly to it. Pollution of surface water intensively depends of water flow. Appearance of the first water wave is the most critical period for the pollution of springs in the area of Loška dolina. The accumulated pollution in catchment area and watercourse quickly rinses and concentration penetrates underground via swallows. Biological research results are in compliance with physical-chemical and bacteriological examinations and show that high organic pollution is the main cause of water quality degradation of this watercourse whose section trough the Prezid has alfa-mesosaprobic water characteristic.

Results are as expected, but since the opportunity to conduct integral qualitative and quantitative monitoring of hydrological, physical-chemical and biological parameters in the same pilot areas is quite rare, the results deserve full attention and can be used for planning further similar researches. These researches were performed within the project 'INTERREG III A Keep Waters clean' proposed by local community – The Town Čabar in the Republic of Croatia and the county of Laška dolina in the Republic of Slovenia (2004 - 2006). It must be pointed out that these results would have a larger impact on protection measures planning in the investigated catchment area in case of equivalent researches in Laško polje, because the main aim of this project was to provide quality data for designing protection of underground water.

References

[1] Šerko A and Michler I 1952 *Postojnska jama i druge zanimljivosti krasa* (Ljubljana: Turistično podjetje Kraške jame Slovenije)
[2] Gams I 2004 *Kras v Sloveniji v prostoru in času* (Ljubljana: Inštitut za raziskovanje krasa ZRC

SAZU)
[3] Zavod za zdravstveno varstvo 2007 *Projekt ohranimo vode čiste – poročilo* RRA (Postojna: Notranjsko-Kraške regije, d.o.o.)
[4] Građevinski fakultet Rijeka 2007 *Monitoring stanja voda – praćenje hidroloških prilika na području Prezida* (Rijeka, unpublished)
[5] RiEKO-LAB 2007 *Monitoring stanja voda – kakvoća podzemnih i površinskih voda na području Prezida* (Rijeka, unpublished)

IOP Conf. Series: Earth and Environmental Science **4** (2008) 012022 doi:10.1088/1755-1307/4/1/012022

Hydrological scenarios of future seasonal runoff distribution in Central Slovakia

K Hlavcová[1], J Szolgay[1], S Kohnová[1] and G Bálint[2]

[1] Department of Land and Water Resources Management, Slovak University of Technology, Radlinského 11, Bratislava, Slovakia
[2] Surface Waters, VITUKI Environmental and Water Research Institute, Budapest H-1095, Kvassay 1, Hungary

E-mail: kamila.hlavcova@stuba.sk

Abstract. The hydrological scenarios of future seasonal distributions of runoff in the upper Hron River basin, which was chosen as a representative mountainous region in Central Slovakia, were evaluated. Changes in the future climate were expressed by three different climate change scenarios developed within the framework of the Central and Eastern Europe Climate Change Impact and Vulnerability Assessment Project (CECILIA). The climate change scenarios were constructed using the pattern scaling method from the outputs of transient simulations made by 3 GCMs - ECHAM4/OPYC3, HadCM2 and NCAR DOE-PCM. A conceptual hydrological balance model calibrated with data from the period 1971-2000 was used for modelling changes in runoff with monthly time steps. The runoff change scenarios for the selected basin in the future time horizons of 2025, 2050 and 2100 show changes in the seasonal runoff distribution.

1. Introduction

The impacts of climate change on hydrological processes are often estimated by defining scenarios of changes in climatic inputs to a hydrological model from the output of general circulation models (GCMs). Two approaches are often used to construct weather series representing a changed climate [1]: an observed weather series is modified by the scenario's parameters [2], or a weather series is produced by a weather generator whose parameters have been modified according to the scenario [1].

Monthly conceptual water balance models are intended to simulate selected hydrological processes, usually by conceptualising a catchment as an assemblage of interconnected storage systems through which water passes from input as rainfall to output as streamflow at the catchment outlet; the controlling equations usually satisfy the requirements of the hydrological balance. In impact studies, the model's parameter values from the representative periods are usually considered to be representative of runoff generation in the future.

Several climate change impact studies have been conducted in recent years on the territory of Slovakia. Usually, three types of climate change scenarios have been used in previous impact studies: analogue scenarios based on an analogy with warmer periods and periods with a specified variability in the climate in the past; regionally downscaled GCMs scenarios with typical time horizons of 2010, 2030 and 2075; and incremental climate change scenarios. Modified model input time series have usually been constructed from the baseline data by adding the differences in mean air temperature and

IOP Conf. Series: Earth and Environmental Science 4 (2008) 012022 doi:10.1088/1755-1307/4/1/012022

precipitation prescribed by the scenarios for the given time step. Observed runoff series from 1931 – 1980 or 1951–1980 have usually been considered as baseline periods. Additional details can also be found in the following studies on Slovak rivers: Danihlík, et al. (2004) [3]; Fendeková (1999) [4]; Fendeková and Némethy (2001) [5]; Halmová (2004a,b, 2005, 2006) [6, 7, 8, 9]; Halmová and Melo (2006) [10]; Hlavčová, et al. (1999, 2000, 2002, 2005, 2006a,b) [11 – 16]; Kostka and Holko (2000, 2001) [17, 18]; Kostka (2003) [19]; Majerčáková (2000) [20]; Majerčáková and Takáčová, (2001) [21]; Pekárová, et al. (1996, 2001) [22, 23]; Pekárová (2000) [24]; Pekárová and Miklánek (2001 [25]); Pekárová and Szolgay, eds. (2005) [26]; Petrovič (1998, 2000) [27, 28]; Szolgay, et al. (1997, 2002, 2003, 2007a,b) [29 – 32] and Takáč (2001) [33].

In this study the potential impact of climate change on river runoff in the upper Hron River basin was evaluated using a conceptual spatially-lumped water balance model. The conceptual water balance model was calibrated with data from the 1971-2000 period. The period of 1971-2000 was assumed to be the reference for impact simulations in the Central and Eastern Europe Climate Change Impact and Vulnerability Assessment project (CECILIA) [34]. Changes in climate variables in the future were expressed by three different climate change scenarios developed within the framework of the CECILIA project. The climate change scenarios were constructed using the pattern scaling method from the outputs of transient simulations made by the ECHAM4/OPYC3, HadCM2 and NCAR DOE-PCM global circulation models [35].

2. Scenarios of climate change

In this study three climate change scenarios developed for the Hron basin within the framework of the 6 FP CECILIA (CECILIA, 2007) were applied. The scenarios were based on transient simulations with three GCMs, i.e., ECHAM (ECHAM4/OPYC3), HadCM (HadCM2) and NCAR (NCAR DOE-PCM), were constructed using pattern scaling techniques from the outputs of these GCMs and representative scenarios for the development of emissions of greenhouse gases and aerosols, and the intermediate climatic sensitivity to the emissions. The climate change scenarios were constructed for three future time horizons, i.e., 2025, 2050, and 2100.

The methodology for constructing the scenarios is described in Dubrovsky, et al. (2005) [35]. In this method, the standardized scenario, which relates the climate's variable responses to a 1°C rise in the global mean temperature (TG), is multiplied by the predicted change (ΔTG). The standardized scenarios were determined from GCM runs and ΔTG values, which were calculated by the simple MAGICC climate model [36, 37] for 3 combinations of conditions that were selected from representative emission scenarios and climatic sensitivities.

The SRES A1, A2, B1, and B2 emission scenarios from the IPCC Third Assessment Report [38] were used in the estimation of future increases in global temperature together with the most likely range of the values for the climate sensitivity factor, i.e., an increase in global temperature by 1.5-4.5°C per doubling of the atmospheric CO_2 concentration [38]. The values of the increases in global temperature for the emission scenarios used were compared for the low (LO), medium (MI), and high (HI) estimates of the climate sensitivity factor, i.e., the most optimistic, the mid-range, and the most pessimistic.

Changes in the mean monthly air temperature for the time horizon 2050 in the Hron basin are illustrated in Table 1.

Table 1. Scenarios of changes in the mean monthly air temperature in the Hron River basin for the horizon 2050 in °C as compared to the 1971–2000 reference period.

Scenario	Months											
	I	II	III	IV	V	VI	VII	VIII	IX	X	XI	XII
ECHAM												
High (HI)	2.6	2.7	2.4	2.9	1.6	2.9	3.2	3.7	3.4	2.7	2.7	2.7
Low (LO|	0.9	0.9	0.8	1.0	0.6	1.0	1.1	1.3	1.2	0.9	0.9	0.9
Medium (MI)	1.7	1.7	1.6	1.9	1.0	1.8	2.1	2.4	2.2	1.7	1.7	1.7

HadCM												
High (HI)	2.3	2.8	2.2	2.3	2.5	2.6	3.6	4.3	3.7	2.7	2.2	3.1
Low (LO|	0.8	1.0	0.7	0.8	0.9	0.9	1.3	1.5	1.3	0.9	0.8	1.1
Medium (MI)	1.5	1.8	1.4	1.5	1.6	1.7	2.3	2.8	2.3	1.7	1.4	2.0
NCAR												
High (HI)	2.3	2.6	1.5	1.3	1.3	1.4	2.9	3.2	2.6	3.0	2.6	2.1
Low (LO|	0.8	0.9	0.5	0.5	0.5	0.5	1.0	1.1	0.9	1.0	0.9	0.7
Medium (MI)	1.5	1.7	1.0	0.8	0.8	0.9	1.8	2.0	1.7	1.9	1.7	1.4

3. Hydrological scenarios of changes in seasonal runoff distribution

The scenarios of the changes in the seasonal runoff distribution were constructed using the following methodology:

- calibration of the conceptual hydrological balance model for the selected pilot basins,
- simulation of the reference mean monthly runoff series using input climate data from the reference period of 1971–2000,
- modification of the climate input data from the reference period (precipitation and air temperature) according to the climate change scenarios for the time horizons of 2025, 2050 and 2100,
- simulation of the monthly runoff series using a hydrological balance model based on the changed input data and parameters of the model from the calibration,
- comparison of the differences between the seasonal runoff distribution for the individual scenarios and the time horizons considered.

3.1. Description of the pilot area and input data

The upper Hron River basin to Banská Bystrica profile was selected as the pilot basin for the impact study. The basin has an area of 1766 km^2; the minimum elevation of the basin is 340 ma.s.l.; the maximum elevation is 2004 m a.s.l.; and the mean elevation is 850 m a.s.l. Seventy percent of the basin's area is covered by forest, 10 % by grasslands, 17 % by agricultural land and 3 % by urban areas. The location of the basin within the territory of Slovakia is shown in Fig. 1.

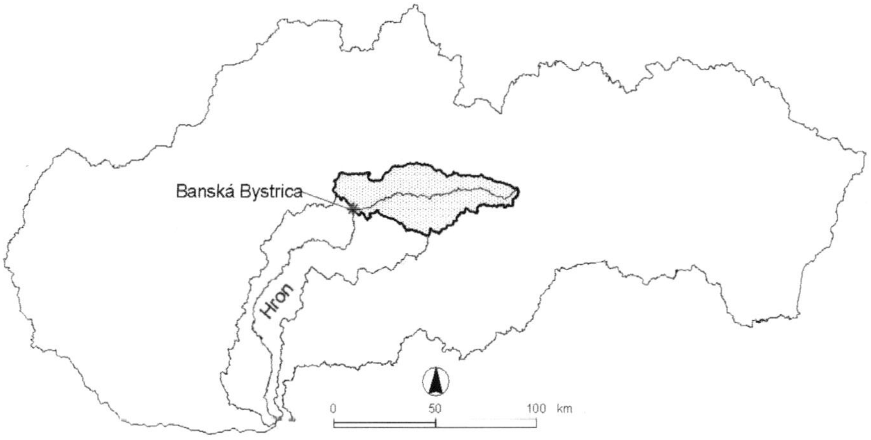

Figure 1. Location of the upper Hron River basin in Slovakia.

The hydrological and climate input data for the hydrological modelling were collected in monthly time steps during the period of 1971-2000. Monthly precipitation totals were available from 23 precipitation stations; the data of the mean monthly air temperature, mean monthly water vapour, cloudiness and snow cover were collected from 5 climate stations in the basin. The values of the mean monthly potential evapotranspiration were calculated by the Tomlain method, based on energy balance equations (Tomlain and Damborská, 1999). The mean monthly discharges were available from the Banská Bystrica gauging station.

The lumped input data for modelling the basin were prepared using the nearest neighbour interpolation method and the elevation gradient method.

3.2. *The monthly water balance model*

For estimating the changes in the seasonal runoff distribution, a conceptual water balance model developed at the Slovak University of Technology was used. This model, which simplifies a river basin into 2 nonlinear reservoirs, simulates water accumulation in the basin, snowmelt, evapotranspiration, runoff from impermeable areas in the basin, surface and subsurface runoff and baseflow. The inputs required for modelling the water balance in a monthly time step are: the mean monthly precipitation for the basin, the mean monthly discharges in the outlet of the basin and the mean monthly potential evapotranspiration (*PET*). For calculating the potential evapotranspiration, various methods can be used (the Tomlain, Thornthwaite, Ivanov and FAO methods), and additional climate data (the long-term mean monthly hours of the duration of sunshine, the long-term mean monthly values of the relative air humidity, and the mean monthly air temperature values) are required.

The basic mass balance equation in the model is written as:

$$S_i - S_{i-1} = \left(P_i\left(1 - drc\right)\right) - Rs_i - Rss_i - Ev_i - Rb \tag{1}$$

where:

S_i, S_{i-1}	-	current water storage in the basin in months i and i -1 [mm],
i	-	time step [month],
P_i	-	basin's average precipitation in the month i [mm],
drc	-	direct runoff coefficient ($0 \leq drc \leq 1$) [–],
Rs_i	-	surface runoff in the month i [mm],
Rss_i	-	subsurface runoff in the month i [mm],
Ev_i	-	basin's average actual evapotranspiration in the month i [mm],
Rb	-	baseflow [mm].

In the calibration procedure of the hydrological balance model, 11 model parameters are optimized (*Smax*, α, γ, ε, *PeffPar*, T_s, T_l, *Rb*, *ActEpar*, *drc* and *Zinitial*). The *Zinitial* parameter is the initial value of the ratio between S_i and S_{max}. In the model a genetic algorithm (GA) is built in to calibrate the model parameters for the at site data, and several criteria (or their combinations) are used as an objective function. The basic optimization criteria are the Nash-Sutcliffe criterion, the sum of the squared differences between the measured and simulated values, the sum of the squared differences between the logarithms of the measured and simulated values, and the Nash-Sutcliffe criterion for the long-term mean monthly values.

The hydrological model was calibrated using monthly data from the period 1971-2000 and validated using data from the period 1961-1970. The values of the Nash – Sutcliffe criterion for the calibration and validation period are listed in Table 2.

Table 2. Values of the Nash –Sutcliffe criterion for the calibration and validation period.

Nash – Sutcliffe	
Calibration 1971-2000	Validation 1961-1970
0.826	0.819

4. Results

The values of the long-term mean monthly discharges in the reference period 1971-2000 and the 2050 time horizon calculated for the ECHAM, HadCM and NCAR climate change scenarios are shown in Fig. 2; the differences in percentages between the long-term mean monthly discharges in the reference period and the future time horizons are compared in Table 3.

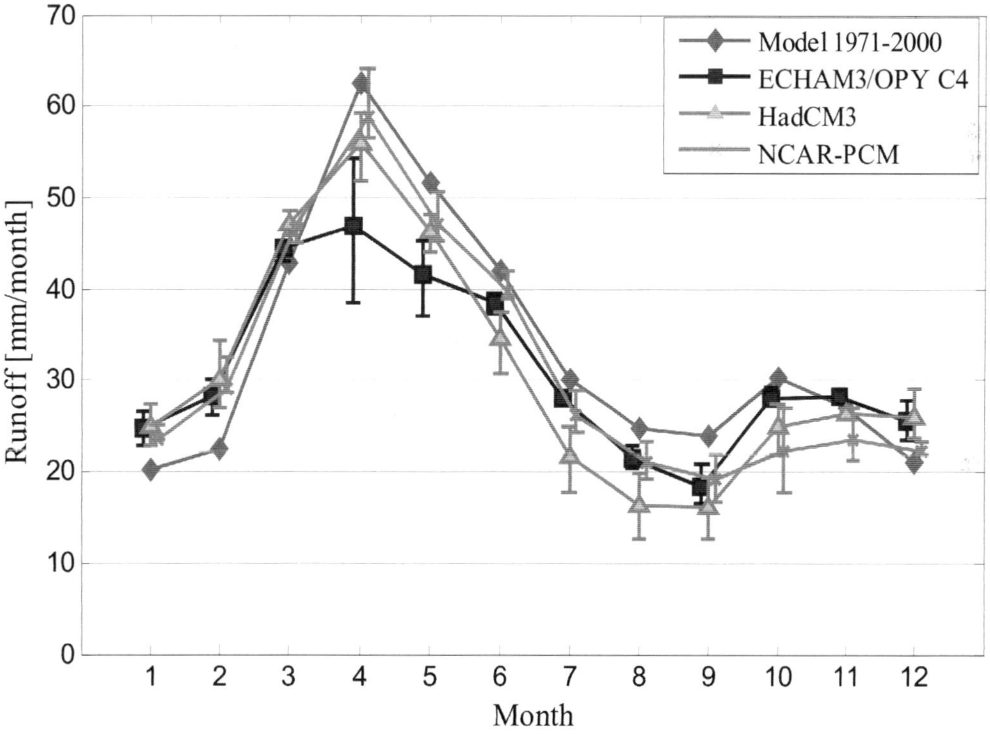

Figure 2. Long-term mean monthly runoff in the reference period 1971-2000 and in the time horizon of 2050 for all considered climate change scenarios.

Table 3. Percentage changes in the long-term mean monthly runoff for all the considered climate change scenarios in comparison with the reference period of 1971-2000

Scenario/	Month											
Horizon	1	2	3	4	5	6	7	8	9	10	11	12
ECHAM												
2025_HI	21	24	6	-18	-15	-6	-4	-9	-17	-4	5	18
2025_LO	8	9	3	-8	-8	-3	-2	-4	-7	-1	3	7
2025_MI	13	16	3	-13	-12	-5	-4	-6	-11	-3	4	11
2050_HI	32	34	0	-39	-28	-11	-8	-17	-31	-10	8	31

2050_LO	13	16	4	-13	-12	-6	-4	-8	-13	-4	4	11
2050_MI	23	25	4	-25	-20	-9	-7	-14	-23	-7	5	21
2100_HI	54	39	-17	-63	-47	-23	-16	-32	-55	-28	8	63
2100_LO	25	24	6	-30	-22	2	8	-12	-38	-8	7	25
2100_MI	32	34	0	-39	-29	-12	-9	-18	-32	-11	8	32
HadCM												
2025_HI	21	30	10	-8	-8	-14	-22	-28	-26	-14	-1	20
2025_LO	8	12	5	-3	-5	-7	-10	-13	-12	-6	0	7
2025_MI	13	19	7	-5	-7	-11	-17	-20	-19	-9	0	12
2050_HI	36	53	13	-17	-15	-27	-41	-49	-47	-28	-4	37
2050_LO	13	20	7	-5	-7	-11	-17	-20	-19	-9	0	12
2050_MI	23	34	10	-11	-11	-18	-28	-34	-32	-18	-2	23
2100_HI	74	93	9	-29	-21	-47	-68	-75	-74	-54	-11	77
2100_LO	28	39	16	-5	-6	-14	-24	-29	-27	-13	3	27
2100_MI	37	54	13	-18	-14	-27	-42	-50	-48	-28	-4	38
NCAR												
2025_HI	14	25	9	-5	-7	-5	-12	-13	-17	-22	-10	5
2025_LO	5	10	4	-2	-3	-1	-5	-6	-7	-9	-3	2
2025_MI	9	16	6	-3	-5	-3	-7	-9	-10	-14	-6	3
2050_HI	26	45	14	-10	-12	-7	-19	-23	-30	-42	-21	7
2050_LO	18	27	15	2	-2	0	-5	-6	-8	-11	0	11
2050_MI	17	29	10	-6	-9	-5	-13	-15	-19	-27	-12	6
2100_HI	53	85	25	-18	-19	-11	-35	-42	-52	-73	-53	6
2100_LO	15	26	9	-5	-8	-4	-11	-14	-18	-24	-10	5
2025_HI	26	46	15	-10	-12	-7	-19	-23	-30	-42	-22	7

The results presented of the modelling of the long-term mean monthly discharges indicate future changes in the seasonal runoff distribution in the upper Hron river basin. According to the pessimistic version of the ECHAM runs, the highest relative increase in runoff in comparison with the reference period can be expected in February, i.e., +24 % in 2025, +34 in 2050; and in December/January, i.e. +63 % in 2100. This increase could be caused by an increase in air temperature during the winter and a shift in the snow melting period from the spring months to the winter period. The most extreme relative decrease in runoff can occur in April, i.e. -18 % in 2025, -39 % in 2050, -63 % in 2100; and in September, i.e., -17 % in 2025, -31 % in 2050 and -55 % in 2100. The decrease in April is caused by an increase in air temperature (+1.5 °C in 2025, +2.9 °C in 2050 and +6.2 °C in 2100) and by a decrease in precipitation (about -7 % in 2025, -13 % in 2050 and -29 % in 2100). The decrease in September is caused by the increase in air temperature (+1.8 °C in 2025, +3.4 °C in 2050 and +7.2 °C in 2100) and by the decrease in precipitation (-10 % in 2025, -18 % in 2050 and -39 % in 2100) if compared to the reference period.

According to the pessimistic version of the HadCM runs, the highest relative increase in runoff in comparison with the reference period can be expected in February, i.e. +30 % in 2025, +53 % in 2050; and in December, i.e., +77 % in 2100. The most extreme relative decrease in runoff can occur from July to September, i.e., -28 % in 2025, -49 % in 2050 and -75 % in 2100. The lower runoff is caused by the increase in air temperature (in August: +2.3 °C in 2025, +4.3 °C in 2050 and +9 °C in 2075), as well as by the decrease in precipitation (in August: -15 % in 2025, -29 % in 2050 and -62% in 2100).

For the pessimistic version of the NCAR scenario, the highest relative increase in runoff can again be expected in February, i.e., +25 % in 2025, +45 % in 2050 and in December, i.e. +85 % in 2100. The most extreme relative decrease in runoff can occur from September to October/November, i.e., -22 % in 2025, -42 % in 2050 and -73 % in 2100. This decrease is caused mainly by the considerable decrease in precipitation (in October: -19 % in 2025, - 35 % in 2050 and -76 % in 2100).

It could generally be concluded for both of the investigated scenarios, that during the winter and early spring periods, an increase in the long-term mean monthly runoff could be assumed. The period of an increase in runoff could occur from November/December to March. This increase could be caused by the increase in air temperature and a shift of the snow melting period from the spring months to the winter period. The period of the decrease in runoff could occur from April to October/November. The most extreme decrease could be expected in April-May and July-September. The increase in winter runoff and the decrease in summer runoff are expected to be more extreme for later time horizons.

5. Conclusions

According to the anticipated changes in the seasonal distribution of the mean monthly runoff, the Hron river basin could become vulnerable to drought in the summer and early autumn. In months with increased water demands for irrigation, domestic and industrial use and tourism, the monthly flows could exhibit a decrease under conditions of climate change. The intensity of the changes could increase towards the time horizon of 2100. A continual general decrease in the utilizable potential of the surface and subsurface water resources is likely to occur, which will have to be taken account of in the planning and management of water resources in the near future.

However, due to several uncertainties, it is rather difficult to present quantitative adaptation guidelines and estimate their cost. Uncertainties in the boundary conditions for the long-term development and management of water resources in Slovakia include: the not-quite-clear priorities for Slovakia's economic transformation, its energy policy, and the gradually changing water legislation in the near future; the changing, but unsettled, priorities for institutional, industrial, agricultural and urban developments; and the possibility of the privatization of the water supply, water power and irrigation sectors.

A "no regret" policy based on environmental protection priorities, which would incorporate adaptive measures for the mitigation of the effects of climate changes, is therefore seen here as a solution for the near future. In order to eliminate the effects of the possible negative trends in the hydrological cycle and development of water resources in Slovakia, the adaptation strategies addressed on the national level will have to incorporate a number of legislative, organizational and technical actions aimed at protecting the country's water resources and their source areas.

Acknowledgement. The research presented in this paper was supported under the VEGA project 2/0096/08. The support is gratefully acknowledged.

6. References

[1] Dubrovský M, Žalud' Z, Šťastná M 2000 Sensitivity of CERES-maize yields to statistical structure of daily weather series *Clim. Change* **46** 447–72
[2] Maytín C E, Acevedo M F, Jaimez R, Andressen R, Harwell M A, Robock A, Azkcar A 1995 Potential effects of global climatic change on the phenology and yield of maize in Venezuela *Clim. Change* **29** 189–211
[3] Danihlík R, Hlavčová K, Kohnová S, Parajka J and Szolgay J 2004 Scenarios of the change in the mean annual and monthly runoff in the Hron Basin *J. Hydrol. Hydromech.* **52** 4 291-302
[4] Fendeková M 1999 Quantitative aspects of groundwater regime *Acta Geologica Universitatis Comenianae* **54** 27–52
[5] Fendeková M, Némethy P 2001 Limiting conditions for groundwater withdrawals utilised for the water supply in the upper part of the Torysa river catchment *International conference proceedings: Water is Life–Take Care of It* (Bratislava: Water Research Institute) pp 375–378
[6] Halmová D 2004a Conflicts between reservoir water demands and climate change inflow *J. Hydrol. Hydromech.* 4 **52** 329-341
[7] Halmová D 2004b Impact of potential climate change upon the water supply observed during

operation of the Vihorlat reservoir *22nd Conference of the Danubian countries on the hydrological forecasting and hydrological forecasting* (Prague: Czech Hydrometeorological Institute) p 11

[8] Halmová D 2005 Impact of climate change on utilization of water storage of the Orava reservoir In: Pekárová P, Szolgay J eds. *Assessment of climate change impact on selected components of the hydrosphere and biosphere in the Hron and Vah basins* (Bratislava: VEDA) 253-282

[9] Halmová D 2006 Impact of expected climate change on storage volume utilisation of important water reservoir of the Slovak Republic In: Stevan Bruk, Tioslav Petkovic, eds.: *23rd Conference of the Danubian countries on the hydrological forecasting and hydrological bases of water management* (Belgrad: NC of Serbia for the IHP of UNESCO) p 10

[10] Halmová D, Melo M 2006 Climate change impact on reservoir water supply reliability *IAHS-AISH Publication* Wallingford **308** 407-12

[11] Hlavčová K, Szolgay J, Čunderlík J, Parajka J, Lapin M 1999 Impact of climate change on the hydrological regime in Slovakia *Publication of the Slovak Committee for Hydrology No 3* (Bratislava: SUT, NC IHP UNESCO) p 101

[12] Hlavčová K, Szolgay J, Parajka J, Čunderlík J 2000 Modelling of climate change impact on the runoff regime in Central Slovakia *Publication of National Climate Programme of the Slovak Republic* 9 (Bratislava: SHMI, ME SR) pp 15−38

[13] Hlavčová K, Kalaš M, Szolgay J 2002 Impact of climate change on the seasonal distribution of runoff in Slovakia *Slovak Journal of Civil Engineering* **X** 2 10−17

[14] Hlavčová K, Kohnová S 2005 Proposals of framework adaptative measures. In: Pekárová P, Szolgay J eds. *Assessment of climate change impact on selected components of the hydrosphere and biosphere in the Hron and Vah basins* (Bratislava: Veda) pp 419-492

[15] Hlavčová K, Szolgay J, Kohnová S, Parajka J 2006a Hydrological scenarios of changes in the mean annual and monthly runoff in Slovakia In: Bruk S, Petkovic T eds. *23nd Conference of the Danube Countries on the Hydrological Forecasting and Hydrological Bases of Water Management* (Belgrade: NC of Serbia for the IHP of UNESCO) p 11

[16] Hlavčová K, Parajka J, Szolgay J, Kohnová S 2006b Grid-based and conceptual approaches to modelling impact of climate change on runoff *Slovak Journal of Civil Engineering* **XIV** 1 19-29

[17] Kostka Z, Holko L 2000 Impact of climate change in a small mountain catchment *Publication of National Climate Programme of the Slovak Republic* 9 (Bratislava: SHMI, ME SR) pp 91−109

[18] Kostka Z, Holko L 2001 Impact of vegetation changes on river runoff in a small mountain catchment *Publication of National Climate Programme of the Slovak Republic* 10 (Bratislava: SHMI, ME SR) pp 82-93

[19] Kostka Z 2003 Changes in hydrological processes in mountainous regions in interaction with a changing environment *Acta Hydrologica Slovaca* **4** 1 190-196

[20] Majerčáková O 2000 Modelling of runoff changes due to climate change in Central Slovakia *Publication of National Climate Programme of the Slovak Republic* 9 (Bratislava: SHMI, ME SR) pp 5−14

[21] Majerčáková O, Takáčová D 2001 Possible impact of climate change on groundwater in alluvial fans *Publication of National Climate Programme of the Slovak Republic* 11 (Bratislava: SHMI, ME SR) pp 31−49

[22] Pekárová P, Halmová D, Miklánek P 1996 Simulation of runoff changes under changed climatic conditions in the Ondava catchment *J. Hydrol. Hydromech.* **44** 5 29−311

[23] Pekárová P, Miklánek P, Pekár J 2001 Analysis of runoff fluctuations in the moderate and subarctic regions *Acta Hydrologica Slovaca* **2** 1 122−129

[24] Pekárová P 2000 Mean annual runoff fluctuation *Publication of National Climate Programme of the Slovak Republic* 9 (Bratislava: SHMI, ME SR) pp 39−57

[25] Pekárová P, Miklánek P 2001 Increase of floods extremality on the Uh river *International*

Conference on Water and Nature Conservation in the Danube-Tisza River Basin (Debrecen: Magyar Hidrológiai Társaság) pp 469–480
[26] Pekárová P, Szolgay J eds. 2005 *Assessment of Climate Change Impact on Selected Components of the Hydrosphere and Biosphere in the Hron and Váh River Basins* (Bratislava: VEDA) pp 419-492
[27] Petrovič P 1998 Climate change impact on hydrological regime for two profiles in the Nitra River basin In: Bonacci O ed. *20th Conference of the Danube Countries on Hydrological Forecasting and Hydrological bases of Water Management* (Osiek) pp 117–122
[28] Petrovič P 2000 Impact of climate change on the hydrological regime in the Nitra River Basin *Publication of National Climate Programme of the Slovak Republic* 9 (Bratislava: SHMI, ME SR) pp 58–73
[29] Szolgay J, Hlavčová K, Parajka J, Čunderlík J 1997 Assessment of the potential impact of climate change on river runoff in Slovakia *Publication of the Slovak National Climate Programme of the Slovak Republic* 6 (Bratislava: SHMI, ME SR) pp 11-110
[30] Szolgay J, Hlavčová K, Kalaš M 2002 Estimation of climate change impacts on a runoff regime *J. Hydrol. Hydromech.* **50** 4 341–371
[31] Szolgay J, Hlavčová K, Lapin M, Danihlík R 2003 Impact of climate change on mean monthly runoff in Slovakia *Meteorological Journal* **6** 3 9–21
[32] Szolgay J, Hlavčová K, Lapin M, Kohnová S, Parajka J 2007 *Impact of changed climate on runoff regime in Slovakia* (Ostrava: KEY Publishing) p 160
[33] Takáč J 2001 Impact of climate change on the water balance in an agricultural landscape *Publication of National Climate Programme of the Slovak Republic* 10 (Bratislava: SHMI, ME SR) pp 16–26
[34] CECILIA 2007 Central and Eastern Europe Climate Change Impact and Vulnerability Assessment. D 5.4: The evaluation of climate change impacts on simulated monthly river flow along the Bohemia/Moravia/Slovakia/Romania geographic gradient
[35] Dubrovský M, Nemešová I, Kalvová J 2005 Uncertainties in climate change scenarios for the Czech Republic *Climate Research* **29** 139-156
[36] Harvey LDD, Gregory J, Hoffert M, Jain A et al 1997 An introduction to simple climate models used in the IPCC Second Assessment Report *IPCC Tech Paper 2, Intergovernmental Panel on Climate Change* Geneva
[37] Hulme M, Wigley T M L, Barrow E M, Raper S C B, Centella A, Smith S, Chipanshi A C 2000 Using a climate scenario generator for vulnerability and adaptation assessments: MAGICC and SCENGEN Version 2.4 Workbook. *Climatic Research Unit* Norwich
[38] IPCC 2001: Houghton J T Y, Ding D J Climate Change 2001 *The Scientific Basis Contribution of Working Group I to the Third Assessment Report of the Intergovernmental Panel on Climate Change*
[39] IPCC 2007 IPCC Fourth Assessment Report. *Working Group I Report "The Physical Science Basis* http://www.ipcc.ch/ipccreports/ar4-wg1.htm p 996

IOP Conf. Series: Earth and Environmental Science **4** (2008) 012023 doi:10.1088/1755-1307/4/1/012023

Estimation of climate change impact on water resources by using Bilan water balance model

Stanislav Horáček, Ladislav Kašpárek and Oldřich Novický

Department of Hydrology, T. G. Masaryk Water Research Institute, Podbabská 30, 160 62 Prague 6, Czech Republic

E-mail: stanislav_horacek@vuv.cz

Abstract. Modelling of water balance under changed climate conditions has been carried out by T. G. Masaryk Water Research Institute in Prague for basins in the Czech Republic since 1990. The studies presently use climate change scenarios derived from simulations by regional climate models. Climate change scenarios are reflected in meteorological time-series for given catchment and subsequently used for simulation of water cycle components by using Bilan water balance model. Results of Bilan model simulations for input meteorological series not affected and affected by climate change scenarios give information for assessing the climate change impacts on output series of the model. The results of the studies generally show that annual runoff could largely decrease. The increased winter temperature could cause an increase in winter flows and a decrease in snow storage, and consequently, spring and summer outflows will decrease significantly, even to their current minimum values. The groundwater storage and base flow could also be highly reduced. The described method has been used in a number of research projects and operational applications. Its typical application is aimed at assessing possible impacts of climate change on surface water resources, whose availability can subsequently be analysed by using water management models of the individual basins. The Bilan model, particularly in combination with Modflow model, can also suitably be used for simulation and assessments of groundwater resources.

1. Introduction

Results of studies of climate change impacts on water resources in basins of the Czech Republic support water management planning at different levels, which include central planning carried out by Ministry of Agriculture, planning at regional level and water management activities of individual water users. The studies are performed by using methods and software developed by T. G. Masaryk Water Research Institute in Prague. These tools are based on water balance modelling by using Bilan model for climate conditions alternatively not affected and affected by climate change. The paper describes principles of the methods, summarises results of a study which was aimed at comparing changes in outflows from basins in the Czech Republic, and shows an example of typical application of the model in assessing surface water resources.

2. Climate change scenarios

Climate change scenarios are used for modification of meteorological time-series. Presently, the results of the simulation by regional climate models (RCM), which replaced those derived from the simulation by global climate models (GCM), are used. The climate change scenarios were derived from the results of PRUDENCE project for two scenarios of the emissions of greenhouse gases, which reflect possible alternatives of human socioeconomic progress. The SRES A2 scenario assumes permanent growth of the emissions while the SRES B2 scenario is based on an assumption of a decrease in emissions during this century.

In terms of climate change models, the climate change scenarios are based on the simulation alternatively by HIRHAM model, which was developed by meteorological institutes of 8 European countries (based on the results of ECHAM model simulation), and RCAO model, which was developed at Rossby Centre in Sweden. All of the simulations were performed for the period 2071–2100 (time horizon of 2085) with reference time period 1960–1990.

Climate change scenarios based on these results were derived by Faculty of Mathematics and Physics of Charles University in Prague. They were prepared in the form of monthly time series of absolute changes in values of air temperature and dew-point temperature and relative changes in precipitation [1]. Space resolution of the used regional models is about 50 × 50 km and the area of the Czech Republic is covered by about 50 points.

3. Methods for assessing climate change impact

The Bilan model, which is used for the studies of possible impacts of climate change on water resources, is a lumped water balance model that was developed by T. G. Masaryk Water Research Institute (described in [2]). This model can be used for assessing water balance components in monthly or daily step. The catchment is schematized as a system (figure 1), whose inputs are basin precipitation, air temperature and relative air humidity and its output is the total streamflow.

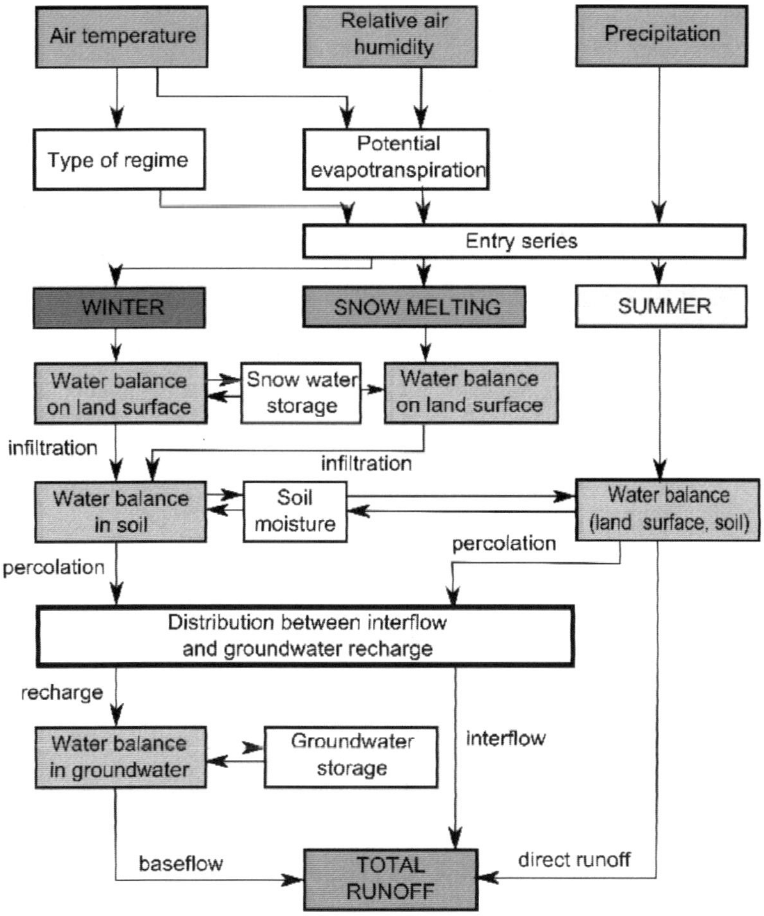

Figure 1. The structure of Bilan water balance model

The model simulates water budget at three vertical levels: on land surface, in soil layer and in groundwater aquifer. Three water balance algorithms that are applied were developed for winter conditions, snow melting and summer conditions. Surface water balance depends on evapotranspiration, which is determined by meteorological conditions by using empirical values that have been derived for different climate zones. Excess water (precipitation minus evapotranspiration) forms direct runoff or infiltrates to deeper zone, where it is divided into interflow and groundwater recharge. Parameters of the model are determined by two-step optimization, which is aimed at attaining a good fit between the observed and simulated river flows. The parameters that significantly affect the total runoff are calibrated in the first step while the second step includes those parameters that affect distribution of the runoff into its components. The calibration by using results of base flow measurements can alternatively be applied if such data are available.

The simulation of climate change impacts on water regime is based on knowledge of the model parameters that have been calibrated for observed input series of meteorological variables and runoff values. These parameters are used in combination with the meteorological series which have been modified according to given climate change scenario.

The changes in the meteorological variables for given catchment are derived from the values simulated by RCM for points, which are located within the catchment area and in its surroundings. The derivation uses inverse distance weighting method based on distances of the points from the centre of the catchment. Precipitation and air temperature time-series are modified by adding the interpolated values, while relative humidity is calculated from interpolated dew-point temperature.

Results of the simulations for affected and not affected input series give information for assessing the climate change impacts on output series of the model, which include potential evapotranspiration, basin evaporation, three components of river flow (surface runoff, interflow and base flow), groundwater recharge and three components of water storage (in snow cover, soil layer and groundwater aquifer).

4. Changes in meteorological variables

Generally, seasonal variability of changes in meteorological values resulting from RCM simulation is more important than spatial variability across the area of the Czech Republic. HIRHAM model predictions are influenced by local orography, particularly by the altitude of given location. Opposed to this, RCAO model puts the accent on simulation of air circulation between ocean and continent and its resulting changes of meteorological variables correspond with east-west gradient across the Czech Republic.

Precipitation is the most important variable in the water balance assessment. Its change in terms of the annual mean is negligible as compared to its seasonal changes. Winter precipitation increases (especially in February, see figure 2) from 10% to 60% dependably on climate model, emission scenario and the locality while summer precipitation decreases, which is particularly significant in August (the decrease between 5% and 50%). HIRHAM predicts the highest winter increase and summer decrease in the northern Bohemia and in the eastern Moravia. RCAO assumes the highest variability in precipitation in the region of southern Moravia. Influence of emission scenario is more important than that stemming from type of the climate model. Models that use SRES B2 scenario predict higher precipitation.

Both models predict significant temperature increase for all months, the changes range from 3 to 8 °C for A2 emission scenario and from 1 to 6 °C for B2 emission scenario. RCAO model predicts lower changes than HIRHAM. Temperature changes in August are shown in figure 3. Temperature seasonality simulated by HIRHAM can be divided into two time intervals. The interval from April to September is typical by higher changes in lowland regions while during the remaining part of a year, higher changes are predicted in mountain areas. In harmony with spatial variability of the results of RCAO model simulation, the temperature changes are higher in the eastern part of the Czech Republic than in its western part.

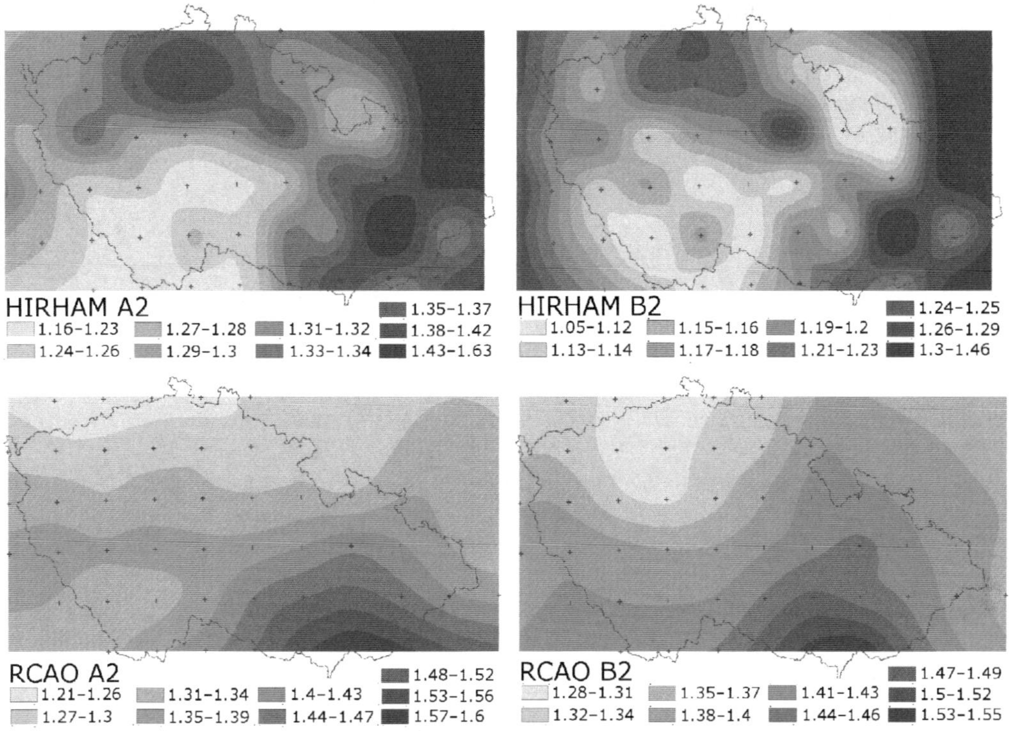

Figure 2. Change in precipitation in February for given climate models and emission scenarios

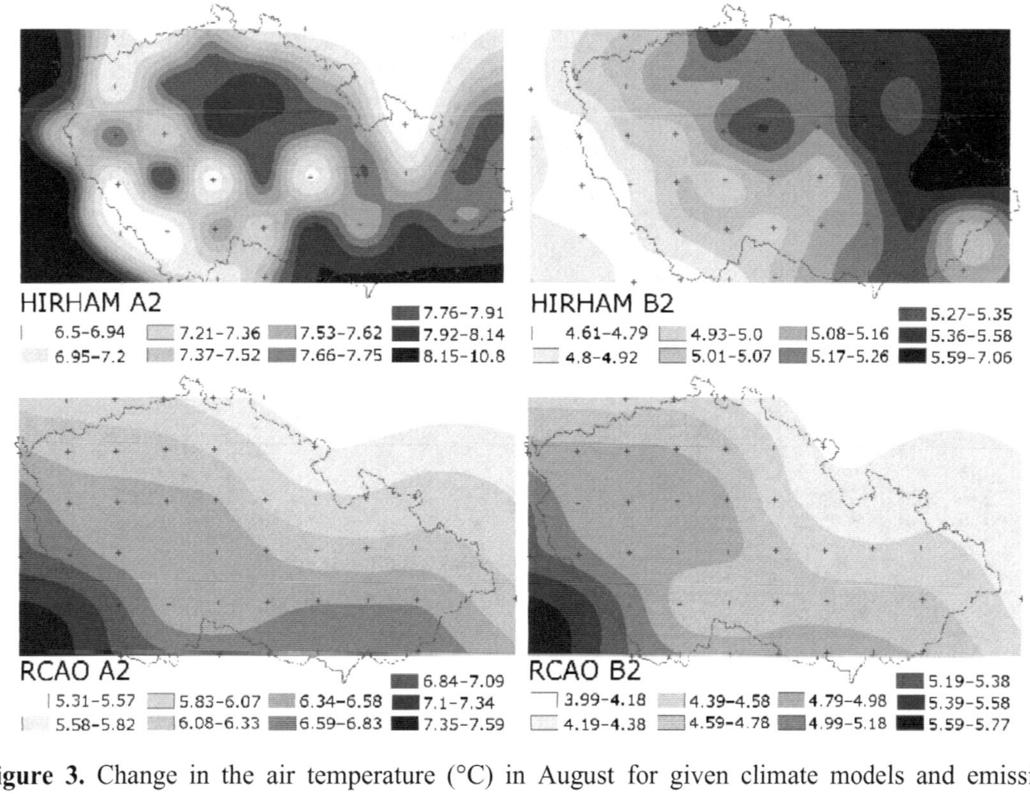

Figure 3. Change in the air temperature (°C) in August for given climate models and emission scenarios

5. Changes in runoff in basins of the Czech Republic

Climate change impact on water resources was studied in monthly time step for Czech parts of the Elbe and Danube River basins [3]. The study was performed for four alternative climate change scenarios that were derived from two types of climate models and two types of emission scenarios.

Parameters of Bilan model were calibrated and subsequent simulations were performed for 61 basins, whose data for the period 1932–2003 were provided by Czech Hydrometeorological Institute. The results of the study showed that the climate change would be mainly reflected in a decrease in mean annual runoff, which applies to all of the basins except for several basins in the southern Bohemia and southern Moravia and simulations of there hydrological conditions by using SRES B2 emission scenario. Consequently to the climate change, the annual runoff (mean value calculated from the results for all of the basins) would drop to 67% of its unaffected value for HIRHAM model and A2 scenario, to 80% for both HIRHAM B2 and RCAO A2 alternatives and to 92% for RCAO model and B2 scenario.

For most of the months the runoff also decreases, except for winter period from December to February when outflows increase or only moderately decrease. This is caused by higher winter temperature (monthly temperature below zero is predicted only for December), which is reflected in an increase in runoff and a decrease in snow storage. Consequently, spring and summer outflows decrease significantly in some basins, even to their current minimum values. An example of typical annual runoff distribution is shown in figure 4.

The results of the Bilan Model simulation also indicate that groundwater storage and base flow could be also highly reduced by the impacts of climate change. It was shown in some other studies carried out by the Institute [4] that for detailed studies of the impacts of climate change on groundwater resources, the Bilan model simulation can suitably be combined with the groundwater flow simulation by using Modflow hydraulic model. In this approach, the groundwater recharge series simulated by Bilan model for conditions not affected and affected by climate change are used as an input into the Modflow model.

Spatial variability in runoff change for different models and scenarios is illustrated in figure 5, which indicates that it can be different even in the same region. However, there is a common trend to smaller decrease in the northern Bohemia (basins of the Upper Elbe and Ohře Rivers).

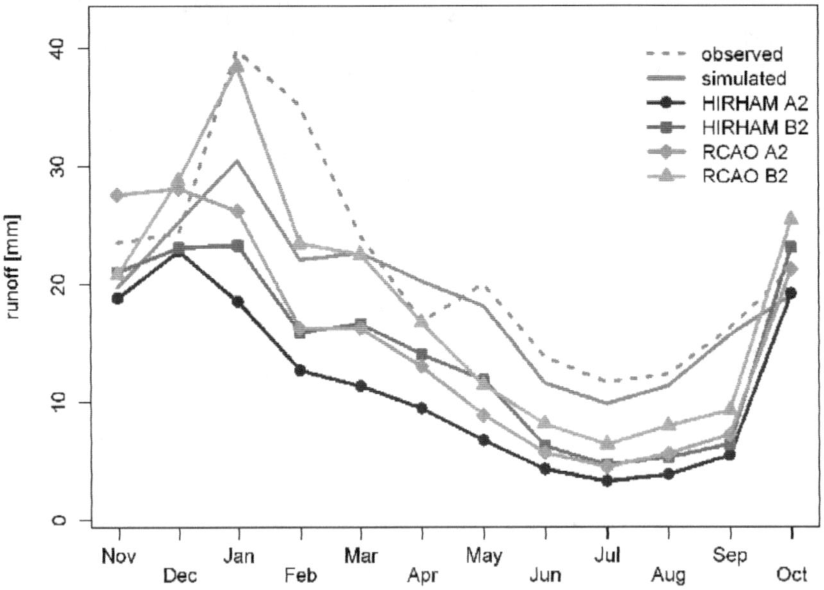

Figure 4. Example of typical seasonal patterns of total runoff – mean values of unaffected and affected runoff calculated for the Morava River at Olomouc – Nové Sady

IOP Conf. Series: Earth and Environmental Science **4** (2008) 012023 doi:10.1088/1755-1307/4/1/012023

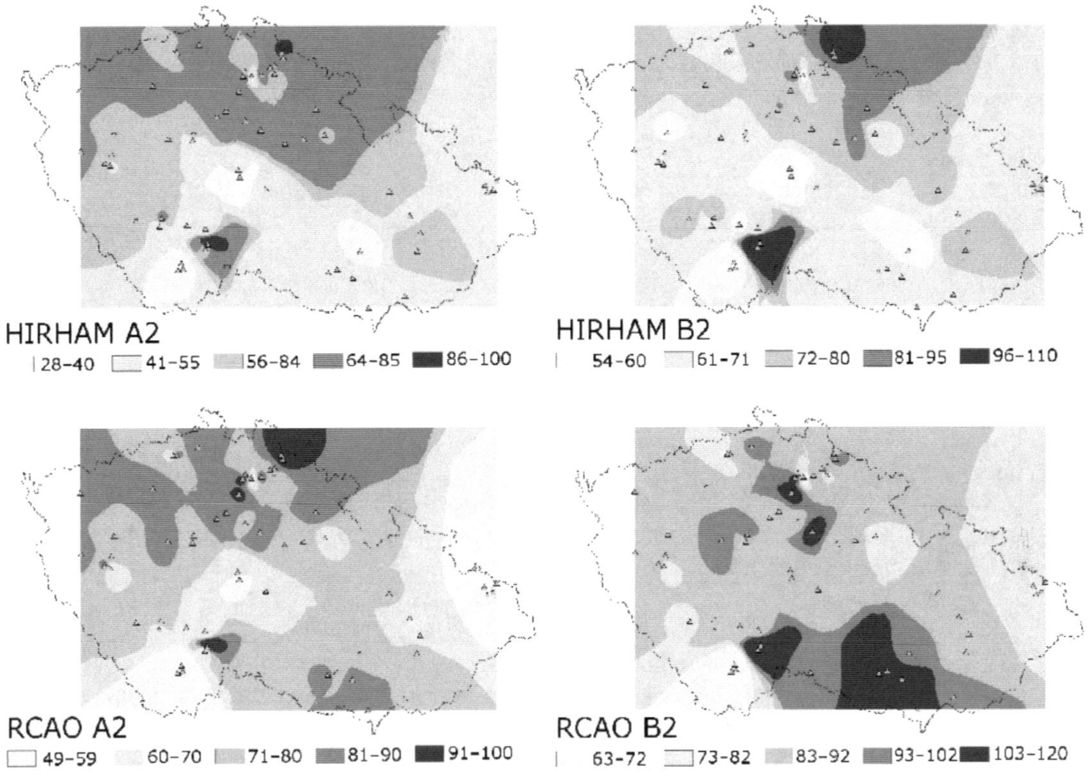

Figure 5. Mean annual runoff for given climate models and emission scenarios (in percent of unaffected runoff)

6. A case study: estimation of climate change impact on water availability for a power plant

The described method has been used in a number of applications. Its typical application is aimed at assessing possible impacts of climate change on surface water resources, whose availability can subsequently be analysed by using a water management model, which includes capacities and functions of reservoirs in the basin.

One of these studies was focused on availability of technological water for Temelín Nuclear Power Plant [5], which is located in the southern Bohemia. Water balance was assessed for three subbasins of the Vltava River (which include the reservoir used for the withdrawals) and for two basins of its tributaries. For the purposes of the study, time series of hydrological and meteorological variables whose length is 44 years were available. The runoff series simulated for observed meteorological data were compared with those derived for climate change scenarios. The scenarios were alternatively based on RCM model simulations for time horizon of 2085 and on short-term outlook derived for the year 2020 from current trends of meteorological variables. Reductions of the annual flows consequently to climate change ranged from 5% to 25% for 2020 or from 5% to 65% for 2085 dependably on the basin and a combination of RCM and emission scenario used for the analysis. In the season from November to March, monthly runoff was highly variable and its mean exceeded that of the remaining period, whose runoff variability was smaller. This flow distribution reflects the change in seasonality of precipitation and earlier snow melting consequently to the increased winter temperature.

The results of the hydrological analysis were used as the input into the model of water management system. This model was applied for analysis of water supply reliability, which was calculated as minimum monthly ratio between available volume of water and that derived from the water supply requirements. The resulting reliability was 100% for all of the scenarios for 2020 and for SRES B2 scenarios for 2085. For SRES A2 scenario, the resulting reliability was 40% for HIRHAM model and 76% for RCAO model. In these cases, maximum duration of the failure was 3 months and 5 months respectively.

7. Conclusions

The lumped hydrologic water balance model (Bilan) is a suitable tool for analyses of possible impacts of climate change on water cycle and its components. The results of the simulation by regional climate change models provide a base for derivation of necessary meteorological data. The described method was applied for general studies of large basins as well as for detailed studies aimed at supporting the activities in water management planning. The results of these studies show that mean annual runoff and monthly runoff in spring, summer and autumn can significantly decrease consequently to climate change. These impacts can be reflected in a risk of insufficiency of the existing retention capacities in water resources for meeting the water supply requirements in future.

References

[1] Kalvová J 2005 Vytvoření klimatických scénářů pro ČR (Preparation of climate scenarios for the CR) *Research and protection of hydrosphere, Part A. Hydrology. Research report* (Prague: VÚV T.G.M.) pp 22–46

[2] Tallaksen L M and van Lanen H A J 2004 *Hydrological Drought – Processes and Estimation Methods for Streamflow and Groundwater* (Amsterdam: Elsevier)

[3] Kašpárek L et al. 2006 Dopad klimatické změny na Českou republiku podle regionálních klimatických scénářů (Climate change impact in the Czech Republic by using regional climate scenarios) *Research and protection of hydrosphere, Part A. Hydrology. Research report* (Prague: VÚV T.G.M.) pp 90–105

[4] Novický O, Kašpárek L and Uhlík J 2007 Possible impacts of climate change on groundwater resources and groundwater flow in well developed water bearing aquifers *Proc. of the Third International Conference on Climate and Water (Helsinki, 3–6 September 2007)* ed M Heinonen (Helsinki: Finnish Environment Institute) pp 358–64

[5] Kašpárek L, Hanel M, Vyskoč P, Zeman V and Picek V 2007 *Studie možnosti zajištění odběrů vody z VD Hněvkovice pro výhledové rozšíření JE Temelín (Possibilities of securing water supply from Hněvkovice reservoir for future production of Temelín Nuclear Power Plant)* (Prague: VÚV T. G. M.)

Slovenian Experience in Applicability of Remote Sensing Data in Hydrology

A Horvat, A Vidmar, S Petan and M Brilly

University of Ljubljana, Faculty of civil and geodetic engineering, Chair of hydrology and hydraulic engineering, Hajdrihova 28, Ljubljana, Slovenia

E-mail: mbrillygg.uni-lj.si

Abstract. Nowadays remote sensing data are great value for many ways of use as for snow hydrology. Snow cover mapping on basis of satellite images is widely spread because it is easier than in-situ measuring and interpolating. To verify the use of satellite imagery (MODIS product) we used all available data of snow cover and satellite images and compared detected snow cover on images and in-situ measured snow cover. The information about clouds and snow cover measured on the Ljubljana climatological station were acquired and analysed. The use of satellite images showed out to be useless because of climate characteristics of Slovenia and low elevation of much area.

1. Introduction

Snow cover maps are important basin for snow hydrology research. A snow depth map can be derived from satellite image or from in-situ measurements and interpolation [1]. Satellite snow cover images have already been avaliable since 1966, first from NOAA (National Oceanic and Atmospheric Administration) and now their use is widely spread mostly over snow covered countries [5]. With combining of snow cover maps and hydrology models we can determine influence of snowmelt on groundwater dynamics and furthermore how much does snowmelt contribute to water availability in mountain environment or what is the impact of groundwater on runoff formation and on the production of flood hydrograph. But for now we still need to validate satellite snow cover maps with measurements because this method is still developing. Some validations show promising results, like Turkish experience when snow cover maps derived from satellite images were used for purpose of determine snow melt runoff in mountain regions with 82% of agreement between measurements and MODIS snow cover maps [11].

Many developed examples of snow pack models all over the world, but most for mountain regions, in Europe for Alps. For Swiss Alps is developed a model that combines snow pack simulation, GIS and satellite images and could finally be used for natural risk management (avalanches, debris flows, flooding), hydropower generation or tourism. [14].

Remote sensing provides many adventages to hydrology science. Mostly it eases the hard work of measuring and colecting data, because with use of satellite images we get large areas covered [10]. But for generalising results and developing fundamental understanding more field studies would be needed [2].

2. Background

2.1. About MODIS

MODIS means Moderate Resolution Imaging Spectroradiometer and is an instrument included in Terra satellite. It is imaging instrument with objectives to study global vegetation, land cover, global land-surface change, vegetation, surface albedo, temperature and snow and ice cover. The temporal resolution is daily or near daily and the spatial resolution of snow maps is 500 m [5].

Problem in snow mapping with MODIS is the discrimination of snow from clouds that have similar spectral reflectance. This would demand changes in snow algorithm that would allow satellite sensor to indentify differences between snow and clouds.

2.2. Snow measurement stations

Snow cover in mountains is very variable due to windy sides, lee sides and the valleys [6]. For this reason it is very important to think carefully before setting up measurement stations or to use data. For validation purpose we used snow measurements from station operated by Slovenian environmental agency (ARSO). Since area of validation is small (about 12 000 km^2) we could rely on one station measurement, that is in Ljubljana.

3. Data and Methodology

Snow cover map is essential basis for many hydrology applications as combined remote sensing, runoff modeling, GIS, DEM, and more [3]. Nowadays more and more snow cover maps are produced with help of computer techniques but still need to be validated on field studies. Comparison of in-situ measurements and MODIS images was also made for Austria [7].

The basis of our study is that Slovenian average duration of snow cover is more than 50 days a year (figure 1), which means that snow runoff contributes to surface or groundwater flow. This can be supported by a fact that Ljubljana, capital of Slovenia, lies on the largest aquifer in the republic that recharges from surface water. For this reason it is important to know the impact of the snow melt on ground water and from basis it is necessary to have snow cover map. According to wide use of satellite snow cover images we decided to explore the possibility of use in Slovenian area.

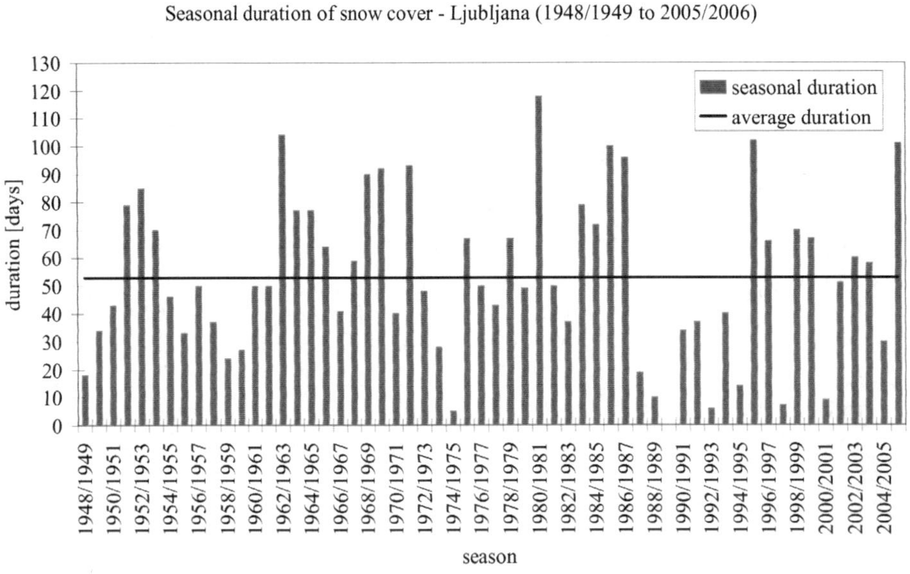

Figure 1. Seasonal duration of snow cover.

IOP Conf. Series: Earth and Environmental Science **4** (2008) 012024 doi:10.1088/1755-1307/4/1/012024

3.1. Available EO Modis data for snow cover area

The release #3 of the SCA product has been performed using the information coming from MOD bands 02, 04, 06 and from Landsat TM images of the project Corine 2000.

3.1.1. Basic Data Description.. Projection: UTM33 (Universal Transverse Mercator, Zone 33 North);

Geo. Datum: WGS84 (World Global System 1984); Raster Format: GEOTIFF (Geo Tagged Image File Format), 4-bit Pixel Depth.

Table 1. List of SCA images data.

ID	SCA Raster File Name	Time	Hour	Date	Qty.SCA
1	mod02hkm_2002299.1050_fuso33_sca3.tif	2002299_1050	10:50	26.10.2002	little
2	mod02hkm_2002324.1045_fuso33_sca3.tif	2002324_1045	10:45	20.11.2002	little
3	mod02hkm_2002363.1050_fuso33_sca3.tif	2002363_1050	10:50	29.12.2002	more
4	mod02hkm_2003014.1050_fuso33_sca3.tif	2003014_1050	10:50	14.1.2003	little
5	mod02hkm_2003055.1045_fuso33_sca3.tif	2003055_1045	10:45	24.2.2003	much more
6	mod02hkm_2003078.1050_fuso33_sca3.tif	2003078_1050	10:50	19.3.2003	more
7	mod02hkm_2003103.1045_fuso33_sca3.tif	2003103_1045	10:45	13.4.2003	more
8	mod02hkm_2003126.1050_fuso33_sca3.tif	2003126_1050	10:50	6.5.2003	very little
9	mod02hkm_2003162.1025_fuso33_sca3.tif	2003162_1025	10:25	11.6.2003	no
10	mod02hkm_2003201.1030_fuso33_sca3.tif	2003201_1030	10:30	20.7.2003	no
11	mod02hkm_2003229.1055_fuso33_sca3.tif	2003229_1055	10:55	17.8.2003	no
12	mod02hkm_2003261.1055_fuso33_sca3.tif	2003261_1055	10:55	18.9.2003	no

Classification scheme:

1	805867	undetermined	
2	19502	lake	
3	131900	sea	
4	65796	snow	
5	239114	clouds	

3.1.2. Overview and calculation of SCA with vector GIS. Due to the GIS vector data processing used to achieve high accuracy of calculation at large scales (1 : 5000 and 1 : 25,000) and compatibility with available existing vector data, the raster content of SCA3 had to be vectorized.

The conversion to vector was performed with the vectorizer tool called "Vextractor" for automatic conversion to the vector format. The software enables »batch« processing of GeoTIFF geopositioned raster images, so that a subsequent vectorizing geopositioning and/or conversion is not necessary. The obtained vectorized topology assumes the content of projection and ratios automatically from the raster and simultaneously for all dates. Only the data indicated in light blue colour were vectorized (Class 4), that is, the data representing snow cover. Upon vectorization, we obtained the vector contours of snow cover.

The vector layer was overlain with the contour of the Republic of Slovenia, which was the basis of the vector SCA polygons in the area of block UTM33&WGS48 covering half of the surface of Slovenia, that is, its western part (Figure 2).

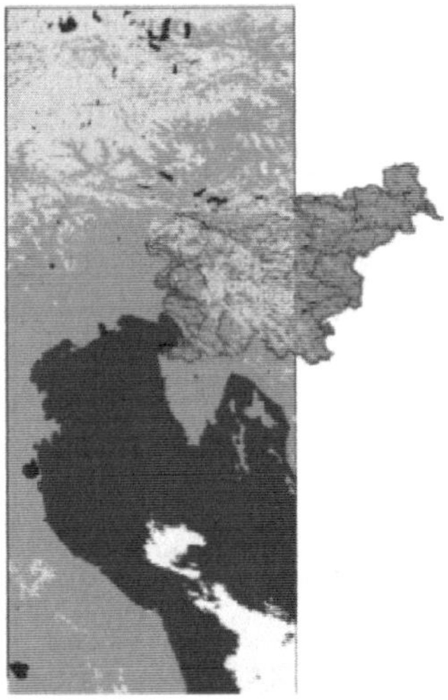

Figure 2. SCA, 24 February 2003 (UTM33-WGS84 whole area over R Slovenia).

The SCA3 polygons were obtained for 8 dates (Figure 3). 4 images show the lack of snow cover in the area of Slovenia (Table 1). In July, June, August and September of 2003 there was no snow, however, in February 2003 there was considerably more snow.

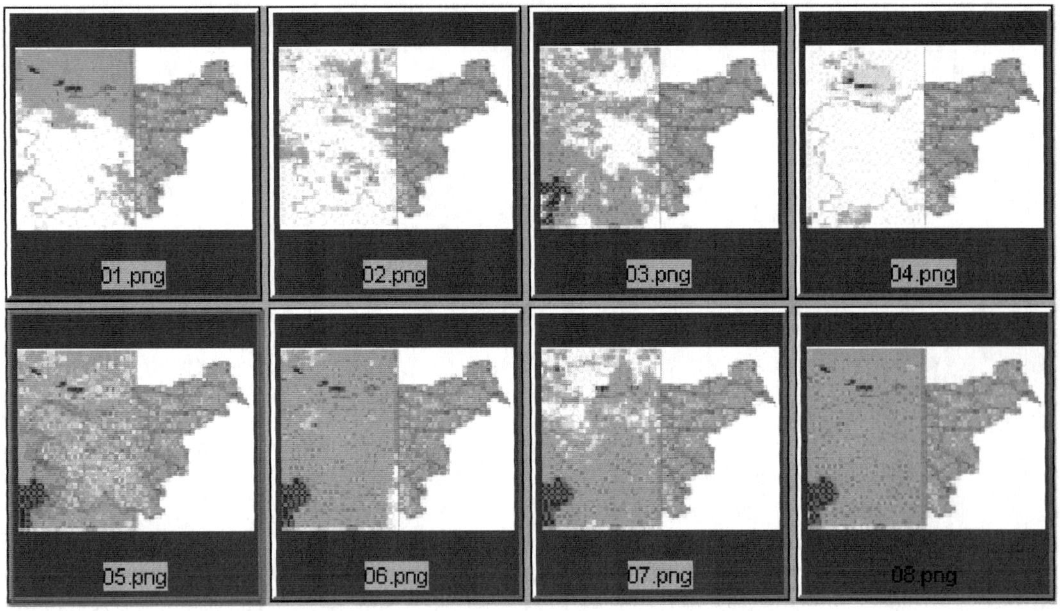

Figure 3. Snow cover of the western part of the Republic of Slovenia, from October 2002 to May 2003.

The obtained vector polygons contained several patches (islands) that needed to be excluded from further processing. A more complex typology had to be elaborated (topology with islands). The nested-in islands (single nested or multi-nested – island within an island) were excluded separately. These were determined based on topological attributes and algorithm, looking at the left and right side of the vectors, helping us to determine whether or not the polygon is an island.

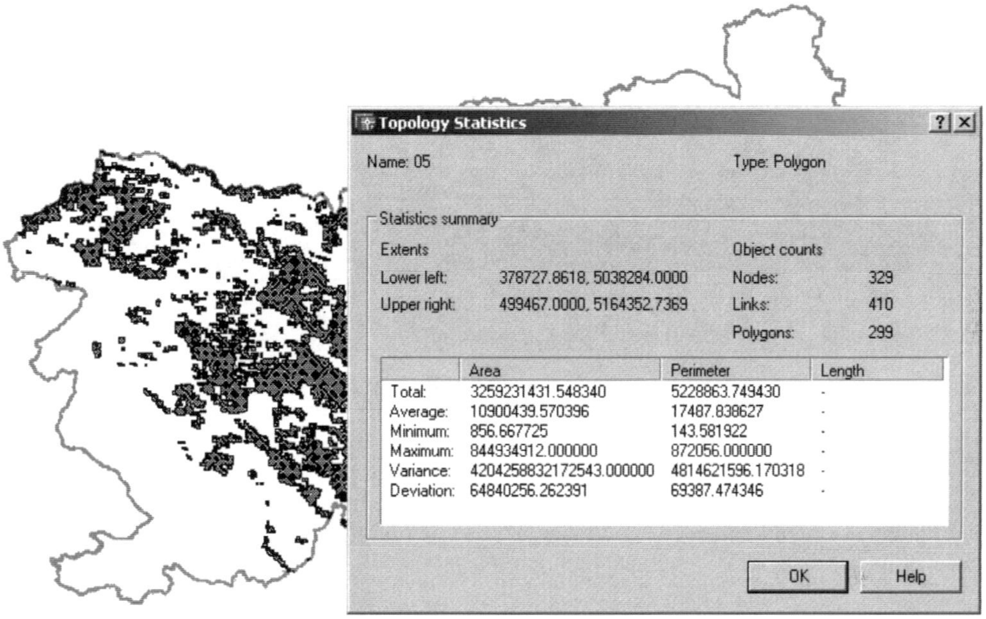

Figure 4. Topology statistic SCA3, 24 February 2003 (3,259.2 km^2).

Then we built a new topology (Figure 4). The islands were subtracted to obtain the »Polygon with Islands«, which provided an accurate overlay of the SCA raster content (Figure 5).

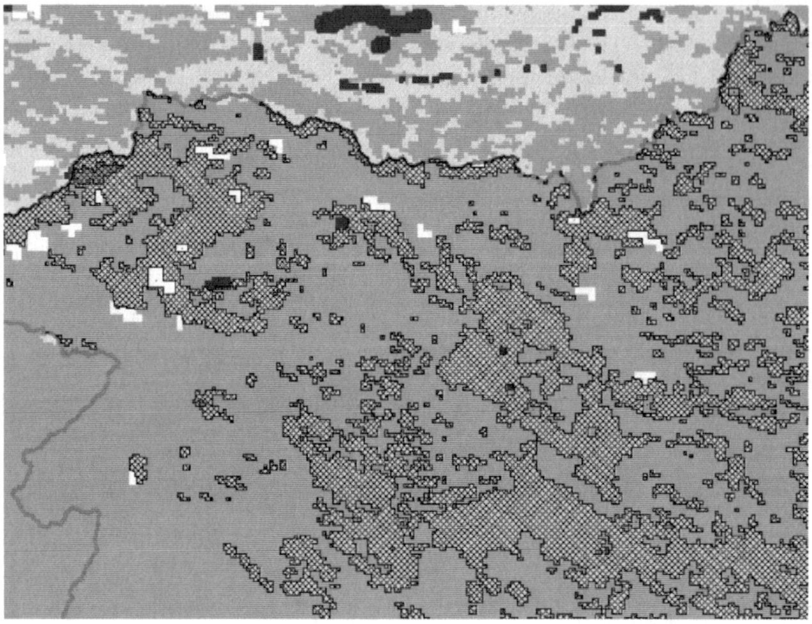

Figure 5. Vectorized areas under snow, 24 February 2003, Republic of Slovenia.

The SCA vectorized polygons could be used for further analysis of overlaying with vector polygons of other content, such as vector polygons of altitude belts, hydrographic areas – catchment areas, administrative areas vectorized CLC (Corine Land Cover) and similar.

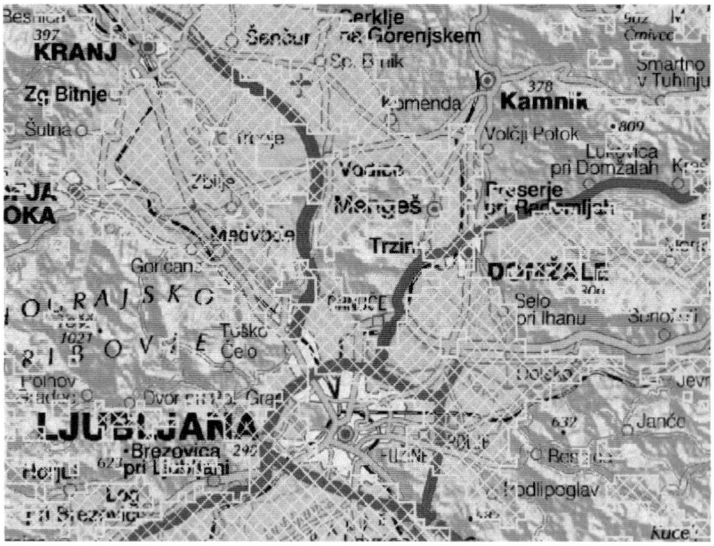

Figure 6. SCA in the Ljubljana area.

4. Results and Disscision

4.1. Analysis of use of the obtained satellite images

The analysis of the possibility to receive useful satellite images for Ljubljansko polje was derived from the data on cloud coverage for the period from 2000/2001 to 2004/2005, Table 1.

The satellite images received for modelling were mainly without snow cover over the Sava River watershed and Ljubljansko Polje aquifer, Table 2.

Table 2. Number of days with snow cover and cloud cover in winter

season	number of days with SC	number of days with SC and		
		clear sky	partly clear sky	no clear sky
2000/2001	11	0	0	11
2001/2002	51	14	2	35
2002/2003	61	19	1	41
2003/2004	59	8	3	48
2004/2005	31	10	1	20
sum	213	51	7	155
	33%	8%	1%	24%

Clear sky…cloud cover less than 10% of sky (useful satellite image)
Partly clear sky… cloud cover between 10% and 20 % of sky (partly useful image)
No clear sky… cloud cover more than 20% of sky (useless satellite image)

Figure 7 shows how much time cloudiness is present above and below 800 ma.s.l. when snow cover is also present. It is seen that below 800 ma.s.l. there is more cloudiness, which means that satellite images are harder to use in lower parts.

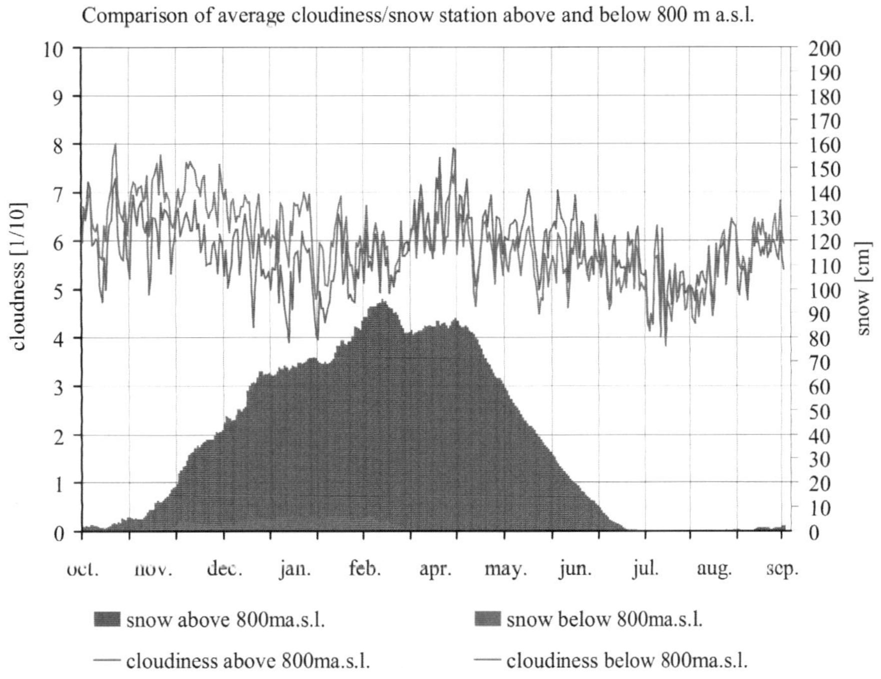

Figure 7. Comparison of cloudiness and snow presence for Slovenia.

IOP Conf. Series: Earth and Environmental Science **4** (2008) 012024 doi:10.1088/1755-1307/4/1/012024

Figure 8: Most common view from mountains in winter in Slovenia.

4.2. Analysis of the obtained SCA data, VER. 3

At validation process it is important to compare data for the same date as for satellite images as for snow cover measurements [4]. That is why we compared the satellite image mod02hkm_2003055.1045_fuso33_sca3 with snow cover data on the same date, this is 24th of February 2003.

According to data in Table 3 we can conclude that snow cover was present in all area even if it was low, except in coastal region.

Table 3: Snow cover heights for meteorological stations on 24th of February 2003.

Station Name	Altitude [m a.s.l.]	Snow cover depht [cm]	Station Name	Altitude [m a.s.l.]	Snow cover depht [cm]
Letališče Brnik	364	8	Novo mesto	220	19
Kredarica	2514	195	Celje	244	10
Rateče Planica	864	18	Šmartno pri Slov.Gradcu	455	18
Vojsko	1067	60	Murska Sobota Rakičani	188	5
Bilje	46	0	Lesce	515	8
Postojna	533	6	Metlika	210	12
Kočevje	467	28	Vogel	1535	90
Ljubljana Bežigrad	299	15	Lisca	943	52
Črnomelj	157	12	Letališče Portorož	2	0

The visual check indicates that much of the area under investigation is not covered in snow, or that the cover is undetermined. However, based on our knowledge and experience, there should be the presence of snow cover in these areas. Namely, these areas are covered in forest with the snow cover under the tree canopy, which, in all probability, the EO satellites failed to detect. These areas include large forest-covered surface of the hills of Polhograjsko hribovje and Škofjeloško hribovje, and the plateaus of Pokljuka and Jelovica. In Pokljuka, especially, snow was in abundance in the time, the Pokljuka nordic ski World Cup being underway. Also, there are several ski areas and snow recreation centres in the area.

Figure 9. Forested areas in R Slovenia.

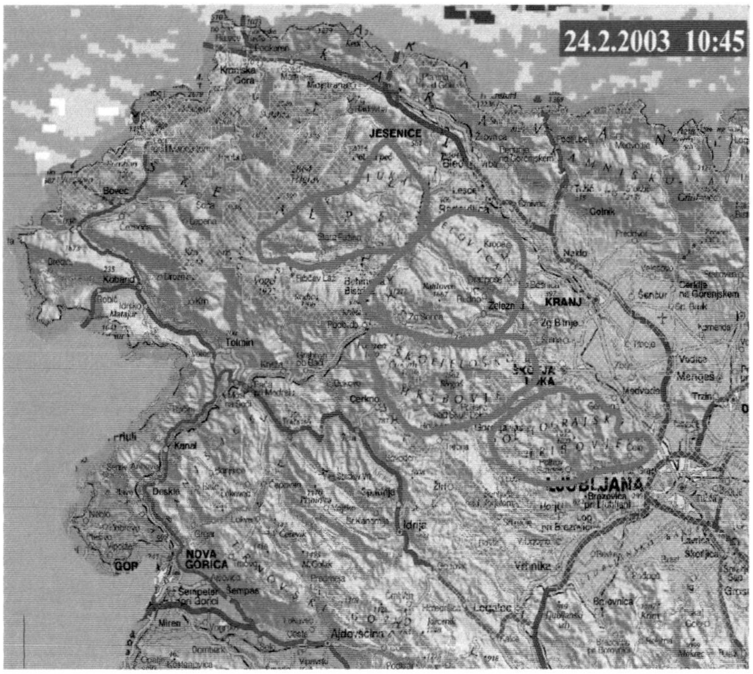

Figure 10. Vectorized area with marked forested areas.

We overlaid satellite image that was vectorized with image with vegetation, Figure 9. The result is on Figure 10, where is clearly seen that between detected snow there is mostly forest area. In this case MODIS was not reliable for detecting snow cover in forest with height below 25 cm. The system detected snow cover only above and below forest height limit.

On satellite image there is also some areas without snow cover betwen Kranj and Jesenice. This area are low altitude and heihgt of snow cover is from 5 to 10 cm, that was measured at 7 a.m. and was already melted when satellite made image at 11:45.

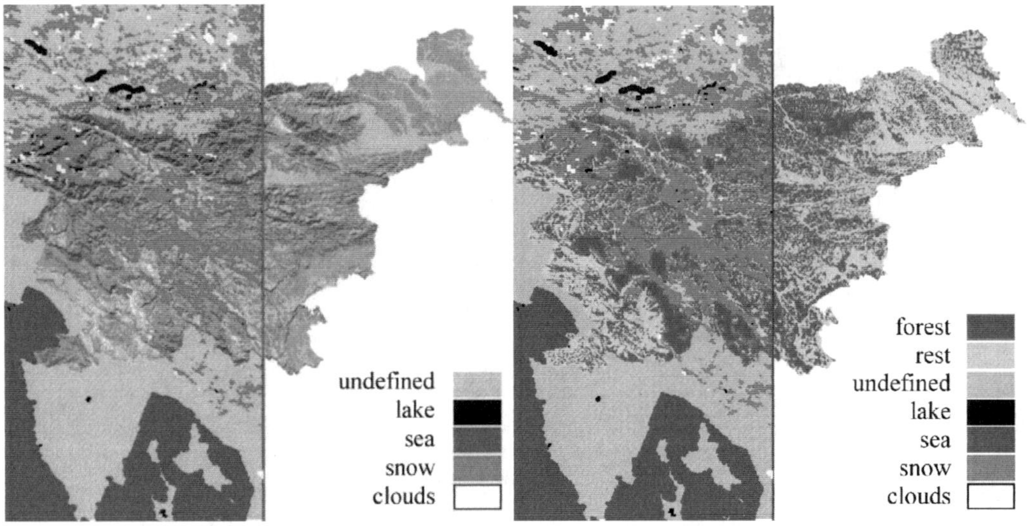

Figure 11. Image mod02hkm_2003055.1045_fuso33_sca3 system MODIS date 24th of Feb. 2003 overlaid with relief of Slovenia (left) and vegetation image (right).

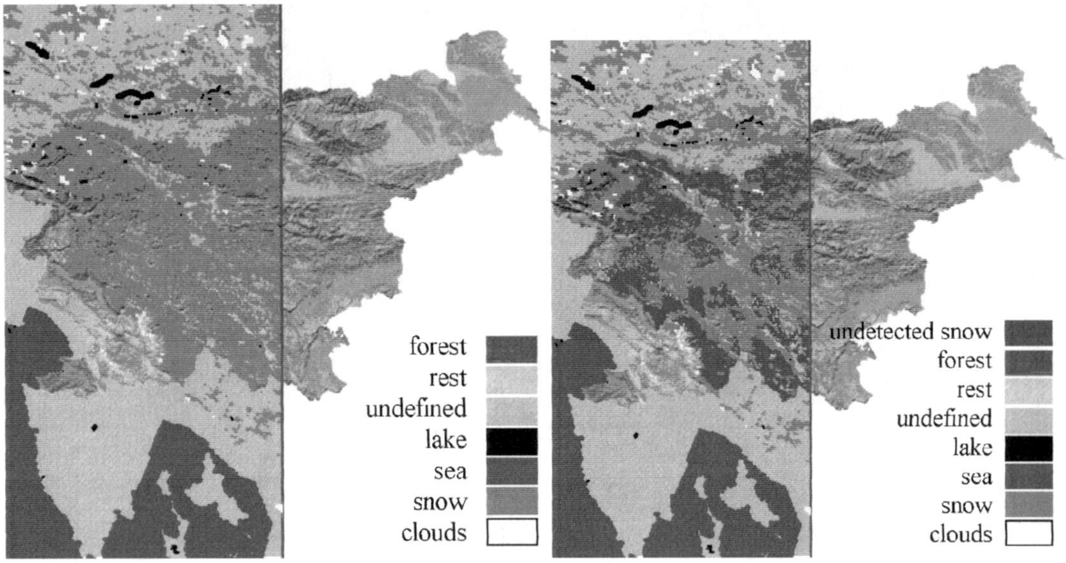

Figure 12. Actual snow cover according to data from climatological station on 24th of Feb. 2003 (left) and red colored areas, where MODIS did not detected snow cover (right).

Figure 13. Data processing, a) SC according to measurement, b) satellite detected SC, c) difference between a and b.

Slovenia is a country abundant in forest, 57.9% of the surface being covered in forest. In terms of snow cover, Slovenia is placed third in the European Union, right after Sweden and Finland. 1,173,847 ha of forests cover more than a half of the surface area of the country (Source: The Slovenian Forestry Service (Zavod za gozdove Slovenije), 2006). Not only that under canopy snow cover is not visible on raw satellite images, but also the snow is melting different under the canopies [10]. That is why detecting snow in forested areas very uncertain.

Figure 14. Snow cover in forested area.

5. Conclusions

It has been established that with large forest cover the EO (Earth Observation) method for SCA detection fails to give satisfactory results, that is, the actual snow cover cannot be determined in a satisfactory manner. The presented method is attractively simple and appears to give results adequate for_mapping daily snow depth over mountainous terrain [1] but now for lower altitudes and forested areas. There has been some solutions of a challenging problem of forests for snow-cover mapping by optical satellite remote sensing developed. One of those investigates different reflectance variability [13].

Weather conditions in wintertime are unfavourable for satellite data collection by images. Only 30% of time with snow cover, less than 20% of the sky is covered by clouds, which is also the reason why only two images were useful for snow cover analysis (Table 2). For useful satellite image is needed clear sky during 11:45 - 12:00 in 13:00 - 13:15 a.m. and that means in average 23% days with snow cover and clear sky. Some improvements on problem of detecting clouds in stead of snow are already being tested.

On basis of only one useful satellite image it is impossible to determine objective error that becomes at snow cover detection. For this purpose we would need more satellite images, processed with GIS application. Later on we could determine the nature and size of errors.

Into consideration shoul also be taken that it is a great possibility that snow cover somewhere is so low that could melted before satellite made images. Satellite images are taken at 11:45 a.m. and measurements are made early in the morning, when snow that fall during the night is still present.

Aknowledgements
Research was derived as part of AWARE project funded by EU in the 6th framework.

References

[1] Foppa N, Stoffel A and Meister R 2006 *Int. J. Appl. Earth Obs. Geoinf.*, **xxx** xxx In Press
[2] Grayson R B, Bloeschl G, Western A W and McMahon T M 2002 *Adv. Water Resour.* **25** 1313
[3] Haefner H, Seidel K and Ehrler H 1997 *Regions. Phys. Chem. Earth* **22** 275
[4] Hall DK, Solberg R and Riggs G A 2002 *Proceedings of 59th Eastern Snow Conference Stowe* Vermont, USA **59** 55
[5] Hall D K, Riggs G A and Salomonson V V 2002 *Rem. Sens.* **82** 181
[6] Ogrin M 2005 *Acta geographica Slovenica* **45** 63
[7] Parajka J and Bloeschl G 2006 *Hydrol. Earth Syst. Sci.* **10** 679
[8] Riggs G A and Hall D K 2002 *Proceedings of 59th Eastern Snow Conference, Stowe* Vermont, USA **59** 205
[9] Schmugge T J, Kustas W P , Ritchie J C, Jackson T J and Rango A 2002 *Adv. Water Resour.* **25** 1367
[10] Suzuki K and Ohta T 2003 *J.Hydromet.* **4** 1181
[11] Tekelia A E, Akyuerekb Z, S¸ormanc A A, S¸ensoyc A and S¸ormana A U 2005 *Rem. Sens.* **97** 216
[12] Vikhamar D and Solberg R 2002 *Rem. Sens.* **84** 69
[13] Vikhamar D and Solberg R 2003 *Rem. Sens.* **88** 309
[14] Weibel D, Wunderle S and Kleindienst H 2002 *Proceedings of EARSeL-LISSIG-Workshop Observing our Cryosphere from Space* Bern **2** 179

IOP Conf. Series: Earth and Environmental Science **4** (2008) 012025 doi:10.1088/1755-1307/4/1/012025

The influence of Arctic and North Atlantic Oscillation on precipitation regime in Serbia

Gordana Jovanović [1], Irini Reljin [2] and Branimir Reljin [3]

[1] Republic Hydrometeorological Service of Serbia, Belgrade, Kneza Višeslava 66, Serbia
[2] ICT College, Belgrade, Zdravka Celara 16, Serbia
[3] Faculty of Electrical Engineering, University of Belgrade, Belgrade, Bulevar Kralja Aleksandra 73, Serbia

E-mail: gordana.jovanovic@hidmet.sr.gov.yu

Abstract. The North Atlantic Oscillation (NAO) is the most significant large-scale mode of natural climate variability on the Northern Hhemisphere and Arctic Oscillation (AO) is its supplement. Both have large impacts on weather and climate in the North Atlantic region and surrounding continents including Europe. NAO and AO phases cause specific already defined changes in distribution of climate elements in wide area. The cross-correlation between daily precipitation data on stations in Serbia and NAO daily index plus AO daily index for the period 1951–2003 is analysed in this paper. Number of cross correlation coefficients have been calculated and the highest values were obtained for daily precipitation in Serbia with the AO index in winter, ranging from –0.44 to – 0.38. The analyses of trend for two periods 1961-2006 and 1980-2006 for the whole year and during winter months only, at 21 stations in Serbia have been performed, too. Very high positive trend for precipitation was obtained for many places in Serbia especially in Vojvodina. The results also indicate to non-homogeneities in the region and influence of NAO during winter.

1. Introduction

Simultaneous, recurrent variations of climate elements on large distances are called teleconnection. That means, beside local influences, local climate is under permanent influence of different phenomena on a large scale. One of the most significant teleconnection patterns on the Northern Hemisphere is the North Atlantic oscillation – NAO. NAO is defined as difference in surface pressures between the center of high pressure above Azores and low pressure stationed above Iceland. The impacts of NAO are felt in the long range from USA to Siberia and from the Arctic to northern Africa, but its consequences are most significant near the activity centers. Although NAO exists during the whole year, the amplitudes of oscillation are greatest during winter, from December to March. NAO comprises two extreme phases, a positive one that means strong activity of both centers and negative phase that denotes weak activity of pressure centers. The changes of NAO phases cause the changes in regime of pressure, precipitation, temperature, and wind on very large areas, especially in boreal winter [1], [2].

The positive phase occurs when pressure is below normal in the northern Atlantic around Iceland, and above normal over subtropical and central Atlantic, i.e., the Azores. The consequences of NAO positive phase are great activity of Azores high and Iceland low-pressure centers causing pronounced western stream over the Atlantic and Western Europe while over Mediterranean, North Africa, and Near East prevails northern flow [3]. Precipitation above normal occurs over northern Europe and Scandinavia, while precipitation deficit dominate above southern Europe and Mediterranean. Negative NAO phase has opposite mark of pressure anomalies that means less pronounced both pressure centers over Iceland and Azores. During negative phase the climate regimes are mostly opposite to positive phase i.e. over southern Europe and Mediteranean, including Serbia it is mostly colder with precipitation sufficit. In order to present current state of essential climate elements between centers of action, climate indices are identified. To represent the status and timing of NAO phase so called NAO Index is completed denoting normalized pressure difference between Lisbon and Reykjavik. Figure 1. [http://www.cgd.ucar.edu/cas/jhurrell/nao.stat.winter.html] shows time series of NAO index for the period 1864-2006 where from it is obvious that NAO phase is mostly positive since 1980.

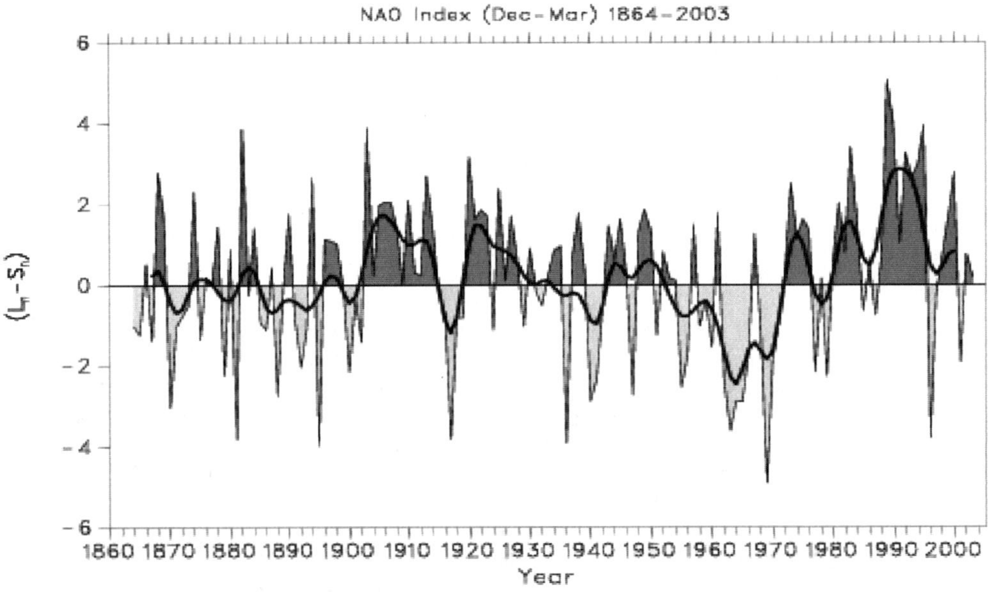

Figure 1. Time series of the NAO index for the period 1864-2003.

The Arctic Oscillation (AO) is treated as an alternate view of NAO, described as a seesaw pattern in which atmospheric pressure fluctuates between positive and negative phases at polar and middle latitudes. The negative phase brings higher-than-normal pressure over the polar region and lower-than-normal pressure at about 45 degrees north latitude. The positive phase brings drier conditions to Spain, Middle East, southern Europe, including Serbia, while negative phase brings opposite conditions. Both phases are shown in Fig. 2. a) and b) [http://nsidc.org/arcticmet/patterns/arctic_oscillation.html].

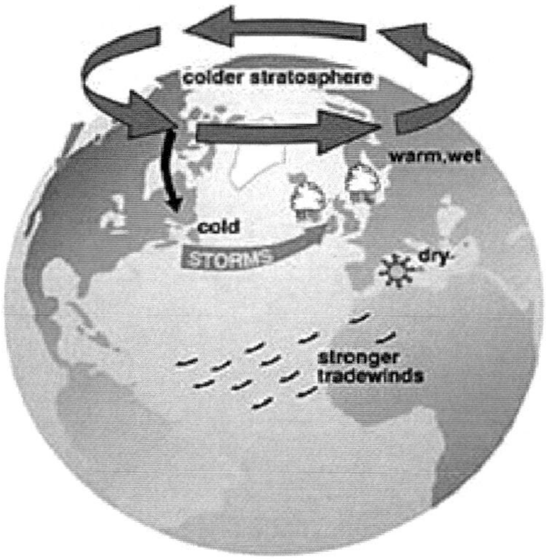

Figure 2a. Positive AO phase
Stronger polar vortex brings cool winds over Canada; North Atlantic storms bring rain and mild temperatures to northern Europe, while droughts prevail in the Mediterranean.

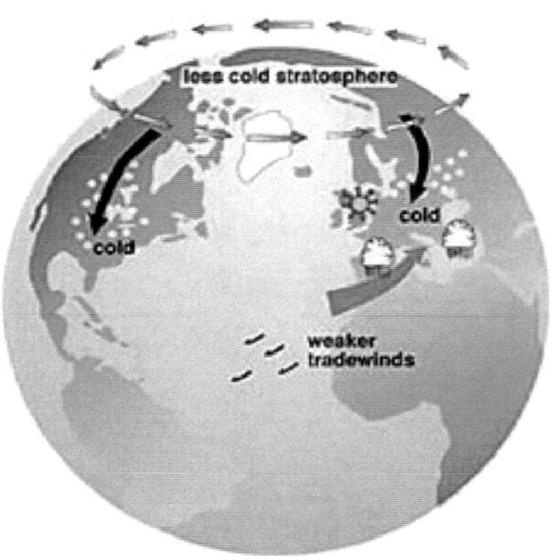

Figure 2b. Negative AO phase
The weaker polar vortex allows cold air to plunge into the mid-western United States and western Europe, while storms bring rain into the Mediterranean.

Wallace [4] argued that the NAO is in fact the part of the AO since AO involves atmospheric circulation in the entire hemisphere and lower stratosphere. During winter the AO extends up through the stratosphere, 6 to 30 miles above the Earth's surface. Some scientists say that Arctic oscillation might be of importance for global warming and of important impact on climate variability. So, the assumption is that AO and NAO represent the same teleconnection mode of climate variability for Northern Hemisphere since there are no meaningful distinctions between them and their consequences.

Time series of AO is presented by the AO index that is very similar to NAO index. Systematic existence of positive phase of both indices since 1980 has coincided with relatively warm conditions in the Arctic and net melting of the Arctic ice pack. During the same period in the Mediterranean surrounding dry conditions prevailed.

2. Results

The previous analyses [5], [6], [7] based on the averaged monthly data of pressure, temperature and precipitation, have indicated to the significant influence of NAO on climate in Serbia and Montenegro. The anomalies of temperature, pressure and precipitation data were calculated in order to explore the NAO impact during winter, for the period 1951-2002 in reference to normal 1961-90. Time series of average pressure (13 stations) and precipitation (32 stations) data in Serbia together with NAO Index are presented in Figures 3 a) and b). The results approved positive cross correlation between NAO index and pressure, as obvious from Figure 3 a), as well as for temperature although it is not presented in graphic form. Particularly prominent and negative as expected, is cross correlation between precipitation in Serbia and NAO Index as seen from Figure 3 b.

The cross-correlation between NAO Index and monthly averaged parameters for Serbia and Montenegro for winter period 1951-2000 have been calculated for a number of stations but the most representative results are the following: correlation coefficient for pressure is 0.60 and temperature 0,64 in Belgrade, while in Niksic is for precipitation –0,53.

Another previously used method is the Empirical Orthogonal Function [8], which are creating the maps of rationalization of pressure, precipitation and temperature field in Serbia and Montenegro. The charts of the first EOF precipitation mode in the Euro-Atlantic region and in Serbia and Montenegro showed great similarity that indicated to the NAO impact, too.

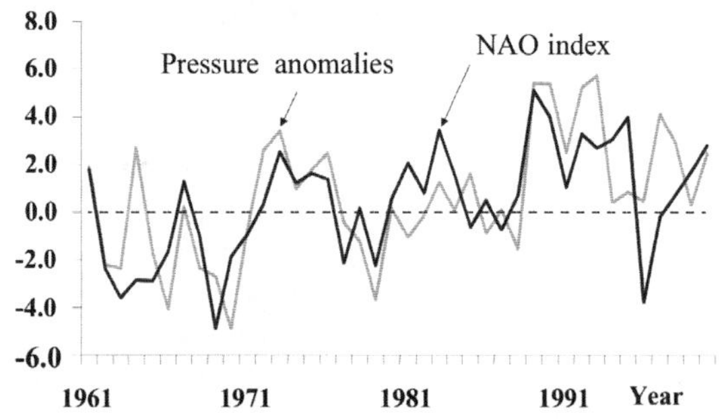

Figure 3a. Time series of pressure anomalies in Serbia in reference to 1961-90 normal and NAO index.

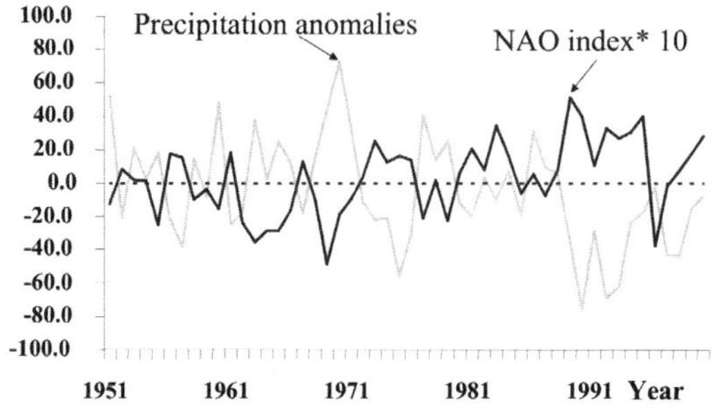

Figure 3b. Time series of precipitation anomalies in Serbia in reference to 1961-90 normal and NAO index.

A new evidence of NAO and AO influences on climate in Serbia has been proved by the calculation of cross-correlation between daily NAO and AO index and daily precipitation data at 12 stations in Serbia for the period 01.01.1950. to 31.12.2003. only during December, January, February. The obtained results gave much higher dependence of daily precipitation with the AO than with the NAO indices, as seen from Table 1. Maximum negative correlation coefficient i.e. dependence of our precipitation with the AO index are ranging from –0.44 in Kikinda to –0.38 in Kragujevac while for NAO it is from –0.14 to –0.11. Three to four times higher correlation coefficients for AO then for NAO suggest to more significance of the AO influence then the NAO in the Serbian region. It seems that AO teleconnection signal has stronger impact then NAO, but it has to be to be explained and these results should be compared with similar ones from surrounding regions.

Table 1. Maximum value of correlation coefficient between daily amount of precipitation and daily AO index and NAO index on each of 12 stations in Serbia

	Station name	Min AO	MinNAO
1	Beograd	-0.39	-0.12
2	Dimitrovgrad	-0.41	-0.13
3	Gradiste	-0.39	-0.14
4	Kikinda	-0.44	-0.12
5	Kragujevac	-0.38	-0.13
6	Kraljevo	-0.35	-0.14
7	Nis	-0.40	-0.11
8	Novi Sad	-0.43	-0.12
9	Sjenica	-0.39	-0.13
10	Sombor	-0.40	-0.13
11	Smed.Palan	-0.39	-0.13
12	Zajecar	-0.40	-0.11

Trends of daily precipitation for the period 1961-2006 and for the period 1980-2006, when positive NAO phase has been dominating, were obtained for 21 stations in Serbia. Precipitation trends were also tested for the winter months to facilitate the AO/NAO influence. The results showed the increase of precipitation on most of the country except on a few stations in the eastern and central parts of the country for both periods during the whole year. Very high positive trends were obtained for Vojvodina, which is in agreement with also, very high positive temperature trends for the same period 1961-2006. Trends are even higher since 1980, and much lower during winter season, when AO and NAO are the most active for both periods. The obtained results have verified significant impact of dominant climate phenomenon AO and NAO on precipitation regime in Serbia.

Table 2. Precipitation trends on the representative stations in Serbia.

	Annual trend 1961-2006	Winter trend 1961-2006	Annual trend 1980-2006	Winter trend 1980-2006
Beograd	y = 0.7x + 671	y = -0.7x + 164	y = 1.1x + 670	y = 0.8x + 126
Nis	y = 0.1x + 587	y = -0.5x + 142	y = 3.4x + 527	y = 0.2x + 121
Novi Sad	y = 2.8x + 544	y = -0.3x + 125	y = 5.2x + 556	y = 0.6x + 106
Sjenica	y = 3.6x + 653	y = 1.0x + 124	y = 8.4x + 628	y = 2.8x + 111
Zajecar	y = -3.1x + 652	y = -0.8x + 147	y = -0.2x + 575	y = -1.0x + 139

In Table 2 are shown trends for five stations typical for each region: Novi Sad–northern part of the country, Beograd-central, Nis-southern, Sjenica – mountain and Zajecar far eastern, in order to represent all 21 stations. Zajecar is example for exceptional high decrease of precipitation in all periods especially during 1961-2006. Winter trends are lower than those during the year due to the NAO/AO activity. It is also evident from the raw data that precipitation amounts have been decreasing after 1980 especially during winter when NAO/AO positive phase was dominating. Winter of 1989 had the highest NAO/AO index and during that winter the lowest precipitation amounts were measured in Serbia.

Climate indicator R95T, which is defined as fraction of annual total rainfall due to events above the 95th percentile, was calculated for the period 1961-2006 in reference to the period 1961–1990 [9]. It has confirmed the above results and shown high positive trend on most of stations in Serbia with especially high trends in the mountain stations.

3. Conclusion

NAO and AO have dominant influence on precipitation regime in Serbia, especially during winter. By comparing the observed data from the stations in Serbia with the NAO and AO indices, the significant influence of these teleconnection pattern have been proved. Higher correlation coefficients were obtained for AO then for NAO index suggesting that AO phenomenon has more prominent impact on the region.

Precipitation trends for two periods 1961-2006 and 1980-2006 for the whole year and during winter showed high positive values for most places in Serbia especially in Vojvodina. The results also indicate the influence of AO/NAO on precipitation regime in Serbia especially during winter.

References

[1] Hurrel J 1995 *Decadal trends in North Atlantic Oscillation and relationships to regional temperature and precipitation* (*Science* vol 269) pp 676-9

[2] Hurrel J 1996 *Influence of variations in extratropical wintertime teleconnections on Northern hemisphere temperature* (*Geophysical Research Letters* vol 23) no 6

[3] Hurrel J et al 2000 *Climate variability over the North Atlantic (Meteorology at the Millennium)* (Hadley Centre for Climate Prediction and Research, UK Met. Office)

[4] Wallace J M and Rigor I G 2004 *Variations in the age of Arctic sea-ice and summer sea-ice extent* (*Geophysical Research Letters* vol 31) L09401

[5] Reljin I, Jovanović and Reljin B 2002 The analyses of FRY climate and NAO teleconnection *Proc. Conf. Information Technologies: Present and Future, IT '02* (Zabljak Montenegro)

IOP Conf. Series: Earth and Environmental Science **4** (2008) 012025 doi:10.1088/1755-1307/4/1/012025

[6] Jovanovic G, Urosev L and Savic T 2002 NAO influence on climate in FRY *Proc. Conf. Information Technologies: Present and Future, IT '02* (Zabljak, Montenegro)
[7] Jovanović G, Reljin I and Savić T October 2002 NAO Influence on Climate Variability in FRY *Proc. 18th Int. Conf. on Carpathian Meteorology (Proc.CD)* (Belgrade, Serbia)
[8] Reljin I, Jovanović G and Reljin B 2001 The Climate Signal Analysis through Linear and Non-Linear Methods *Proc. Int. Symp. on Theoretical Electrical Engineering (ISTET)* (Linz, Austria)
[9] Djordjevic S 2008 Temperature and precipitation trends in Belgrade and indicators of changing extremes in Serbia (Novi Sad: Geographica Pannonica)

The effect of flow regulation on the distribution and dynamics of channel geomorphic units (CGUs) and implications for Marble Trout (*Salmo marmoratus*) spawning habitat in the Soča River, Slovenia

Ian Maddock[1,3], Natasa Smolar-Žvanut[2] and Graham Hill[1]

[1] Institute of Science and the Environment, University of Worcester, Henwick Grove, Worcester, WR2 6AJ, UK
[2] Institute for Water of the Republic of Slovenia, Hajdrihova 28c, 1000 Ljubljana, Slovenia

Email: i.maddock@worc.ac.uk

Abstract. A river channel survey was completed along three reaches (totalling 14.3km), i.e. an unregulated stretch and two regulated reaches (with reduced flows) of the Soča River to assess the spatial pattern of CGU type, size, hydraulics and distribution. In addition, one regulated reach was re-surveyed at different discharges to investigate the dynamics of CGUs and their relationship with flow. CGUs were classified and mapped using visual assessment and physical measurements of the hydraulic characteristics (velocity and depth) in each CGU. The effect of flow regulation on the hydraulic character of the river becomes apparent by highlighting differences in the types of CGUs present between the regulated and unregulated reaches. Reduced flows from river regulation also significantly reduces the size of CGUs, alters their hydraulic character, and affects the longitudinal distribution of types by creating greater habitat fragmentation. Hydraulic preferences for spawning habitat of marble trout (Salmo marmoratus) were obtained from previous research. The hydraulic character of CGUs were analysed at different discharges and combined with the hydraulic preferences of the species to evaluate the impact of flow regulation on habitat availability. Analysis shows that intermediate measured flow provides increased spawning habitat availability in the chosen reach for this target species.

1. Introduction

The quantity and quality of physical habitat in rivers is determined by the interaction of geomorphology and hydrology. It plays an important role in determining 'river health' and influencing the structure and function of aquatic communities [1]. Traditional assessment of both physical habitat and biotic communities has tended to focus on sampling at individual points, cross-sections or along short (i.e. <200m) stretches of river ('micro' or small scale). Subsequent extrapolation to the sections of river inbetween ('upscaling') provides catchment assessments (at the 'macro' or large scale). It has been argued [2] that river habitat assessment should concentrate on assessing complete reaches at the 'meso' or 'intermediate' spatial scale in order to recognise the river

[3] Author to whom any correspondence should be addressed.

landscape as a spatially continuous longitudinal and lateral mosaic of habitats. This approach underpins the use of physical habitat mapping.

A range of different river habitat mapping methods and classification systems have been developed to facilitate this approach. Habitat mapping surveys are normally completed for one of two main reasons:-

1) For survey purposes, i.e., to provide an inventory of the physical habitat present as part of a basic resource assessment in it's own right [3].

2) For impact assessment purposes, i.e., as part of aquatic habitat modelling studies. Habitat mapping data are used to either a) model the relationship between physical habitat availability and flow directly from the mapping results without the need for any further fieldwork (e.g. MesoHABSIM [4]), or, b) to identify representative cross-sections and reaches for subsequent and more detailed study. This use of habitat mapping is also often completed as part of PHABSIM Studies [5].

River habitat mapping aims to identify the types and spatial configuration of geomorphic and hydraulic units. The terms used to describe these units differ between authors and include 'channel geomorphic units' (CGUs) [6], 'mesohabitats' [7], 'physical biotopes' [8] and 'hydraulic biotopes' [9]. [10] provide a review of the use of some of these terms and the differences between them.

In addition to the need to assess rivers at the most appropriate scale and along continuous reaches, others have called for the translation of key concepts that are well established in landscape ecology to be translated to riverine environments [11]. These key concepts include patch dynamics, habitat connectivity, complexity and fragmentation, and the importance of understanding river ecosystems at a range of spatial scales. This requires a shift in traditional ways of conceptualising and sampling river habitats. A recent study examining macroinvertebrate assemblages has demonstrated the importance of this new approach [12]. Habitat heterogeneity has also been successfully quantified using spatial heterogeneity measures to examine the influence of relative sediment supply [13]. River habitat mapping is likely to underpin an understanding of the links between physical habitat dynamics and instream biota in general, and particularly for fish species.

The aims of this research were two fold. Firstly, we utilised a rapid habitat mapping approach in it's traditional sense to examine the impact of flow regulation on the spatial distribution and temporal dynamics of physical habitats or channel geomorphic units (CGU) in the Soča River, Slovenia. Secondly, we characterised the hydraulic preferences of spawning marble trout using field data from previous research [14] and an approach developed by [15], and attempted to determine the impact of flow regulation on habitat availability for this important target species.

2. Site Details

The Soča River rises in the Slovenian Alps, flowing for 95km through Slovenia before crossing into Italy and discharging into the Adriatic Sea. It has a catchment area of 1576 km^2 and is predominantly underlain by limestone, but the lower parts the river run over flysch and quaternary gravels. The Soča has a flashy flow regime, with high flows occurring at any time of year. The lowest flows are experienced both in summer and winter months with generally higher snow-fed flows in spring and rain fed flows in autumn. The Soča River is well known for the presence of marble trout and recreational (white-water rafting) opportunities.

The river is regulated for hydro-power production at the Podsela Dam and Ajba Dam. Water is abstracted from the impoundment upstream from each dam. It then flows along a bypass channel to the hydropower plant and is subsequently augmented back to the river channel further downstream. Therefore, bypassed sections with reduced flows exist below each dam.

No long term flow records are available to describe the pre- and post river regulation flow regimes exactly, but it is clear that the hydro-power scheme abstracts the vast majority of water for long periods of time, leaving by-passed sections of river with greatly reduced flows. Prior to 2001, the highest possible abstraction rate at Podsela Dam was 96 m^3s^{-1} and the measured flow below the

IOP Conf. Series: Earth and Environmental Science **4** (2008) 012026 doi:10.1088/1755-1307/4/1/012026

Podsela Dam for most of the year is 0.2 m^3s^{-1}. The highest possible abstraction rate at the Ajba Dam is 75 m^3s^{-1} whilst flow releases until 2003 were normally 0.5 m^3s^{-1}.

In order to assess the impact of these reduced flows on physical habitat type, size and fragmentation, three reaches of river were assessed. **Reach 1:** an unregulated 5.14km stretch of the river between Volarje and Tolmin flowing through a broad open floodplain; **Reach 2:** on a 4.20km by-passed section of the river affected by abstraction below the Podsela Dam that flows through a confined river valley bordered by bedrock walls; and **Reach 3:** another regulated part of the river below the Ajba Dam (4.95km long) with a relatively intermediate-sized valley floor. The three reaches are illustrated in figure 1.

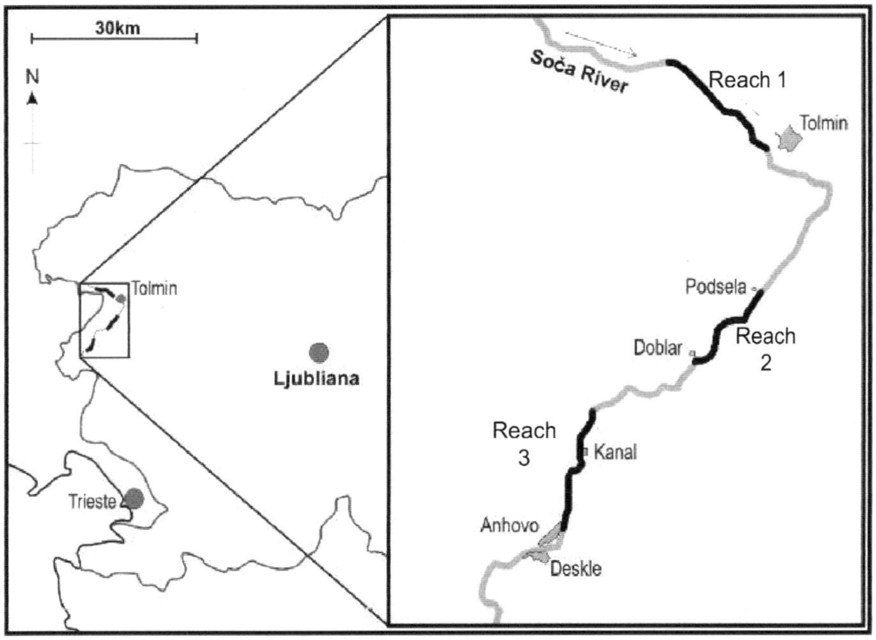

Figure 1. Site location

3. Method

3.1. Habitat Mapping

In order to assess the impact of flow alteration on CGU type, size, hydraulics and distribution, habitat mapping surveys were completed during July 2004 along the three reaches, and at two further flows along the regulated Reach 2 during July 2005. The three flow surveys along Reach 2 were conducted at 0.940 m^3s^{-1}, 1.546 m^3s^{-1} and 3.496 m^3s^{-1}. These represent relatively small variations in discharge compared to variation associated with the natural flow regime, but in order to distinguish them for the purposes of this research have been labelled as 'low', 'medium' and 'high' respectively.

CGUs were classified using a modified version of the [6] approach and mapped on foot and from a boat using a combination of visual assessment and physical measurements in each CGU. Channel width and water width were recorded to the nearest metre within each CGU. Water depth was measured to the nearest cm and the average water column velocity was recorded at a representative point in each CGU. Substrate sizes present (based on the Wentworth classification) were identified and assigned to 'dominant', 'subdominant' and 'present' categories. The proportion of the surface area of each CGU taken up with instream cover (e.g. instream macrophytes, large woody debris) and overhanging cover (e.g. from overhanging trees and boughs) were visually estimated to the nearest 10 percent. The presence of lateral-, point- and mid-channel bars, their location (e.g. left or right bank), and whether they were vegetated (>50% of surface covered) or unvegetated (<50%) were also noted. Photographs were taken of each CGU. Mapping-grade GPS was used to locate CGU boundaries to sub-metre accuracy, and the application of GIS (MapInfo) enabled the description and analysis of the longitudinal distribution of CGUs along each reach.

Using MapInfo 7.5 software, GPS data were combined with digitised maps at 1:50,000 scale and data recorded during the survey to create maps showing the CGU locations. The measured width and length data were used to calculate total water area in each reach and for individual CGU types in each reach. Physical habitat heterogeneity, defined as the spatial complexity of CGUs within a stream reach [13] was quantified using Shannon's Diversity Index (H) and Shannon's Equitability / Eveness Index (E).

3.2. Development of 'Fuzzy Logic' habitat preference curves

The use of habitat suitability curves to describe the depth, velocity and substrate preferences of individual species and life stages is well known in habitat modelling studies. These curves can be created by the use of professional judgment or detailed field observations at the site of interest, or a neighbouring unimpacted site [16]. The latter has the advantage of providing site specific information but may be flawed for a range of reasons, including the influence of flow conditions and the range of habitats available during sampling, species may select different habitats depending on the time of day or season, and the influence of competition for space by other species / life stages [17].

However, we recognise that even these curves are susceptible to the limitations outlined above, and therefore developed 'fuzzy logic' curves. Fuzzy logic is designed to be applied to situations where only imprecise or even ambiguous information is available and therefore is ideally suited to the application of habitat suitability criteria [15]. Fuzzy models are particularly valuable for situations where noise exists in the data and is not necessarily quantifiable or in other words, where the situation is ambiguous [18]. Thus, fuzzy analysis commonly maps onto linguistic definitions (good, medium, poor) rather than numerical definitions. The ecological uncertainty surrounding habitat preference, which arises from the methods obtained to derive the curves may be large. By developing fuzzy habitat models it is possible to explicitly retain the ambiguity that is implicit in habitat preference curves.

We based the development of 'fuzzy' habitat suitability curves and subsequent analysis of habitat availability on that used by [15]. In order to examine the influence of flow regulation on marble trout spawning habitat, information on hydraulic requirements was extracted from a report on the habitat preferences for the chosen target species and life stage produced by [14]. These depth and velocity requirements were interpreted into three classes, i.e. poor, medium and good for both parameters (see table 1 below).

Given that water depth and velocity influence fish habitat suitability in combination with one another rather than acting as independent variables, we examined combinations of these two variables. Combinations of poor, medium and good habitat classes for each parameter can produce nine types of habitats. These nine possible combinations were classified into six groups ranging from unsuitable (a combination of poor depth and poor velocity), very poor, poor, intermediate, good and finally to excellent (a combination of good depth and good velocity). Each of the six groups were assigned a score from 0-5 as show in table 2 below.

Table 1. Habitat preferences for marble trout spawning habitat adapted from [14].

	Poor	Medium	Good
Water Velocity	< 0.15 m/s > 0.90 m/s	0.15 - 0.40 m/s 0.70 – 0.90 m/s	0.40 - 0.70 m/s
Water Depth	< 0.24 m >0.60 m	0.48 – 0.60 m	0.24 – 0.48 m

Table 2. The rule set used to define habitat classes for marble trout spawning and the corresponding 'score' associated with each combination [15].

	Velocity poor	Velocity medium	Velocity good
Depth poor	Unsuitable habitat 0	Very poor habitat 1	Poor habitat 2
Depth medium	Very poor habitat 1	Intermediate habitat 3	Good habitat 4
Depth good	Poor habitat 2	Good habitat 4	Excellent habitat 5

For Reach 2, and at each of the three flows, the depth and velocity recorded within each CGU was converted into a poor, medium or good class based on the values given in table 1. Where the rating was poor or medium, it was also noted whether the velocity and depth was either higher or lower than the preferred range ('good') for further analysis outlined later. Each CGU was then assigned a habitat class and given the corresponding habitat score based on the combination of depth and velocity classes outlined in the rule set in table 2. These were used to assign 'weightings' to each habitat class, and weightings were multiplied by the total surface area of each CGU (m^2) to obtain a Weighted Usable Area (WUA). Finally, WUA values were divided by reach length to obtain a value of WUA in m^2 per 1000m of stream reach.

4. Results and Discussion

4.1. CGU composition between reaches and the effects of flow regulation

Results demonstrated significant differences in the CGU composition between the unregulated and regulated reaches (figure 2a). The unregulated stretch (Reach 1) was dominated by glides (55%) with the rest of the reach consisting of fast-flowing and turbulent features (runs, riffles and rapids). The dominant feature of the regulated reaches were the slow flowing pool CGUs occupying 44% of Reach 2, and 76% of Reach 3, with glides, runs, riffles and rapids forming the remainder of the CGUs. This highlights the effects of reduced flows on habitat composition in the regulated reaches.

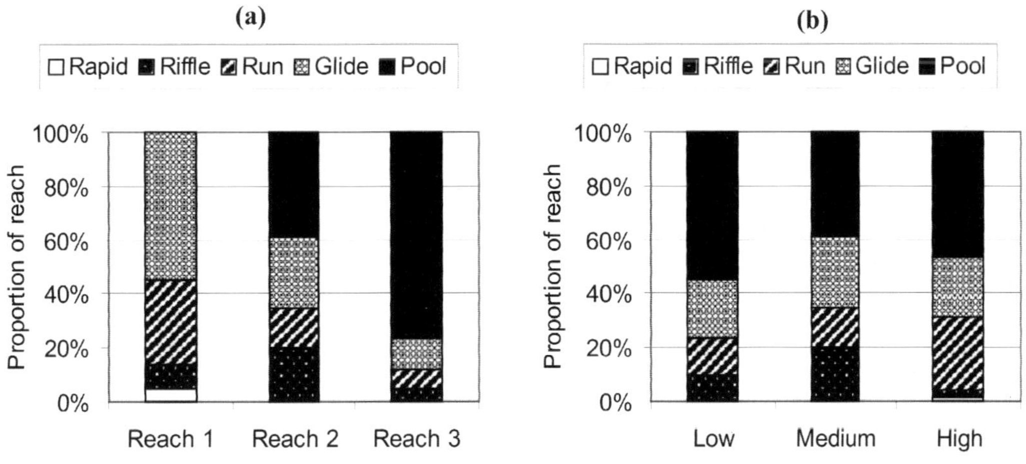

Figure 2(a) Variations in CGU composition between reaches determined by flow regulation, and **Figure 2(b)** effects of flow alteration on CGU composition within a single reach (Reach 2).

The repeat surveys in Reach 2 (figure 2b) highlighted differences (albeit smaller variations than the spatial changes) in habitat composition caused by changing discharges within the reach. At the relatively higher flow, the proportion of runs increased (from 14.1% to 26.9% of reach area) and the proportion of riffles decreased (from 8.6% to 2.8% of reach area). This change is created by riffles becoming submerged or 'drowned out' during the relatively higher flow and becoming run type CGUs.

Habitat mapping data were analysed to evaluate the effect of flow regulation on the size of CGUs (figure 3). The unregulated reach is dominated by larger CGUs (wider and longer units). As the flows decline, CGUs become shorter and narrower and hence fragmentation increases. This is supported by CGUs becoming progressively shorter and narrower in the three habitat mapping surveys conducted at different flows along Reach 2 (shown in figure 3).

Data were also analysed to examine the spatial complexity of CGUs using two indices, i.e. Shannon's Diversity Index (H) and Shannon's Eveness Index (E). The former describes the range of CGUs present, with a higher index indicating greater diversity of CGUs. The latter describes the spread of CGUs with a lower index indicating an uneven spread in the CGU composition (e.g. one or two types dominate the overall CGU composition). Results of calculating the two indices and comparing between reaches, and examining the influence of flow changes within a reach are plotted in figure 4a and b.

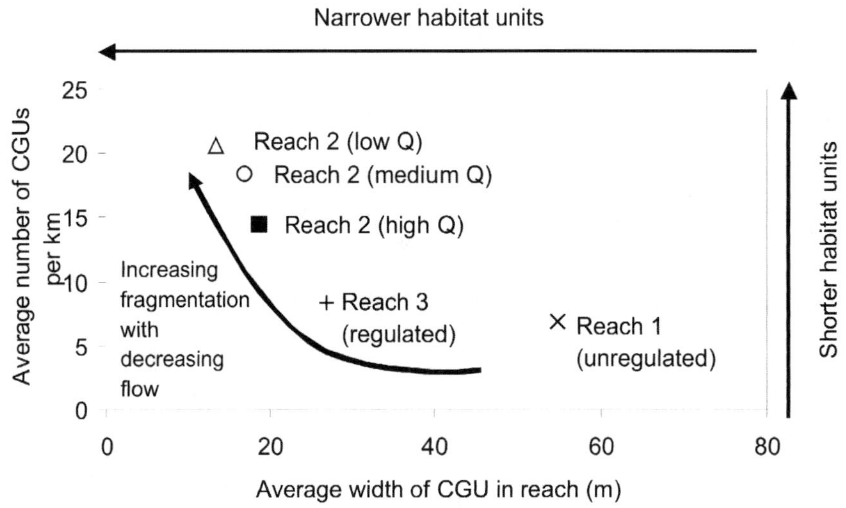

Figure 3. Comparison of average CGU width and no. of CGUs per km between reaches and within the same Reach (2) at three different flows. Points in the upper left represent reaches dominated by large numbers of short and narrow CGUs, i.e. fragmented habitats.

Figure 4a indicates that the diversity of CGUs is lowest Reaches 1 and 3, and these stretches of river are also characterised by a more uneven spread in the CGU composition of the reach. Reach 2 had the highest diversity and more even spread of CGU types, but smaller habitat units (indicated by the smaller symbol in figure 4a, and previously described in figure 3). Figure 4b highlights that Reach 2 has greater diversity of CGU types, and a more even spread of CGU composition at the intermediate measured flow compared to the low or high flow.

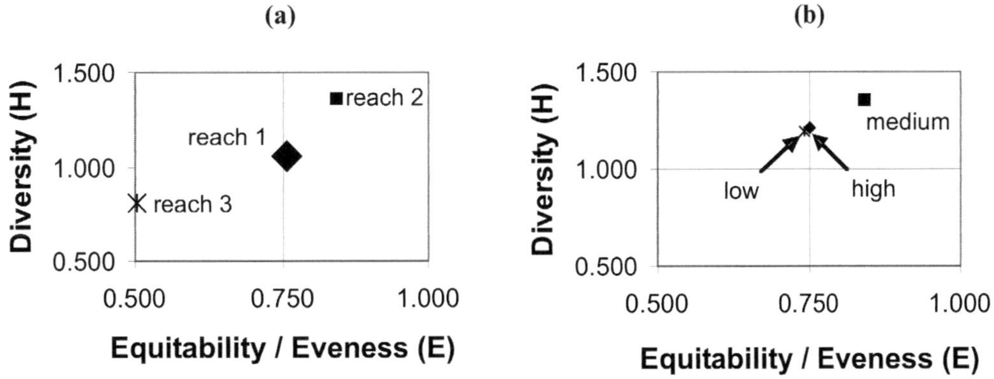

Figure 4(a) (left). Diversity and Eveness of CGU composition compared **between reaches.** Note larger symbols indicate larger habitat units. **Figure 4(b)** (right). Diversity and Eveness of CGU composition in reach 2 compared **between the three different measured discharges** (low =0.940 m^3s^{-1}, medium = 1.546 m^3s^{-1} and high = 3.496 m^3s^{-1}).

4.2. The influence of flow regulation on Marble Trout spawning habitat

Figure 5 illustrates the relationship between WUA and discharge for Reach 2. This highlights that the intermediate discharge provides a greater amount of suitable habitat compared to that available at the low and high measured flows.

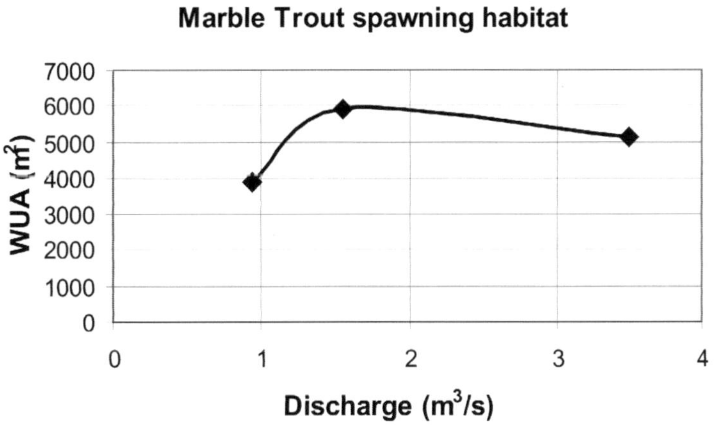

Figure 5. Flow versus habitat relationship for marble trout spawning habitat in Reach 2 based in CGU depth and velocity.

Further analysis of the habitat suitability of each CGU was completed to determine the limiting factors that inhibit habitat availability and which combinations of depth and velocity are most common within this reach and at the three measured discharges for this target species and life stage. These are illustrated in figure 6.

The diagrams illustrate that there is a significant proportion of the reach at all measured flows that has a velocity that is too low and depth that is too high for marble trout spawning (i.e. 'pool' habitat, shown in bottom left of each diagram). It also illustrates that at the lowest measured discharge, there is a significant proportion of CGUs that are limited by insufficient water depth (three light grey squares across the top row), but as flows increase, then a greater proportion contain sufficient water depth, but spawning habitat becomes limited by high velocities (note the dark grey square with 'good' water depth but 'poor (high)' velocity at the highest measured flow).

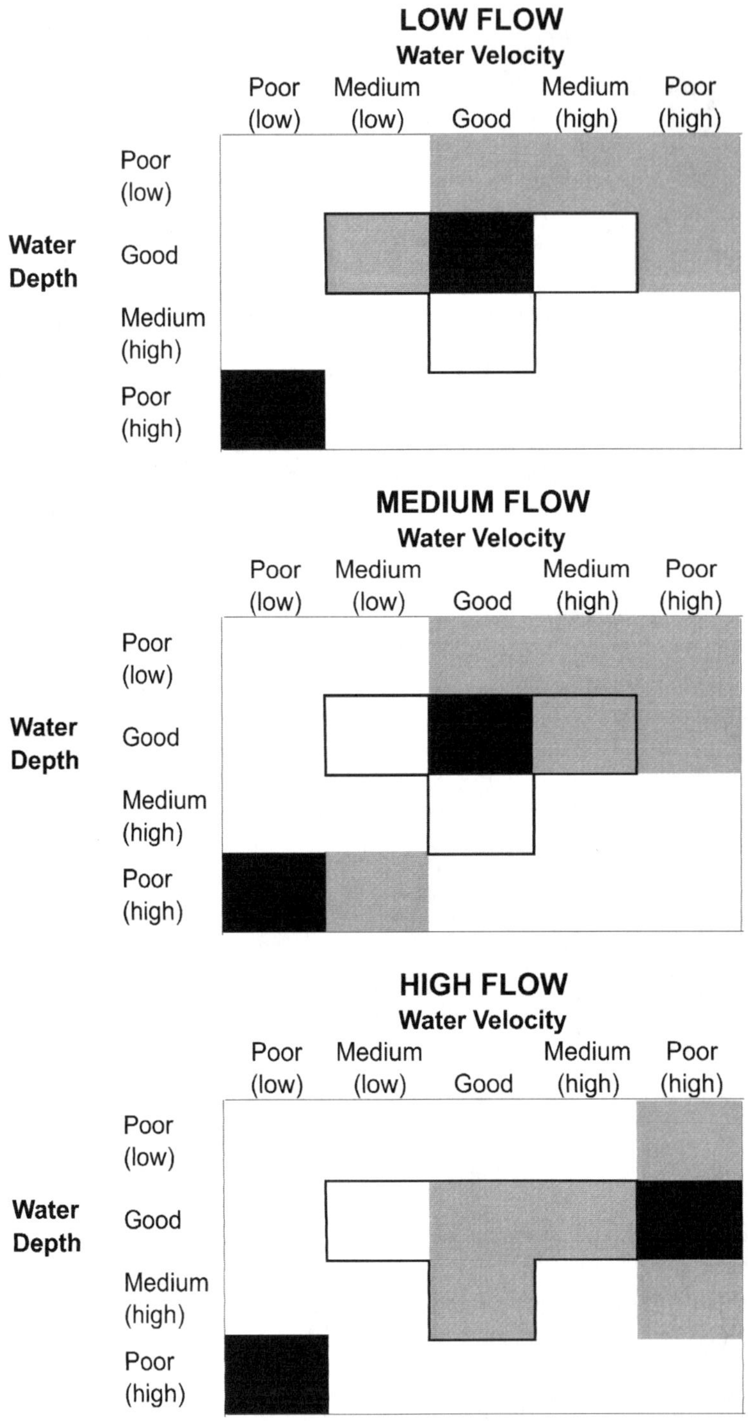

Figure 6. Combinations of depth and velocity habitat classes in Reach 2 at low, medium and high flow. Light grey shaded areas indicate between 5-15% of the CGU's had this combination of depth and velocity. Dark grey shaded areas indicate >15% of the CGU's have this depth and velocity combination (and hence are common). Unshaded areas indicate <5%. The highlighted 'T shaped area in the centre represents the combinations of depth and velocity that have habitat classes 4 and 5, i.e. good and excellent.

We acknowledge there are limitations to this approach. Other factors, in addition to depth and velocity influence habitat availability, such as water temperature and substrate size. We intend to build substrate size into our future analysis as this was also recorded in the field when CGUs were assessed. Separate monitoring of water quality variables (including temperature), and modelling studies to incorporate the influence of these parameters are still needed. Additional analysis, using the hydraulic preferences of other life stages, and/or other target species would help move towards a more holistic evaluation of the impacts of flow changes o the instream ecology. Our approach has not incorporated modelling at flows inbetween the three measured flows, and if completed, this additional work would enable us to determine the nature of the relationship between measured flows and assess optimum discharges with more confidence.

5. Conclusion

The effect of flow regulation on the physical and hydraulic character of the river becomes apparent by using habitat mapping results to highlight differences in the dominant types of CGUs present. Flow regulation reduces discharge, and habitat mapping is an effective tool to highlight increasing proportions of slow flowing types evident under these conditions. Declining flows also reduces the size of CGUs, and affects the longitudinal distribution of types thus creating greater habitat fragmentation. Habitat availability for the spawning requirements of marble trout were greatest at the measured intermediate flow for this particular river. The habitat mapping applied here was sufficiently sensitive to detect changes in the hydraulic and physical character of CGUs in a regulated reach with relatively small changes in flow.

We believe that this type of approach that considers the river system as a set of interconnected habitats, analyses the entire reach rather than at individual and disparate points, and combines field data with information on habitat requirements using habitat classes rather than absolute depth and velocity habitat suitability curves provides a more pragmatic approach to assessing the impact of flow changes in physical habitat availability. These tools can then be used to make recommendations for environmental flows with greater confidence.

Acknowledgments
The authors wish to thank the COST626 Aquatic Habitat Modelling Network, the European Science Foundation and the British Council Slovenia for financial assistance that enabled this collaborative research to take place. We would also like to acknowledge the assistance and hard work of Dušan Rebolj and Barbara Breznik during the field data collection.

References

[1] Maddock I P 1999 The importance of physical habitat assessment for evaluating river health *Freshwat.Biol.* **41** 373–91
[2] Fausch K D, Torgersen C E, Baxter C V and Li H W 2002 Landscapes to riverscapes: bridging the gap between research and conservation of stream fishes *BioScience* **52** 483-98
[3] Halwas K L and Church M 2002 Channel units in small high gradient streams on Vancouver Island, British Columbia *Geomorphology* **43** 243-56
[4] Parasiewicz P 2001 MesoHABSIM: A concept for application of instream flow models in river restoration planning *Fisheries* **26** 6-13
[5] Maddock I P, Bickerton M A, Spence R and Pickering T 2001 Reallocation of compensation releases to restore river flows and improve instream habitat availability in the Upper Derwent catchment, Derbyshire, UK *Regul. River.* **17** 417-41
[6] Hawkins C P *et al.* 1993 A hierarchical approach to classifying stream habitat features *Fisheries* **18** 3-12
[7] Tickner D, Armitage P D, Bickerton M A and Hall K A (2000) Assessing stream quality using information on mesohabitat distribution and character *Aquat. Conserv.* **10** 179-96
[8] Padmore C L 1997 Biotopes and their hydraulics: a method for determining the physical

component of freshwater habitat quality *Freshwater Quality: Defining the Indefinable* ed P J Boon and D L Howell (Edinburgh: HMSO) pp 251-57

[9] Wadeson L A 1994 A geomorphological approach to the identification and classification of instream flow environments *S. Afr. J. Aquat. Sci.* **20** 1-24

[10] Newson M D and Newson C L 2000 Geomorphology, ecology and river channel habitat: mesoscale approaches to basin-scale challenges *Prog. Phys. Geog.* **24** 195-217

[11] Wiens J A 2002 Riverine landscapes: taking landscape ecology into the water *Freshwat. Biol.* **47** 501-15

[12] Heino J, Louhi P and Muotka T 2004 Identifying the scales of variability in stream macroinvertebrate abundance, functional composition and assemblage structure *Freshwat. Biol.* **49** 1230-9

[13] Yarnell S M, Mount J F and Larsen E W 2006 The influence of relative sediment supply on riverine habitat heterogeneity *Geomorphology* **80** 310-24

[14] Povž M 2008 The habitat requirements of Salmo marmoratus. *Unpublished Report,* Institute for Natural Sciences, Ljubljana, Slovenia 4 p.

[15] Conlan K, Wade T, Ormerod S, Lane S, Durance I and Yu D 2007 Preparing for climate change impacts on freshwater ecosystems (PRINCE). *Environment Agency Science Report* SC030300/SR (Bristol: Environment Agency)

[16] Bovee K D 1982 Guide to Stream Habitat Analysis Using the Instream Flow Incremental Methodology. *US Fish and Wildlife Report No. FWS/OBS-82/26* (Fort Collins: US Fish and Wildlife Service)

[17] Vismara R, Azzellino A, Bosi R, Crosa G and Gentili G 2001 Habitat suitability curves for brown trout (Salmo trutta fario L.) in the River Adda, Northern Italy: comparing univariate and multivariate approaches *Regul. River.* **17** 37-50

[18] Negnevitsky M 2005 *Artificial Intelligence: A Guide to Intelligent Systems* (2nd ed) (Harlow:Addison-Wesley)

Climate Change and Water Resources: Scenarios of Low-flow Conditions in the Upper Danube River Basin

W Mauser[1,2] , T Marke[1], S Stoeber[1]

[1] Department of Geography, Ludwig-Maximilians University, Munich, Germany

E-mail: w.mauser@iggf.geo.uni-muenchen.de

Abstract. Global Climate Change will have regional impacts on the water resources and will force water resources managers and farmers to adapt. Both low-flow and its duration are critical hydrological parameters, which strongly influence the state of aquatic ecosystems as well as power production, reservoir management and industry. Impacts of future climate change is analysed using scenarios for the change of meteorological drivers and regional hydrological simulation models. The project GLOWA-Danube (www.glowa-danube.de) develops integrative modelling techniques combining process knowledge from both natural and social sciences to examine the sustainability of regional water systems as well as water management alternatives in the Upper Danube watershed (A = 77000 km²). Special emphasis is given to changes in low-flow condition. DANUBIA describes the regional water cycle both physical and spatially distributed. It consists of a collection of tightly coupled models, which strictly preserve energy and matter and are not calibrated to maximise their overall predictive abilities. The paper demonstrates that DANUBIA can reproduce the daily discharge for the time period from 1971-2003 with a Nash-Suttcliffe coefficient of 0.84 (gauge Achleiten). Based on a statistical climate simulator 12 realisations of the IPCC A1B climate scenario were used to investigate impacts of climate change during the simulation period of 2011–2060. The change in discharge and frequency of occurrences of low-flow in the watershed for the scenario ensemble were analysed for the outlet gauge. The analysis shows that strong changes were simulated in the frequency of occurrences of low-flow conditions. The changing climate gradually reduces a 50-years NM7Q discharge of today to less than half of its discharge in the year 2060. These results clearly indicate that the expected climate change will strongly alter the low-flow conditions in the Upper Danube watershed.

1. Introduction

Conditions of low-flow are an important limiting factor for the utilization of water resources. Low-flow is characterized by a prolonged below average discharge in a river system. The reasons for low-flow can be manifold. They range from reduced rainfall, elevated evapotranspiration, reduced water storage or cold temperatures with freezing soils and water or natural water storage in mountain snow packs. Usually only a combination of reasons creates severe low-flow conditions. Severe low-flows can impose limitations on those water dependent infrastructures, which use river water, e.g. hydropower production, cooling in the process of conventional power production and navigation. Low-flow can therefore cause substantial financial losses since the installed structures either shut down or have to adapt their operation to the limitations created by the missing flow.

[2] To whom any correspondence should be addressed.

Low-flow augmentation can be achieved through reservoirs, which collect runoff and release it during times of droughts. Reservoirs and their operation therefore are an important component of the overall water infrastructures in a basin.

From a global perspective climate change is usually perceived as an increase in average air temperature, which is assumed to be well predictable both in its magnitude and spatial distribution. When it comes to the expected changes in rainfall predictions are much less certain. For the case of Central Europe IPCC expects no significant changes in annual rainfall but significant increases in rainfall amounts in winter and corresponding decrease in summer [1]. How serious will be the effects of changing climatic drivers on the regional water cycle and what are suitable adaptation strategies for the water infrastructures are important questions in the context of regional impact studies of Global Change?

A case study with the aim to analyse the expected changes in low flows was conducted within the interdisciplinary research project GLOWA-Danube (www.glowa-danube.de) to exemplify both the power of DANUBIA as a scenario tool and the regional impacts of climate change on the complex, large, mountainous watershed of the Upper Danube in Central Europe. The water cycle components of DANUBIA are set up as a spatially distributed, physical hydrological model, which conserves mass and energy and which is not calibrated against measured historical runoff [2].

The paper is set up as follows: first the Upper Danube watershed and the trends in climate change, which have already occurred there are introduced, second the general approach of GLOWA-Danube and the integrated Global Change decision support system DANUBIA is described and checked against measured historical stream flows, third a statistical climate change simulator, which produces possible future time series of meteorological drivers using already measured data is introduced and fourth the impact of an ensemble of climate realisations for the next 50 years on the low-flow conditions of the Upper Danube watershed is demonstrated.

2. The Upper Danube Watershed

2.1. The Geography of the Upper Danube

In order to validate DANUBIA it is applied to the Upper Danube catchment. The Upper Danube catchment covers an area of 76 653km² and is situated in parts of Southern Germany, Austria, Switzerland and Italy (figure 1). The catchment is characterized by its Alpine topography, the relief stretching from altitudes of 287 m.a.s.l. at the discharge gauge Achleiten up to 4049 m.a.s.l. at Piz Bernina in its Alpine head-waters. The Upper Danube catchment shows strong meteorological gradients with annual precipitation ranging from 550 to more than 2000 mm, an annual mean temperature from -4.8 to 9°C, evapotranspiration from 100 to 700 mm per year and annual discharge from 150 to 1750 mm per year. Geology is dominated by the northern rim of the Alps with a sequence of mountain ridges composed of granite, lime-stone to sedimentary rocks. The lowlands north of the Alps are composed of moraine material deposited during the last ice ages and a sedimentary basin, which is up to 3000 m deep. Most rivers emerge in the Alps and cross the lowlands towards the North to feed the Danube, which flows in W-E direction in the northern part of the watershed. Soils are very heterogeneous ranging from coarse soils in and close to the Alps to deeply weathered fine-grained soils in the low-lands of the Danube. As a result of all heterogeneities the land use and land cover pattern is highly diverse: it ranges from glaciers in the upmost headwaters through large conifer and deciduous forests and meadows in the Alpine valleys and on the moraines to intensively managed meadows and agricultural areas (maize, cereals, potatoes, sugar beet) in the valleys of the lowlands [3]. North of the river Danube the catchment is framed by the mid-altitude mountains of the Bavarian Forest and the Swabian Alb.

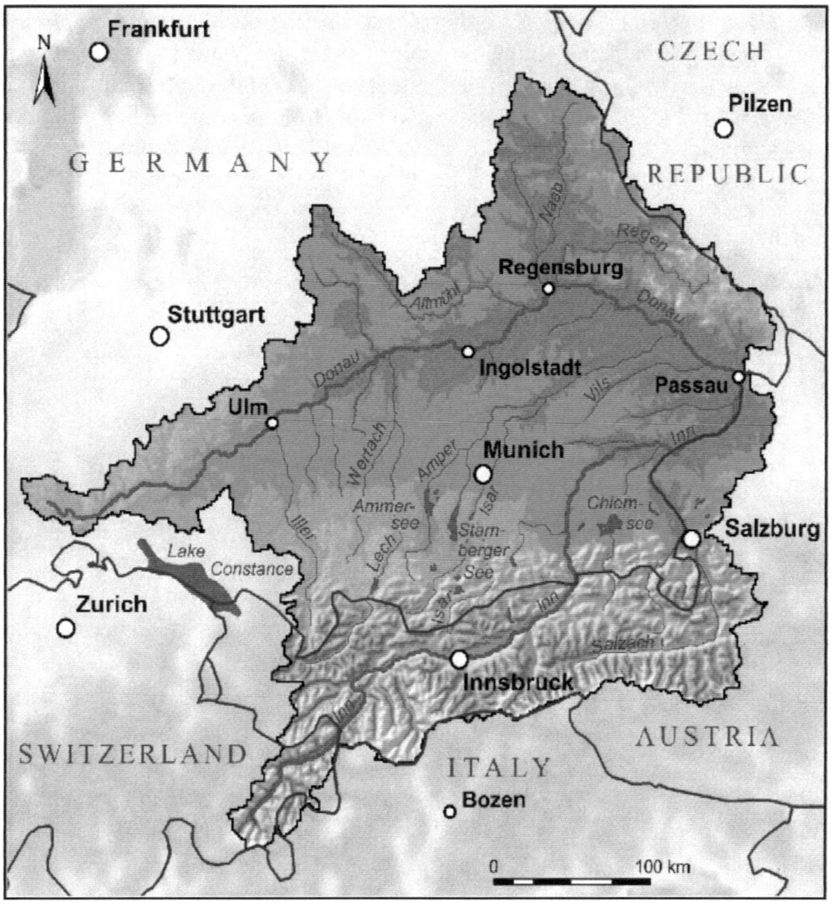

Figure 1. The Upper Danube Watershed

The water resources in the Upper Danube are intensively used by 5 countries. Most river runoff in the catchment is managed either for hydropower production through run-of-river power stations, water diversions, cooling of power plants or large reservoirs, which in the Alps are mainly used for low-flow augmentation during the winter period. The water resources of the Upper Danube, which are not consumed within the watershed through evapotranspiration or through water diversion into other watersheds (e.g. Main) are released to the downstream Danube countries. They strongly depend on the water surplus of the Upper Danube for power production, irrigation and navigation.

2.2. Climate Change in the Upper Danube Watershed
The Upper Danube has already undergone considerable regional climate change. A thorough analysis of the temperature increases, which have already occurred, was conducted in the GLOWA-Danube project in order to compare global climate trends as forecasted by the IPCC [1] with regional effects. The analysis is based on the record of 277 meteorological stations in the watershed from the German and Austrian Weather service and uses the record from 1960 to 2006. Figure 2 shows the result of the analysis. It can be seen clearly that air temperatures in the watershed have already increased from 7.2 °C in 1960 to 9.2 °C in 2006. This strong increase can only be explained by the IPCC A1B scenario if one assumes, in accordance with IPCC, that the regional increase in air temperature can deviate strongly from the global mean. For the Upper Danube the assumption that the regional increase in air temperature is larger than the global increase by a factor of 1.7 explains the observed curve in the past and leads to the regression line in figure 2. Figure 2 shows that the temperature

increase also applies to the annual monthly minima and maxima. Using the assumption of figure 2 a statistical climate generator produced future possible climate time series (red), which follows three conditions: 1) it is statistically equivalent to the measured data from 1970-2006 in that it considers the temporal development of the statistical relation between mean weekly temperature and rainfall sums over the year, 2) it considers coupled standard deviations of both temperature and rainfall through a random number generator, 3) it considers the IPCC A1B temperature trend over the next 55 years multiplied by a factor of 1.7.

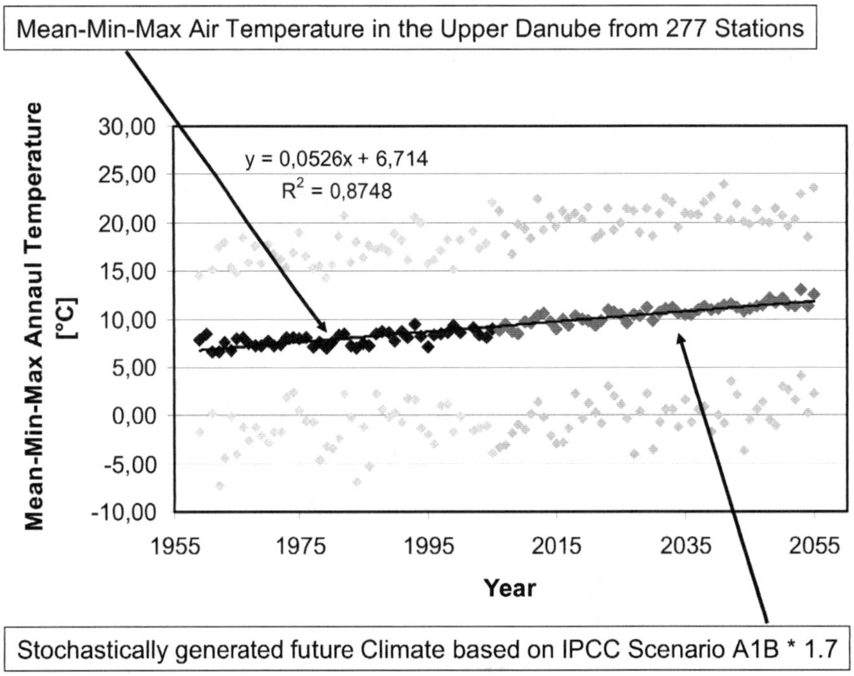

Figure 2. Past and future trends in air temperature for the Upper Danube catchment based on the IPCC A1B scenario.

The way the climate scenario information is used within DANUBIA to evaluate the consequences of a changing climate on the low-flow conditions in the Upper Danube watershed is shown in principle in figure 3. The figure shows that two main drivers affect the future development of water management in the Upper Danube watershed. First there is climate change. It acts upon DANUBIA through climate scenarios originating from different sources. Currently results from regional climate models like MM5 and REMO can be assimilated into DANUBIA. The second driver is regional development, which includes demographic change, economic development, changing EU-regulations as well as external (global) boundary conditions. Regional development influences the decisions of actors in DANUBIA. Actors are self-contained entities, which make decisions. They receive information on their environment, have plans for actions and experiences from past actions. Actors can be different groups of society, the decisions of which are simulated. Currently DANUBIA contains farmers, households, water suppliers, tourists and industrial branches. The decisions of the actors as well as changing climate simultaneously influence the water flows in the Upper Danube watershed leading to a complex web of interactions, which result in a future change of the water balance in the watershed. Key factors for change are characterized by changing land use (through farmers' decisions), melting glaciers, reduced snow cover, changing reservoir operations, etc.

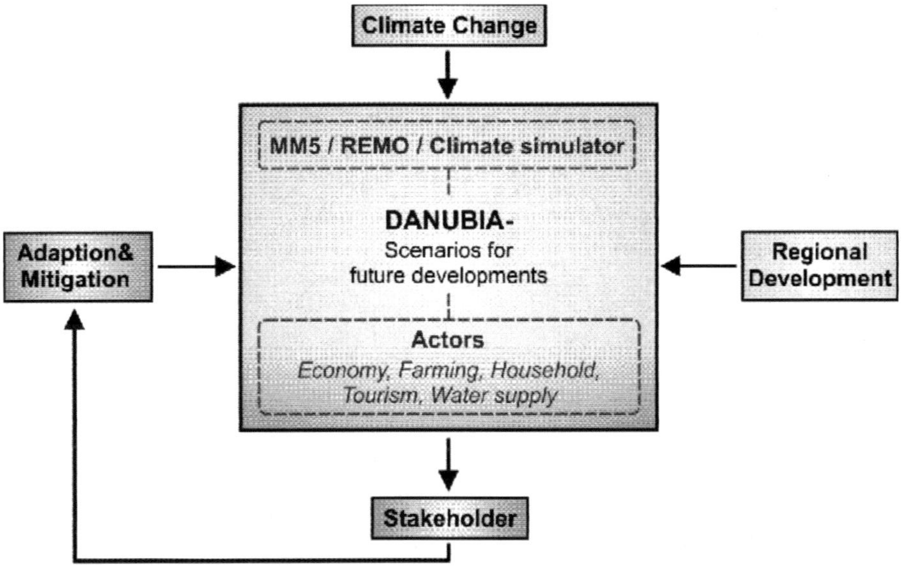

Figure 3. Schematic diagram of the utilization of the Global Change Decision Support System DANUBIA for the stakeholder dialogue within the GLOWA-Danube project.

The presented paper concentrates on the influence of changing climate alone and takes the first order assumption that land use, demography, water consumption etc. will not change during the modelling period from 2011 to 2060.

As Figure 3 shows the results of the scenario simulations are presented to interested stakeholders. In an interactive process the results are discussed with the stake-holders and ways to both regionally adapt to climate change and to mitigate causes of the changing climate are elaborated by the stakeholders. They are implemented in DANUBIA and lead to a new scenario run. The efficiency and effectiveness of the proposed decision alternatives are then discussed with the stakeholders in a second iteration.

3. PROMET - The hydrological model behind DANUBIA

The investigation of the impacts of climate change on the regional hydrology needs predictive hydrological models, which include physical descriptions for all compartments of the regional water cycle. PROMET was developed as a raster model with 1x1 km raster elements. Figure 4 shows a schematic diagram of the compartments represented in PROMET. The basic principles, which were followed in the design of PROMET [2] were:

1. a fully physical and physiological description of the water fluxes in the compart-ments of complex watersheds of medium size (A~100 000 km²). The different components of PROMET should be fully coupled. The components are:
   a. meteorological drivers either from regional climate models or from station data,
   b. land-atmosphere energy and mass exchange based on plant physiologic control of gas exchange (interception of rainfall, evapotranspiration, sensible heat exchange, carbon uptake or release, long- and shortwave radiation balance as well as momentum exchange),
   c. snow and ice accumulation and ablation,
   d. vertical and lateral unsaturated and saturated flows including infiltration
   e. channel flow and flow through lakes,
   f. flow through man-made reservoirs and water transfers.

2. PROMET should strictly conserve mass and energy as a whole and within as well as throughout all its components and feedbacks.
3. PROMET should be spatially explicit in all process descriptions. In order to be most compatible with models from other disciplines (e.g. regional climate models, carbon models, groundwater models, multi-agent raster-based decision models) the spatial representation in PROMET is based on an isotropic grid. A watershed is composed of a set of raster elements, which exchange data with their surrounding and all processes treated within PROMET are consequently described based on the same set of raster elements.
4. Physical consistency and predictive power should not be diminished or lost in the model calibration process. Therefore the values of the model parameters of PROMET should not be calibrated using measured discharge at gauges in or at the boundary of the considered watershed. Instead we use literature sources and/or measurements (both in the field and from remote sensing sources) and a detailed analysis of the digital terrain model to define the initial values of all model parameters.

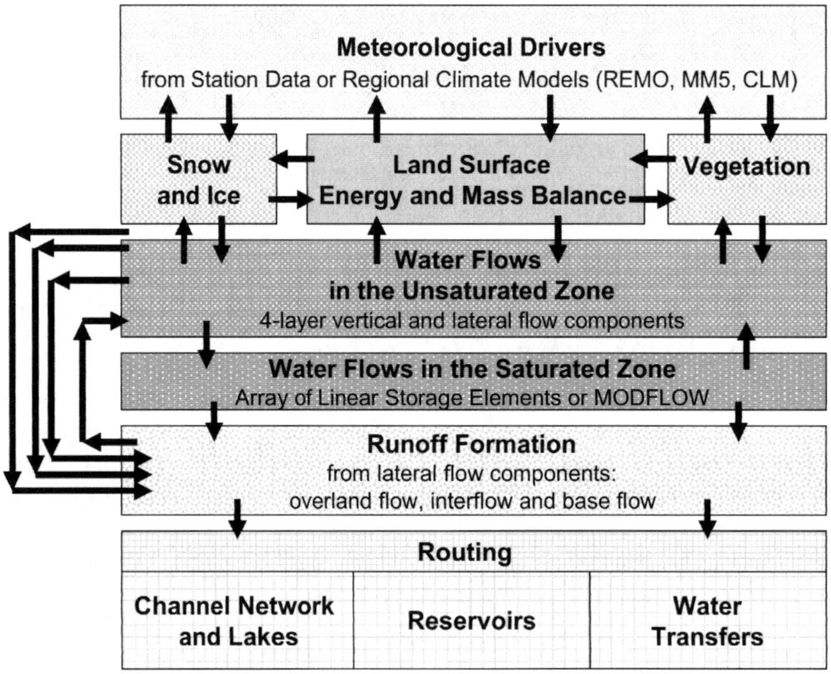

Figure 4. Schematic diagram of the compartments of PROMET.

4. Validation of PROMET

Figure 5-1 shows the resulting daily discharge modelled for the period between 1971 and 2003 for the Achleiten gauge close to the outlet of the 77000 km² watershed. As can be seen clearly the daily discharge is modelled with a high accuracy. Figure 5-2 shows the measured and modelled annual low-flow, which is defined as the annual mini-mum 7-days average flow for the same gauge for the same period. The annual variability of low-flow is well captured by the model for historical periods.

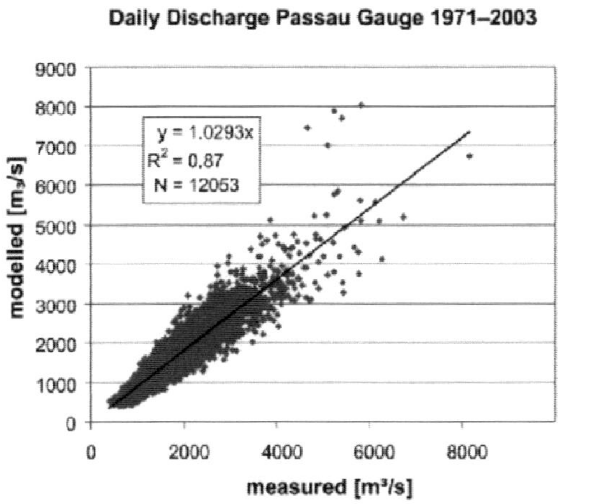

Figure 5-1. Measured and modelled daily discharge for the reference period of 1971-2003 at Achleiten gauge, Upper Danube watershed.

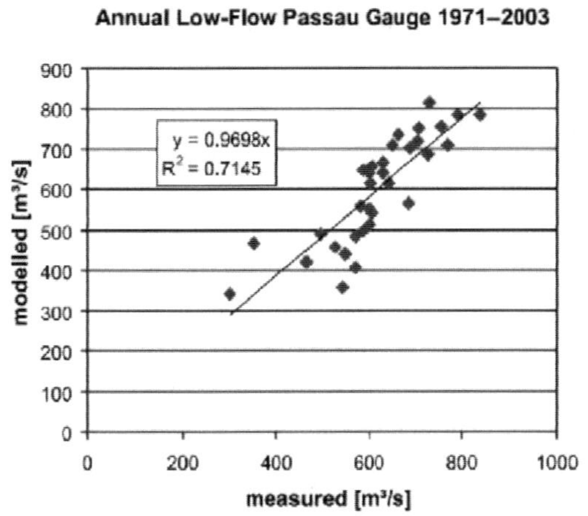

Figure 5-2. Measured and modelled annual low-flow (minimum 7-days average discharge) for the reference period of 1971-2003 at Achleiten gauge, Upper Danube watershed.

Table 1 shows the results of the validation runs for selected gauges in the Upper Danube watershed representing watersheds of different areal extents and hydrological conditions. The Salzach as well as the Inn are two tributaries of the Upper Danube which are strongly influenced by the hydrological regime of the Alps. Under current climate conditions this means that discharge is low in winter due to snow storage and large in spring and early summer due to snow melt. The Naab and Ammer are tributaries, which are not or hardly influenced by Alpine hydrology. Table 1 therefore covers a broad range of watersheds with very different hydrologic conditions. As can be seen slopes of the 1:1 regressions, as shown in figure 5-1, are close to 1.0 and the coefficient of determination R^2 is high throughout all watersheds. So is the Nash-Sutcliffe coefficient with the one exception of the Isar. There the Sylvenstein reservoir, which is operated in a manner, which is not transparent to the authors is strongly affecting discharge.

Table 1. Selection of gauges within the Upper Danube watershed. Achleiten is at the outlet of the Upper Danube near Passau.

Gauge	River	Upstream Area [km²]	Slope [-]	R^2	Nash-Sutcliffe Coefficient [-]
Achleiten	Danube	76 673	1.03	0.87	0.84
Hofkirchen	Danube	46 496	1.11	0.87	0.81
Dillingen	Danube	11 350	1.13	0.84	0.72
Oberaudorf	Inn	9 715	0.94	0.81	0.80
Plattling	Isar	8 435	1.08	0.75	0.47
Laufen	Salzach	6 112	0.86	0.85	0.80
Heitzenhofen	Naab	5 431	0.99	0.78	0.79
Weilheim	Ammer	607	0.98	0.63	0.69

5. Climate impact on low-flow – scenario investigations

5.1. The Synthetic Climate Generator

Based on these results the period from 2011 to 2060 was modelled using the output of a stochastic climate generator. It uses the IPCC-A1B scenario [1] with the empirical multiplication factor of 1.7 for the regional trends of temperature increase in the Upper Danube watershed. This regional adaptation of the IPCC A1B scenario results in a 3 degree temperature increase in the Upper Danube watershed until the year 2060. A random number generator, which is based on the covariance matrices of weekly mean temperatures and rainfall sums generates pairs of deviations from a mean temperature and rainfall sum for each week in the period between 2011 and 2060. These values together with the IPCC temperature trend produce pairs of statistical mean temperature and rainfall sum for each week from 2011 to 2060. For each future week a minimum distance algorithm then searches for the most similar meteorological data set in the measured data from 1960 to 2006. Through this procedure a synthetic time series of meteorological input data is produced, which 1) confirms to the statistical behaviour of the past (one necessary but critical assumption) and 2) reflects the regional temperature trend of IPCC. This synthetic climate generator can easily be used to create a large number of statistically equivalent future climate data sets by changing the seed for the random number generator an rerunning the procedure. In order to evaluate the uncertainty of the estimated change in low-flow an ensemble of 12 different realisations of the same IPCC A1B climate scenario was modelled and analysed. The following strategy was applied for the realisations described in table 2:

1) For each of the 12 realisations a different story was developed to select the valid realisation from a large number of simulated meteorological time series. A story in this context is a set of assumptions about the future, which can be converted into rules to select a certain realisation of a scenario.
2) Based on the assumption 5000 meteorological data sets were produced with different seeds for the random number generator.
3) The 5000 meteorological data sets were analysed statistically according to the underlying assumptions of the stories formulated under 1) and the realisation, which fulfilled the assumption with a probability of 1% was selected.

Table 2. List of the realisations 1-12 together with the zero-hypothesis that temperature will not change in the future. Column 2 expresses the story, columns 3 and 4 give values for total rainfall or mean temperature for the period under consideration.

Reali-sation	Story	1% Rain-fall [mm]	1% JJA-Temp. [°C]
0	No change in air temperature until 2060		
1	Min. rainfall sum of 5 consecutive years between 2011-2035	4015	
2	Min. rainfall sum of 5 consecutive years between 2036-2060	3883	
3	Min. rainfall sum of 3 consecutive years between 2011-2035	2517	
4	Min. rainfall sum of 3 consecutive years between 2036-2060	2387	
5	Min. rainfall sum of 1 year between 2011-2035	791	
6	Min. rainfall sum of 1 year between 2036-2060	762	
7	Average rainfall between 2011-2035	1027	
8	Average rainfall between 2036-2060	922	
9	Max. aver. JJA temp. of 5 consecutive years 2011-2035		20.15
10	Max. aver. JJA temp. of 5 consecutive years 2036-2060		20.35
11	Max. aver. JJA temp. of 1 year between 2011-2035		21.62
12	Max. aver. JJA temp. of 1 year between 2036-2060		21.7

5.2. Effects of Climate Change Scenarios on low-flow

For all selected scenarios together with one scenario, the zero-hypothesis, which assumes that no further change in temperature will occur in the future, the water flows were modelled on an hourly basis for the period from 2011 to 2060 using DANUBIA. The resulting hourly runoff data was aggregated to daily values for the outlet gauge in Achleiten. The resulting discharge records were analysed for the lowest annual mean 7-days discharge (NM7Q), which was selected as criterion for low-flow conditions. The time series of annual low flows is shown in figure 6. The left part of figure 6 shows the modelled and measured historical low flows from 1971 to 2003. As can be seen clearly there is considerable scatter both in the measured and modelled data. The right hand side of figure 6 shows the development of the annual low flow for the period from 2011 to 2060 for realisation 0 to realisation 12 in table 2. As can be seen clearly the zero-hypothesis of no temperature increase in the future leads to no significant changes in the simulated future annual low-flow. On the other hand the assumed temperature increase leads to a considerable reduction of simulated annual low-flows for all selected stories of the IPCC A1B scenario. As can be expected there also is a considerable annual scatter in the development of low-flow. This means that it may well be that in 2050 there could be an annual low-flow occurring, which is as high as today's. Nevertheless the general trend of decreasing low-flows is obvious.

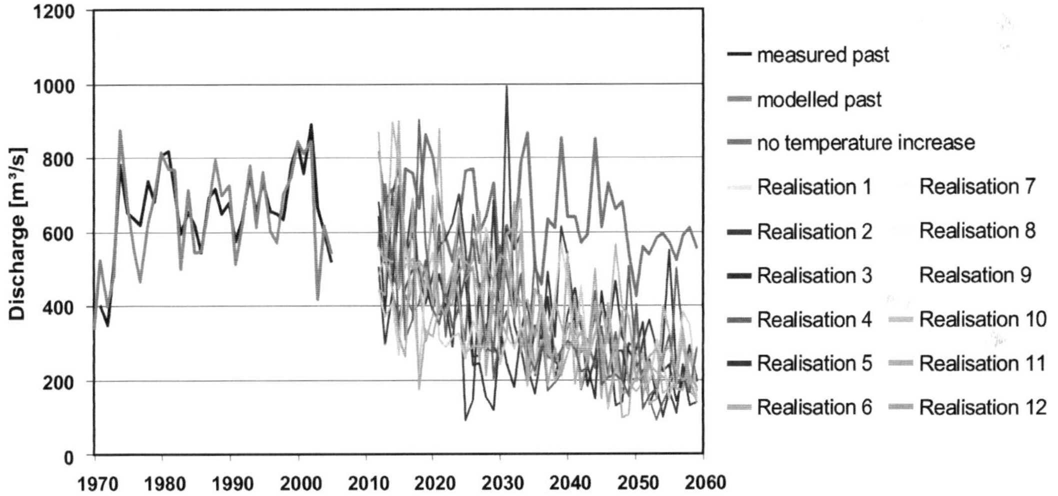

Figure 6. Annual low-flow (min. 7-days mean discharge) from 1971-2003 and from 2011-2060 at gauge Achleiten using the 12 realisations of the IPPC A1B scenario as listed in table 2.

Figure 7 shows the result of an analysis of the course of the 50-years return period low-flow condition at gauge Achleiten. The blue, green and red curves were determined using a 25 years window, which was shifted over the measured and modelled data. For each window the 50-years low-flow was determined by fitting a log-normal distribution to the annual low-flow data and determining the minimum annual 7-days average discharge with 98% excess probability (NM7Q50). The blue curve in figure 7 represents the change in measured data, the green curve stands for the temporal development of the NM7Q50 when no temperature increase occurs and the red curve represents the change in NM7Q50 when assuming the average development of the annual low-flows of all realisations shown in figure 6. As can be seen clearly 1) the simulated 50-years return period low flow hardly changes in the future when assuming no change in the air-temperature and the related rainfall, 2) NM7Q50 decreases sharply and is reduced to half of its present value by 2030 and to one third of its value by 2060 when investigating the effects if a series of realisations of the IPCC A1B climate change scenario. This indicates a change in flow regime in the discharge of the Upper Danube since low-flows with such a discharge have not yet been measured.

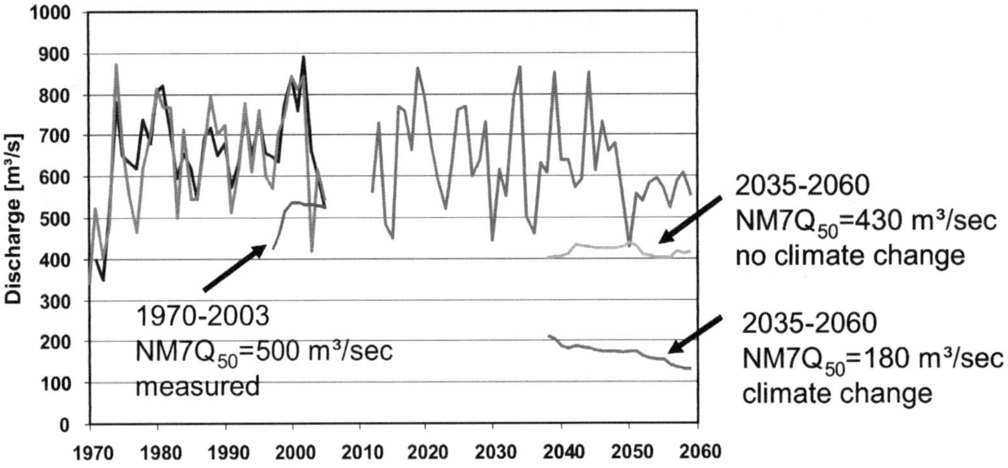

Figure 7. Development of the 50-years return period NM7Q low-flow condition at gauge Achleiten; blue = determined from measurements, green = no climate change, red = average from realisation 1-12 from figure 6.

6. Conclusions

The simulations of the effects of an ensemble of different statistically equivalent realisations of the IPCC A1B climate change scenario on the low-flow conditions have shown that, following the assumptions of the statistical climate generator described above, severe changes in the low-flow conditions in the Upper Danube can be expected in the future. They range from a reduction of the expected low-flows to half of their present discharges by 2030 to a reduction to one third of their present values by the year 2060. The reasons for the steep reductions are manifold and interrelated in a complex manner. The main reasons are: 1) the changes induced in the snow cover in the Alps. With increasing temperatures an increasing fraction of today's snowfall will fall as rain, will run off during the winter season and not be stored in the Alps; 2) the decreasing summer rain fall, 3) the increasing evapotranspiration. All three factors work towards a reduction of low-flow in summer and amplify each others.

It can be expected that the reduction of low-flows will have a notable impact on water resources management both within the Upper Danube watershed and downstream. More frequent and severe low-flows will mainly affect power-production and transportation. Future investigations will concentrate on possible adaptation measures to compensate for the changes in low-flows in the Upper Danube. Measures like changes in land-use or the increase of managed storage capacity in the watershed as well as changes in the operation of existing reservoirs will be investigated.

Acknowledgement

The authors wish to thank the German Federal Ministry of Education and Research (BMBF), the Free State of Bavaria, the State of Baden-Wuerttemberg and the Ludwig-Maximilians University Munich for funding the GLOWA-Danube project. Thanks also go to the German and Austrian Weather Services as well as the Bavarian State Authority for the Environment (LfU) for the supplied data.

References

[1] IPCC 2007 *Climate Change 2007: The Physical Science Basis. Contribution of Working Group I to the Fourth Assessment Report of the Intergovernmental Panel on Climate Change,* ed. S Solomon, D Qin, M Manning, Z Chen, M Marquis, K B Averyt, M Tignor and H L Miller (Cambridge and New York: Cambridge University Press) p142

IOP Conf. Series: Earth and Environmental Science **4** (2008) 012027 doi:10.1088/1755-1307/4/1/012027

[2] Mauser W and Bach H 2008 PROMET – a Physical Hydrological Model to Study the Impact of Climate Change on the Water Flows of Medium Sized Mountain Watersheds *J. Hydrol.* (submitted)

[3] Ludwig R, Probeck M and Mauser W 2003 Mesoscale water balance modelling in the Upper Danube watershed using sub-scale land cover information derived from NOAA-AVHRR imagery and GIS-techniques *Physics and Chemistry of the Earth* **28** 1351-1364

Detection of changes in flow variability of the upper Danube between 1876–2006

Pavla Pekarova[1], Peter Skoda[2], Pavol Miklanek[1], Dana Halmova[1], Jan Pekar[3]

[1] Institute of Hydrology SAS, Racianska 75, 831 02 Bratislava, Slovakia
[2] Slovak Hydrometeorological Institute, Jeseniova 17, 833 15 Bratislava, Slovakia
[3] Department of Applied Mathematics and Statistics, FMPI CU Bratislava, Mlynska dolina, 842 48 Bratislava, Slovakia

E-mail: pekarova@uh.savba.sk

Abstract. It is widely acknowledged that global warming will result in an increased extremality of hydrological phenomena. As for streamflow it is anticipated that duration of low–flow conditions and flood frequency will both increase. Approving this hypothesis by statistically evaluating daily runoff characteristics depends on the availability of long and homogeneous data series. The aim of this paper is to analyze the change of the average daily flows of the Danube at Bratislava for different subsets of the period 1876–2006. The statistical tests did not reveal any increase in the extremality of daily runoff over the period 1876–2006. On the contrary, variability of daily flows of the Danube River decreases. Over the last 30 years streamflows increased in the winter-spring season and decreased in the summertime, compared to the past periods. This can be explained by the higher air temperature in the headwaters. A warmer climate causes an earlier onset of snowmelt in the winter-spring season, thus less runoff is observed in the summer with higher rates of precipitation.

1. Introduction

The impact of anthropogenic activity on runoff processes can be assessed by analyzing statistical characteristics of mean daily streamflows. In general, an opinion prevails that as the atmospheric temperature increases, precipitation and runoff extremality tend to rise as well – leading to higher peaks and lower minimum values. However, such an opinion can be either confirmed or disapproved through a detailed statistical analysis of the various runoff characteristics. Statistical analyses of observed data series are often hindered by short data series, non-homogeneity of observation series, gaps in data sets, etc. There is only one gauging station in the Slovakia's part of the Danube (at Bratislava) with a continuous time series of discharge observations over the 130 years. Therefore an effort was made to find (through statistical methods) any long-term changes in the characteristics of the mean monthly flows, mean daily flows, and extremes (*Qmax*) in the historical record.

This paper is organized to provide answers to the following issues:
1. Filling in gaps in the daily flow records of the Danube at the Bratislava station (1876–1890);
2. Analysis of changes in the distribution of daily flows in 1876–2005;
3. Analysis of changes in the temporal pattern of mean monthly flows;
4. Analysis of variability of the several flow characteristics and their long-term trends.

IOP Conf. Series: Earth and Environmental Science **4** (2008) 012028 doi:10.1088/1755-1307/4/1/012028

2. Results

2.1. *Filling in gaps in the daily flow records of the Danube at the Bratislava station*

Water stage measurements on the Danube River at Bratislava were first made in 1823. At that time the gauge datum was at 131.08 m J (Adria system). After 1876 the average daily stages were recorded in Hungarian yearbooks [13]. In 1942, the Bratislava gauge datum was lowered by two meters, down to 129.08 m J (Adria system). After 1964, the gauge datum was set at 128.46 m B.p.v (Baltic system).

First discharge observations on the Danube at Bratislava, based on measurements of flow velocities, were conducted as early as 1882 [9, 10]. These observations revealed that the river channel at Bratislava was subject to scouring long time before the river was dammed at the town of Cunovo due to construction of the Gabcikovo Hydro Project in 1992. Deepening of the river's channel bottom can be readily seen in the shape of the rating curve (Fig. 1) [3, 4, 5].

The historical rating curve, valid before 1903 (according to [6] and [15] was approximated by two third-order polynomials. Relations between discharge Q and water stages WS for the range of stages below 480 cm and above 480 cm, are shown in Figure 2. Using these two relationships, the average daily stages were converted into average daily flows for a period of fifteen years (1876–1890). Adding these data to the existing records allowed us to create a series of 130 years (Fig. 3) suitable for a thorough statistical analysis of changes in the runoff series [8].

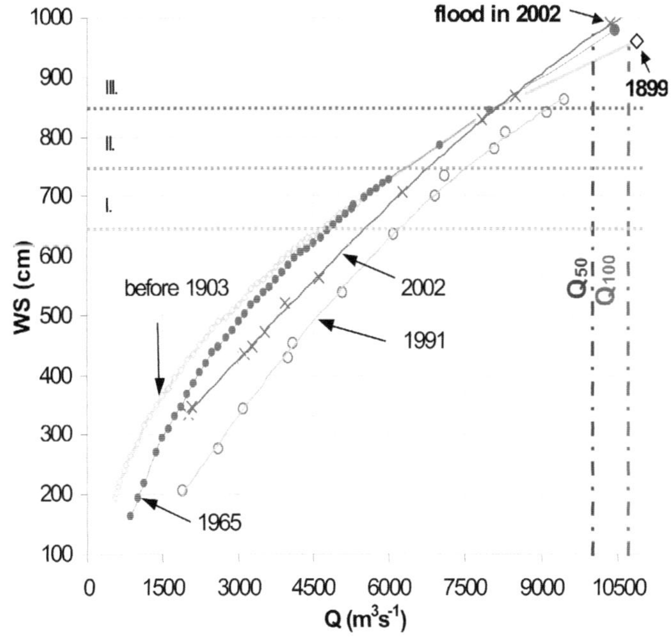

Figure 1. Changes of the Danube rating curve at Bratislava gauge (related to the present gauge zero level).

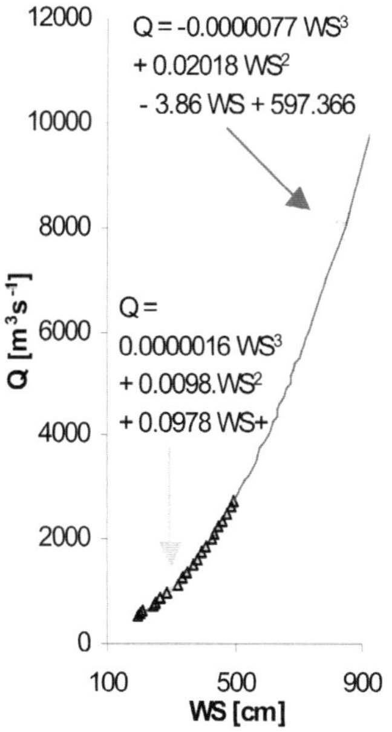

Figure 2. Relationship of the discharges Q to water stages WS, below and above the 480 cm level before 1903.

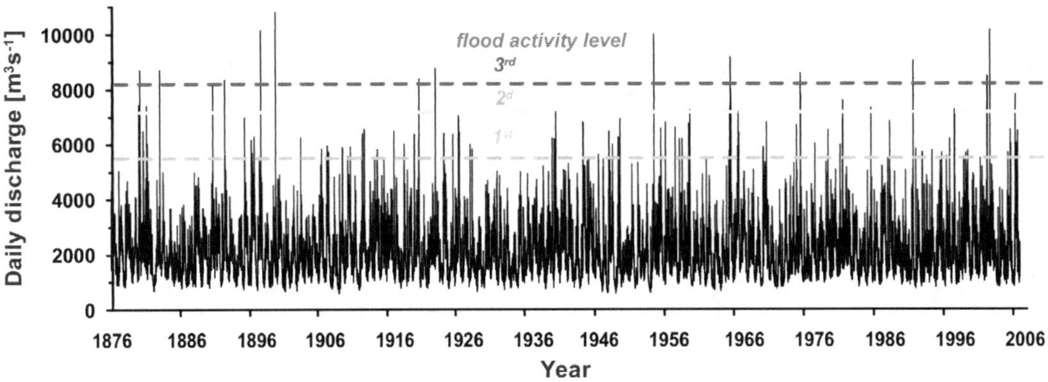

Figure 3. Mean daily flows of the Danube at Bratislava, flood activity level (1876–2006).

2.2. *Analysis of distribution of the daily flows*

The long-term average annual regime (365 days) of the average daily flows for the two periods (1876–1940, and 1941–2005) is graphed in Figure 4. The arrows point at the most significant differences between the periods in question. In late March, values from the period 1941–2005 exceeded the values from 1876–1940 by 492 m^3s^{-1}. In contrast, in the middle of September, average flows dropped by 541 m^3s^{-1} below those of the previous 65-year period. This time shift in the average daily flows was also analyzed by cross-correlations. The highest correlation coefficient between the average daily flows of the both periods (1876–1940 and 1941–2005) was found at the shift of +13 days. This may explain why the snowmelt in the Danube basin shifted earlier by 13 days due to the increased average daily temperatures. Nevertheless, this fact can also be attributed to the earlier atmospheric precipitation over the basin [1, 14, 16]. However, the latter alternative cannot be confirmed unless an analysis of rainfall amounts over the Danube River basin is done.

The probability of excedance curves of the mean daily flows at the Bratislava gauging station for the four 30-years periods (1886–1915, 1916–1945, 1946–1975, and 1976–2005) are shown in Figure 5.

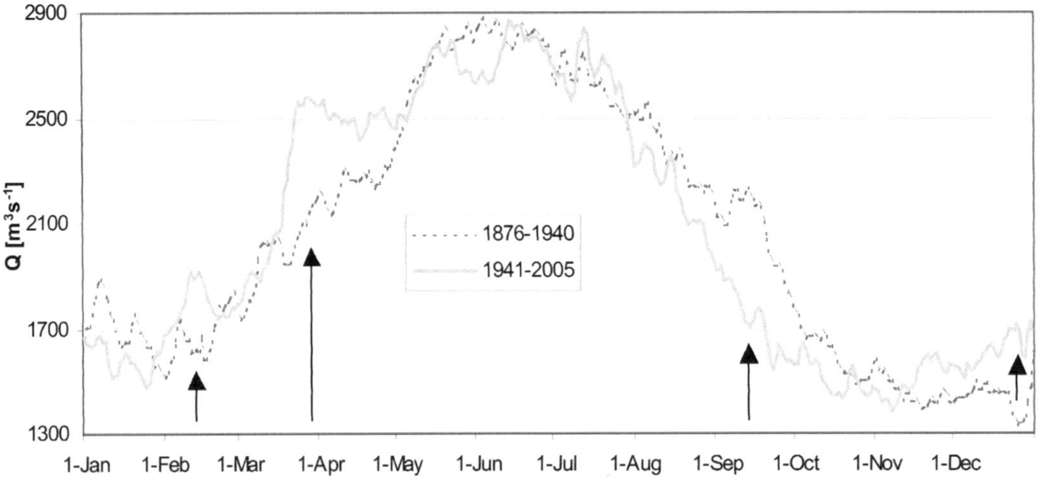

Figure 4. Long-term of the average daily flows of the Danube at Bratislava for periods: 1876–1940, and 1941–2005.

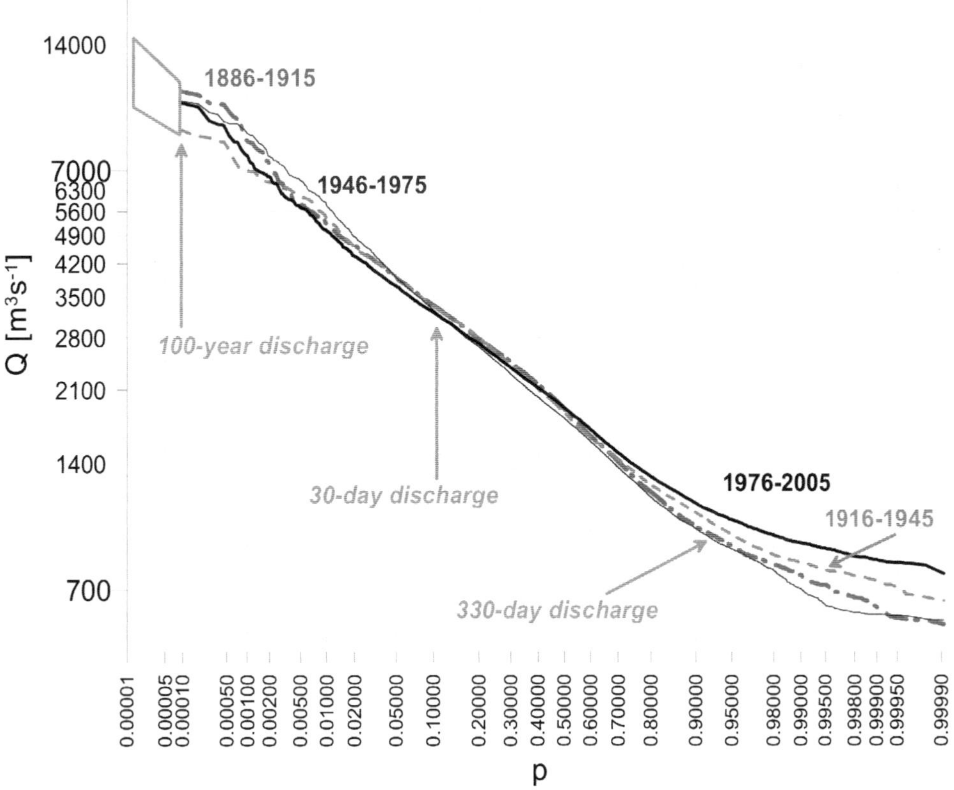

Figure 5. Empirical probability of exceedance curves of the mean daily streamflows for four periods, logarithmic-probabilistic scale.

This frequency curve may be used to determine the *M*-day discharge, that was reached or exceeded in *M*-days during the selected period (*M*-day discharge). The daily flows are more smoothed and the long-term mean daily flow was lower in last three decades. Comparing with the reference period 1930–1980, the *30*-day flow during 1975–2005 was lower by more than 300 m^3s^{-1}, while the *364*-day flow was higher by 160 m^3s^{-1}.

2.3. *Changes in the monthly runoff distribution over a year*

Intra-annual distribution of runoff is an important characteristic of a river's streamflow. The annual runoff distribution of the upper Danube reflects the high-mountain conditions in the headwaters. Maximal monthly streamflow (7324 m^3s^{-1}) was observed in June 1965 (Figure 6), while the absolute minimal monthly flow was observed in October 1947 (633 m^3s^{-1}).

Certain changes in the distribution of average monthly flows were identified by a simple determination of the 30-year average values from the 1876–2005 gauging records (Figure 6). A comparison of the averages from the two 30-year periods: 1901–1930, and 1976–2005 is depicted in Figure 7a. In the years 1901–1930, the wintertime streamflows were moderately below, while in June and July a little above the average. Over the last period of 30 years, the intra-annual runoff distribution changed markedly. In the winter-spring season (November through April) the streamflow remains above the average, while in the summertime (June through September) a decline in the flow regime is evident.

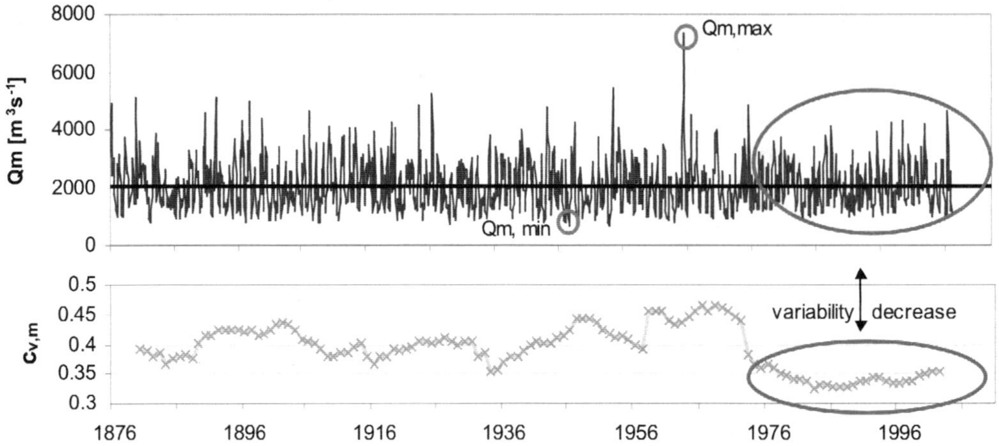

Figure 6. a) Monthly flows; b) 30-year moving averages of coefficient of variations $c_{v,m}$ of the monthly flows.

This can be explained by the higher air temperature in the upper Danube basin. A warmer climate causes an earlier onset of snowmelt in the winter-spring season, leading to lower runoff in the summertime with high precipitation rates. Therefore, discharge waves resulting from snow-melt and extreme summer rainfall may not occur coincidently. Nonetheless, the phenomenon can be attributed to changes in the patterns of annual precipitation over the Danube upper basin (Figure 7b). But we cannot prove this, since records with monthly areal precipitations for the Danube upper basin over the period 1876–2005 are not available.

2.4. *Long-term trends and the multi-annual oscillation of the stream flow characteristics*

In this section of the paper the annual Danube runoff characteristics will be evaluated at the Bratislava gauging station. The following time series covering 130 years were computed from average daily discharge for each year:

1. annual peak flows (Q_{maxd});
2. *30*-day flows (Q_{30}, flows that exceeded 30 days in a year – indicator of elevated flows);
3. average annual flows (Qa);
4. *330*-day flows (Q_{330}, flows that exceeded 330 days in a year – indicator of lower flows);
5. annual minimum flows (Q_{min});
6. annual variation coefficients (c_v).

Temporal evolution of the analyzed series is shown in Figure 8.

As can be assumed from these graphs, no substantial changes in the mean annual runoff within the last 130 years occurred. Minimum flows Q_{min} and *330*-day flows Q_{330}, have risen only moderately, approx. by 0.6 m^3s^{-1} per year. On the other hand, the *30*-day flows Q_{30} have dropped moderately, approx. by 0.8 m^3s^{-1} per year. The mean daily maxima Q_{max} series have been rising only slowly (by 1.2 m^3s^{-1} in a year), which compared with the order of the Danube peak flow values is not a substantial figure. The annual series of the variation coefficient c_v for the mean daily flows dropped as well, $b = -0.0002$. The statistical tests did not support the general belief that the daily runoff extremality of the Danube between 1876–2006 would increase.

Finally, we compared the 15-years moving averages of the *30*- and *330*-day flows, (Figure 9). Visually, the trend of the *30*-day flows declines; whereas in the case of the *330*-day flows it rather increases. Significance of these trends was detected using the CPTA software [7, 8, 9]. From the statistical point of view, no significant trends were confirmed for both of the above discharge characteristics within the analyzed period of 130 years.

a)

b)

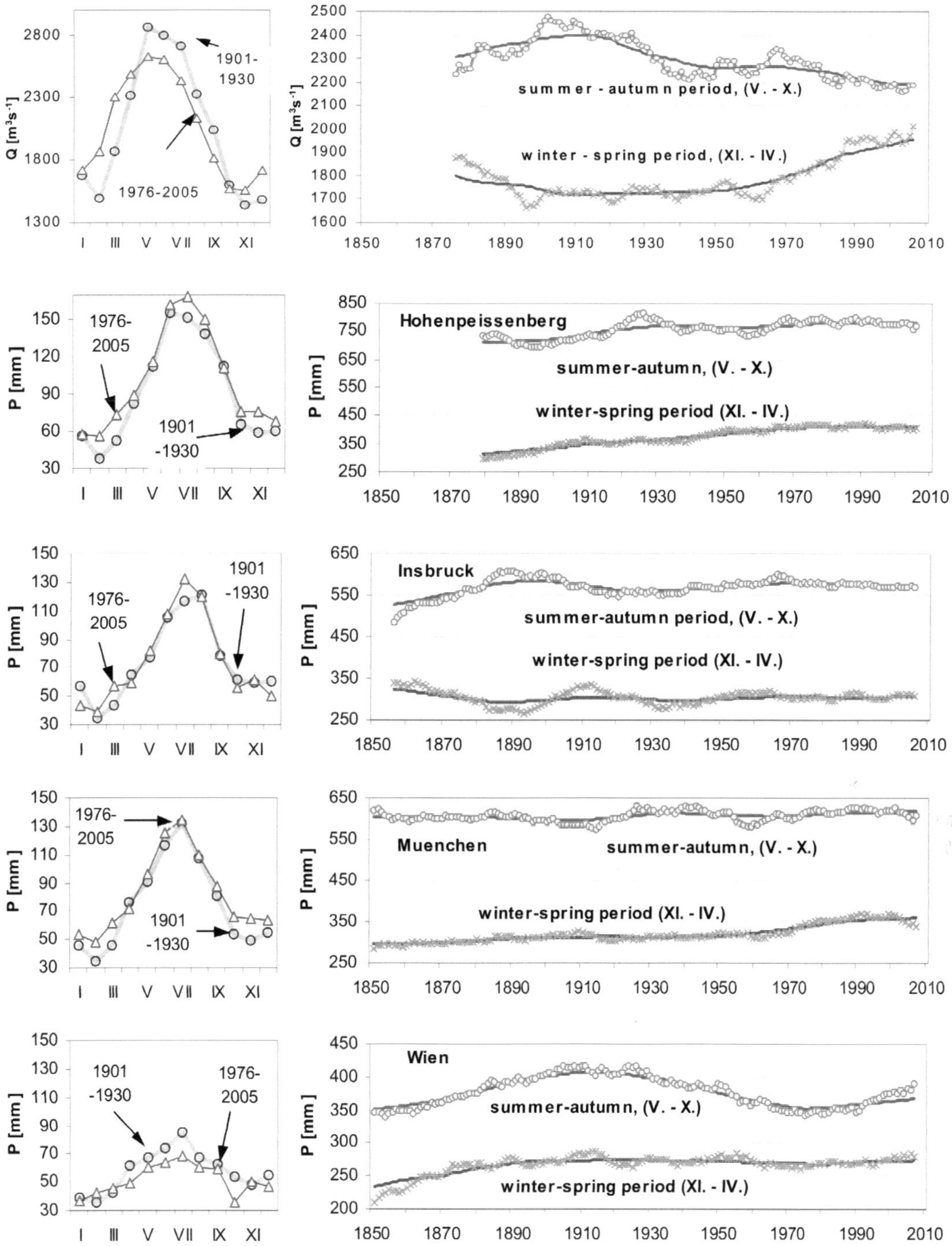

Figure 7. a) Intra-annual changes in the distribution of discharge Q over two 30-year periods: 1901–1930, and 1976–2005 (left side); Long-term runoff variability (1876–2005), increase of winter-spring discharge and declined summer-fall discharge over the period of 1970–2005 (right side).
b) the same for monthly precipitation totals, stations Hohenpeissenberg, Insbruck, Muenchen and Vienna (Austria), precipitation data series according to [2].

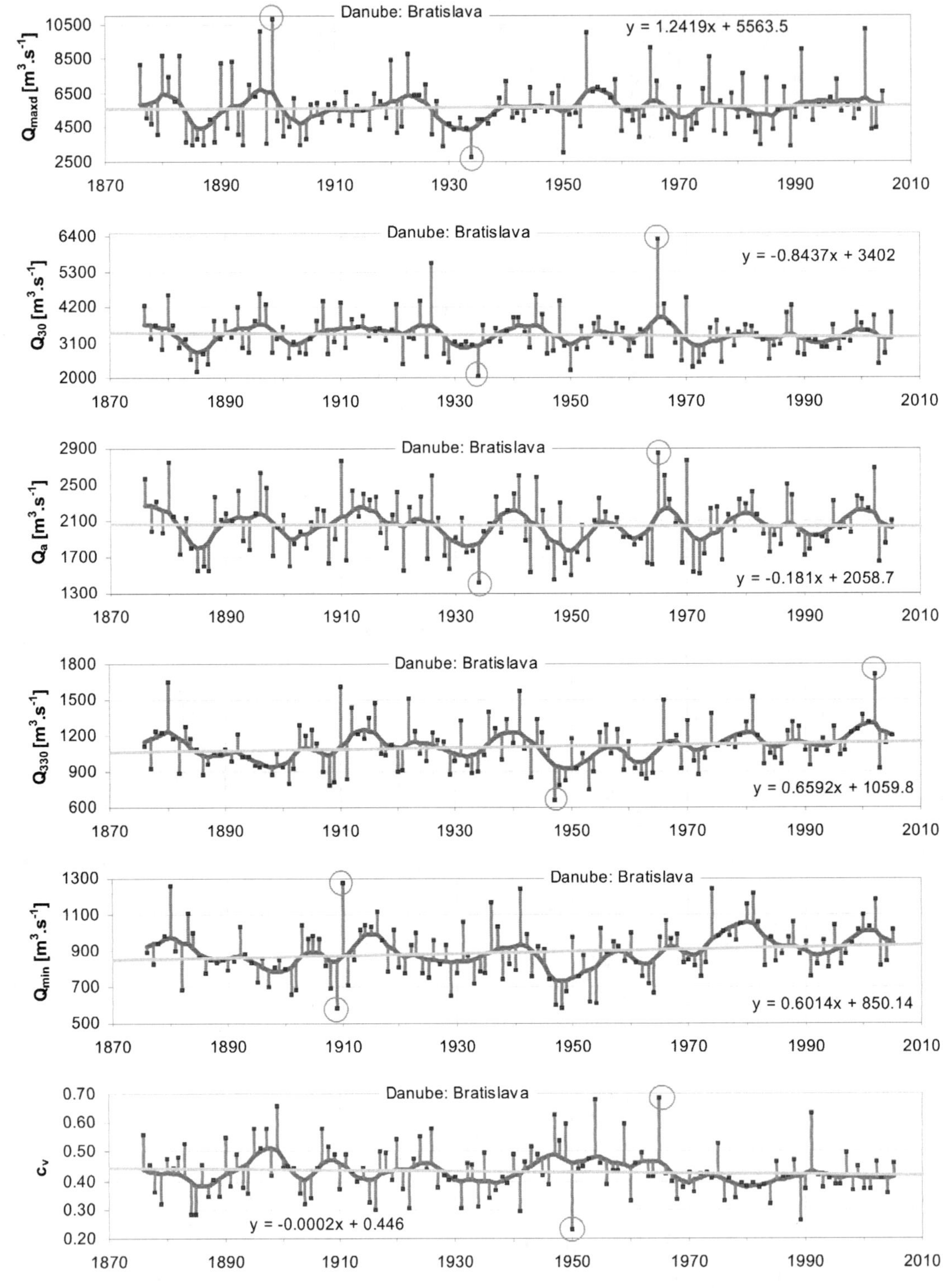

Figure 8. Variation of the peak annual flows Q_{maxd}, *30*-day flows Q_{30}, mean annual flows Q_a, *330*-day flows Q_{330}, minimum flows Q_{min}, and annual variation coefficients c_v series, of the Danube at Bratislava. Deviations from the 5-year moving averages, (1876–2005).

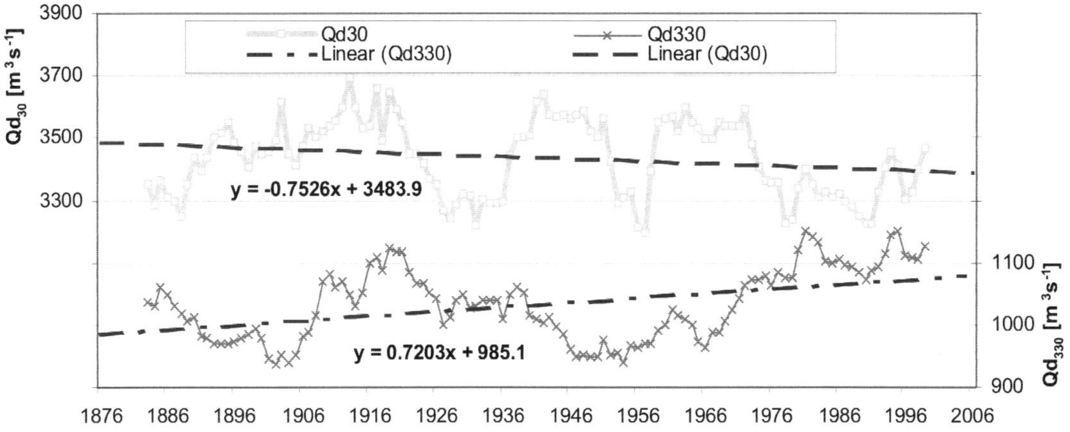

Figure 9. Trends in the 15-year moving averages of *30*-day and *330*-day flows.

3. Conclusion

Following the obtained results it can be said that the intra-annual distribution of runoff of the Danube at Bratislava changed considerably. Over the last 30 years, in the winter-spring season (November through April), the streamflow remains above its long-term average, while in the summertime (June through September) a decline is observed. The higher air temperature in the upper Danube basin can explain this fact. A warmer climate causes an earlier snowmelt in the winter-spring season, generating less runoff in the summertime when precipitation rates are high. Therefore, discharge waves resulting from snow-melt and extreme summer rainfall may not occur coincidently. Nonetheless, this finding can be attributed to changes in the annual precipitation in the Danube headwaters. Yet there is no evidence to prove this explanation, for records with monthly areal precipitations for the Danube headwaters over the period 1876–2005 are not available.

The underlying results reveal that the variability of daily flows at Bratislava did not rise in the 1876–2005 period whatsoever. It would rather decrease, as documented by the series of the variation coefficients for the period 1876–2005. Even if the intra-annual runoff distribution changed (higher winter-spring runoff), this seems not to influence neither Q_{30}, nor the Q_{330} values, since these do not depend on the intra-annual distribution of discharge.

Acknowledgement
This study was supported by projects MVTS „Flood Regime of Rivers in the Danube River Basin" APVV 0443-07 and VEGA-0096/08.

References

[1] Andrade-Leal RN, Bachhiesl M, Drabek U, Gutknecht D, Haiden T, Holzmann K, Hebenstreit K, Kirnbauer R and Nachtnebel HP 2002 Hydrological prediction models in operational use in the hydro-power industry *Österreichische Wasser- und Abfallwirtschaft* 54 **9-10** pp 129–134

[2] Klein Tank AMG, Wijngaard JB, Konnen GP, Bohm R, Demar′e G, Gocheva A, Mileta M, Pashiardis S, Hejkrlik L, Kern-Hansen C, Heino R, Bessemoulin P, Muller-Westermeier G, Tzanakou M, Szalai S, Palsdottir T, Fitzgerald D, Rubin S, Capaldo M, Maugeri M, Leitass A, Bukantis A, Aberfeld R, van Engelen AFV, Forland E, Mietus M, Coelho F, Mares C, Razuvaev V, Nieplova E, Cegnar T, Antonio Lopez J, Dahlstrom B, Moberg A, Kirchhofer W, Ceylan A, Pachaliuk O, Alexander LV and Petrovic P 2002 Daily dataset of 20th-century surface air temperature and precipitation series for the European Climate Assessment *International Journal of Climatology* **22** pp 1441–1453

[3] Miklanek P, Mikulickova M, Mitkova V and Pekarova P 2002 Changes of floods travel times on upper Danube CD from XXI Conference of Danube countries Romania 12 pp
[4] Mitkova V 2002 Changes of the travel times of flood waves for Danube River (In Slovak) *Acta Hydrologica Slovaca* 3 **1** pp 20–27
[5] Mitkova V, Pekarova P, Miklanek P and Pekar, J 2005 Analysis of flood propagation changes in the Kienstock–Bratislava reach of the Danube River *Hydrological Sciences Journal* 50 **4** pp 665–668
[6] Pacl J 1955 Danube floods in July 1954 (In Slovak) Hydrological study HMI Praha 26 pp
[7] Pekarova P, Miklanek P and Pekar J 2007 Long-term Danube monthly discharge prognosis for the Bratislava station using stochastic models *Meteorologický casopis* 4 **10** pp 211–218
[8] Pekarova P and Pekar J 2007 Multiannual variability of Danube runoff characteristics in Bratislava gauge (In Slovak) *Acta Hydrologica Slovaca* **1** pp 12–21
[9] Prochazka M, Deyl M and Novicky O 2001 Technology for Detecting Trends and Changes in Time Series of Hydrological and Meteorological Variables - Change and Trend Problem Analysis (CTPA) User's Guide CHMI Prague 25 pp
[10] Skoda P and Turbek J 1995 Monitoring of the Hydrological Regime of Danube (In Slovak.) Proc Int Conf "Danube – the artery of Europe" ZSVTS Bratislava pp 194–206
[11] Svoboda A, Pekarova P and Miklanek P 2000 Flood hydrology of Danube between Devin and Nagymaros SVH – IH SAS 96 pp
[12] Svoboda A and Szolgay J 1988 Impact of Anthropogenic Activity upon the River Runoff (In Slovak.) Report of Phase KE 01 Determination of runoff regime changes in the abandoned river bed, using mathematical runoff model ÚHH SAV Bratislava
[13] Vizallasok 1890 Vizallasok a Dunaban Hornyanszky Viktor Könyvnyomdaja Budapest
[14] Webb BW and Nobilis F 2007 Long-term changes in river temperature and the influence of climatic and hydrological factors *Hydrological Sciences Journal* 52 **1** pp 74–85
[15] Zatkalik G 1965 The 1965 Danube Flood (In Slovak) Vodni hospodarstvi 12 pp 519–526
[16] Zsufa I 1999 Impact of Austrian hydropower plants on the flood control safety of the Hungarian Danube *Hydrological Sciences Journal* 44 **3** pp 363–371

IOP Conf. Series: Earth and Environmental Science **4** (2008) 012029 doi:10.1088/1755-1307/4/1/012029

Influence of vegetation cover on air and soil temperatures in the Šumava Mts. (Czech Republic)

M Tesař[1], M Šír[1], M Krejča[2] and J Váchal[2]

[1] Institute of Hydrodynamics, v.v.i., Academy of Sciences of the Czech Republic, Pod Paťankou 30/5, 166 12 Praha, Czech Republic
[2] Institute of Technology and Business in České Budějovice, Okružní 10, 370 01 České Budějovice, Czech Republic

E-mail: miroslav.tesar@iol.cz

Abstract. Air temperature at heights of 5 and 200 cm above soil surface, and soil temperature at depths of 15, 30 and 60 cm, were measured in the cold climatic zone at three localities (catchments) under different plant cover during the growing seasons 2002–2007. The catchments Kout (dead forest), Doupě (clearing) and Stolec (mature spruce forest) are situated in the National Park of the Šumava Mts. (Czech Republic) at an elevation of 1105–1330 m a. s. l. This region is part of the metamorphic complex – Moldanubicum. It is formed mainly by metamorphosed rocks (paragneiss with smaller anomalies). The three catchments have very similar natural conditions, but differ significantly in vegetation cover. The catchments are covered with acid brown soil developed on paragneiss. The clearing and dead forest catchments both exist as small islands inside an extensive spruce forest. The following conclusions were obtained from the interpretation of measured data: (1) The air temperature of the catchment covered by the dead forest had a greater daily fluctuation compared to the stands covered by clearing or mature forest. (2) The air temperature of catchments covered by mature forest had the smallest daily fluctuations. (3) The more extreme air temperatures in the dead forest caused the systematic raising of the soil temperature at a depth of 15 cm compared to soil in the clearing and mature forest. These conclusions are valid for a cold climatic zone during a vegetation season in which plant transpiration plays a governing role in solar energy dissipation.

1. Introduction

Plant transpiration modifies climatic conditions at the land surface and in the soil during the vegetative season [1], [2], [3], [4]. In mountainous areas of Central Europe transpiration recycles about 50% of the rain water during the vegetative season. Evaporative cooling induced by transpiration reduces the maximum plant surface temperature by about 15 °C on a hot and dry day [5].

Water exchange between the soil and plants is driven by heat from solar radiation, which, in combination with the air temperature, is the cause of plant heating. The plants protect themselves against heating up over a certain temperature by transpiration [6]. In this way, heat input is divided into two parts – latent heat (used for water vaporisation) and sensible heat. Sensible heat irradiated from plant cover is the dominant cause of heating of the lower atmosphere layer and the near-surface soil horizon. Water for transpiration is imbibed from the soil root zone. In cases of water scarcity, transpiration ceases, the plants do not cool (the latent heat is equal to zero), and therefore, the plant cover, atmosphere, and soil are overheated by solar radiation. This relation between plant transpiration, soil water and income of solar energy is central to the co-evolution of plants, soil cover and climate [7].

The data describing the coupling between soil water tension (tensiometric pressure), transpiration, and temperature in cold climatic conditions can be summarised as follows: (1) plant temperature does not rise above the maximum temperature (about 25 °C) if there is enough water for plant transpiration, (2) on sunny days, scarcity of water for plant transpiration results in an increase in plant temperature to about 35 °C and air temperature to about 30 °C, and (3) the limiting value of tensiometric pressure, below which water uptake for plant transpiration is impossible, is –60 kPa for grass, dwarf pine and the Norway spruce stand in the course of the vegetation season [8].

Between 1992–2000, large areas in the National Park of the Šumava Mts. were deforested due to attack by bark beetle. These areas are located in extremely severe natural conditions. The natural regeneration of the forest is currently in progress. The influence of deforestation and recurrent afforestation on the hydrologic and climatic conditions is being studied. The air temperature (5 and 200 cm above the soil surface), the soil temperature (15 and 60 cm below the soil surface), precipitation and runoff are continuously measured in three stands differing in vegetation cover: clearing covered by herbs, mature spruce forest, and dead spruce forest with a herb undergrowth. One study [9] concluded that in conditions with sufficient soil moisture to prevent any reduction in plant transpiration: (1) both soil and air temperature are influenced by plant cover, (2) the extremes in daily and night-time air temperatures are a function of transpiring vegetation height on hot and dry days, with higher daily maximums and lower night-time minimums for areas with lower vegetation, (3) over the entire growing season, the mean air temperature was independent of plant cover, but the magnitude of the dispersion variance was ranked in ascending order: mature forest – clearing – dead forest.

The aim of this study is to demonstrate how different plant cover modifies temperature at the land surface during the vegetation season.

2. Experimental catchments and methods

The three experimental catchments lie in the National Park of the Šumava Mts. (Czech Republic). These catchments have very similar macroclimatic conditions, but differ significantly in vegetation cover (table 1). Soil cover of the catchments consists of acid brown soil developed on paragneiss. All

Table 1. Characteristics of the experimental catchments.

	Kout	Doupě	Stolec
vegetation cover	dead spruce forest with herb undergrowth	clearing covered by herbs	mature spruce forest
age of former forest (years)	150	150	130
age of new vegetation cover (years)	0–5	0–5	none
catchment area (km^2)	0.1	0.17	0.07
elevation (m a. s. l.)	1210–1275	1180–1330	1105–1251
precipitation total June–September (mm)	361.1	342.4	322.4
runoff depth June–September (mm)	25.6	40.9	28.1
runoff coefficient (%)	7.1	11.9	8.7

catchments have a northfacing aspect. The Kout catchment is covered by dead trees and herb undergrowth (figure 1). Originally it was a typical Calamagrostis villosa-rich spruce forest. The height of the herbs is about 40 cm. The dead trees have a height of 5 to 10 m and occur at a density of 200 to 300 per hectare. The Doupě catchment is a clearing of a Calamagrostis villosa-rich spruce forest and is nowadays covered by herbs (figure 2). Trees are virtually absent, and the height of the herbs is about

Figure 1. Dead spruce forest with herb undergrowth in the Kout catchment.

Figure 2. Clearing covered by herbs in the Doupě catchment.

30 cm. The Stolec catchment consists of an acidophilous mountain spruce-beech forest (figure 3). The majority of trees are older than 140 years (with a height of about 29 m), and around 20 % of the trees are younger than 40 years (with a height of about 6–10 m). Herb height ranges between 10 and 20 cm. The clearing and the area covered with dead trees form small "islands" inside an extensive spruce forest.

Air temperature at heights of 5 and 200 cm above soil surface, and soil temperature at depths of 15, 30 and 60 cm, were measured in every catchment using automated monitoring stations.

Figure 3. Mature spruce forest in the Stolec catchment.

3. Results and discussion

Air temperature at a height of 5 cm above the soil surface in the clearing, dead forest and mature forest in August 2002 is presented in figure 4. On cloudy days (e. g. 12 to 14 August) the midday temperature in the dead forest was about 2 °C and 9 °C higher than that in the clearing and mature forest, respectively. The lower temperature in the mature forest can be explained by the fact that it is mainly the tree crowns that are heated by solar radiation. As their distance to the soil surface is about 29 m, the heat emitted by the tree crowns is attenuated markedly during its passage downward. On the other hand the distance between the plant canopy and soil surface is about 30–40 cm in the clearing and dead forest. As a result, the heat emitted by plant cover is less attenuated during its passage to the soil surface, and the air temperature at a height of 5 cm is higher than that in the mature forest.

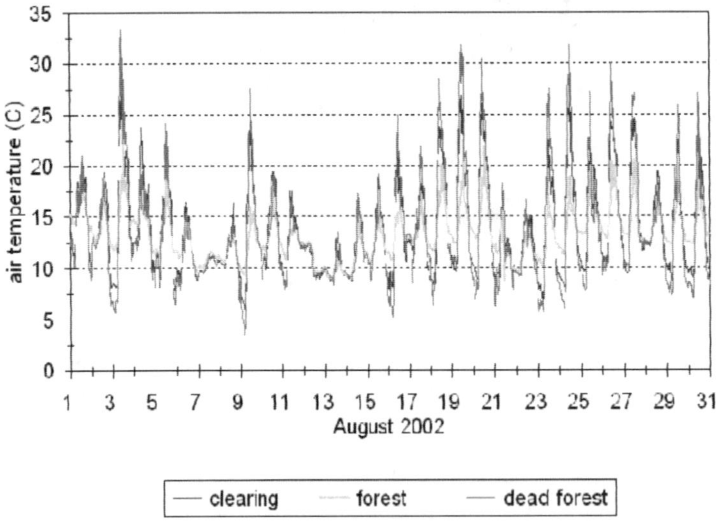

Figure 4. Air temperature at a height of 5 cm above the soil surface in clearing, dead forest and mature forest in August 2002.

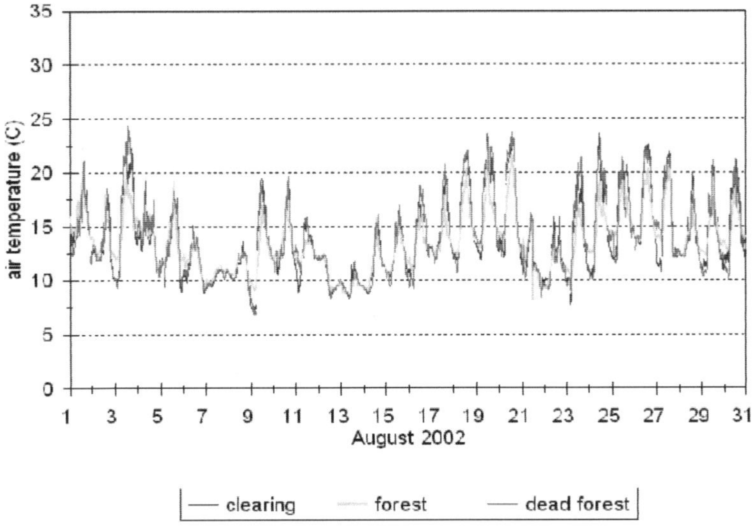

Figure 5. Air temperature at a high of 200 cm above the soil surface in clearing, dead forest and mature forest in August 2002.

On sunny days (e. g. 18 to 20 August) transpiration was about 4–5 mm per day. The midday air temperature in the dead forest was over 30 °C whilst in the clearing and mature forest it was about 23–25 °C and 22 °C, respectively. This temperature difference (between dead forest on the one hand and the clearing and mature forest on the other hand) is the result of transpiration cooling.

Figures 4 and 5 show the more extreme air temperatures (an increase of the temperature throughout the day and a decrease at night) in the dead forest compared to the clearing. The increase in daily temperature results from the fact that the dead non-transpiring stems are more greatly heated during sunshine than the transpiring plant cover. The heat emitted from the dead stems is the cause of the peak temperature in the midday hours on the days with a large income of solar energy. The drop in the night temperature is the cause of the lower greenhouse effect of the atmosphere. The dead stems cause a drop in the green vegetation area, and therefore, a drop in the total transpiration. The lower total transpiration results in the lower content of vapour in the air. This is the cause of greater emission of heat from the soil surface at night, accompanied by a drop in night temperature. It is obvious from table 2 that the night-time greenhouse effect attenuates emission of heat from the soil (the attenuation is approximately equal to the difference in air temperature at the height of 5 cm and 200 cm).

Figure 6 demonstrates that the more extreme air temperatures in the dead forest caused the systematic raising of the soil temperature at a depth of 15 cm compared to the clearing and mature forest.

Table 2. Maximum air temperature differences at the heights of 5 and 200 cm above soil surface in August 2002.

vegetation cover	maximum air temperature differences at the heights of 5 and 200 cm (°C)	
	in the day-time	at night
forest	1	-1
clearing	4	-2
dead forest	8	-4

4. Conclusions

The following conclusions were obtained by the interpretation of measured data: (1) The daily air temperature of the catchment covered by the dead forest fluctuated more than in the stands covered by clearing or mature forest. (2) The air temperature of catchments covered by mature forest had the smallest daily fluctuations. (3) The more extreme air temperatures in the dead forest caused the systematic raising of the soil temperature at a depth of 15 cm compared to the clearing and mature forest. These conclusions are valid for a cold climatic zone during a vegetation season in which plant transpiration plays a governing role in solar energy dissipation.

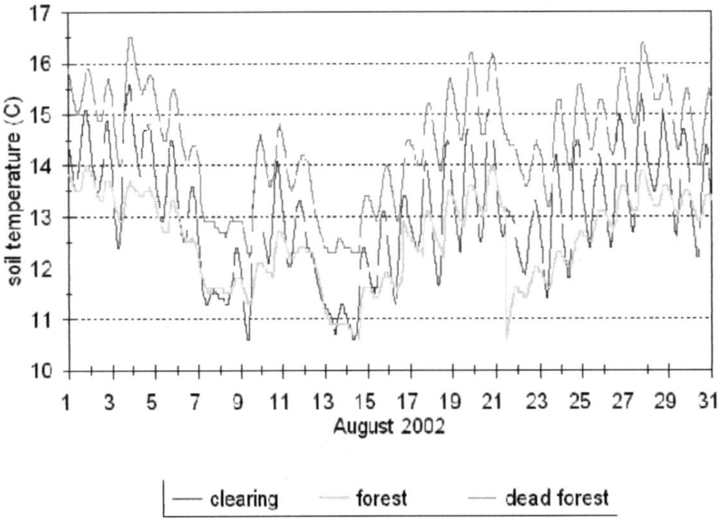

Figure 6. Soil temperature at a depth of 15 cm below the soil surface in the clearing, dead forest and mature forest in August 2002.

These results are consistent with the experimental findings gained in other mountainous catchments in the Czech Republic [10], [11],[12], [13],[14], [15], [16].

Acknowledgements
This research was supported by the Academy of Sciences of the Czech Republic (Institutional Research Plan AV0Z20600510 and Project No. 1QS200420562), Grant Agency of the Czech Republic (Projects No. 205/06/0375, 205/08/1174 and 526/08/1016), Ministry of Education, Youth and Sports (Project No. MEB 0808114) and Ministry of the Environment of the Czech Republic (Project No. SP/1a6/151/07).

References

[1] Eagleson P S 1978 Climate, soil, and vegetation *Water Resources Research* **14** 705–776

[2] Eagleson P S 2005 *Ecohydrology: Darwinian Expression of Vegetation Forms and Function* (Cambridge: Cambridge University Press) p 483

[3] Kabat P, Claussen M, Dirmeyer P A, Gash J H C, de Guenni L B, Meybeck M, Pielke sr. R A, Vörösmarty C J, Hutjes R W A and Lütkemeier S [eds.] 2004 *Vegetation, water, humus and the climate A new perspective on an interactive system* (Berlin: Springer-Verlag) p 566

[4] Pokorný J 2001 Dissipation of solar energy in landscape – controlled by management of water and vegetation *Renewable Energy* **24** 1641–1645

[5] Novák V and Havrila J 2006 Method to estimate the critical soil water content of limited availability for plants *Biologia* Bratislava **61**/Suppl. **19** S289–S293

[6] Larcher W 2003 *Physiological plant ecology (Ecophysiology and stress physiology of functional groups)* 4th edition ISBN 3-540-43516-6 (Berlin Heidelberg New York: Springer-Verlag) p 504

[7] Kleidon A 2006 Quantifying the biologically possible range of steady-state soil and surface climates with climate model simulations *Biologia* Bratislava **61**/Suppl. **19** S234–S239

[8] Tesař M, Šír M, Dvořák I J and Lichner Ľ 2004 Influence of vegetative cover changes on the soil water regime in headwater regions in the Czech Republic *IHP/HWRP-Berichte* **2** 57–72

IOP Conf. Series: Earth and Environmental Science **4** (2008) 012029 doi:10.1088/1755-1307/4/1/012029

[9] Tesař M, Šír M, Lichner and, Zelenková E 2006 Influence of vegetation cover on thermal regime of mountainous catchments *Biologia* Bratislava, **61**/Suppl. **19** S311–S314

[10] Chlebek A and Jařabáč M 1984 Analysis of peak runoffs in the Beskydy Mts. experimental catchments (in Czech) *Lesnictví* **30** 383–396

[11] Chlebek A and Jařabáč M 1988 Consequences of the forest renewal on the runoff from the experimental catchments in the Beskydy Mts (in Czech) *Zprávy lesnického výzkumu* **33** 7–12

[12] Chlebek A and Jařabáč M 1994 40 years of hydrologic research of forest in Beskydy Mts. (in Czech) *Vodní hospodářství* **44** 21–24

[13] Hais M, Brom J, Procházka J and Pokorný J 2006 Effect of water drainage on the forest microclimate; case study of two small catchments in the Šumava Mountains *Ekológógia* Bratislava **25**/Suppl. **3** S18–S26

[14] Hojdová M, Hais M and Pokorný J 2005 Microclimate of a peat bog and the forest in different states of damage in the National Park Šumava *Silva Gabreta* **11(1)** 13–24

[15] Jařabáč M and Chlebek A 1989 Methodological topics on the measurements in the experimental catchments (in Czech) *Zprávy lesnického výzkumu* **34** 13–17

[16] Tesař M, Šír M, Syrovátka O, Pražák J, Lichner Ľ and Kubík F 2001 Soil water regime in headwater regions – observation, assessment and modelling *J.Hydrol. Hydromech.* **49** 355–375

IOP Conf. Series: Earth and Environmental Science **4** (2008) 012030 doi:10.1088/1755-1307/4/1/012030

Trends of River Discharges in Slovenia

F Ulaga[1], M Kobold and P Frantar

Environmental Agency of the Republic of Slovenia, Vojkova 1b, SI-1000 Ljubljana, Slovenia

E-mail: florjana.ulaga@gov.si

Abstract. Trends of river discharges are very important in recognition of climate changes and also in water management. In last years droughts and floods are more and more frequent, therefore monitoring and investigating of hydrological situation became one of the guidance scientific approach. In analysis of mean annual discharges made in Agency of the Republic of Slovenia for the Environment in last years, statistical significance trend was find out only by rivers of northwest alpine region [1] [2] [3]. In recent research of time changing discharges, presumption of generally decreasing of water quantity all over Slovenia was confirmed. Our basic conclusion – in Slovenia it is well recognized decreasing of water quantity in long period of observation. This fact and also worsening of water quality became alarming problems of our future, especially in dry months of the year.
Keywords: trend, river discharge, statistical significance, decreasing of water, Hydrospect

1. Introduction

According to the results of analyses performed in previous years by the Environmental Agency of the Republic of Slovenia, the average annual runoff in Slovenia over the last fifty years is diminishing. A similar tendency is exhibited by the environmental indicator of the annual river balance. The analyses of extremes carried out up to now have found no statistical significance of the regional assessment of discharge trends. The present discharge trend study represents an expansion of the analysis of the temporal variability of data sets on the discharges of Slovenian rivers using a larger number of representative water gauging stations. Trends are important indicators of the temporal variability of phenomena (discharges). By analysing the time sequence of the discharges, we assess the magnitude and significance of the temporal variability. In recent years, when floods and droughts as a result of climate change have become more frequent, the monitoring and study of hydrological states and events is becoming increasingly topical.

2. Methodology

In the analysis of the trends of discharges on Slovenian rivers streams data on the low, mean and high discharges were examined. The analysis was carried out for 22 water gauging stations with sufficiently long and reliable continuous data sets and are which are representatively distributed across Slovenia. In analysing the trends of short data sets, the results can be affected significantly by climatic variation [4] which is why only the water gauging stations with at least 50-year data sets were included into the selection. The data from the entire period of observations available in the EARS archives were used,

[1] To whom any correspondence should be addressed.

up to and including the year 2005. The Laško station on the Savinja River has the longest data set of the selected water gauging stations. There, the water stages and discharges have been measured since 1907 with the exception of an interval in 1940-1945. That is 93 years. The average length of the data sets from the selected water gauging stations is 62 years.

Changes in the time series can occur as sudden incidental changes, cyclically repetitive changes or as a continuous change with a distinct trend. The type of the trend can be most easily established analytically using linear regression analysis and graphic depiction. The simple test statistic for the depiction of a trend is the slope of the regression line that shows how distinct the trend is. If there is no trend, the value of the regression gradient is close to 0. If the trend is significant, the value of the regression gradient is (considerably) different from 0 - positive for an increasing trend and negative for a decreasing one. The statistical significance of the discharge trends was assessed with a significance rate that adopts a ten percent confidence level and the assumption that the distribution of the data is normal [4]. Values above 90% were adopted as a statistically significant phenomenon for changing discharges or for some selected hydrological indexes.

The analysis of the value trends and discharge indexes was carried out based on the time series of mean daily discharges. Using the HYDROSPECT program, developed under the auspices of the World Meteorological Organization [5], various indexes were calculated, such as the derived time series of low, mean and large discharges [6] [7]. The trend of the indexes treated was determined using linear regression analysis, where time was the independent variable.

3. The Rate of the Statistical Significance of Trends

3.1. Index of mean annual discharges
Statistical estimates of the mean annual discharges (Qs) indicate the average annual quantity of water. By analysing the trend of these time series, we can assess the magnitude and characteristics of changes in the quantity of water and the envisaged trend of this in the future.

3.2. Index of low discharges
Based on the mean daily discharges, the minimum annual discharges lasting for 1 (Qmin1), 7 (Qmin7) and 30 (Qmin30) days were analysed. The analysis of minimum annual discharges of various durations also shows the characteristics of the drought periods and the occurrences of hydrological droughts, which were frequent over the last decade [8] [9].

3.3. Index of high discharges
Five different indexes were used to assess the temporal variation in the high discharges. Studies of high discharges usually focus on the trends of the maximum annual discharges (Qvp), meaning that only one event is taken into account each year, irrespective of whether there were several high-water waves or no high-water waves at all. A more representative way of describing the frequency of high waters is the use of the threshold method (the POT – 'Peak Over Threshold'). When using this method, all events with discharges in excess of a set threshold are selected under the condition that the selected events are not interdependent. Therefore we can have several high discharges recorded in a year or none at all. High threshold values with a frequency (on average) of one (POT1-mag) and three (POT3-mag) values per year have been chosen. In addition to assessing the trend of the magnitude of high river flows, the trend of the frequency of high flows by calculating the number of values above the threshold (POT) - namely POT 1-f and POT3-f - for each year and the trend of these data sets were also assessed. The data sets POT1 (POT1-mag and POT1-f) describe the size and frequency of the extreme high discharges, while the sets POT3 (POT3-mag and POT3-f) describe the variability of more moderate high-water discharges.

Table 1 shows for each analysed station the significance of the trends of the nine flow index series estimated using linear regression and the ten percent confidence level. When the trend is negative, the significance level is also shown as negative.

Table 1. The rate of the statistical significance of trends

River - Water gauging station	Q s	Qvp	Low discharges			Magnitude of the extreme		Frequency of the extreme	
			Qmin1	Qmin7	Qmin30	POT1 mag	POT3 mag	POT1- f	POT3 - f
Mura - Gornja Radgona	-37.86	29.99	83.66	71.40	18.70	-31.14	-22.90	13.92	-0.70
Ščavnica - Pristava	-99.20	55.69	55.60	27.13	-17.55	99.88	99.97	27.00	-75.52
Ledava - Polana	-99.55	-96.55	-38.00	-69.30	-81.98	-38.10	-90.00	-99.15	-99.64
Meža - Otiški Vrh	-98.76	-92.64	-96.90	-95.30	-98.75	18.66	-7.99	-99.38	-99.94
Dravinja – Videm *	-64.95	81.54	84.76	77.37	85.47	18.16	96.18	69.27	61.75
Pesnica - Zamušani	-96.48	35.14	-27.40	-18.10	-18.93	44.62	94.24	10.46	-69.21
Radovna – Podhom *	-99.87	-99.99	-98.80	-98.30	-88.60	-99.84	-99.99	-99.99	-99.93
Sava Bohinjka - Sv. Janez	-99.81	-85.90	-98.60	-98.50	-99.10	-66.73	-22.49	-28.93	-53.18
Sava - Radovljica	-98.88	-7.96	-98.20	-99.90	-99.90	-62.68	6.52	65.38	64.78
Sava - Litija	-99.41	-84.56	-39.90	-42.50	-82.53	59.45	-75.68	-99.87	-99.59
Sora - Suha	-62.16	16.67	86.44	71.80	-18.16	-65.94	-38.60	-23.88	-36.22
Kolpa - Radenci	-99.15	-97.94	-96.80	-97.50	-95.58	-73.86	65.30	-81.13	-99.99
Ljubljanica - Moste	-93.20	-62.45	-86.60	-99.50	-99.54	-56.11	-92.91	-77.57	20.59
Unica - Hasberg	-91.74	-99.88	88.27	92.60	71.11	-99.99	-95.60	-98.95	-99.99
Savinja - Laško	-99.94	-91.10	-93.60	-94.20	-97.34	39.83	-63.13	-99.88	-99.88
Krka - Podbočje	-96.79	-98.80	92.76	75.86	54.87	-93.73	-98.30	-99.83	-99.99
Soča - Kršovec	-91.80	-98.90	-92.10	-95.30	-98.67	-91.22	-88.17	-98.64	-99.52
Soča - Log Čezsoški	-99.81	-97.39	27.39	10.50	-63.73	-94.44	-83.36	-89.47	-98.28
Soča - Solkan	-79.24	43.67	-12.60	-64.0	-94.94	32.43	69.78	53.20	-21.45
Idrijca - Hotešk	-91.72	-70.21	-21.20	-61.70	-72.40	-68.21	34.83	-12.80	-83.89
Vipava - Miren	-69.99	-10.83	-28.70	-49.90	-87.26	-75.38	-19.43	-18.61	49.44
Reka -Cerkvenikov mlin	-92.37	6.36	97.84	99.58	96.54	-9.64	90.74	-2.66	-93.45

* rivers with high and low significante rate – use as examples

3.4. Trend of mean annual discharges – Qs

Qs is a decreasing one at all the analysed water gauging stations. With the exception of the Mura, Dravinja, Sora, Soča at Solkan and Vipava rivers, which do not exhibit a statistically significant decrease in the mean annual discharges, all the other streams exhibit a statistically significant decrease in the quantity of water, irrespective of the length of the set used. The trend of the mean annual discharges indicates a decrease in the annual quantity of water in all the landscape units of Slovenia. The decrease in discharges is first and foremost a result of the decrease in the precipitation, which directly reduces the runoff, and the increase in temperature, which in turn increases evapotranspiration.

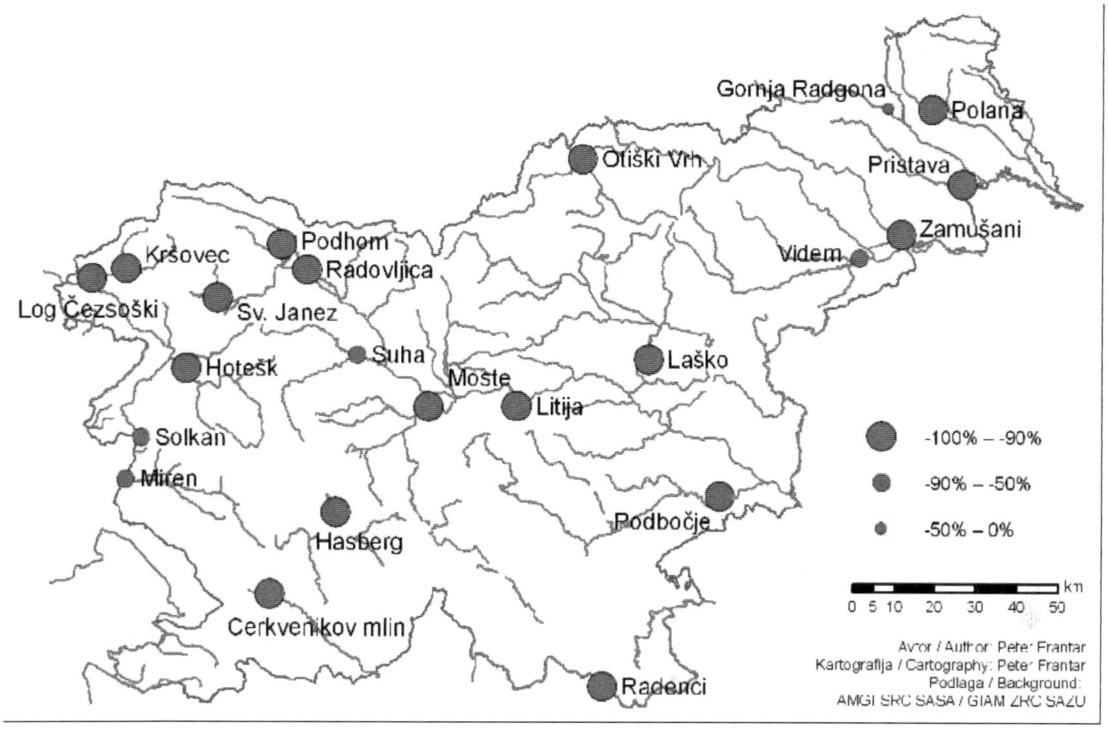

Figure 1. Gauging stations on main rivers and the significance rate of the trend of mean annual discharges.

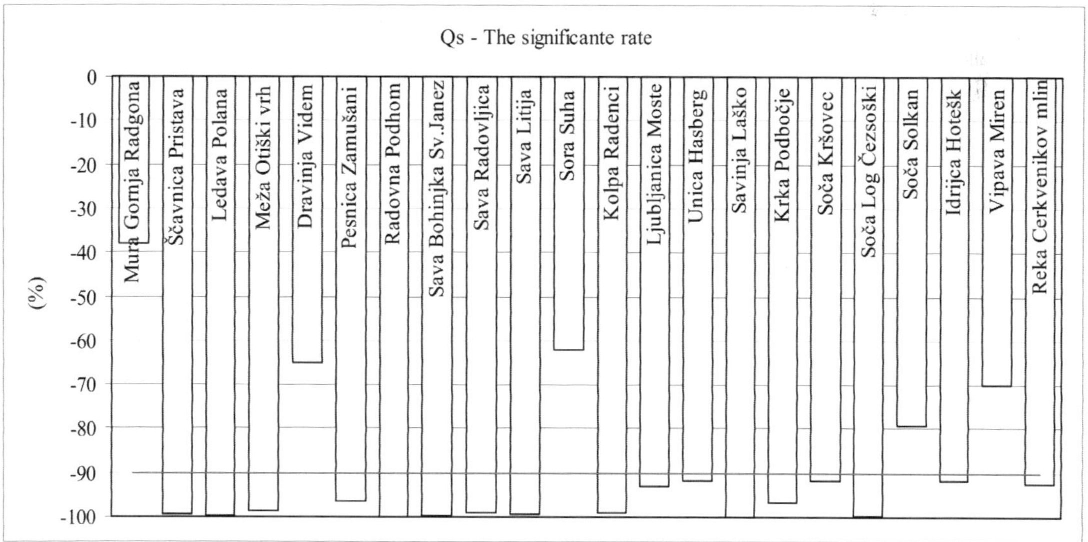

Figure 2. The significance rate of the trend of mean annual discharges.

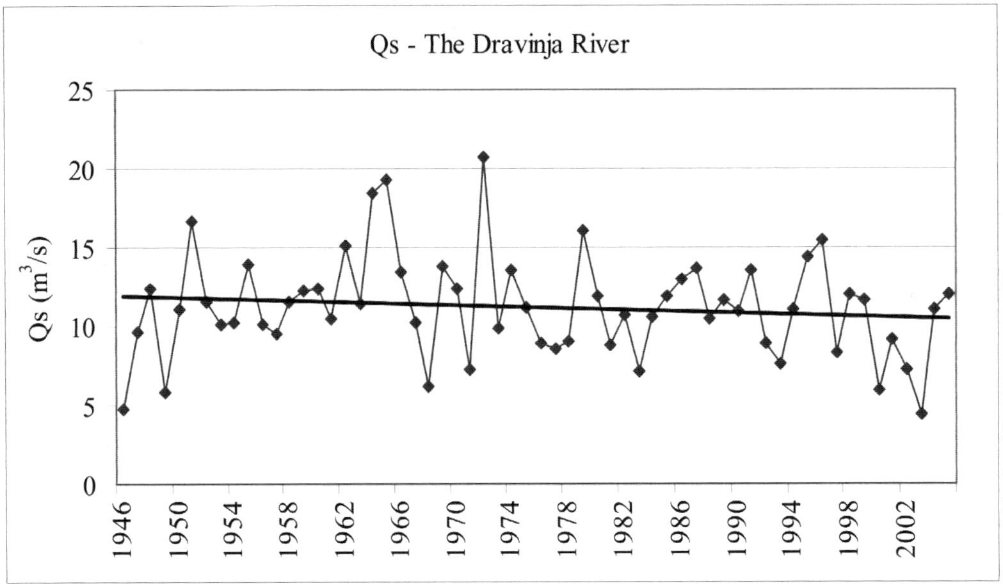

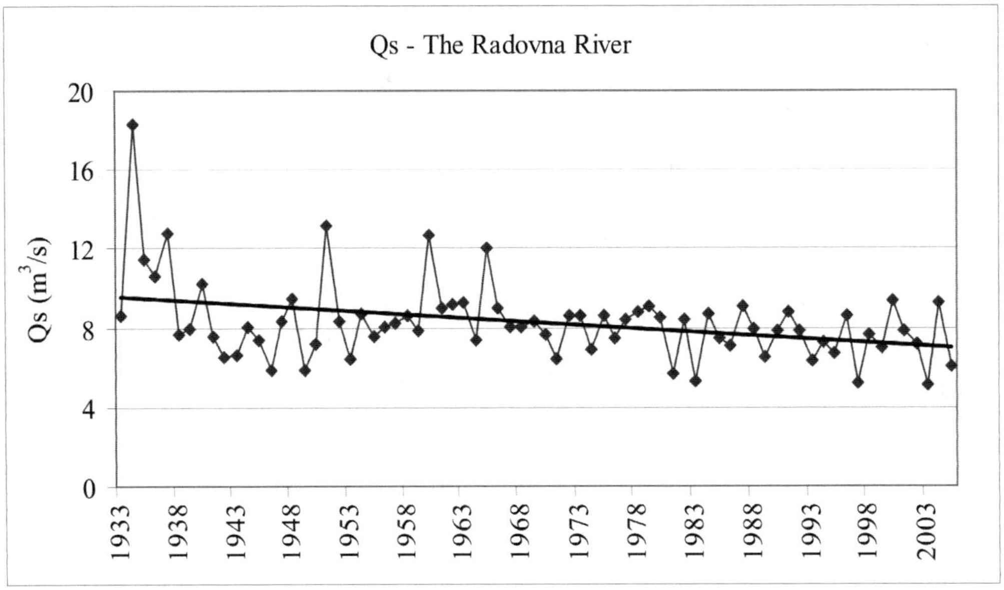

Figure 3. Mean annual discharges of Radovna and Dravinja rivers.

3.5. Trend of the minimum annual discharges – Qmin

Qmin were examined for the minimum annual mean daily discharges (Qmin1), minimum 7-day (Qmin7) and minimum 30-day (Qmin30) discharges. Analysis of the minimum mean daily discharges (Qmin1) indicates a statistically significant decreasing trend at the selected water gauging stations for mountain catchment areas. An increasing trend was observed in these discharge data sets on the karstic and eastern areas (in the east and south-west) of Slovenia, and is statistically significant on the Krka and Reka rivers. A similar geographical distribution is also exhibited in the longer duration minimum discharges (Qmin7 and Qmin30). The number of water gauging stations with an increasing trend is getting smaller though, while the number of those with a decreasing trend is ever greater. The Radovna River has a prominently decreasing trend in the minimum annual discharges Qmin1 and Qmin7, while these are Qmin7 and Qmin30 for the Ljubljanica. In the rivers with an increasing trend, the prominent values are exhibited by the Unica River for the Qmin7, the Krka River for the Qmin1 and the Reka River for all the low discharges, most probably caused by the artificial impact of the Molja and Klivnik water reservoirs in its catchment area.

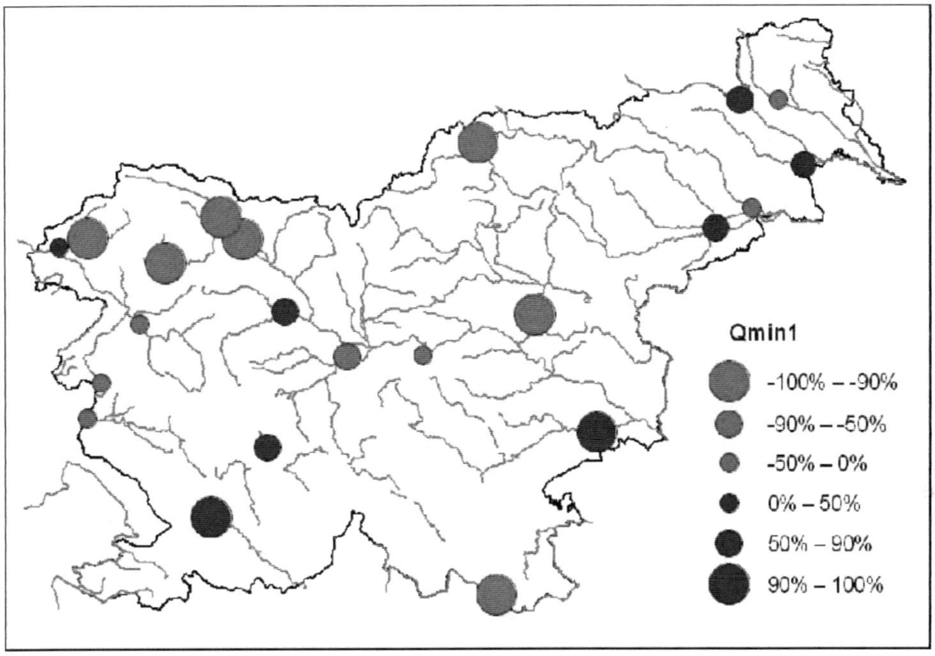

Figure 4. The significance rate of the trend of minimum daily discharges (Qmin1).

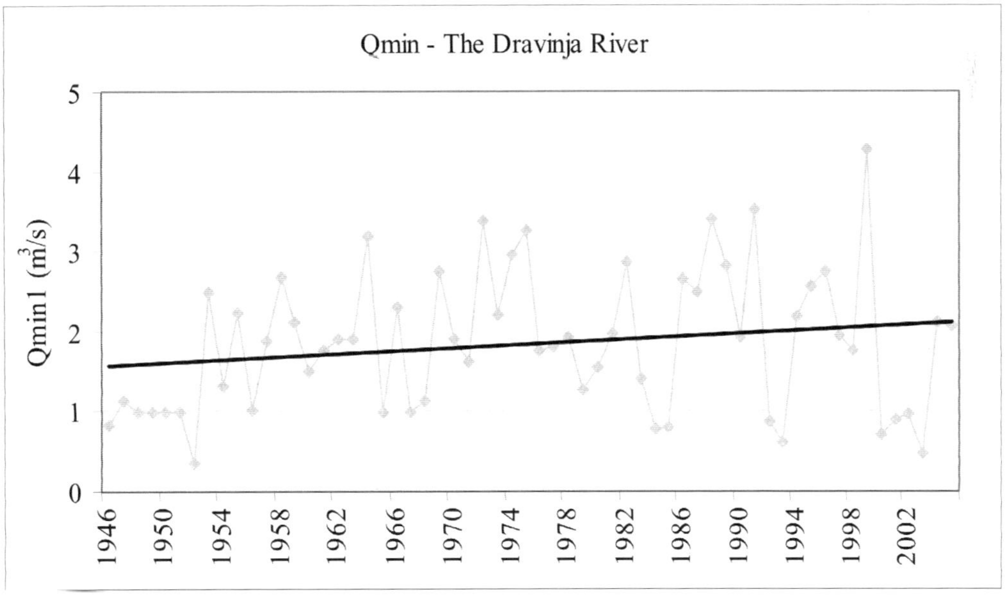

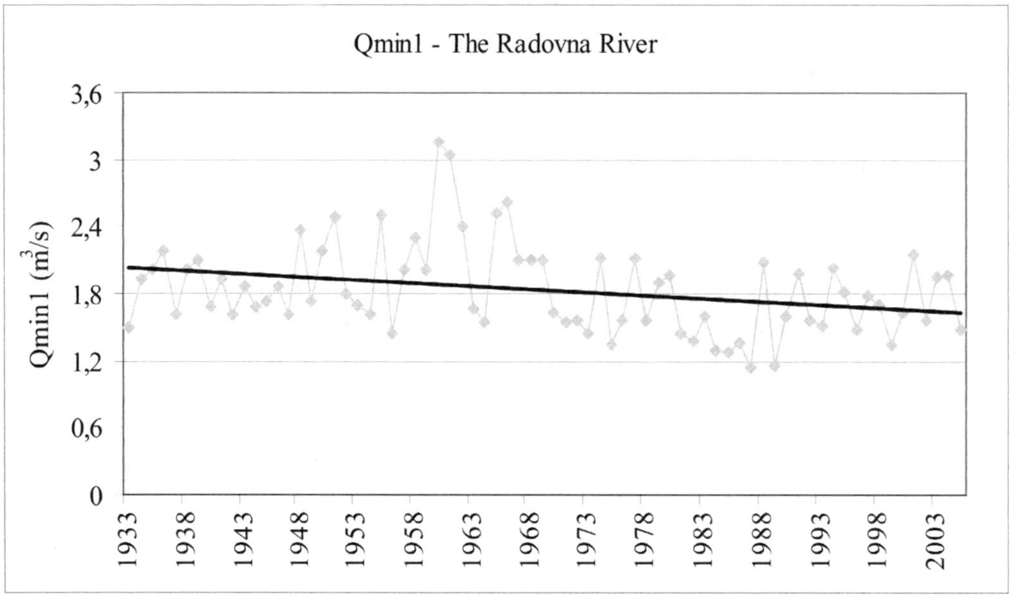

Figure 5. Minimum daily discharges of Radovna and Dravinja rivers.

3.6. Trend of the maximum annual discharges - Qvp

Qvp indicate a lower statistical significance than the trends of the mean daily discharges. The majority of the water gauging stations analysed indicate a falling trend, which is statistically significant on the Ledava River and Meža River, on tributaries of the Sava River (on the Radovna, Unica, Savinja, Krka and Kolpa rivers) as well as on the water gauging stations of the upper Soča River (Kršovec, Log Čezsoški). A non-significant increasing trend is only exhibited by the Dravinja and Ščavnica rivers. The trends on other rivers are not statistically significant.

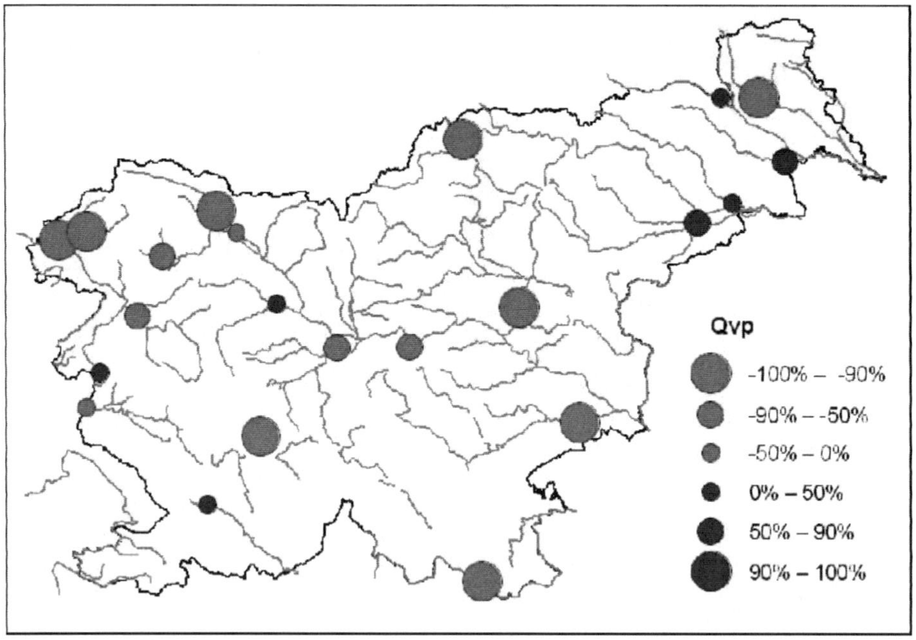

Figure 6. The significance rate of the trend of the maximum annual discharges (Qvp).

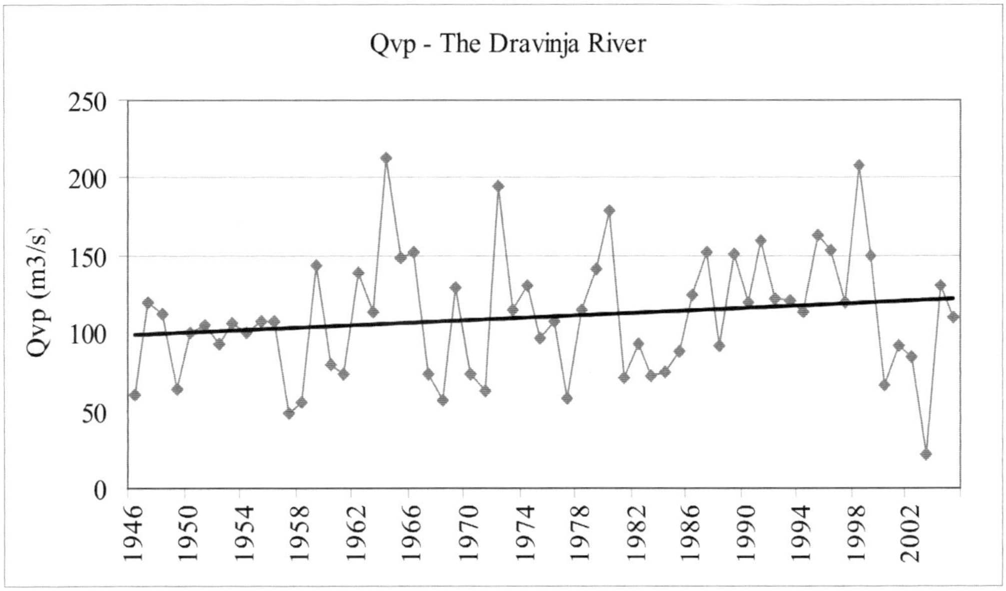

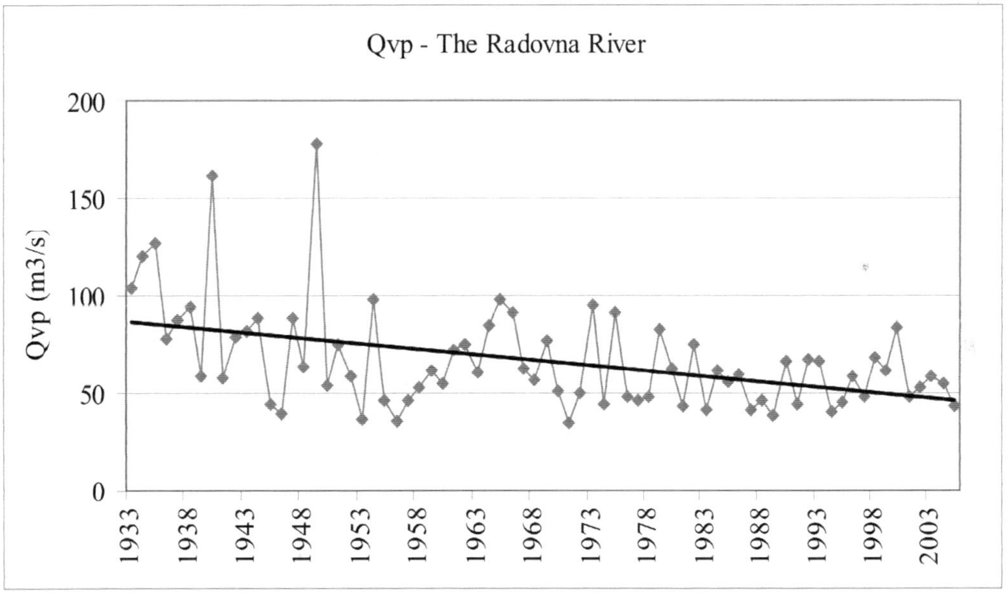

Figure 7. Maximum annual discharges of Radovna and Dravinja rivers.

3.7. The index of the magnitude

The index of the magnitude of the occurrence of high extremes (on average, one value per year - POT1 mag) indicates a similar regional diversity to the trends of the maximum annual discharges (Qvp) with a predominantly decreasing trend. A statistically significant decreasing trend is exhibited by the Radovna and Krka rivers as well as the Soča River in its upstream part, while only the Ščavnica River exhibits an increasing trend. The other analysed rivers do not exhibit significant trends in prominent extremes. With the significance index of the occurrence of high extremes three times a year (POT3 mag), the picture changes. The number of water gauging stations with an increasing trend increases, with the Ščavnica, Dravinja, Pesnica and Reka rivers being statistically significant.

IOP Conf. Series: Earth and Environmental Science **4** (2008) 012030 doi:10.1088/1755-1307/4/1/012030

Statistically significant decreasing trends were observed on the Ledava, Radovna, Ljubljanica, Unica and Krka rivers. The other rivers analysed do not exhibit significant trends in prominent extremes.

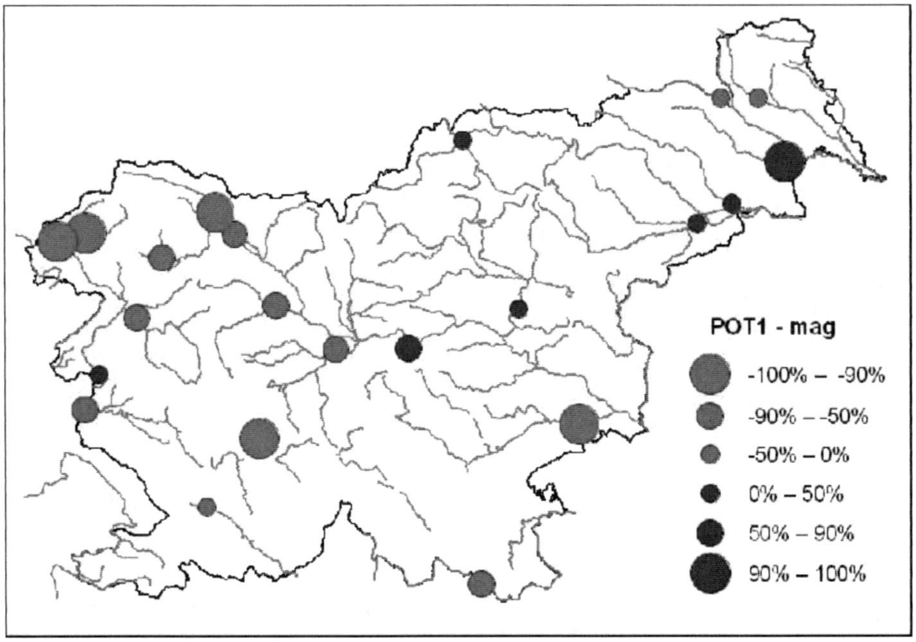

Figure 8. The significance rate of the trend of the magnitude of the occurrence of high extremes – on average once a year.

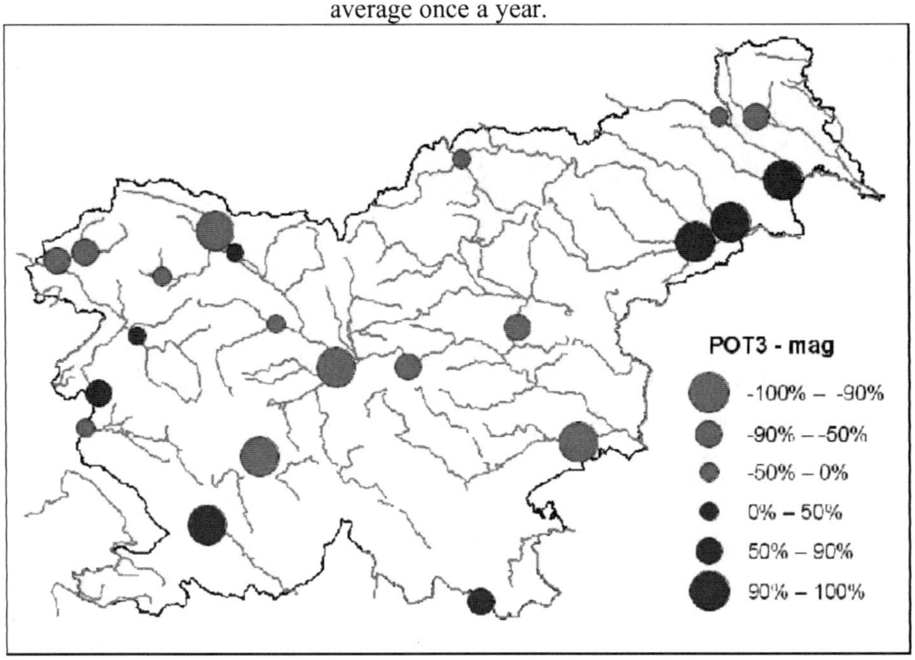

Figure 9. The significance rate of the trend of the magnitude of the occurrence of high extremes – on average three times a year.

To a considerable extent, the frequency of the high extremes of the POT1-f and POT3-f indexes indicates a decreasing trend that is statistically significant for almost a half of the stations. The phenomenon of high extremes occurring three times a year has 11 significant trends, while the phenomenon of extremes occurring once year has 8 decreasing trends, showing that the number of extreme high-water events occurring in a year is decreasing.

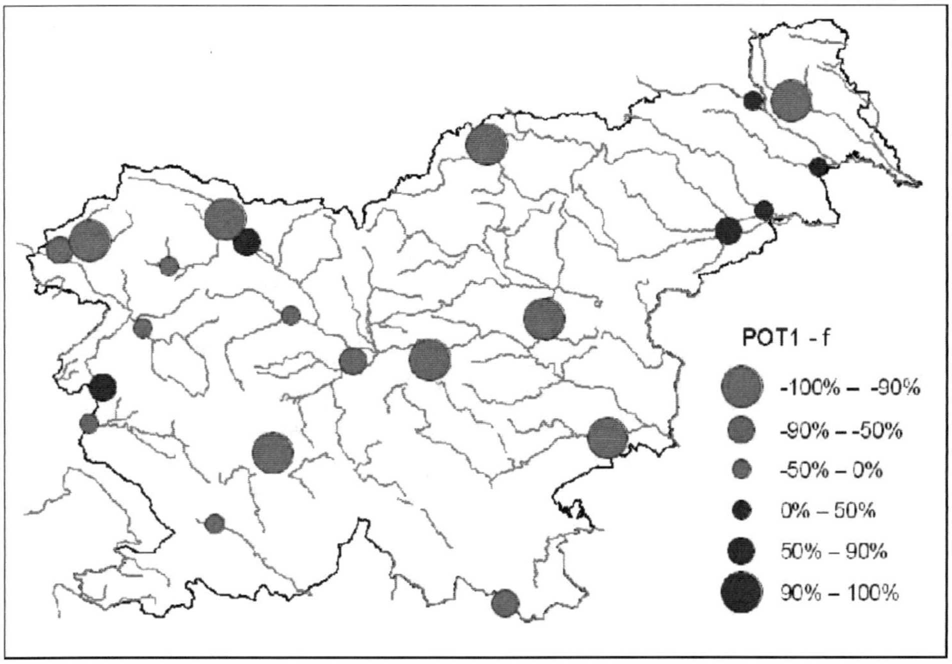

Figure 10. The trends of the frequency of the occurrence of high extremes – on average once a year.

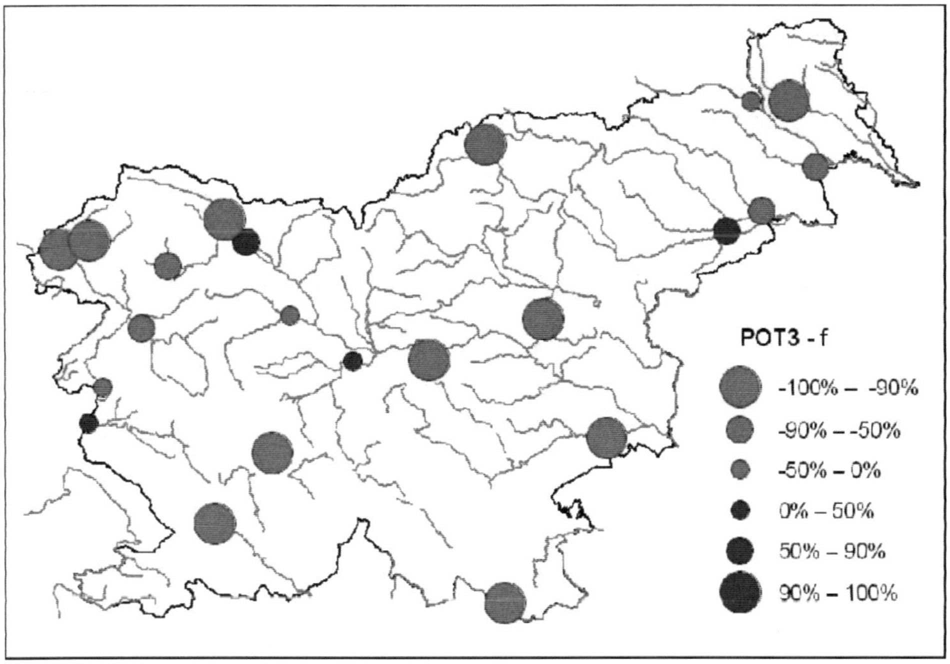

Figure 11. The trends of frequency of the occurrence of high extremes – on average three times a year.

IOP Conf. Series: Earth and Environmental Science **4** (2008) 012030 doi:10.1088/1755-1307/4/1/012030

4. Overview of the Trends by River Basins

4.1. The Mura River Basin

Overview of the trends by river basins shows us that discharge trends of the largest River Mura in the Mura river basin, are statistically non-significant. This is presumed to be the result of the use of the water from the Mura for energy generation in Austria. The Ščavnica and Ledava exhibit a statistically significant decreasing trend in the mean annual discharges, while the Ledava exhibits a decreasing trend in the maximum annual mean daily discharges. In the Ledava, the trends of the magnitude and frequency of high extremes are also decreasing. The discharges of the Ščavnica are exhibiting a statistically significant increasing trend in the time series of the statistical indexes POT1-mag and POT3-mag.

4.2. The Drava River Basin

The Meža River has the most statistically significant trends of the rivers of the Drava river basin that were analysed. Except for the two indexes of the magnitude of the occurrence of high extremes (POT1-mag and POT3-mag), the Meža exhibits statistically significant decreasing trends. Of the other selected rivers of this area, the Pesnica River has a statistically significant decreasing trend in the mean annual discharges (Qs) and an increasing trend in the magnitude of a high extreme exceeding the POT3-mag threshold. The trends of other hydrological indexes in Zamušani are not statistically significant. The Dravinja River at Videm does not exhibit a statistically significant trend, except for an increasing trend in the magnitude of the occurrence of the POT3-mag high extremes.

4.3. The Sava River Basin

Analysis of the time series on the mean daily discharges (Qs) for all the selected water gauging stations in the Sava river basin is showing a decrease that is statistically significant everywhere except on the Sora River. The Sava exhibits a statistically significant decreasing trend in its low discharges Qmin1, Qmin7 and Qmin30 at Radovljica. In Litija, the decreasing frequency trends of the high extremes (POT1-f and POT3-f) are non-significant. The significance of the decreasing trend for the low discharges decreases, when moving down the Sava River. The frequency of the occurrence of high water extremes has a non-significant trend at Radovljica, though the trend at Litija is a significantly decreasing one. The Qvp trend at the selected water gauging stations on the Sava River is not statistically significant. The tributaries of Sava River, Radovna and Sava Bohinjka rivers exhibit a decreasing trend in the majority of the hydrological indexes analysed. The mean annual and low discharges of both rivers have exhibited a statistical significance exceeding 98%. The decreasing of the trends covering Qvp, frequency and the magnitude of the high extreme are statistically significant on the Radovna River, while they are slightly less so on the Sava Bohinjka River (86%). The Alpine tributaries of the Sava River are therefore characterised by more or less prominent decreases in the water quantities, especially of the hydrological high extremes.

The Sora River, with its catchment area in the pre-Alpine area, does not exhibit statistically significant trends in changing discharges, with even the annual discharge only exhibiting a mild and statistically non-significant decrease. A tendency towards statistically non-significant increases can be observed in the low discharges. The Savinja River, with its catchment area in the pre-Alpine world, is exhibiting prominent decreasing trends in the mean, high and low discharges, as well as in the frequency of the occurrence of high extremes. The magnitude of the hydrological extreme is exhibiting a statistically non-significant decrease. The hydrological picture given by the almost 100-year long data set has been confirmed in recent years during the high waters that occurred several times in the Savinja river basin.

The tributaries of the Sava River, with their karstic catchment area (with the exception of the Ljubljanica River, which is exhibiting a mild and non-significant negative trend), exhibit a statistically significant decrease in the mean discharges (Qs), as well as a decrease in the high discharges (Qvp).The trend of the low discharges is exhibiting a statistically significant decrease on the Kolpa and Ljubljanica River, while the Unica and Krka rivers are exhibiting a prominently increasing trend in the 7-day and 30-day low discharges. The Unica and Krka rivers are also exhibiting a statistically significant decreasing trend in the magnitude and frequency of the high water extremes. The decreasing trend of the frequency of the POT3-f hydrological high extremes is statistically significant on the Kolpa River, while on the Ljubljanica River the trend of the magnitude of the POT1-mag high extreme is exhibiting statistical significance.

4.4. Adriatic River Basin

The significance of the trend of the Adriatic river basin is it possible to combine in two groups. On the Soča River we analysed the hydrological data from three water gauging stations with long data sets. The most prominent decreasing trend of all the analysed hydrological indexes is exhibited by the discharge of the Soča River at Kršovec, where we can see the most prominent decreasing trends in mean, low and high discharges, as well as in the magnitude and frequency of the high extremes among all the analysed water gauging stations. At Log Čezsoški, the decreasing of the trends Qs and Qvp, the magnitude of the high extremes POT3-mag and the frequency of the high extremes are statistically significant. At Solkan, only the decreasing trend of the minimum discharges with a duration of 30 days exhibits statistical significance. The Qvp trend and the trends of the magnitude and frequency of the occurrence of extremes are increasing, but remain statistically non-significant. In the majority of its indexes analysed, the Idrijca River does not exhibit a high level of statistical significance in its trends. Only the decreasing trend of the mean discharges (Qs) is statistically significant. The Vipava River also does not exhibit statistically significant trends. The data from the Cerkvenikov mlin water gauging station on the Reka River exhibits a statistically significant decreasing trend in the Qs discharges as well as in the magnitude (POT3-mag) and frequency of the occurrence of high extremes (POT3-f). On the Reka River, a statistically significant increasing trend was also observed in the low discharges: the minimum annual mean daily, 7-day and 30-day discharges, though this is most probably the result of the already mentioned impact of the water reservoirs in the catchment area.

5. Conclusions

Our conclusions about discharge trends on Slovenian rivers are as follow. When analysing the temporal variation of discharges, the assumption of a general decrease in water quantities in the rivers was confirmed. The discharges at the majority of the water gauging stations analysed have a decreasing trend in the mean daily discharges. Even in the high water trends, we can observe a decreasing trend to a large extent, though with a greater spatial variation in the statistical significance of the established trends. Decreasing trends with a greater statistical significance were seen in data sets for high waters, observed at water gauging stations with predominantly high-mountain and karstic catchment areas. An increasing trend is only exhibited by some of the water gauging stations analysed, predominantly in eastern Slovenia. The spatial distribution of statistically significant trends in the data sets for low discharges is similar. Northern, western and central parts of Slovenia exhibit decreasing low water discharges, while the western and southern parts exhibit an increasing of the same. A significant spatial variation in the statistical significance of the trends was also observed here. In general, we can confirm that the quantities of water in Slovenia are diminishing.

References

[1] Ulaga F 2002 *Trendi spreminjanja pretokov slovenskih rek* (*Geografija in njene aplikativne možnosti)* ed M Špes (Oddelek za geografijo Filozofske fakultete Univerze v Ljubljani) pp 93-114

[2] Uhan J 2002 *Spremenljivost hidroloških ekstremov* (*Nesreče in varstvo pred njimi)* ed B Ušeničnik (Uprava RS za zaščito in reševanje Ministrstva za obrambo Ljubljana) pp 252-255

[3] Frantar P, Ulaga F, Uhan J and Sušnik M 2006 *Trends of the annual mean river discharges in Slovenia* (*Poster for the XXIII Conference of the Danube Countries on the Hydrological Forecasting and Hydrological Bases of Water Management)* Belgrade

IOP Conf. Series: Earth and Environmental Science 4 (2008) 012030 doi:10.1088/1755-1307/4/1/012030

[4] Kundzewicz Z W and Robson A 2000 *Detecting trend and other changes in hydrological data* (WMO Report WMO/TD-No 1013 Geneva)
[5] Radziejewski M 2007: http://www.staff.amu.edu.pl/~maciejr/hydrospect/ (25.03.2007)
[6] Svensson C, Kundzewicz Z W and Maurer T 2004 *Trends in flood and low flow series* (GRDC Report 33 Koblenz)
[7] Kundzewicz Z W, Graczyk D, Maurer T, Przymusinska I, Radziejewski M, Svensson C and Szwed M 2004 *Detection of change in world-wide hydrological time series of maximum annual flow* (GRDC Report 32 Koblenz)
[8] Kobold M 2004 *Hidrološka suša slovenskih vodotokov v obdobju 2000-2002* (Ujma 17/18 Ljubljana)
[9] Kobold M and Sušnik M 2004 Analiza nizkovodnih razmer slovenskih vodotokov leta 2003 (Ujma 17/18 Ljubljana)

Integrated modelling and the impacts of water management on land use

W Dorner, K Spachinger and R Metzka

University of Applied Sciences Deggendorf, Edlmairstr. 6+8, D-94469 Deggendorf, Germany

E-mail: wolfgang.dorner@fhd.edu

Abstract. River systems and the quantity and quality of water depend on the catchment, its structure and land use. In central Europe especially land is a scarce resource. This causes conflicts between different types of land use, but also with the interests of flood protection, nature conservation and the protection of water resources and water bodies in the flood plain and on a catchment scale. ILUP - Integrated Land Use Planning and River Basin Management was a project, funded by the European Union, to address the problems of conflicting interests within a catchment. It addressed the problems of conflicting land use from a hydrological perspective and with regard to the resulting problems of water management. Two test river basins, Vils and Rott, both with a catchment size of about 1000 square kilometres, were considered for the German part of the project. Objective of the project was to identify means of managing land use with regard to water management objectives and adapt planning strategies and methodologies of water management authorities to the new needs of catchment management and planning. Catchment models were derived to simulate hydrological processes, assess the safety of dams and improve the control strategy of detention reservoirs with regard to land use in the lower system. Hydrodynamic models provided the basis to assess flood prone areas, evaluate flood protection measures and analyze the impacts of river training and discharge on morphology. Erosion and transport models were used to assess the impacts of land use on water quality. Maps were compiled from model results to provide a basis for decision making. In test areas new ways of planning and implementation of measures were tested. As a result of model scenarios in combination with the socio economic situation in the catchment new methods of land management and land use management were derived and implemented in model areas. The results of the project show that new ways of managing land use in river catchments are necessary to fulfil quantitative and qualitative criteria of new European legal regulations. The paper presents means to use hydrological models and data as a basis of land use planning and river basin management.

1. Introduction and scientific background

In the past water management and hydraulic engineering provided the spaces for new forms of land use in the floodplain through river training or the protection of agricultural areas against floods through detention structures and levees. Nowadays authorities for water management are facing the consequences because of high maintenance costs, water quality problems and other qualitative and quantitative problems including conflicts between different stakeholder groups [2]. It can be stated that the project costs were covered by the public while the incomes and gains were privatized.

In parallel future strategies for water management must deal more and more with problems in the catchment [1,3]. Besides classical tasks in the river and the floodplain several severe impacts regarding water quality (agriculture and erosion) but also the development of floods (land use and loss of detention structures) occur in the catchments. As a result the classical hydrological perspective becomes of importance for water management because hydrological functional dependencies define therefore the borders. Instead of providing the space for new forms of land use through diking and river training, water management authorities have to deal with existing land use in the catchment and must define restrictions or even ban land use from targeted spaces to achieve defined legal and societal objectives.

The European Water Framework Directive (WFD) and the new Flood Risk Directive (FRD) are requiring new planning approaches and methodologies. While till today water management was focused on the water bodies themselves, new concepts , also formulated in both directives, are asking for a catchment perspective. Therefore quantitative and qualitative hydrology can be called lead disciplines, defining the frame for planning not only in the river and the floodplain, but also in the catchment. This means that hydrological model results have to provide the data and information necessary for decision making.

The WFD and FRD are defining new objectives or are setting strict limits for existing ones and show the need for an integrated management of rivers and their catchments. Today main water related problems result from activities outside the river or on property in the floodplain which is privately owned. But till today the water body itself defines the classical focus of water management authorities. Agriculture, urban and infrastructure development, but also recreation, tourism and the changes induced for example by the subsidies for renewable energy sources are threats for fulfilling the objectives of new European Directives. There is a main gap between the public interest in landscape functions and private property and land use rights.

With regard to the catchment perspective new institutions and administrations, classically not involved in water management like municipalities (land use planning and urban planning), rural development (land allocation and land consolidation) or offices for agriculture and farming, will draw responsible for sectoral implementation of river basin management plans. Because of this new approaches in applied hydrology are necessary to identify impacts and drivers on a catchment scale and prepare the basis for decision making.

Classical functional segmentations like the responsibility of water management authorities for rivers and agricultural administrations for agriculture and farming are becoming irrelevant. The interactions of these disciplines are of importance to achieve protection goals on the one hand side and define frameworks for land use and economic development on the other hand side. Another conflict exists between public and private interests. Land use and ownership are private, but land provides a lot of functions and services which can be defined as public goods or public services. The loss of these functions results in increased flooding, problems with water quality and also with river morphology.

There is a need for solutions how to handle these conflicts and find solutions to fulfil the requirements of EU directives, protect our resources and establish new, alternative or adapted forms of land use. Hydrological models can help to find solutions for this resource based conflict through providing design values for the status quo, evaluating possible solutions and their environmental/hydrological impacts.

2. Approach

2.1. Project structure

The idea behind River Basin Management was the integration of process management and rapid prototyping as a dynamic component in contrast to actual static and linear planning processes. A regular planning process as a top down approach defines long term objectives. New scientific results, changing framing conditions and learning during implementation can not be considered. Therefore a lot of planning documents are quickly outdated and need to be completely reviewed after short times.

Rapid prototyping, as an industry approach to always integrate newest knowledge and experimental results into product development, does not claim to achieve perfect results at each level of development. As a developing, planning, and implementation method it is able to provide a flexible and updateable instrument to follow changing framing conditions. The idea behind a dynamic river basin management was to integrate institutional learning and an updateable approach to follow trends, scientific achievements and political developments. Three topics or pillars were identified to cover all relevant aspects (Fig. 1):

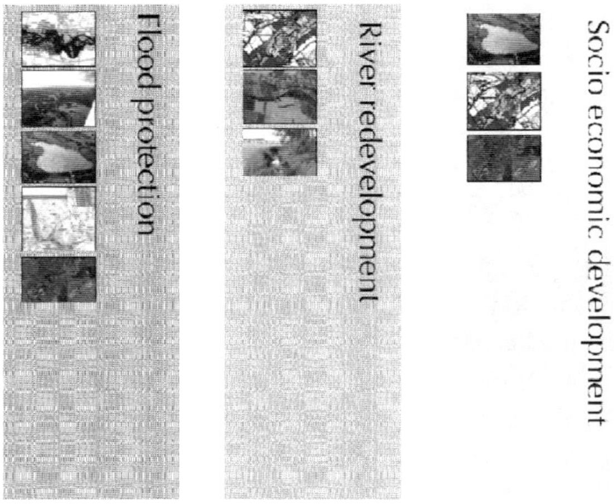

Figure 1. The main three thematic pillars of the project.

Flood protection is a standard task for water management authorities. Already during the design phase of the project a link between the WFD and the flood management was identified although a flood directive was not foreseeable. River development, derived from the German word Gewässerentwicklung, was defined as a task to provide the necessary space and framing conditions to fulfil the qualitative conditions of the WFD. As a third factor the socio-economic development of the region was seen as an important driver but also threat to water and water bodies. This pilar consisted of studies to identify relevant developments and societal needs, e.g. a module agro-economics, but also the definition of interfaces towards the population through the integration of touristic and recreational objectives in water management projects as well as stakeholder participation.

Besides new water management planning, the interdisciplinary idea behind river basin management also asks for a distinct process management. Like experiences from the past showed, different authorities and planning bodies start activities in parallel, often pushing conflicting objectives. To achieve a maximum of synergies and avoid contradictory goals disciplinary plans need to be parallelized. In a next step inter-sectoral implementation projects can combine e.g. the strengths of different funding and implementation measures and achieve a maximum benefit from a societal and stakeholder perspective.

At this point of the planning process first results of hydrological models have to be available to allow an analysis of the status quo and provide a basis to simulate different planning scenarios. The following types of models were used to provide this background information in a catchment scale but also for decision making at a local scale and the preparation of implementation projects:

- Hydrological models: for flood forecasting, providing design values for flood protection works, improved control of reservoirs.
- Hydrodynamic models (2D): Evaluation of flood plains and endangered areas, providing the data basis for flood hazard maps, evaluation of different planning scenarios for levee constructions and river renaturalisation, evaluation of shear stress in the river bed

- Erosion model (based on modified Universal soil loss equation - USLE): Estimation of spatial erosion on agricultural sites regarding different land use scenarios, calculation of sediment delivery [4]

The European Partners in the project were located in Austria, Czech Republic and Hungary. In Bavaria main partner was the State Ministry for the Environment, Health and Consumer Protection, represented by four local Water Management Offices in Landshut, Deggendorf, Pfarrkirchen and Passau. Over 80 subcontracts including modelling and planning tasks were given to engineering and consulting companies as well as universities.

Before the start of the project a first evaluation of problems and drivers resulted in a lot of measures and tasks ranging from applied quantitative hydrology via erosion modelling towards agro-economic evaluation of the status quo in the catchment and the assessment of measures of water management.

Major tasks were:

- Hydrology for flood forecasting and providing design values for further hydraulic assessments
- Erosion modelling and evaluation of diffuse pollution
- Agro-economic evaluation of current agricultural practices and assessment of impacts through projects and measures
- Hydrodynamic modelling, providing the basis for flood protection studies and assessment of processes in the flood plain
- Evaluation of habitat conditions and measures for habitat development and management
- Hydromorphological studies especially to identify measures for heavily modified river sections and artificial flood channels
- Evaluation of flood risk management practices ranging from technical measures towards hazard mapping and forecasting
- Evaluation of sustainable land use practices

2.2. Project area

In Bavaria two neighbouring catchments were identified as ideal project areas. The rivers Vils and Rott and its smaller tributaries are characteristic for a lot of water related problems in Bavarian catchments. Both areas are located in southern Bavaria in the tertiary hilly landscape (Fig. 2). As right hand side tributaries of the Danube and Inn they are closely interlinked with the project areas of the other partners.

In general the catchments are dominated by intensive agriculture. Partially the influences of the urbanisation of Munich and the Danube axis led to fast urban and industrial development.

The Vils catchment has a size of 1200 km^2 and the main river reach a length of 120 km. Main urbanisations are Vilsbiburg in the head and Vilshofen in the tailwater. The main river reach and all its tributaries are dominated by a rural structure, small villages and intensive agriculture. During the past 50 years a lot of measures changed the hydrologic characteristic of the main river reach. The Vilstalstausee was built 20 km east of Vilsbiburg as a detention reservoir to protect the downstream agricultural areas and villages. The following sections of the Vils have been trained and along 50 kilometres the main river was bypassed by a flood channel. In contrast to this the upper reach of the Vils in the headwater of the reservoir gives an ideal picture of an almost untouched riverine landscape (Fig. 3).

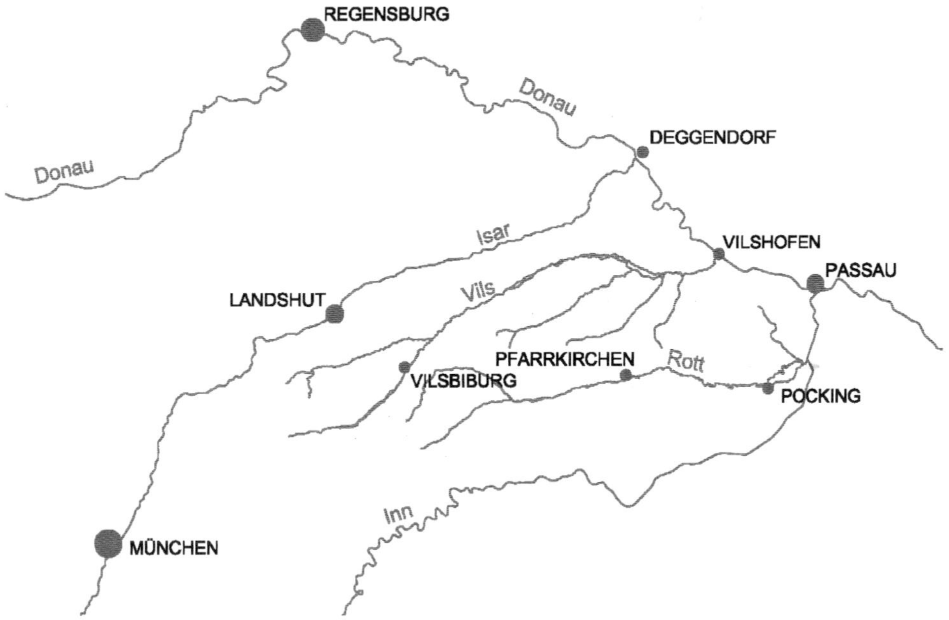

Figure 2. Situation of the project area of Vils and Rott within in the Danube catchment.

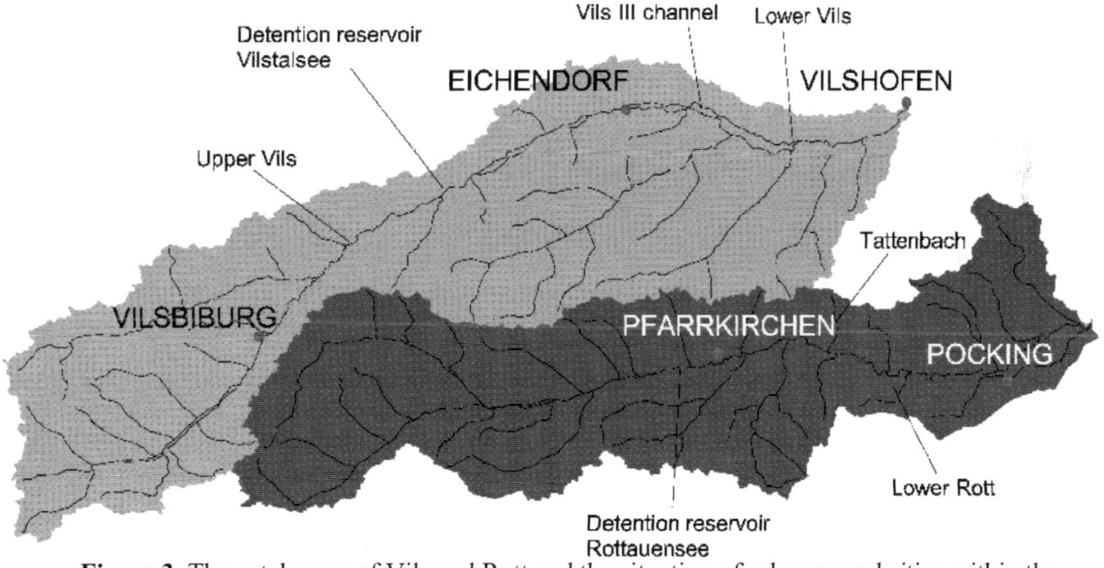

Figure 3. The catchmens of Vils and Rott and the situation of subareas and cities within the catchments.

The situation in the Rott is almost similar to the Vils catchment. Also at the Rott in the head water of Pfarrkirchen a detention reservoir is located, intended to protect downstream agricultural and urban areas. Agriculture is the dominant form of land use. Because of the extreme hill slopes in the eastern part of the catchment in combination with fertile loessy soils and maize cultivation erosion and diffuse pollution is a major problem. In contrast to the Vils a lot of urban areas like Eggenfelden, Pfarrkirchen and Pocking are located in the main river valley. Especially close to Pocking industrial areas are predominant in the wide and flat flood plain and are a major concern from a flood risk management perspective.

3. Results

ILUP had a focus on different types of actions in river basin management: Generation and computation of basic data, strategy development on a catchment scale, bargaining and joint planning with stakeholders and other planning disciplines, preparation of implementation and joint implementation.

The results show very well, that the sectoral instruments of planning and implementation are well understood. Within the process from strategy development on a catchment scale towards implementation the bargaining process with stakeholder groups and even other planning disciplines becomes the major bottleneck or even obstacle, retarding successful implementation. The actual focus of planning theory and practice misses to address the interests, objectives and fears of people or groups affected by plans. As a result the emphasis for the future must be the management of processes including the steps data gathering, stakeholder participation, spatial management, planning and implementation.

River basin management can take place at many scales. ILUP had a closer look on three of these levels: The catchment of a river system (1000 km^2 like Vils and Rott), subcatchment or sections (100km^2 or 10 km like Upper Vils or Lower Vils [7]) and local scale (Vilsbiburg or Sulzbach).

The catchment level is of relevance for general data collection, the definition of strategic objectives and the identification of focal areas. Using hydrologic, geomorphologic or land use data, a catchment can be subdivided in characteristic subunits. Within these subunits (subcatchment or section) often identical land use practices, hydraulic circumstances or quality problems can be found. The bargaining process with stakeholders or other authorities about the need for measures, availability of land or the change of land use practices can therefore set up on a defined and regional common level.

On a local scale the willingness to participate in a planning and implementation process can differ. The definition of measures and the preparation of implementation projects must take into account professional needs, the availability of land and the willingness to implement measures resulting from a subcatchment analysis and identification of individual objectives. Here the major task is to prioritise professional needs, identify the willingness and available space and to use other implementation projects like urban development or land reallocation projects to integrate water related objectives and measures.

River basin management is more a management than a planning task. Major activities are therefore to bring together technical or professional requirements (project idea), the necessary space, funding and the willingness of different parties to participate in a planning and implementation project. This means to define the framing conditions and professional requirements for a subcatchment, having available the necessary data from a catchment perspective and always to take into account the overall strategy and objectives.

3.1. Hydroeconomics

The analysis of the subprojects shows clearly, that economic interests can be a main driver or obstacle. Therefore these aspects can be both a risk but also a chance in river basin management, if they are recognised. Some of the subprojects dealt directly with the economic situation of stakeholder groups or regions whereas others were confronted during a planning or implementation process with business interests.

From the beginning the interests of farmers were monitored as a part of the agro economic evaluation [8,9]. This meta-project served as service for other implementation oriented subprojects. The economic evaluation of farms delivered important hints, how interests of individual businesses as well as groups of farmers can be considered.

During the project ILUP the discussion about renewable energy sources and the necessity to improve the ecologic situation of water bodies (Water Framework Directive) grew. This was reflected in two subprojects. The cattail project dealt with the technical and economic feasibility of cattail growing in the project area and its processing [10]. The study delivered important indicators in what regions and under which framing conditions in general cattail growing can be an agricultural alternative in the flood plain and which renewable energy and resource stock it could provide.

The second study covered the conflicting area of hydro power and river ecology, identified long ago as a problematic field [6]. For the main river reaches the need to establish fish ways, to allow again fish migration, was analyzed from the technical and ecological point of view. In a second step for each facility the economic situation before and after the installation of a fish pass was calculated. The result was a system of incentives including compensation from the Renewable Energy Sources Act.

It was recognized that economic evaluation must go beyond classic market theory and deal with environmental economic questions. Within a river basin river systems can work as transfer-mechanisms for unidirectional externalities. Rivers link head with tail water. Activities in the upper catchment, like for example farming, have a lot of consequences. Erosion and sediment delivery, transport of fertilizers and pesticides as diffuse pollutions impact downstream riparian owners or bodies. Public costs like the maintenance of river systems are only an indication of the total externality of upstream land use.

Future research must therefore concentrate on the environmental economic interrelations in river basins and use hydrological modelling as a mean to simulate different environmental outcomes to derive relevant steering mechanisms. Also consequences for the implementation of the Water Framework Directive can be indentified. The Directive addresses at the moment only water services, but not other forms of environmental economic relations, which could be of even greater importance. Hydrological quantitative and qualitative models as well as for example erosion, sediment delivery and transport models.

3.2. Catchment segmentation regarding hydrological "zones"

In the history of hydraulic engineering projects often "created" new space for different forms of land use through flood protection, river training and draining. In contrast modern water management strategies create restrictions for land use and need space to develop detention, avoid the development of flood risks and renaturalize river systems. Measures do not only affect the closer flood plain but also have to take place in the wider catchment, to reduce surface run-off , diffuse pollution and develop networks of habitats along water courses.

This need for space along the water courses, the optimised allocation of land use types in the flood plain and the preparation of spatial measures in the wider catchments requires a spatial planning instrument. Four types of activities with spatial impacts can be identified:

- The purchase of land for river renaturalisation and the building and realignment of flood protection works
- Reallocation projects to optimise land use structures in the flood plain from a land users and water management point of view to reduce flood damages and diffuse pollution and allocate available land for river renaturalisation along the water courses
- Restrict land use and urban development in the flood plain to reduce flood damages
- Influence agriculture to reduce diffuse pollution, the delivery of erosive material and surface run-off

At the Upper Vils [5] a land use plan was developed together with the farming community and representatives of the municipalities. The main objective was to allocate available areas along the main water courses, to protect and develop habitats, protect the river against diffuse pollution and renaturalize trained sections. The project resulted in a joint reallocation project, which in addition also provided an optimised allocation of intensive and extensive agriculture in the wider and farer flood plain. This project and also other measures showed the need to visualize and communicate the spatial objectives of water management authorities.

These analysis of sub projects like this one resulted in a hydrologic based spatial management plan and a zoning concept (Fig. 4), with four different zones indication a rising importance of the areas from a water point of view. For the definition of the zones established or well known technical and legal criteria were used. Therefore the results of GIS-calculations, erosion models and hydrodynamic flood models were combined.

Zone 1 (dark blue), as the core zone of water management authorities, includes the water bodies and a section of the closer flood plain restricted by the border of the one year flood and settled areas located in the 100 year flood plain.

Zone 2 (medium blue) is defined by the 5 year flood plain and settled diked areas to identify residual risks.

Zone 3 (light blue) 100 year flood plain and Natura 2000 areas with a water context

Zone 4 (brown) the area of the probable maximum flood and agricultural fields with a high erosion risk.

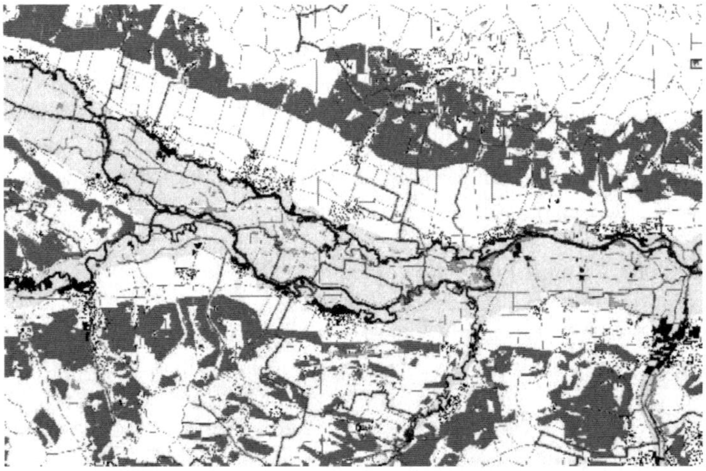

Figure 4. Representation of the zoning system for a 10 km section of the Lower Vils.

A hydrological oriented spatial plan can be used to integrate water related objectives in other spatial planning activities like rural development, land reallocation projects, urban development or the development of agricultural policies. Within the different zones water management authorities will purchase or rent land to influence the land use. In other zones it will bargain with land users to motivate them to change land use practices or reallocate property or use.

3.3. The role of municipalities The role of municipalities

Spatial management is an important topic for municipalities, because they are in contact with land users and land owners and are responsible for different types of spatial plans on a local a level. Therefore they are the key for success for an implementation in the whole catchment. A general implementation of actions can only be achieved if water objectives can be integrated in policies, strategies, plans and implementation projects of other sectors. Rural development, urban planning and different stakeholder groups can therefore become important partners for implementation.

As a result of ILUP a lot of municipalities now picked up the idea of catchment management. This was a consequence of the bandwidth of topics addressed by ILUP, which included 35 lead projects and over 100 sub-actions. It is also a result of the publication and participation strategy, which addressed politicians as well as citizens in the catchment.

Municipalities are the main administrative bodies dealing with spatial questions and problems. Because of their close contact to citizens and responsibilities for the small river systems and water courses they are the perfect organisational structure to implement spatial activities. ILUP addressed a lot of these aspects integrating communities in subprojects.

The main idea was to motivate communities to show responsibility for their water courses on the one hand side and raise awareness for the implementation of actions initiated by the water management authorities on the other hand side.

Three types of measures of ILUP were of importance for these developments:

- Tourist and recreational projects
- Integration of municipalities in water management projects
- Initiation of subprojects to address municipal problems

In a first step councils and political representatives were invited to discuss planned projects. To give a better impression of the local situation and problems, meetings were often organised as excursions. A second step was the integration of municipal interests, like cycle paths, urban development or tourist attractions, in water management projects. The close integration of local administrations had an multiplying effect. The started participative process now initiated an implementation of water related projects on a local and municipal level.

River basin management on a catchment scale, recommended by the water framework directive, must be accompaigned by a local process, a municipal river basin management (MRBM). As an implementation oriented planning and management it could be the driving force to address and solve spatial problems. Municipal alliances on a subcatchment level offer the possibility to follow the catchment approach. Therefore not only water management related issues, but also local economic, social and environmental problems and interests need to be addressed from a water perspective:

- Flood protection
- Water protection and water quality management
- Recreation and tourism
- Habitat management

MRBM can be seen as the implementation oriented downscaling of ILUP. In parallel to urban development plans and spatial planning MRBM could integrated water related responsibilities in the municipal decision making processes.

In Germany municipalities draw responsible for land use planning and also the management of small rivulets. With regard to water management municipalities must be enabled to deal with hydrological issues. Besides modelling on a large catchment scale and in contrast on a plot scale there is a growing need to apply hydrological models also on a municipal level. This also means that hydrological models must become easy to handle tools to allow an assessment of municipal activities.

4. Conclusions

As part of the project hydrological, hydrodynamic and other type of scientific computer models provided the basis for decision making. Model results were compared with economic and socio-economic data and resulted in participatory planning processes.

River basin management sets a new boundary for water management. The jump into the catchment confronts engineers and planners with the needs of different stakeholder groups affected by planning or impacting the hydrological cycle. The classical instrument of sectoral planning and a linear process from planning towards implementation must be substituted by the definition of a catchment strategy and a new form of process management for different planning levels. Modelling and simulation of different scenarios in combination with different planning techniques must result in concrete actions and especially the reallocation of land ownership and land use. Basic models like hydrological model, hydrodynamic models and erosion sediment transport models are necessary scientific prerequisite and the core of any decision making process.

Most water related problems, like flood risk or water protection, have a spatial character and are located in the flood plain or the catchment. Solving them requires changes or restrictions of land use and the reallocation of land parcels or types of land use. Therefore water management needs first a planning instrument to define spaces with a water context and second a set of instruments to influence, avoid or change negative activities with spatial character. Land property and maintenance costs for public areas also require new strategies for a sustainable and optimized land use of these environmental or functional areas. For an effective and efficient catchment management the available instruments of all public bodies (water management, municipalities, nature conservation, rural development, …) have to be combined. Parallelization of different planning processes helps to avoid conflicts because of differing knowledge of partners with different interests. Good sample is the parallel investigation of hydro-morphology and river development, land use optimization and habitat management at the Upper Vils section resulting in a joint management and development plan of water management authorities, farmers and nature conservation authorities. With regard to conflicting interests of public interests and private land ownership and land use additional measures are necessary. Not only the use of our water bodies but also intensive land use are all representations of economic activities. River basin management as a driver to improve the situation within a catchment and with respect to the water bodies can only work, if we recognise and integrate economic interests. Therefore economic evaluation methods have been part of subprojects. The results showed clearly, that we understand well the analysis of individual businesses as well as the evaluation of implementation projects for example through cost comparison calculation. A general economic assessment on a catchment scale and the integration of environmental economic theory is still missing in practice and even needs to be the topic of future research.

The broad implementation of water protection objectives can only be achieved, if municipalities foster spatial measures. As the main responsible bodies for spatial plans and the representatives of citizens they can not only offer support, but be the main drivers of a change process.

5. Acknowledgements

The full reports can be found at www.interreg-ilup.de. Transnational reports are at www.ilup.org. The project ILUP Integrated Land Use Planning and River Basin Management was co-financed by the European Union, Interreg IIIB CADSES. The paper presents results of a work financed through the project ILUP and the Bavarian Ministry for the Environment, Health and Consumer Protection.

References

[1] Dorner W, Spachinger K, Riedel B, Dullinger F and Metzka R 2007 ILUP *Integrated Land Use Planning and River Basin Management - Erfahrungen und Empfehlungen Wasser und Umwelt* (Fachhochschule Deggendorf/University of Applied Sciences) www.interreg-ilup.de, 23.3.2008

[2] Dorner W, Spachinger K and Metzka R 2007 Integrated land use planning and river basin management *10th International Riversymposium & Environmental Flows Conference* (Brisbane)

[3] Metzka R, Dorner W and Spachinger K 2006 Flussgebietsmanagement im Spiegel der integrierten ländlichen Entwicklung *Beitrag anlässlich der Fachtagung der Bayerischen Verwaltung für Ländliche Entwicklung* (31. Mai 2006, Passau) http://www.fachtagung-le.bayern.de/passau/htdocs/pdf/vortraege/05_Prof_Metzka.pdf, 23.3.2008

[4] Pirkl A and Riedel B 2007 *Maßnahmenkonzept zum Stoffrückhal* www.interreg-ilup.de, 23.3.2008

[5] Riedel B and Schacht H 2007 *Gewässer- Aueentwicklung Obere Vils* www.interreg-ilup.de, 23.3.2008

[6] Roland F and Lang T 2007 *Restwasser und Durchgängigkeit* www.interreg-ilup.de, 23.3.2008

[7] Schacht H 2006 Drei unter einem Hut *Landwirtschaftliches Wochenblatt, 2006*

[8] Schätzl R and Hoffmann H 2006 Was kostet die Umnutzung-Teil1 *Landwirtschaftliches Wochenblatt, 2006*

[9] Schätzl R and Hoffmann H 2006 Was kostet die Umnutzung-Teil2 *Landwirtschaftliches*

Wochenblatt, 2006
[10] Schätzl R and Hoffmann H 2007 *Agrarökonomische Untersuchungen 2007* www.interreg-ilup.de, 23.3.2008

IOP Conf. Series: Earth and Environmental Science 4 (2008) 012032 doi:10.1088/1755-1307/4/1/012032

Influence of spreading urbanization in flood areas on flood damage in Slovenia

B Komac[1], K Natek[2] and M Zorn[1]

[1] Anton Melik Geographical Institute, Scientific Research Centre of the Slovenian Academy of Sciences and Arts, Gosposka ulica 13, 1000 Ljubljana, Slovenia
[2] Department of Geography, Faculty of Arts, University of Ljubljana, Aškerčeva cesta 2, 1000 Ljubljana, Slovenia

E-mail: blaz.komac@zrc-sazu.si, karel.natek@guest.arnes.si, matija.zorn@zrc-sazu.si

Abstract. Damage caused by natural disasters in Slovenia is frequently linked to the ignoring of natural factors in spatial planning. Historically, the construction of buildings and settlements avoided dangerous flood areas, but later we see increasing construction in dangerous areas. During the floods in 1990, the most affected buildings were located on ill-considered locations, and the majority was built in more recent times. A similar situation occurred during the floods of September 2007. Comparing the effects of these floods, we determined that damage was always greater due to the urbanization of flood areas. This process furthermore increasingly limits the "manoeuvring space" for water management authorities, who due to the torrential nature of Slovenia's rivers can not ensure the required level of safety from flooding for unsuitably located settlements and infrastructure. Every year, the Environmental Agency of the Republic of Slovenia issues more than one thousand permits for interventions in areas that affect the water regime, and through decrees the government allows construction in riparian zones, which is supposedly forbidden by the Law on Water. If we do not take measures with more suitable policies for spatial planning, we will no long have the possibility in future to reduce the negative consequences of floods. Given that torrential floods strike certain Slovene regions every three years on average and that larger floods occur at least once a decade, it is senseless to lay the blame on climate change.

1. Introduction

The potential for natural disasters is an important component of landscapes and therefore it is essential to consider this when planning the use of space. Using geographical research methods we can identify the presence of potential natural processes in areas, define their effects on nature and society, and using this information in cooperation with other relevant expert input we can look for possibilities for "coexistence" with them through the instruments of spatial and other forms of planning (Natek, 2008, 148). However, spatial planning is subject to the interests of capital and local communities, and agriculture suffers the blind destruction by urbanization of the highest quality farmland that could produce food in future crisis periods while at the same time placing ever greater demands on limited natural resources (primarily on water for irrigation).

IOP Conf. Series: Earth and Environmental Science **4** (2008) 012032 doi:10.1088/1755-1307/4/1/012032

Figure 1. Frequency of floods in Slovenia (1750–2000).

On the other hand, much has been done in recent years in the field of protecting and safeguarding people and property. Good preparation, effective rescue services, and comprehensive help in dealing with the consequences of natural disasters all reflect the high level of government management, but they cannot be an alternative to preventive action oriented toward preventing or at least mitigating the consequences of future natural disasters.

2. Preventive limitation of urbanization in flood areas

In the past, settlement avoided dangerous flood areas. However, even before World War II, there were permanent flourmills, sawmills, and similar operations and a handful of flood-threatened settlements in flood areas, and later entire residential neighbourhoods grew in such areas (on the southern edge of Ljubljana the population grew by several thousand to more than 30,000). Major public infrastructure projects were built in dangerous areas that now require protection against flooding. This is a major challenge for water management authorities, but it will never be possible to control all dangerous flooding completely.

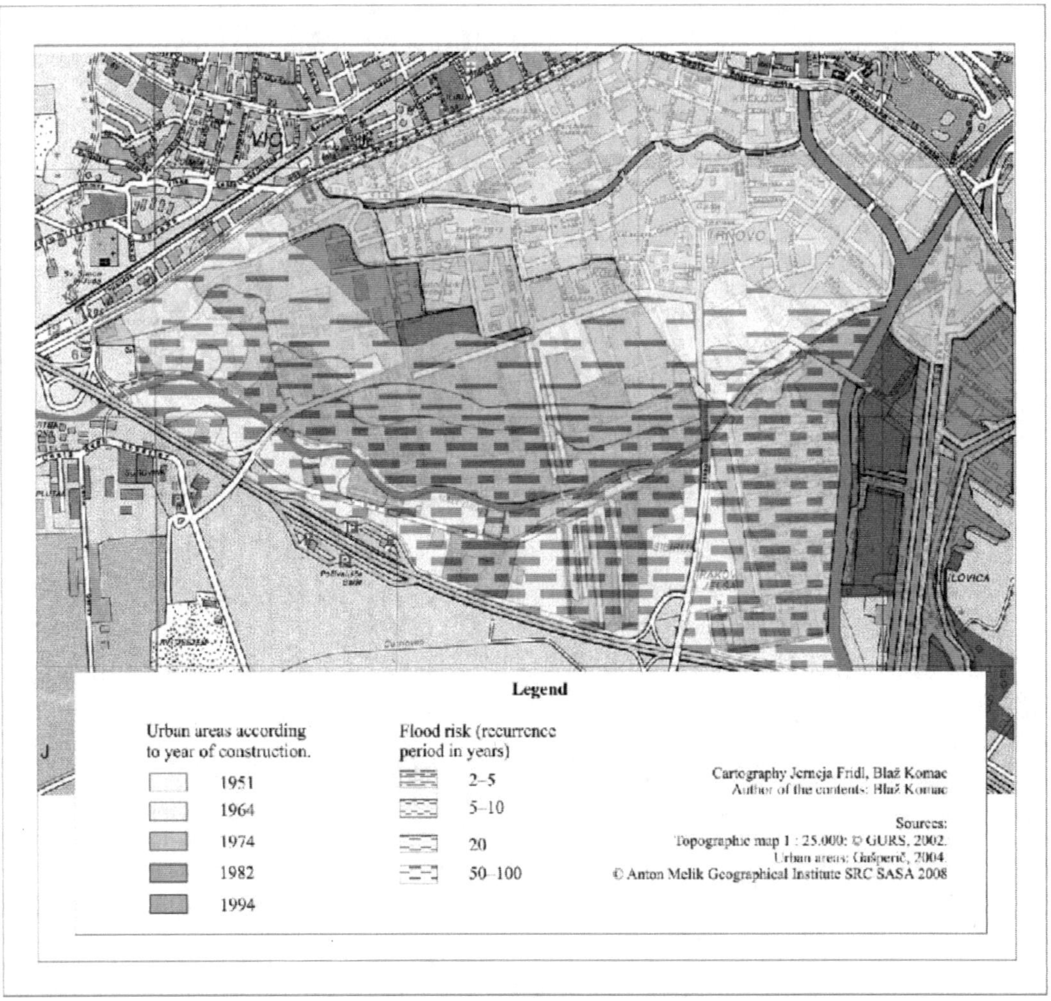

Figure 2. In recent decades, Ljubljana has spread southward [1] where flooding threatens a considerable part of the settled area.

Because the ever-increasing pressure from investment and construction in flood areas shows no regard for the threat of future floods or their inevitable intensification due to global climate change, the government must regulate this urbanization through long- and medium-term spatial planning policies and the professional cooperation of the water management authorities.

Declarative references to the paradigm of sustainable development in a transitional society are very frequent. A corresponding approach to "coexistence" with nature is concluded in the *Strategy for the Spatial Development of Slovenia*; however, due to different development priorities, its provisions are not realized in practice. The Strategy clearly states, "*Natural processes that can threaten settlement and human activity must be respected as limitations in planning uses and activities in space... Potential risk should be reduced through preventive planning, certainly through the distribution of activities in space outside areas of potential disasters, with appropriate management of primary activities in dangerous and threatened areas, and with the monitoring of activities that can cause natural and other disasters*" [2].

A European Union document entitled "Best practice on flood prevention, protection and mitigation" [3] states, "*It is necessary to support and coordinate the transformation of water policies and the use of space ... that will improve the control of floods in the framework of the overall management of riparian areas*" (page 6).

The *National Program of Protection against Natural and Other Disasters* [4] outlines measures that contribute to greater security and are necessary to consider in spatial planning. The *Law on Water* [5] is also quite clear: "*The use and other interventions in water, riparian, or coastal land areas and land in protected or threatened areas ... it is necessary to program, plan, and execute so as ... to provide security from damaging activity of water, the preservation of natural processes, the natural balance of water and riparian ecosystems*" (Article 5). On the basis of these starting points, "*interventions in space are not allowed on water or riparian land* [including flood plains] *or in the area of disappearing lakes*," except in some explicitly determined cases.

By signing the *Alpine Convention* [6], Slovenia bound itself to strive for suitable measures in the field of spatial planning, including relative to natural disasters.

In spite of such clear legislation, we can find in Slovenia hundreds of cases of legalized new construction of residential, business, and other buildings on flood areas. This also conflicts with the *Law on Spatial Planning* [7], according to which a key goal of spatial planning is "*to enable harmonious spatial development ... in the fields ... of protecting natural resources and defence and safety from natural and other disasters*" (Article 3).

This lack of respect for the legislation is unacceptable from the viewpoint of sustainable development or environmental ethics, since "*spatial planning should not look merely at satisfying the needs for space of the existing society but also must from the viewpoint of environmental ethics consider intergenerational responsibility*" [8].

A lack of consideration of natural disasters reduces the capacity of the environment and its self-regulation capability, which is reflected in ever greater damage due to natural disasters and in social instability or injustice. The consequence of an acquisitive attitude toward land is greater liability of natural processes that subsequently produce effects in other areas or in other forms [9].

3. Problems in changing existing practices in spatial planning

3.1. Municipalities of Ljubno, Luče, and Nazarje

In the case of Ljubno and Nazarje, Privšek [10] investigated to what extent the terrible consequences of the flood of November 1, 1990, influenced changes to the spatial planning documents of the two municipalities. The flooding most affected the new parts of settlements that spread onto the flood areas after 1970 (Boltinov travnik and Log near Luče and Prihova and the Glin factory area in Nazarje along the Savinja River, and Dobletina and the lower part of Nazarje along the Dreta River).

The situation around Luče has not changed significantly since 1990. Four houses and a small factory have been built near the Lučnica and Savinja rivers, and the bad flood experience has only been partly considered in the municipality's spatial planning for Luče since although new residential construction is anticipated within the existing settlement, replacement construction has been allowed on farmland on the right bank of the Lučnica River. They intend to provide greater flood safety for the lower parts of Luče by constructing a bypass on the Radmirje-Luče regional road and deepening the bed of the Savinja River. This will cause greater pressure on the areas along the Savinja River just above Luče and along the Lučnica River, which in the current spatial plan are defined as mixed use areas [10].

In Nazarje, the flood protection issue is even more contentious. During the flood in November of 1990, extensive areas were affected but small factories and larger industrial plants were subsequently built on both sides of the Savinja River, and a gymnasium, a store, and a restaurant along the Dreta River. The municipality's spatial plan anticipates the continued development of the small manufacturing and industrial areas, which should be protected by the existing dike. To the south, areas of mixed use are expanding onto farmland on the left bank of the Dreta River that are protected by a low dike; however, flood waters poured over this dike during a smaller flood in 1998.

The spatial development of Nazarje could be reoriented from the threatened flood plains to an unthreatened forest area east of Nazarje and above Dobletina a kilometre away, which would certainly reduce the threat to the settlement.

3.2. Celje and Laško

Flooding in Celje is a geographical inevitability, the consequence of locating a large city at the confluence of the Voglajna, Ložnica, Hudinja, and Savinja rivers. Major floods occurred here in 1550, 1772, 1813, and 1850 [11]. In the 20th century alone, Celje experienced devastating floods in 1901, 1923, 1925, 1926, 1933, 1954, and 1990 [12].

Before the efforts to manage floods by regulating the Savinja between 1876 and 1893, there were almost forty square kilometres of flood areas in the Savinja Valley that were sparsely settled and functioned as an enormous reservoir for flood water. The regulation works reduced the extent of this "natural reservoir" to a few square kilometres, and the construction of dikes on both sides of the Savinja right along the riverbed sealed the fate of Celje, which now has almost no possibility of defending itself from flooding [13].

Because of its location on the floor of a narrow valley, Laško has frequent problems with flooding. The Savinja River, for example, achieved a discharge rate of 1,406 m3/s on November 1, 1990, and rose by six meters (the average discharge between 1961 and 1990 was 41.5 m^3/s). Although the natural conditions in Laško make larger preventive measures impossible, the government allowed by decree the expansion of the health spa on riparian flood land. The consequences were predictable: during the floods in September 2007, the new buildings suffered 1.8 million euros worth of damage. The government has promised to provide financial help since its decision was in complete contradiction to the standards of sustainable development and the applicable legislation [14].

3.3. Škofja Loka region

Lives were lost and great material damage caused unnecessarily during the flooding of the Selška Sora River in September 2007 since ominous precedents were readily available. Previously, the river swept away a bridge in Železniki and five kilometres of road in the Davča Valley farther upriver in 1990, and it flooded again in 1992, in 1993, and in 1995 when three hundred meters of road were washed away in the Davča Valley as well as a bridge and a road in Železniki. In September 2007, the water level of the river rose two meters in half an hour in Železniki. The neighbouring Poljanska Sora River overflowed its banks in 1901 and in 1926 when it destroyed twelve bridges between Žiri and Škofja Loka as well as in 1964, 1965, and 1982 [15].

Similar events occurred in Kropa, which developed around ironworking workshops along the river at the bottom of a narrow ravine, and in Podbrdo, where a textile mill was once built beside the Bača River.

4. Preventive limitation of urbanization in flood areas

The idea that flood defences can guarantee complete security is common, even in professional circles, which results in numerous proposals for the construction of reservoirs for flood waters. However, due to the great space requirements and the irregularity of flooding, this solution is not suitable for the torrential rivers in Slovenia's hilly and mountainous regions. Furthermore, reservoirs do not always provide effective protection against flooding as demonstrated by the floods in Prague and other cities along the Vltava River in August of 2003 and the many times smaller floods along the Drava River below Maribor and the flooding of the northern part of Celje in spite of the Šmartinsko jezero reservoir along the Koprivnica stream on September 18, 2007.

Such "ultimate" solutions to the problem of flooding not only require (too) large interventions in the functioning of natural processes but also create a subjective feeling of complete security from future flooding that encourages the spread of urbanization onto flood-threatened areas. They also conflict with the provisions of the 2007 directive of the European Parliament and the Council of Europe on the assessment and control of flood risk [16] valid since November 26, 2007, that requires of European Union members that they "*must in planning, to the extent possible, consider the conservation and/or renewal of flood areas*" (Preamble, point 14).

The basic problem in areas such as those described above is not floods as a natural phenomenon per se but rather their influence on human activity. Valley floors are actually channels for high water or flood plains. We have exploited these narrow and valuable valley bottom areas by constructing buildings and roads and in a way ignoring natural conditions. The number of buildings of public importance such as factories and hospitals as well as infrastructure in flood areas has grown, and the risk to threatened and vulnerable settlements has increased due to their growing number of buildings and the rising value of real estate.

Regardless of the diverse views on the use of flood areas, natural processes and human activity are always tightly interwoven. While humans adapt more or less to natural conditions, the natural conditions are often greatly altered. While the thought therefore prevails that the land belongs only to humans, relative to controlling natural functions we must identify flood areas as special. In this way we will acknowledge the role of natural processes in the landscape, curtail the ambitions of spatial planners, and reduce the burden on water management authorities.

Figure 3. A flood affected new dwellings in Nomenj on September 18, 2007 (photograph: Aleš Smrekar).

5. Conclusion

Slovenia has almost three thousand square kilometres of flood areas, almost 15% of its national territory. The majority of these flood areas (2,370 km^2) occur in narrow, torrential valleys [17; 18]. Almost ten square kilometres of built-up land and more than 3% of all buildings in Slovenia are located in flood areas.

Figure 4 shows the flood risk to Slovenia's municipalities relative to the number of buildings on flood areas in each municipality and the number of all buildings in each municipality expressed as a percentage. The data on flood areas used to create the map is inaccurate in some cases [19; 20], as evident in the case of the Municipality of Bohinj, for example, which is shown as unthreatened but suffered major flooding on September 9, 2007. Municipalities in the Savinja Valley (Celje, Nazarje) are most at risk, the municipality of Trzin stands out as well, and the municipalities of Luče, Prevalje, Domžale, and Ljubljana are greatly threatened. If the number of residents of individual municipalities is considered, the municipalities most at risk are Ljubljana, Celje, Domžale, Žalec, Brežice, Mozirje, Grosuplje, Slovenske Konjice, Nova Gorica, and Kranj.

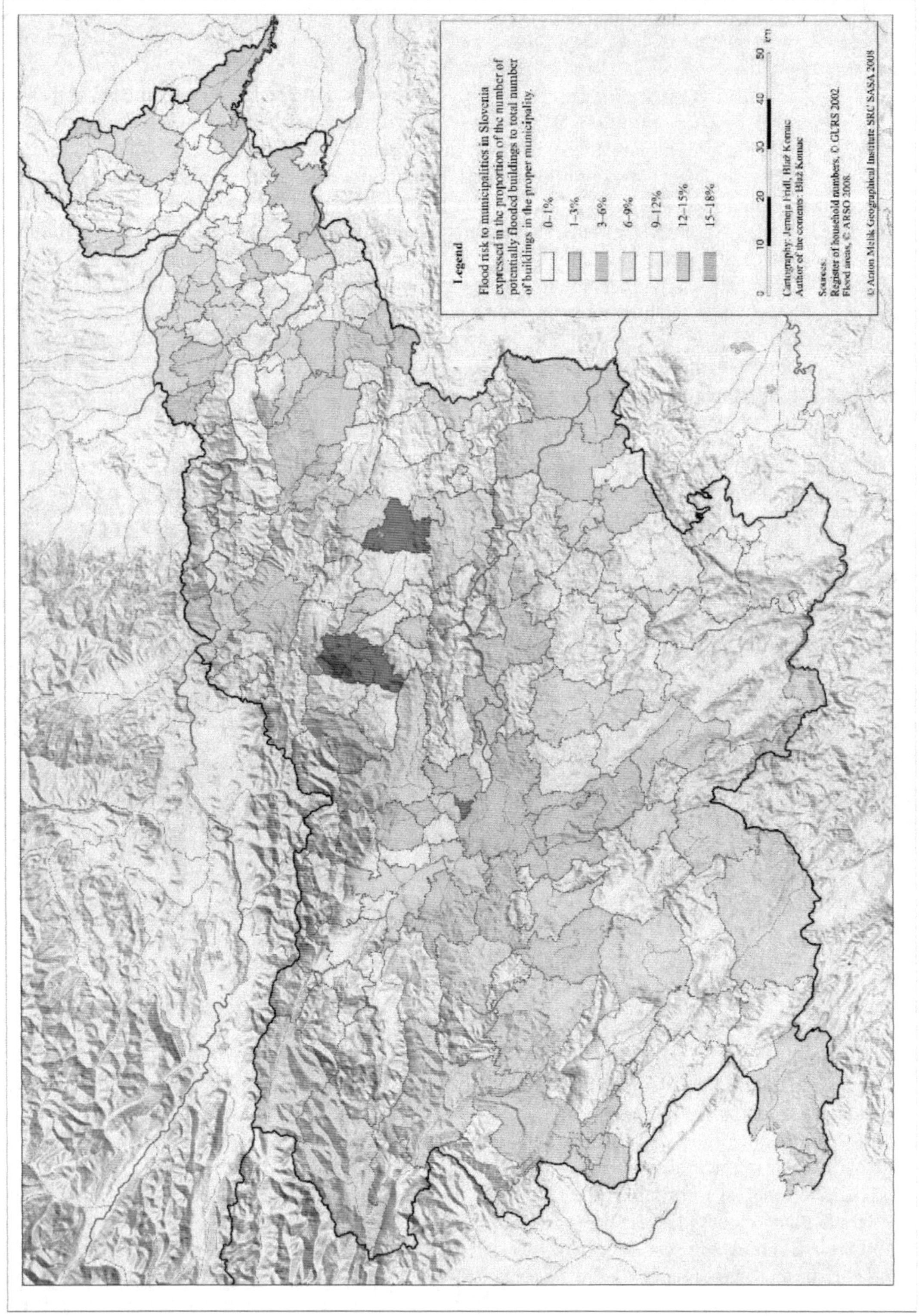

Figure 4. Flood risk to municipalities in Slovenia.

The greatest defect in the management of space in these areas is the lack of consideration of natural factors. We therefore spend much greater sums on cleanup and rehabilitation than on preventive measures [21]. Among the most important reasons for this situation are the lack of adaptation to natural conditions and forgetfulness [22]. Furthermore, it is senseless to lay the blame on climate change; torrential flooding in Slovenia is a completely normal natural process and society must constantly adapt itself to the changing environment.

In dealing with natural threats, we must change our thinking and shift from trusting defensive measures to controlling risk and live with the fact that floods are inevitable [3]. Seeing nature as an obstacle to progress that can be overcome using modern technology rather than as a constantly changing complex of processes is prevalent misconception. Because we think that we can control nature, we wrongly believe we can solve the flood problem just by building dikes. Instead of trying to understand the nature, features, effects, and extent of natural processes and adapt to them, we try at all costs to satisfy our immediate desires. These involve primarily the desire for profit, which is reflected in building infrastructure and industrial objects in flood areas, and the desire for comfort, which is reflected in the wish to live in beautiful but dangerous areas such as along river banks.

Planning the use of space must consider natural limitations and employ different standards for different needs (e.g. the construction of expressways or settlements). The basic paradigm for dealing with these problematic areas should be sustainability, working responsibly with a long-term orientation. It is not good enough merely to take measures after natural disasters but we must above all think preventatively. This would lead to respecting natural processes and the applicable legislation. If the laws had been respected, a study of the 1954 and 1990 floods would not say "*that certain buildings (hospital) were equally affected in both disasters*" [23] or another study after the floods in 1990 that "*the most affected were those residential and business buildings erected on ill-considered sites. Most of them were built in modern times*" [24].

References

[1] Gašperič P 2004 The expansion of Ljubljana onto the Ljubljansko barje moor *Acta geographica Slovenica* **44(2)** 23

[2] *Strategija prostorskega razvoja Slovenije* 2004 (Ljubljana: Ministrstvo za okolje, prostor in energijo)

[3] *Najboljše prakse za preprečevanje poplav, varstvo pred njimi in ublažitev njihovih posledic* 2003 http://www.dhd.si/datoteke/strategija_poplave_koncni_prevod.pdf (15. 3. 2008)

[4] Nacionalni program varstva pred naravnimi in drugimi nesrečami 2002 *Uradni list Republike Slovenije* **44**

[5] Zakon o vodah 2002 *Uradni list Republike Slovenije* **67**

[6] *Alpska konvencija* 1991 http://www.convenzionedellealpi.org/page1_slo.htm (7. 10. 2005)

[7] Zakon o prostorskem načrtovanju 2007 *Uradni list Republike Slovenije* **33**

[8] Plut D 2005 Teoretična in vsebinska zasnova trajnostno sonaravnega napredka *Dela* **23** 62

[9] Natek K 2008 Geografske dimenzije naravnih nesreč in varstva pred njimi *Dela* **28** 158–159

[10] Privšek A 2007 *Geografski učinki poplave 1. 11. 1990 v Zgornji Savinjski dolini in njihovo upoštevanje pri načrtovanju prostorskega razvoja* (Ljubljana: University of Ljubljana)

[11] Trontelj M 1997 *Kronika izrednih vremenskih dogodkov XX. stoletja* (Ljubljana, Hidrometeorološki zavod Republike Slovenije)

[12] Aristovnik B 2005 *Mesto v objemu voda: poplave v Celju v 20. stoletju* (Celje: Zgodovinski arhiv)

[13] Natek K 1992 Geoekološke značilnosti in grožnje prihodnjih katastrof v porečju Savinje *Poplave v Sloveniji* ed M Orožen Adamič (Ljubljana: Republiška uprava za zaščito in reševanje)

[14] Mekina B 2007 Narava? Ne, minister Podobnik *Mladina* **39** http://www.mladina.si/tednik/200739/clanek/slo-tema--borut_mekina (10. 3. 2008)

[15] Orožen Adamič M and Kolbezen M 1984 Neurja in poplave Poljanske Sore v letu 1982

IOP Conf. Series: Earth and Environmental Science **4** (2008) 012032 doi:10.1088/1755-1307/4/1/012032

Geografski zbornik **23**

[16] *Direktiva 2007/60/ES Evropskega parlamenta in sveta z dne 23. oktobra 2007 o oceni in obvladovanju poplavne ogroženosti* http://eur-lex.europa.eu/LexUriServ/site/sl/oj/2007/l_288/l_28820071106sl00270034.pdf (15. 3. 2008)

[17] Orožen Adamič M 1998 Naravne nesreče *Geografski atlas Slovenije* ed J Fridl, D Kladnik, M Orožen Adamič and D Perko (Ljubljana: DZS) p 318

[18] Fridl J, Kladnik D, Pavšek M, Perko D, Repolusk P and Urbanc M 2007 *Slovenia in focus* (Ljubljana: Založba ZRC)

[19] *Enotna evidenca hišnih številk* 2002 (Ljubljana: Geodetska uprava Republike Slovenije)

[20] *Zemljevid poplavnih območij v Republiki Sloveniji* 2008 (Ljubljana: Agencija Republike Slovenije za okolje)

[21] Zorn M and Komac B 2006 Geomorfologija in prostorsko planiranje *Urbani izziv* **17(12)**

[22] Komac B and Zorn M 2007 Pobočni procesi in človek *Geografija Slovenije* **15** (Ljubljana: Založba ZRC)

[23] Jesenovec S 1995 *Pogubna razigranost – 110 let organiziranega hudourničarstva na Slovenskem 1884–1994* (Ljubljana: PUH) p 34

[24] Horvat A 1995 Ujma 1. novembra 1990 na območju Zgornje Savinje *Pogubna razigranost – 110 let organiziranega hudourničarstva na Slovenskem 1884–1994* ed S Jesenovec (Ljubljana: PUH) pp 50–53

IOP Conf. Series: Earth and Environmental Science **4** (2008) 012033 doi:10.1088/1755-1307/4/1/012033

Hydro power plants on the Middle Sava River section

A Kryžanowski, A Horvat and M Brilly

University of Ljubljana, Faculty of civil and geodetic engineering, Chair of hydrology and hydraulic engineering, Hajdrihova 28, Ljubljana, Slovenia

E-mail: andrej.kryzanowski@geateh.si

Abstract. Construction of a chain of hydro power plants is planned on the Sava River from Medvode to the Slovenian-Croatian border which will, apart from the chain of HPPs on the Drava River, represent the linchpin of renewable energy production within the Slovenian power system. The mentioned chain of HPPs will also be one of the country's main renewable energy sources that can still be developed for power generation. Three hydro power plants, Moste, Mavčiče and Medvode, are already operating on the Upper Sava River section. Construction of the chain is underway in the lower part of the stream where Vrhovo and Boštanj HPPs are already in operation; HPP Blanca is under construction and the site planning procedures are taking place for Krško, Brežice and Mokrice HPPs. The planned HPPs on the Middle Sava River section between Medvode and Zidani most will connect the HPPs on the Upper and Lower Sava River into a closed chain which will operate on the principle of run-of-river type power plants with daily storage. Completion of all stages will enable optimal development of available hydro potential. Apart from the energy effects, also other beneficial effects of hydro power plant construction in the region can be expected: flood protection; better water supply; waste water treatment; development of transport and energy networks as well as positive economic and social effects.

1. General

Water is the only indigenous natural resource that is in abundant supply in the resources impoverished Slovenia. The highlands in the north-western part of the country prevent penetration of warm and moist air masses from the Mediterranean. The rivers in the western and central Slovenia receive water from the rainfall in the Sub-alpine region. The most important among them is the Sava River with its tributaries which receives water from a catchment area of 10,872 square kilometers, i. e. 53.6 per cent of the country's total area. The Sava River is characterized by changing flow regime: in the upper part the snowy-rainy regime prevails which turns into rainy-snowy regime in the lower stretch. Maximum annual discharges appear during the spring and autumn months and the minimum discharges during the summer and winter months. The Sava River with its tributaries has a torrential character along the entire river stretch. The ratio between minimum and maximum discharges amounts to 1:100; in exceptional circumstances even to 1:250. Unlike other main streams, the Sava river basin is the least exploited for power generation.

Of the available hydro potential of the Sava River with its tributary Ljubljanica, 18.0 per cent have been utilized so far, i.e. 12.2 per cent of the total hydro energy generated in Slovenia (Table 1). The available hydro potential has been fully utilized on the Drava River and 34 per cent on the Soča River with its tributary Idrijca.

From the point of view of energy, the Sava river basin is divided into three sections: Upper, Middle and Lower Sava River (Figure 1). The idea of utilizing the Sava for hydro power generation dates back to early last century when the first studies about the chain of hydro power plants have been undertaken. A closed chain of hydro power plants along the entire stream has already been proposed. Head storage reservoirs (HPP Moste and HPP Radovljica) were planned in the upper stream with seasonal storage capacity to enable operational capability of the chain during the dry season of the year. Construction of hydro power plants on the Upper Sava River: HPP Moste (1952) and HPP Medvode (1953) started immediately after the War. Later the works were abandoned for some time and only in 1986, the construction of Mavčiče hydro power plant started. The plans to construct storage reservoirs on the Upper Sava River (Radovljica, Radovna) were completely abandoned at that time for nature protection reasons. [1]

Table 1. Hydro Potential of Slovenian Rivers (Porenta, 2002 with supplements)

WATERCOURSE	GROSS POTENTIAL	TECHNICALLY UTILIZABLE POTENTIAL	UTILIZED ENERGETICAL POTENTIAL	PORTION OF HYDROPOWER UTILIZATION
	(GWh/year)	(GWh/year)	(GWh/year)	(%)
SAVA	4,134	2,794	502	18.0
DRAVA	4,301	2,896	2,833	97.8
SOČA & IDRIJCA	2,417	1,442	491	34.0
MURA	928	690	5	0.7
KOLPA	310	209	0	0.0
REST	7,350	1,114	284	25.5
TOTAL:	19,440	9,145	4,115	45.0

By abandoning the head storage reservoirs in the Upper Sava River section, the chain of 15 run-of-river plants from Medvode to the Slovenian-Croatian border at Mokrice is given consideration again. The role of the head storage reservoirs is taken over by the storage reservoirs of Mavčiče and Medvode HPPs, and the role of compensating reservoirs by the storage reservoir of Brežice and Mokrice HPPs. The storage reservoir of the Vrhovo HPP and partly of Jevnica, Kresnice and Ponoviče HPPs are foreseen for intermediate compensation of flows. Construction of the chain of HPPs on the Sava River has started on the Lower Sava River section as the most promising sites due to the aquiferous tributaries: the Savinja and the Krka. Vrhovo HPP, the first in the chain has been operational since 1993, Boštanj HPP has started operation this year, the third in the chain, Blanca HPP, is under construction and it is scheduled for completion by the year 2009; the HPP Krško dam structure is under construction and it is due to be completed by the year 2012; HPP Brežice and HPP Mokrice projects are in the stage of site planning. The chain on the Lower Sava River is scheduled for completion by the year 2018 at the latest.

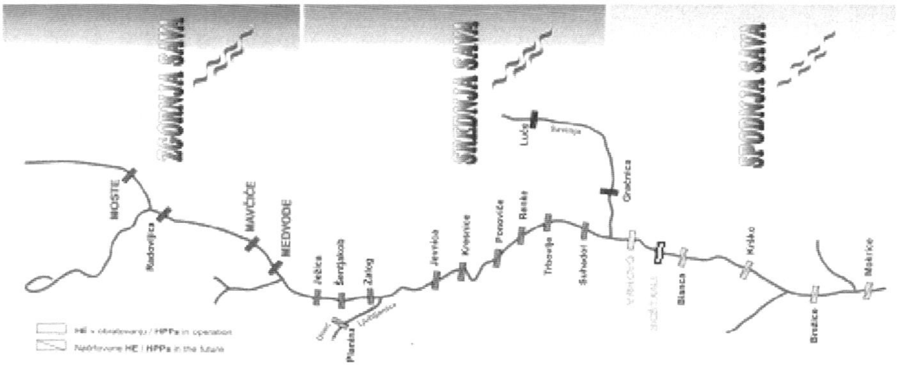

Figure 1. Planned Hydro Power Plants on the Sava River (Source: Savske elektrarne)

2. Hydro Power Plants on the middle Sava River

The hydro power plants on the Middle Sava River will connect the stages on the Upper and Lower Sava River into a closed chain. Nine power plants are envisaged on the section which is from the geographical point of view divided into two parts: the Ljubljana Basin with Ježica, Šentjakob, Zalog, Jevnica, Kresnice and Ponoviče HPPs; and the canyon part of Zasavje with Renke, Trbovlje and Suhadol HPPs. The conditions for hydro development of the Middle Sava River section are regulated by Concession Acts [3]; concession has been granted to Holding Slovenske elektrarne. Construction of the chain of HPPs on the Middle Sava River will probably last 20 years, its completion is expected in 2030. The project is now in the stage of preparing expert groundwork to start the site planning procedures.

The rated discharge of the chain on the Middle Sava River has been determined in a manner which enables power generation in time with the hydro power plants on the Upper and Lower Sava River. It amounts to 260 m3/s on the river section to Zalog, and to 400 m3/s at the confluence with the Ljubljanica River. The gross potential of the Middle Sava River section amounts to 1,185 GWh/year. It has been assessed on the basis of the hydrological data from the period between 1961-90 by taking into consideration the average annual flows and a gross head of 117 meters which represents the difference between the downstream water level at Medvode and the upstream water level at Vrhovo. The average annual output of the chain on the Middle Sava River is estimated at: 994 GWh/year; net electrical output: 295.4 MW. Characteristic data on individual power plants are shown in Table 2.

Table 2. Hydro Potential of Slovenian Rivers (Source: Regulation on the Concession, 2004)

	RESERVOIR VOLUME	NET HEAD	INSTALLED CAPACITY	ANNUAL INPUT
	(10^6 m^3)	(m)	(MW)	(GWh/year)
HE JEŽICA	0.4	26.0	54.5	151.4
HE ŠENTJAKOB	1.9	7.3	15.9	52.4
HE ZALOG	1.1	7.2	15.7	52.8
HE JEVNICA	2.4	6.9	22.9	101.4

HE KRESNICE	1.5	8.3	27.7	96.4
HE PONOVIČE	1.4	18.8	63.0	190.7
HE RENKE	0.4	8.5	28.6	97.9
HE TRBOVLJE	0.3	8.3	27.8	97.6
HE SUHADOL	0.7	11.7	39.3	153.4
TOTAL:			295.4	994.0

In the conceptual design of the chain on the Middle Sava River section, two types of hydro power plants are proposed: run-of-river plants and head water storage reservoirs. With run-of-river type plants, the powerhouse is a constituent part of the dam structure, and the hydraulic head is represented by the geodetic altitude between the water level in the storage reservoir and the tail water. With head water storage reservoirs, the hydraulic head is realized by a headrace tunnel connecting the dam with the powerhouse located downstream. According to the description of the power plants from the Concession Act [3], Ježica and Ponoviče hydro power plants are foreseen as head water storage reservoirs and the other plants as run-of-river type plants. At present, conceptual designs are being prepared and, in order to reduce the construction costs and to minimize the environmental impacts, it is foreseen to replace the head water storage reservoirs by run-of-river plants.

3. The role of the Hydro Power Plants on the Sava River within the Slovenian electrical power system

By changing the structure of electrical energy supply (a constant increase in percentage of thermal energy) and attainment of independence of the Slovenian electrical power system within the European interconnection system, also the role of the hydro power plants within the system has changed. Their role in covering the constant part of daily load diagram has been minimized and the need for peak energy generation and the capability of taking over primary and secondary regulation have increased. The ratio between peak load and base load has been increasing together with the restructuring of the country's industry due to the smaller number of large consumers. Thermal or imported energy are intended to cover the needs of base load, while the energy in the peak portion of the diagram (particularly the power) has to be produced as near to the major consumers as possible.

It is also necessary to underline the conditions that have to be fulfilled for the incorporation of the Slovenian electrical power system into the European interconnection of networks: activation of the required reserve power in the event of failure of the largest generator unit (more than 300 MW) and availability of power reserve for frequency and power regulation (more than 180 MW). Association with the European Union and the introduction of energy market require from the Slovenian power system to provide ancillary services from its own production capacities, such as: own spare peak power, a power reserve during selected periods of the day for secondary regulation, fast activating of reserve peak power, reserve reactive power for voltage compensation and black start in the event of system breakdown. In respect to the envisaged role within the Slovenian electrical power system, the chain of hydro power plants on the Sava River will operate on the principle of run-of-river power plants with daily storage, with Mavčiče and Medvode HPPs taking over the role of the head storage reservoir, and Jevnica, Kresnice, and Ponoviče HPPs on the Middle Sava River and Vrhovo HPP on the Lower Sava River, serving as afterbay and compensation reservoirs as long as the whole chain has not been completed. Upon completion of the chain, the mentioned storage reservoirs will function as intermediate afterbay reservoirs on individual sections, while Brežice and Mokrice storage reservoirs will take over the role of compensation reservoirs for the whole chain.

The possibility to operate as a run-of-river power plant will increase with the degree of completion of the chain. At the beginning, the power plants will mostly operate as run-of-river plants. By increasing the number of stages, the capability of the chain for peak energy production and power regulation will also increase. The operation will depend on the conditions and operational limits required by other beneficiaries.

4. Effects of Power Plant construction on the Sava River

Apart from the energy benefits, also other useful effects of hydro power plant construction on the Middle Sava River can be expected: greater protection against detrimental action of water; improved water supply; waste water treatment; regulation of transport and energy networks; economic and social effects; opportunities for development of recreational facilities and so on.

It is also necessary to emphasize the role of the hydro power plants in ensuring stable run-off conditions of the watercourse [2]. Encroachments on the environment may not in any way deteriorate the conditions of the watercourse and every solution selected shall represent a new value of higher quality in space. In planning the power plants on the Middle Sava River, special protective measures have to be taken, such as: flood protection, protection of groundwater quality and quantity, protection of water regime and river dynamics. At the same time, it is also necessary to ensure passage of aquatic organisms, biodiversity conservation and autochthonous habitats as well as preservation and protection of valuable natural features in the area.

Construction of hydro power plants undoubtedly means positive national economic effects, as the Slovenian industry is capable to ensure the greatest part of the investment, both by equipment supply and project implementation as well. Investments into the construction of hydro power plants will help to maintain the competitiveness and the capabilities, particularly in the field of civil engineering and building industry, mechanical industry and, last but not least, electromechanical equipment industry. Investments into the hydro power plants on the Middle Sava River will enhance the conditions for creation of jobs in the region and industrial production for a period of more than ten years. Storage reservoirs as multipurpose projects will solve the problems of drinking water supply to the population, irrigation of agricultural land, protection against detrimental actions of water and provision of aquatic and riparian environments for tourism and recreational activities. The arrangements in the influential area of the storage reservoirs also include the development of public and local infrastructure in the region, such as: regulation of tributaries, development of transport infrastructure, passage of aquatic organisms through dam structures, development of public utility and energy infrastructures as well as construction of waste water treatment plants.

References

[1] Kryžanowski A 1999 *Hydrotechnical Concretes – Proceedings*, IRMA, Ljubljana, 77
[2] Porenta M 2002 *Hydro energy, Expert Groundwork* Ljubljana
[3] RS 2004 *Official Gazette of the RS* **121/04** Ljubljana

IOP Conf. Series: Earth and Environmental Science **4** (2008) 012034 doi:10.1088/1755-1307/4/1/012034

Temperature and oxygen conditions peculiarities of the cooling ponds of nuclear and thermal power plants

N Mostova

Ukrainian Hydrometeorological Institute
Nauky 37, Kyiv, 03028, Ukraine

E-mail: nimos@uhmi.org.ua

Abstract. On materials of laboratory and expeditionary researches on the cooling pond of the Zaporizhya nuclear power plant and adjoining area of the Kachovka reservoir are found out features of temperature and oxygen regime of the cooling ponds and condition and factors, that them determine. Carried out researches testify to significant thermal loading on ecosystem of the cooling pond which determines a course of hydrobiological and physico-chemical processes. Differences of the hydrochemical regime of cooling ponds from similar reservoirs in natural conditions are caused, first of all, by features of their thermal regime. Because of constant artificial heating, character of hydrobiological, physico-chemical and hydrodynamical processes in cooling ponds are formed peculiar conditions of formation of the dissolved oxygen regime.

1. Introduction

Water temperature belongs to the universal ecological factors, which, first of all, determine a course of hydrobiological and physical-chemical processes in reservoirs. Differences of hydrochemical and hydrobiological regimes of cooling ponds from similar reservoirs with natural conditions are caused, first of all, by features of their thermal regime.

In natural conditions the temperature regime of reservoirs is determined by their geographical position (quantity of solar radiation, climatic conditions), morphometric characteristics of the reservoir, features of water masses circulation, etc. Within the limits of the reservoir, usually, is observed change of water temperature with depth, and surface layers test the influence not only seasonal, but also daily changes of solar radiation intensity, especially in the summer.

Cooling ponds of nuclear and thermal power plants belong to water objects of especial type, ecosystems that test significant thermal loading.

The temperature regime of the cooling ponds, except for natural factors, is determined by volume and temperature of waste water, that is quantity and capacity of working power units.

There can be a threat of thermal pollution at thermal discharge in the cooling pond, owing to ecological equilibrium can be broken and sanitary condition become worse. Researches have shown that at water temperature up to 26 °C is not observed harmful influence on phyto- and zooplankton. However, at water temperature 33-35 °C become weaker photosynthesis, and at 40-45 °C – absolutely stops. For the most fishes, processes of vital functions suppress at temperature 26-30 °C [1].

2. Objects of research

For the initial information are used results of long-term expeditionary researches on the cooling pond of the Zaporizhya nuclear power plant (Zaporizhya NPP) and adjoining water area of the Kachovka reservoir.

Kachovka reservoir is located in the steppe zone of Ukraine and is the last in the chain of the Dnipro Basin.

Zaporizhya NPP – is the biggest nuclear power plant in Ukraine and Europe. Consisting of six power units producing 1000 MW unit^{-1}, it is situated on the left bank of the Kachovka reservoir. The cooling pond surface is 9.5 km^2 and water volume is 43×10^6 m^3, average depth is 4-5 m, maximum – 16 m. Water is constantly fed to the cooling pond from Kachovka reservoir for compensating water losses owing to evaporation.

3. Results and discussion

3.1. *Temperature regime peculiarities*

In the cooling pond because of constant artificial heating, there are especial conditions of temperature regime formation. Comparing to 1996-2005 water temperature in the coldest zone of the cooling pond of the NPP and adjoining area of the Kachovka reservoir, it is evident, that in the cooling pond water temperature on 5-18 degrees above compared with natural conditions during all year (figure 1).

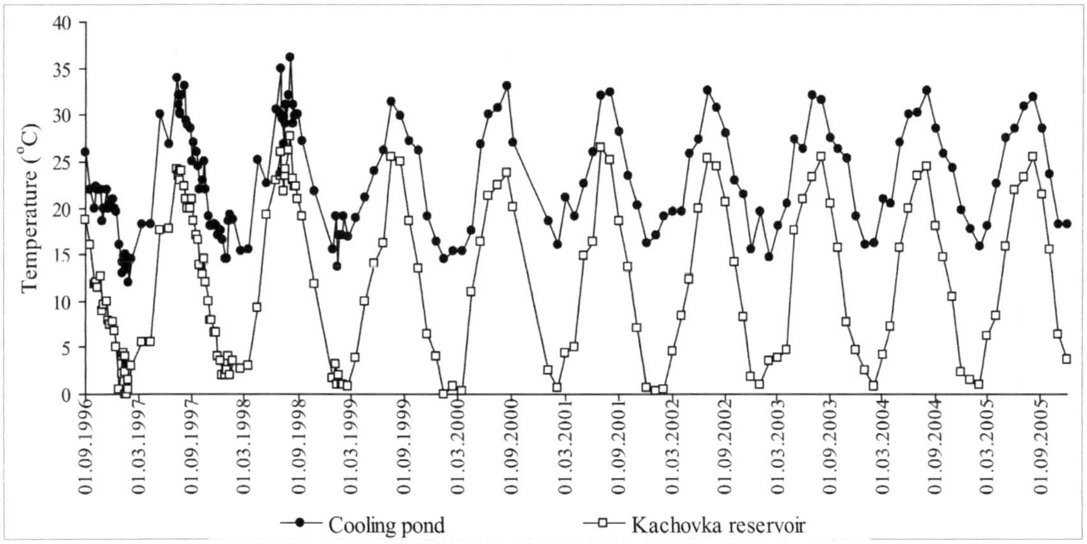

Figure 1. Water temperature dynamics of the cooling pond of the Zaporizhya NPP and Kachovka reservoir in the years 1996 to 2005.

The average data testify to substantial increase of water temperature in the cooling pond of Zaporizhya NPP. So, average long-term (1996-2005) water temperature in the cooling pond makes 23.1 °C, and in Kachovka reservoir – 12.2 °C.

When we compare with monthly average values of water temperature in the researched water objects, it is obviously, that in natural conditions water temperature on the average is less on 13 degrees (table 1).

Researching daily course of water temperature of the cooling pond it is marked, that during the summer period at high air temperature (28-32 °C) the maximal heating of surface layers of the reservoir is observed about 20 hours. The water temperature of surface layer from 6 a.m. till 20 p.m.

can increase on 4.0 °C while in the Kachovka reservoir this value changes maximum on 1.8-2.0 °C [2].

Table 1. Monthly average water temperature of the cooling pond of the Zaporizhya NPP and Kachovka reservoir.

Period	Water temperature, °C		Temperature difference (Δ t), °C
	cooling pond	Kachovka reservoir	
January	13.9	2.1	11.8
February	14.5	3.0	11.5
March	18.2	5.5	12.7
April	18.2	5.5	12.7
May	30.0	17.8	12.2
June	26.8	17.8	9.0
July	31.5	23.7	7.8
August	29.9	20.8	9.1
September	24.9	16.5	8.4
October	22.3	12.3	10.0
November	18.0	7.3	10.7
December	16.0	2.8	13.2

3.2. *Peculiarities of the dissolved oxygen regime*

The content of the dissolved oxygen in water is one of the major hydrochemical parameters, which determine physico-chemical conditions of natural waters, intensity of renewing and oxidizing processes.

The regime of the dissolved oxygen in the cooling ponds, as well as in natural reservoirs, is formed, first of all, according to the temperature regime, character of hydrobiological, physico-chemical and hydrodynamical processes.

For the cooling pond is characteristic that after water passing through heat exchangers its temperature rises almost on 10 °C, as a result in the discharge channels there is water oversaturating by oxygen and its evasion to the atmosphere [1].

Because of the permanent artificial warming up in the cooling pond intensify physico-chemical and hydrobiological processes of different direction. Water of the cooling pond is enriched by oxygen owing to water aeration and strengthening of photosynthesis by phytoplankton, and during intensification of biochemical processes organic substances of destruction and oxidation-reduction processes, its content diminishes.

Weather is also the determinative condition of the oxygen regime forming. It is necessary to note, that during the period of high water and air temperatures, at calm, windless weather and as a result – absence of wind-wave interfusion of water masses – in the cooling pond is observed clear vertical stratification of the oxygen content. At saturation of upper water layer by oxygen (up to 220 %), that is during the concentration of dissolved oxygen to 17 mg/dm^3, beginning from a 4-meter depth, is observed the sudden diminishing of its content to 1.5 mg/dm^3 (19 % saturation) and complete (or almost complete) absence on the 6-meter depth (table 2).

Table 2. Dynamics in depth of water temperature and dissolved oxygen in the cooling pond of the Zaporizhya NPP in July.

Depth (m)	Water temperature (°C)	Dissolved oxygen	
		mg/dm^3	% saturation
0	30.5	17.1	219
1	30.0	16.8	210
2	29.5	13.2	172

3	29.5	8.1	106
4	29.5	1.5	19
5	29.5	0.8	9
6	29.0	0.3	4
7	29.0	0.1	1
8	29.0	0.1	1
9	28.5	0.1	1
10	28.5	0	0
11	28.5	0	0
12	28.3	0	0
13	28.0	0	0
14	28.0	0	0

During the autumn-winter period such phenomena was not observed, that, in our opinion, arise from decrease in water temperature and substantial increase thus oxygen solubility.

Consequently, in the spring-summer period in the cooling pond is formed natural water layer of significant capacity, which does not interfuse with a superficial circulation stream. On a temperature gradient, it is impossible to notice, as the thermocline is absent in the normal sense.

3.3. *Thermal pollution estimation of adjoining area of the Kachovka reservoir*

The researches lead on water objects near Zaporizhya nuclear power plant have shown, that at use on the nuclear power plant of the combined system of power units cooling, influences of the nuclear power plant on thermal pollution of adjoining area of the Kachovka reservoir is not revealed. Thermally enriched water that come in from the waste channel, are completely localized in the cooling pond.

Since 1994 is carried out periodic water exchange of the cooling pond with Kachovka reservoir, so-called "expulsion".

At operation of the cooling pond in the regime of planned "expulsion" the aureole of temperature pollution of water area of Kachovka reservoir with slightly raised temperature (over 2 °C), compared with a natural background, makes less than 1.0 km^2.

Depending on speed and direction of wind aureole of pollution can change a configuration, however total area of Kachovka reservoir with slightly raised temperature in any case makes less than 1.0 km^2.

4. Conclusions

Carried out researches testify to significant thermal loading on ecosystem of the cooling pond which determines a course of hydrobiological and physico-chemical processes.

Differences of the hydrochemical regime of cooling ponds from similar reservoirs in natural conditions are caused, first of all, by features of their thermal regime.

Because of constant artificial heating, character of hydrobiological, physico-chemical and hydrodynamical processes in cooling ponds are formed peculiar conditions of formation of the dissolved oxygen regime.

References

[1] Romas' M 2002 *Hydrochemistry of water objects of nuclear and thermal power engineering* (Kyiv: Kyiv State University Press)

[2] Mostova N 2006 Temperature and oxygen regime peculiarities of the reservoirs in conditions of thermal loading Proc. *Ukrainian Hydrometeorological Institute* **255** 117–26

IOP Conf. Series: Earth and Environmental Science **4** (2008) 012035 doi:10.1088/1755-1307/4/1/012035

River archaeology – a new tool for historical hydrology

Attila J. Tóth (PhD)

National Office of Cultural Heritage, Budapest, Hungary

E-mail: attila.toth@koh.hu

Abstract. River archaeology consists of underwater research on the rivers themselves. It is also concerned with the archaeology of the valleys/floodplains with special attention to human-environmental interactions (reconstructing landscape, the environment, economy and society from material culture and traces of human impact on their surroundings). As historical hydrology is concerned with similar questions, from the hydrologist's point of view, the combination of different approaches offers the possibilities for fruitful cooperation for both disciplines. The intent of this paper is to present the type, nature and limitations of this part of the archaeological record through recent work in the Drava River basin.
Keywords: river archaeology, historical hydrology, underwater survey, GIS, Drava River

1. Introduction

River archaeology is a new field of research [1,2,3] consisting of underwater research on the rivers themselves, and also the archaeology of the valleys/floodplains with special attention to human-environmental interactions (reconstructing the cultural landscape, the environment, economy and society based on material culture and traces of human impacts on their surroundings). As historical hydrology is concerned with similar questions from the hydrologist's point of view [4] this combination of different approaches offers possibilities for fruitful cooperation between both disciplines. Geology (geomorphology, sedimentology, geophysics), dendrochronology etc. should be taken into consideration as partner disciplines [5, 6, 7, 8, 9, 10].

2. The nature of the data and time perspective

Due to extreme flooding events around the turn of the millennium, the study of paleofloods dominates research themes [11, 4, 12]. Historical hydrology may take advantage of the accurate instrumental observations from the 19th-20th c., as well as pre-instrumental written sources of variable nature and accuracy. Extreme flood events have been registered since antiquity, but continuous, more accurate records only begin from the end of the Middle Ages and beginning of the Early Modern Age (15th c.).

Underwater archaeological research and the topographical survey of river valleys offer a wider perspective over the last 8,000 -10,000 years - since the beginnings of agriculture and the intensification of human impacts upon their natural surroundings. The archaeological record could be divided into direct and indirect information. Direct information consists of traces of floods (such as a layer of sterile river sediment lying over archaeological features) or erosion. The Governor's palace, seat of the governor of the Roman province of *Pannonia Inferior* was discovered in the mid-19th c. on an island (Hajógyári-sziget) in the Danube in Budapest, Hungary. The complex had been constructed for the first governor of the province (Hadrian, the later emperor) in the early AD 2nd century. The building was abandoned sometime around the second half of the AD 3rd century It has been suggested that this abandonment was related to the rising of the Danube water level [13]. The Roman *limes*, the river frontier consisting of a system of watch towers, fortifications, forts, and bridge heads, was also influenced by Danube water levels. A bridge head at Bölcske (Tolna County, south Hungary) formerly sitting on the right bank of the river was found in what is now the middle of the river as the result of the movement of the river bed towards the west since the AD 4th century [14]. Bridges, ships, ports, fords and remains of fisheries also represent direct evidence for the nature and use of the rivers.

Indirect information derives from topographical survey: the geographical position of settlements and changes in settlement patterns reflect conditions in the river environment, the average level of floodings, and changes in river beds. Rivers themselves functioned as trade routes, boundaries, sources of energy and water etc.

3. Underwater archaeology and river archaeology

Underwater archaeology is treated as a synonym for maritime archaeology not only by the public, but even by many archaeologists. This situation is due to the hostile river environment. Strong currents and low visibility makes diving dangerous, and survey as well as documentation a time-consuming activity. Prospecting and photographing in the "Great Blue" is much more spectacular than in the "Great Brown" and easier to get sponsors for as well. It is largely forgotten that unlike the marine environment, organic material is much better conserved in sweet water (rivers and lakes).

Archaeological mapping has become one of the most urgent tasks for Hungarian archaeology. Although ca. 55,000 sites are known in the country, it is estimated that there are actually another 15,000-20,000 sites awaiting discovery. In the case of underwater sites and other riverbank sites (covered by dense forest or bush) the situation is even more difficult: probably 90-95% of the underwater sites has not yet been discovered! Naturally, only known sites can be studied and protected. The National Office of Cultural Heritage in Hungary has begun a topographic project in the Drava River valley (in southern Hungary). Dozens of new sites have been identified on the high bank and on the floodplain. These sites date frpm prehistoric times to Early Modern times and include earthwork-fortifications and an outstanding logboat "cemetery" discovered in 2005 [15, 16].

3.1. The Drávatamási logboat site and its environmental context

Drávatamási is a small village on the north bank of the Drava in southwestern Hungary (Somogy County). The river is approximately 100 m wide at this point with an average depth of between 4-7 m, the current is strong, and the visibility is good (up to 1 m and sometimes even better). First investigations at this site was carried out by the Dráva Museum of Barcs in 1992, when the low water level made it possible to photograph some dugouts lying beneath the water. An Ottoman period (17th c.) copper jug was also collected from a ship. Based on these discoveries, we were able to organise an international expedition including Austrian, Croatian, French and Hungarian archaeologists and divers.

Between 2005 and 2007, three surveys were organised by the National Office of Cultural Heritage and the Hungarian Archaeological and Art-historian Society. Altogether 30 very well preserved dugouts were found in a limited area of the river. This is the largest logboat cemetery in Europe, and one of the largest in the world! The boats were carved from a single oak trunk. They are very big (10-13 m long, 1.2-1.5 m wide). All the boats share the same characteristics of a round cross-section and absence of interior reinforcements. Fragments of Ottoman period ceramics as well as some earlier Arpad period 11th-13th c. ceramics also came to light close by. We concluded that the site extends to both the Croatian and the Hungarian sides of the river. One of these so-called monoxyls was documented in detail by the French team. We have also collected wooden samples for dendrochronological dating. One of the great surprises of this survey was the discovery of an almost intact copper cauldron which could be dated to the Ottoman period like the aforementioned jug. The cauldron was found in the immediate vicinity of two of our logboats.

All of the boats have pointed, flat bows. The aft however was straight and open. A notch with treenails is visible in this part suggesting that a wooden plank had been fixed there to close off the aft.

There are oval or circular holes along the edges of the side-wall that probably held some attachments. A hexagonal axel from a watermill wheel was also found together with two large pieces of timber with attachment holes. The morphological homogeneity and the traces of attachment suggest that the monoxyls were not individual objects that had been deposited at this spot from time to time but were part of a larger structure. Although dendrochronology did not provide us with a precise date due to the lack of a "master chronology" for the Drava River, it still revealed that all the trees had been cut down during the same period.

The last, 2007 campaign presented us with new discoveries. We also found a millstone under one of our logboats. The most exciting finds included two planks found some meters upstream of logboat 20. There are hand made iron nails and iron sintels on the planks, which attest that they derive from a plank-built wreck, probably covered by the sand dune. The way the iron nails were used was quite unusual. They connected the planks together in a manner similar to the ancient "shell first" concept. There are only a few traces of the nails that connected the planks with the timbers. The only known parällel comes from Altenwörth, in the region of the Austrian Danube. That ship was dated to the early 19th c. The elongated oval plate of the sintels is identical with the figures from a catalogue of ironworks published in Gräz (Austria) in 1825. The Austrian catalogue referred to this kind of sintel as "Turkish-sintels". The main purpose of the 2008 research has been to excavate and document this strange wreck.

The question of interpretation has brought paleo-environmental problems into the focus. The study of 18th-19th c. maps, historical texts and the observations mentioned above give way to two hypotheses. The first one is the water mill hypothesis. It is based on the watermill axel the millstone found at the site, and ethnographic descriptions mentioning floating mills set on boats and logboats commonly used on the Drava. An 18th c. map shows a group of such mills just 1 km upstream from our site. Ottoman period sources mention a number of ship-mills in the environment of Barcs (nearest town to Drávatamási). This hypothesis raises the question of site formation: 30 logboats (or more) represent a large number of floating mills, destroyed at the same time in the same area. These mills were usually based in suitable places in groups, called mill-villages. The site may have been such a village. A possible hypothesis explaning the destruction is a flood which broke away the ropes of the mills.

The second hypothesis is supported by G. Petthő (a 17th c. historian), who described a Hungarian surprise attack against two small fortifications and a bridge at Drávatamási in 1603. The Ottoman period find material, a timber with rectangular holes (for attachment of the boats) and the absence of closing-planks in the aft of the boats (were they destroyed?) may also support this hypothesis. There is also a possible explanation for the question of watermill elements: the Turks may have confiscated all floatable objects, including boats and mills, on the conquered territory to build a military bridge.

Archive maps show that the river itself has changed a lot since the 18th century, mostly due to river regulations that regularized the bed of the river by cutting off meanders.. Unfortunately, we do not have accurate maps for the Ottoman period or the period before although the river environment should have changed since those times. The site needs to be interpreted in its original context, something that requires a broader, interdisciplinary perspective.

Figure 1. The copper cauldron *in situ.* (Photo: K. Zubčič, Croatian Conservation Institute)

3.2. First river archaeological observations

Parallel to the underwater survey we organised a short field survey east of Drávatamási and within a large sample area on the high bank of the Dráva west of Vízvár. Archaeological information from this border zone between former Yugoslavia and Hungary is scarce due to former border tensions related to the political situation during the Cold War situation when the river and its banks were not accesible to scientists. These preliminary surveys resulted in the discovery of a large number of new sites from different periods. An abundant source of information has been collected from archive aerial photos. The Museum and Institute of Military History has a large collection of archive cartographical ortophotos from the 1950's to early 1980's. A series of photos from the Drava area has been scanned and some of them georeferred. A large number of meanders and river channels are still visible (Figure 2.). Many of them could be identified on 18th century maps, although some were not shown because they were already inactive and filled up at the time of mapping.

IOP Conf. Series: Earth and Environmental Science **4** (2008) 012035 doi:10.1088/1755-1307/4/1/012035

Figure 2. Mosaic of georefered aerial photos of the Dráva valley, near the logboat site. Archaeological sites marked with in pink. Meanders (oxbow lakes) cut during the 19th c. are clearly visible. By Péter Szőcs, National Office of Cultural Heritage

Figure 3. Topographic map with the paleo-channels (blue) and archaeological sites (pink) marked in the area of Drávatamási

Combining archaeological observations and paleo-channel data resulted in interesting paleo-hydro-environmental information. West of the logboat cemetery (id.no. 19519) an S-shaped meander was cut off during 19th century river Regulation work. Our site was however, already in the active riverbed at that time. A large east-west running channel has been identified east of our site (Figure 3, blue). We documented archaeological sites both on its north and south banks (Figure 3, pink). It is evident that the ancient Drava was the main reason for human activity there.

It has been possible to link thecreation, life and abandonment of these sites, not only to historical events, but also to the life on the river. The earliest finds come from the Late Bronze Age. Roman and Medieval finds have been also collected. The latest feature is an Ottoman period small earth-fortification, a guard-post, which served to control that stretch of the river. It may mean that at one time there was a river channel running there from the Late Bronze Age to the Ottoman period (sometime during the 17th c.) although this channel had disappeared by the end of the 18th century. The formation of the logboat site and the closing of the east-west channel occurred (archaeologically) at the same time. Relation of historical and natural events require futher research.

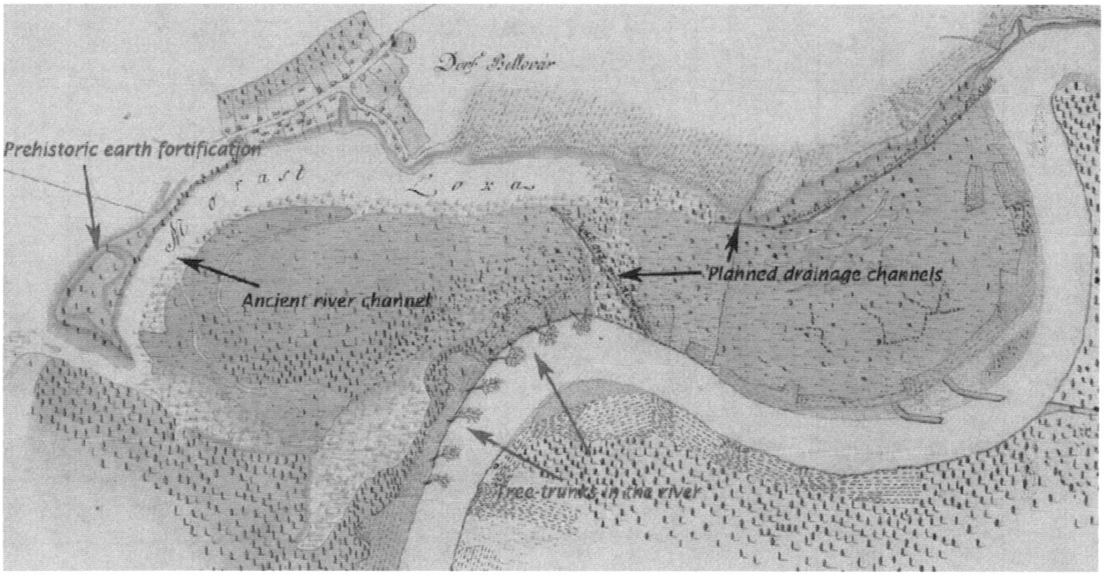

Figure 4. Archaeological and hydro-environmental features at Bélavár (Somogy County). Detail from a manuscript map,1795. Collection of the National Széchenyi Library, Budapest

In the sample area between Őrtilos and Vízvár the survey shows us that fortifications and fortified centres were built on the high (loess) bank on the north edge of the floodplain. Natural loess "peninsulas" cut by paleo-channels were preferred for fortified sites. The river is in a reasonable distance from this high bank nowadays, but surface morphology, aerial photos and in some cases archive maps testify to the former connections between the Drava and these sites. In the case of the prehistoric fortified site (re-used in Ottoman period) at Bélavár, the late 18th century map shows the last phase of the transformation of this historical-hydrological situation. Large settlement sites were also found on the low floodplain in some cases but these date to the Árpád period (11th-13th c.) which suggest that the average flood levels were low over a considerable period, or some kind of adaptation to this humid environment.

Further questions

Although a number of historical, archaeological studies have discussed the special problems connected to working in a riverine environment, even in the Danube Basin, complex, interdiscilinary studies are still awaiting. At this point, it is enough to refer to the close connections between Early and Middle

Neolithic sites and the Tisza River, running though the Great Hungarian Plain [17] and studies dealing with the history of navigation on the Danube [18] or the studies of the Drava River [19] based on images and written sources. Underwater archaeology came into the picture much later, but studies carried out on some European rivers are now accessible (Saône: [1]; Charente: [20]; Ljubljanica: [21]). The idea for and first example of studies combining underwater archaeology and the natural sciences came from France [2] within the framework of international cooperation programmes. The circle of research fields and disciplinas are broadening. Thepurpose of this paper is to inform hydrologists about this new field of research.

Proposed fields for cooperation:

- Paleochannel study (formation, evolution, characteristics).
- Question of the driving forces: extreme, violent events (floods) versus gradual changes.
- Traces of the human impacts on river environment (e.g. built elements like bridges or ports as part of the long-term artificial landscape along the Roman period Tiberis River: [22].
- River and sediment types and strategies employed by archaeological cultures to exploit the riverine environment (e.g.[6, 7]).
- Islands as special places in culture and history (Roman and Medieval towns were often founded near river islands, and the life of these islands influenced developments on the settlements).
- Land use, changes in vegetation cover, and climate change and their influence on rivers and local communities.
- GIS used to combine interdisciplinary data.

References

[1] Bonnamour L 2000 *Archéologie de la Saône* (Paris)

[2] Dumont A Ed 2006 *Archéologie des lacs et des cours d'eau* (Paris)

[3] Tóth J A 2006a River archaeology - a new field of research *New technologies in Archeology, International Symposium* Hungarian National Museum 2005 Archeometriai Műhely 2006/**1** 61-66 (http://www.ace.hu/am)

[4] Brázdil R and Kundzewicz Z W 2006 Historical hydrology – Editorial *Hydrol. Sci. J.* **51(5)** 733-8

[5] Brown A G 1997 *Alluvial geoarchaeology: floodplain archaeology and environmental change* (Cambridge)

[6] Howard A J and Macklin M G 1999 A generic geomorphological approach to archaeological interpretation and prospection in British river valleys: a guide for archaeologists investigating Holocene landscapes *Antiquity* **73** 527-41

[7] Hudson-Edwards K A, Mark G, Macklin M G and Taylor M P 1999 2000 years of sediment-borne heavy metal storage in the Yorkshire Ouse basin, NE England, UK *Hydrol. Process* . **13** 1087-102

[8] Timár G, Sümegi P and Horváth F 2005 Late Quarternary dynamics of the Tisza River: Evidence of climatic and tectonic controls *Tectophysics* **410** 97-110

[9] Tóth T, Vida R, Horvath F and Simpkin P 1997 Shallow-water single and multichannel seismic profiling in a riverine environment *The Leading Edge* **16** no. 11 1691-5

[10] Grynaeus A 2003 *Dendrochronology and Environmental History* People and Nature ed Laszlovszky J and Szabó P (Budapest) pp 175-93

[11] Benito G 2003 *Paleoflood hydrology in Europe* In: Thorndycraft et al. 2003 1-24

[12] Glaser R and Stangl H 2003 *Floods in Central Europe since 1300* In: Thorndycraft et al. 2003 93-98

[13] Kérdő K H 1997 Test excavations at the Aquincum Governor's Palace on the Óbuda Hajógyár Island *Aquincumi Füzetek* **3** 27-39

[14] Soproni S 1989 Eine Spátrömische Festung im Donaubett bei Bölcske. III. *Internationaler Kongress für archaologische Erforschung der Binnengewasser* (Zürich, 1990)

[15] Tóth J A 2006b The Drávatamási Ship Project *International Logboat Conference* North Carolina Maritime Museum (Beaufort, NC, USA)
[16] Tóth J A 2007 The HERMA and the Argonauts Programs: Practices and Problems of Heritage Mapping in River Environment (Hungary) *Web Journal on Cultural Patrimony* **2/1** 97-102. (http://www.webjournal.unior.it)
[17] Kosse K 1979 *Settlement ecology of the Körös and Linear Pottery Cultures in Hungary* (BAR IS 64, Oxford)
[18] Neweklowsky E 1952-1954-1964 *Die Schiffart und Flösserei in Raume der oberen Donau* I-III. Linz
[19] Leskoschek F 1972 Schiffahrt und Flösserei auf der Drau *Zeitschrift des Historischen Vereines für Steiermark* **63** 115-52
[20] Dumont A, Marioti J-F and Pichon M 2003 *La Charente à Taillebourg-Port d'Envaux* (France, dép. Charente-Maritime) Premiers résultats d'une prospection thématique subaquatique. *Archäologisches Korrespondenzblatt* **33 (4)** 585-96
[21] Gaspari A 2003 Archaeology of the Ljubljanica River (Slovenia): early underwater investigations and some current issues *International Journal of Nautical Archaeology* **32 (1)** 42–52
[22] Graham S 2005 Of lumberjacks and brick stamps: working with the Tiber as infrastructure ed . MacMahon A and Price J *Roman working and urban living* (Oxford) pp 106-24
[23] Newson M 1994 *Hydrology and the river environment* (Oxford)
[24] Thorndycraft V R, Benito V G, Barriendos M and Llasat M C ed 2003 Palaeofloods, Historical Floods and Climatic Variability: Applications in Flood Risk Assessment *Proceedings of the PHEFRA Worksh,(Barcelona, 16-19th October, 2002)*

IOP Conf. Series: Earth and Environmental Science **4** (2008) 012036 doi:10.1088/1755-1307/4/1/012036

The UN Convention on International Watercourses and Integrated Water Management: a bridge built

Vasiliki-Maria Tzatzaki

PhD Candidate, University of Athens, Athens, Greece

E-mail: tz_vicky@yahoo.gr

Abstract. The UN Convention on the Law of the Non Navigational Uses of International Watercourses incorporates principles regarding the management of international water resources. The most important principles are the duty of the riparian states to cooperate, not to cause significant harm, to protect the aquatic environment and to utilize the watercourses reasonably and equitably. The lack of hierarchy between these principles signifies that the necessary step for the sound management of shared natural resources is an integrated approach, which takes into account economic development, human needs and environmental protection. Moreover, the UN Convention proved to be useful for the International Court of Justice (hereinafter ICJ) in the settlement of the Gabcikovo- Nagymaros dispute between Hungary and Slovakia for the Danube River. The Court highlighted the importance of the Convention by reminding the riparian states of their obligation to abide by its principles. On the other hand, the ICJ has used the principles of the Convention in the pending case of Pulp Mills between Uruguay and Argentina. This paper is going to show that the UN Convention is an international legal framework with general guidelines in order to create regional conventions, which promotes integrated water management as a solution to the emerging challenges of international water law and potential future conflicts.

1. Introduction

Until 1997 international law had made little progress on the management of international watercourses. The management of transboundary rivers and lakes had been the topic of attention of either riparian states or regional organizations. Treaties and agreements have been concluded, signed and still in effect but they only govern the relationship between states sharing a watercourse or states that are members of an organization. These legal regimes have been regional until recently, given not only the great number of watercourses but also the unique characteristics of each of those watercourses. [1] The international arena lacked a legal text to set the guidelines for the riparian states for the management of shared rivers and lakes.

Soon the international community realized that the management of international rivers and lakes includes the priorities of different users and since most of these watercourses flow between two or more states or form boundaries, the priorities give rise to regional conflicts. [2] Therefore, in May 1997, the General Assembly (hereinafter GA) of the United Nations (hereinafter UN) adopted the Convention on the Law of the Non Navigational Uses of International Watercourses. It is the first international legal instrument on the management of transboundary rivers and lakes; it is the first time that the international community has recognized the necessity to deal with this issue and the

Convention is an attempt to create an international framework with rights and obligations that should be respected by the riparian states.

This paper is going to show that the purpose of the UN Convention is the integrated management of international watercourses. In other words, it aims to provide the states with rules that oblige them to take into account all those factors that are necessary for the management of a shared watercourse. The riparian states when concluding agreements should put under consideration geographical, hydrological, economic, social and environmental factors that are going to assist them in managing their fresh water resource in a sound and sustainable manner.

The first part of the paper is going to present the characteristics of the Convention on International Watercourses and it is going to explain how a cumulative application of the obligations set forth by the articles of the text leads to integrated water management. On the other hand, the second part of the paper is going to deal with water related international disputes before the International Court of Justice (hereinafter ICJ). Specifically, it is going to examine the Gabcikovo-Nagymaros case and the Pulp Mills case and how the ICJ has tried to settle the disputes according to the integrated approach provided by the UN Convention. The second of the disputes is still pending before the ICJ.

2. The Characteristics of the UN Convention on International Watercourses

This part of the paper is going to attempt a brief overview of the characteristics of the UN Convention on International Watercourses and the obligations that it sets to the riparian states. It is going to demonstrate that the rules of the Convention aim at creating a legal basis for the integrated management of international watercourses by the riparian states.

2.1. The UN Convention on the Law of the Non-Navigational Uses of International Watercourses

The Convention of 1997 is the result of the work of the International Law Commission and it codifies and evolves the law of international watercourses. The definition of an international watercourse, given in article 2 of the Convention, includes both surface and groundwater; this definition is in accordance with the hydrological reality and calls for the riparian states to focus on the relationship between all the parts of the system, not only the parts of the surface but also the groundwater parts, which comprise an international watercourse. Hence, any effect caused in a part of the system is going to affect the freshwater system in *toto*. [3] The articles of the UN Convention provide the states with flexible rules for the management of shared watercourses only if one or more of the states that are members to the UN Convention have not agreed otherwise. [4] For future agreements between member states the rules of the Convention are guidelines, allowing thus the states to keep their contractual flexibility. [5]

The UN Convention has seven parts: the first part is the Introduction (articles 1-4), the second part includes the General Principles (articles 5-10), the third part is the Planned Measures (articles 11-19), the fourth part refers to the Protection, Preservation and Management (articles 20-26), the fifth part comprises of the Harmful Conditions and Emergency Situations (articles 27-28), the sixth part is the Miscellaneous Provisions (articles 29-33) and the seventh part has the Final Clauses (articles 34-37). The major obligations for the riparian states that the international legal text adopts are four: the obligation to cooperate, not to cause significant harm, to utilize their freshwater resources in an equitable and reasonable manner and to protect their ecosystem.

2.2. The obligation to cooperate

The obligation of the riparian states to cooperate is incorporated in article 8 of the UN Convention and it has the title: "General Obligation to Cooperate". The article says: "Watercourse States shall cooperate on the basis of sovereign equality, territorial integrity, mutual benefit and good faith in order to attain optimal utilization and adequate protection of an international watercourse." Therefore, the obligation set by the article is general. [6]

The obligation of the states that share a freshwater resource is not standing alone in the text of the Convention. On the contrary, it is accompanied by other obligations, which are enhancing the

IOP Conf. Series: Earth and Environmental Science **4** (2008) 012036 doi:10.1088/1755-1307/4/1/012036

cooperation between riparian states and could prevent damages to the watercourse and conflicts between the users of the resource.

So, article 9 obliges the states to exchange data and information regularly in order to facilitate the management of international watercourses. In article 12 the riparian states find the obligation to notify each other for planned measures with possible adverse effects, while articles 13-16 elaborate on the reply of the notification, the period for that and the absence of reply. According to article 17, the states sharing a river, a lake or an aquifer should consult and negotiate concerning planned measures. Finally, article 33 includes a form of cooperation, which is the peaceful settlement of disputes between the riparian states.

2.3. The obligation not to cause significant harm

The second important obligation of the Convention on International Watercourses is article 7, the obligation not to cause significant harm to the other users of the resource. This rule does not express an absolute obligation. In other words, the cause of damage is not entirely forbidden. It is only forbidden to cause significant damage, while whenever such damage occurs the states should cooperate for its mitigation and should take all the necessary means for its control, as it is expressed in the second paragraph of article 7.

2.4. Equitable and reasonable utilization

In article 5 we find the cornerstone rule for the management of international watercourses; it is the rule of equitable and reasonable use and participation. The principle of equitable and reasonable utilization obliges the states to certain behaviour and at the same time provides them with a right to act. In other words, a state should use its international watercourse in an equitable way towards the rest of the riparian states, omitting to proceed in actions that could disturb this equity, while on the other hand it has the right to an equitable share of the uses and benefits of the international freshwater resource.

The equitable and reasonable use of shared rivers and lakes can be achieved more efficiently through the cooperation of the riparian states. [7] In accordance with article 5, article 6 of the Convention of 1997 defines the factors that play an important role for the equitable and reasonable use of international watercourses. These factors can be geographic, hydrographic, hydrological, climatic, ecological, social and economics, while the list provided by article 6 is not exhaustive.

2.5. The obligation to protect and preserve the ecosystems

Finally, articles 20-26 regulate the protection and the preservation of the ecosystems of international watercourses. The UN Convention follows an integrated approach, which is necessary for environmental security, as the states have the obligation to preserve and develop their natural resources. [8] Such integrated approach can be achieved only through legal regimes that regulate the ecosystem as a whole and not just the watercourses. The obligations and the rights of the riparian states must be settled according to the region and the needs of the ecosystem. [9]

The lack of hierarchy between these principles of international water law within the UN Convention signifies that the necessary step for the sound management of shared natural resources is an integrated approach that takes into account economic development, human needs and environmental protection. The path curved by the UN Convention is the most appropriate, efficient and aligned with the notion of sustainable development.

3. International Case Law on Transboundary Watercourses Disputes:

This part of the paper is going to focus on two examples of case law that regard the management of international watercourses and have been brought before the ICJ for their adjudication. The first is the case of Gabcikovo-Nagymaros between Slovakia and Hungary for one of the major European rivers, the Danube. [10] The second case is the Pulp Mills case between Argentina and Uruguay and concerns the Uruguay river and its pollution due to the construction of a paper mill on one side of the river. Contrary to the first one, the second case has not been yet adjudicated by the ICJ and it is the first environmental dispute that the Court has to settle. Both of these cases are going to demonstrate how the ICJ has tried to apply the rules of the UN Convention in order to guarantee the sustainable development of the rivers and their integrated management.

3.1. The Gabcikovo-Nagymaros case

In the Gabcikovo-Nagymaros case the Court highlighted the importance of cooperation between the nine riparian states of the Danube in order to develop commercially and economically. What is necessary to be done for the Danube is also necessary to be done for every international watercourse.[11] Moreover, the ICJ also underlined the obligation to negotiate in good faith; in other words, the Court in its decision of 25/09/2997 said that both Hungary and Slovakia had violated their international obligations and invited both states to negotiate in good faith in order to ensure the purpose of the UN Convention, taking into account the rules of international environmental law and international law on watercourses.

In this case, the ICJ enhanced the importance of adopting preventive measures for the relationship and the sustainable development of the common water resources. One of such measures is environmental impact assessment, which could be considered as one of the 'appropriate measures' of article 7 of the UN Convention in the obligation of the states to avoid causing significant harm. In reality, the conduct of environmental impact assessment allows the state that initiates the project to move the burden of proof to the states that have been affected. [12]

In the Gabcikovo-Nagymaros case the Court seems to adopt the rule of equitable and reasonable utilization, on the grounds that this is the cornerstone rule in the law of international watercourses. According to MacCaffrey, this signifies that the Court is doubtful of the priority of the obligation not to cause significant harm for the settlement of complicated issues of allocation of the uses and benefits of international sources of freshwater. [13]

Finally, the ICJ highlighted the obligation of ecological protection of international watercoursesby saying that the development of international law has strengthened the principle of community of benefits on a navigational river for its uses except navigation, as it is demonstrated in the UN Convention. In other words, the principle of equitable and reasonable utilization of an international watercourse reinforces the obligation to protect the ecosystems. [14]

3.2. The Pulp Mills case

In the recent case of Pulp Mills the obligation to negotiate is set by the ICJ. Argentina claims that Uruguay has not abided by the Statute of River Uruguay of 1975 and the obligation to negotiate. The Court, in its decision of 13/06/2006 on the provisional measures that Argentina requested said that the states should proceed with negotiations in good faith, as it is set out in the Statute of 1975 and that the Administrative Committee for River Uruguay is the appropriate forum for their realization.

Moreover, the ICJ referred to the obligation of the riparian states to notify each other on their planned measures. Argentina has initiated the proceedings before the ICJ claiming the violation of the Statute of River Uruguay of 1975, when Uruguay gave its unilateral permission for the construction of a mill near the town of Fray Bentos, without abiding by the obligations to notify and negotiate the project, as they are set out in the Statute of River Uruguay. Argentina also requested by the Court to decide and declare that Uruguay had violated the procedural obligations of the Statute and among others the obligation to prepare environmental impact assessment. As for the settlement of the dispute, in order to ground the jurisdiction of the Court, Argentina refers to the first paragraph of article 60 of the Statute of River Uruguay, which allows that any dispute that regards the interpretation and the application of the Statute, which cannot be settled with direct negotiations can be put before the ICJ by one of the two member states. This is a recent example of the way to settle a dispute that arises over a common watercourse, when at least one of the two parties –Argentina in this case- considers that the negotiations have not resulted in a solution.

Finally, as it is mentioned above, the Pulp Mills case regards the pollution of an international watercourse. Argentina requested the Court to rule on provisional measures for the suspension of the construction of the two mills on the river on behalf of Uruguay, claiming imminent danger. The Court recognized Argentina's concern for the protection of the environment and connected the present case with the Gabcikovo-Nagymaros case. However, it did not order provisional measures since, according to its opinion the facts on the present dispute have not proven that the construction of the mills threatens the ecosystem of the river and the interests of the citizens. The Court said that Argentina could not prove that the suspension of the mills will be capable of reverting or reconditioning the economic and social consequences that blames Uruguay for. In other words, it is exceptionally difficult to prove the existence of an imminent danger regarding issues of international watercourses. This is probably due to the technical issues, as well as to the legal parameters in analyzing this technical data.

4. Conclusions

This paper has made an attempt to give a brief overview of the international legal framework that governs the management of international watercourses and to present the two most important international disputes that regard transboundary freshwater resources. Its aim was to provide the reader with the necessary information on the principles governing prevailing in international freshwater law and the attempts of the international community to set a regime for sound, sustainable and most of all integrated management of shared rivers and lakes.

The first part of the paper presented the legal text, which has been adopted under the auspices of the UN, the Convention on the Law of the Non Navigational Uses of International Watercourses. The nature of the Convention and its parts have been presented; the obligations of the riparian states have been analyzed, while special attention have been given to the obligation of cooperation, not to cause significant harm, to utilize the watercourse in an equitable and reasonable manner and to protect the ecosystem.

The second part of the paper focused on two international disputes over watercourses; the Gabcikovo-Nagymaros case and the Pulp Mills case. Both of them regard important rivers, such as the Danube and Uruguay. The ICJ tried to apply the principles of the UN Convention and to illuminate their importance for the management of the rivers. Without any doubt, the Court ruled –in the first case with a final judgment and in the second with a judgment on provisional measures- that riparian states should manage their international rivers in an integrated way.

Therefore, according to the international legal text and to international case law, the riparian states should coordinate for the management of shared freshwater resources. They should cooperate, take into consideration the interests of the others, apply equity and preserve the ecosystems. The bridge has been built for the integrated management of our valuable freshwater resources, as the only path for their protection.

References

[1] Butts K H 1997 Parameters, US Army War College Quarterly 66
[2] Benvenisti E 1996 *American Journal of International Law* **90** 384
[3] MacCaffrey S C and Sinjela M 1998 *American Journal of International Law* **92** 97
[4] Hey E 1998 *RECIEL* **7/3** 291
[5] Calfisch L *1998 World Bank Technical Paper* **414** 1
[6] MacCaffrey S C 2000 *Journal of Land Resources and Environmental Law* **20** 57
[7] MacCaffrey S C 2000 *Journal of Land Resources and Environmental Law* **20** 57
[8] Dupuy R J 1991 *Colorado Journal of International Environmental Law & Policy* **2** 201
[9] Brunnee J and Toope S J 1997 *American Journal of International Law* **91** 26
[10] Liska M B 1995 *The Peaceful Management of Transboundary Resource* 175
[11] MacCaffrey S C 2001 *The Law of International Watercourses- Non-Navigational Uses* 400
[12] Farrajota M M 2005 *Water Resources and International Law* 281
[13] MacCaffrey S C 2001 *The Law of International Watercourses- Non-Navigational Uses* 356
[14] MacCaffrey S C 2001 The Law of International Watercourses- Non-Navigational Uses 392

IOP Conf. Series: Earth and Environmental Science **4** (2008) 012037 doi:10.1088/1755-1307/4/1/012037

Public or Private Water Management: Experience from different European Countries

Johann Wackerbauer

Ifo Institute for Economic Research, Poschingerstraße 5, 81679 Munich, Germany

E-mail: wackerbauer@ifo.de

Abstract. Faced with liberalisation proposals and an increasing internationalisation of water resource management, the question arises as to how a change of the regulatory framework would affect the market structure and the supply conditions in this area. While the term "privatisation" relates to the ownership structure of the providers, the term "liberalisation" implies extensive free market ideas. Privatisation involves the outsourcing of public services from the public authorities to a privately organised organisation. Through this, however, nothing needs to change in terms of the market or the intensity of competition for the commodity in question. Within the framework of privatisation it can also occur that the public monopoly is only transferred to a private monopoly. The term "liberalisation" in addition refers to the basic regulatory constraints: liberalisation signifies the cessation of limitations to competition and supply monopolies, and open competition between several suppliers for the consumers. In the EU-15, the only country where the provision of operational services in the water supply has been totally passed to the private sector is the United Kingdom, but this is only true for England and Wales. Another singular case is France, where there is a mix of mainly private operating companies and municipalities which have divided the regional supply areas among themselves. In six other EU-15 countries where some privatisation took place, either the municipalities or (majority) publicly owned companies are controlling water supply. In the remaining seven countries, the water supply is organised by municipality companies only. In an international comparison, there are three basic models for the regulation of natural monopolies in the public water supply: the Anglo-Saxon, the French and the German model. The delimitation between supervisory bodies and operations in the water supply is strongest in the first model and weakest in the last. This has led to three basic types of privatisation: "full privatisation", "privatisation through delegation" and "privatisation with regulation by the supervisory bodies". These have led to three clearly distinguishable forms of competition: substitute competition simulated by the regulation authorities between private supply-enterprises, competition between private operators for the right to the temporary provision of water supplies, and competition in the product and service markets in the provision of water. In this paper, these different forms of privatisation and liberalisation are described and their impact on competition and market structures is analysed.
Keywords: water management, natural monopoly, privatisation, liberalisation.

1. Introduction

While the term "privatization" relates to the ownership structure of the providers, the term "liberalization" implies extensive free market ideas, especially the removal of barriers to entry the markets for water services, which are supplied by regional monopolies up to now. Privatization means the outsourcing of public services from the public authorities to a privately organized enterprise [1]. This, however, does not necessarily mean a change in the shape of the market or the intensity of competition for the commodity in question. Within the framework of the privatization, it is also possible that the public monopoly is only transferred to a private monopoly.

In the supply of drinking water, the pipe network represents a natural monopoly but not the production of drinking water. As drinking water is provided in different qualities, this is not a homogenous commodity such as, for example, electricity. Operation of the network and production of drinking water can be separated from one another only with difficulty. The high fixed-cost component in the supply of water makes the laying of parallel networks by the competing bidder unprofitable; here we have the classical case of a natural monopoly. This is characterized by subadditivity (i.e. a monopolist can supply the relevant market more cost-effectively than two or more companies) as well as by the irreversibility of investments (so-called "sunk costs"). With the presence of "sunk costs" free entry into and departure from the market are not possible, the relevant market is then no contestable market in the sense of the theory of "contestable markets"[2]. In the case of a natural monopoly, the alternative to competition in the market would be competition for the market, for example by licence bidding.

Therefore, competition in the supply of drinking water is hard to introduce, as the transport of water from competing providers is problematic since it involves the toleration of the mixing of various qualities of water. Operation of the network and production of drinking water are therefore difficult to separate. The public water supply in Germany, in contrast to other infrastructure areas, is thus still an exception in terms of competition law. Despite isolated privatization of municipal water supply companies, no competition in the sense of a liberalization of the market exists.

2. Market Structures

Within the former 15 countries of the European Union, only in France and England/Wales is the majority of inhabitants supplied by private water companies (Figure 1, Population served by type of operator (ownership) in Europe, 2003). In the other EU-15 countries as well as in Switzerland, water supply is predominantly public. Private ownership is also considerably important in Denmark, Italy and Spain. In terms of size (Figure 2, Population served by type of operator (size) in Europe, 2003), in the majority of countries the population is served by either a small (less than 100,000 customers) or a medium sized (100,000 to 10 million customers) operator. Only in France are large operators (more than 10 million customers) dominant, which can be seen as a result of the local model of delegated management to private companies. Apart from France, large operators are also present in the United Kingdom, and in Germany. In England and Wales exclusively private, in most cases medium sized suppliers are active, in Germany contrarily mainly public and small to medium sized suppliers.

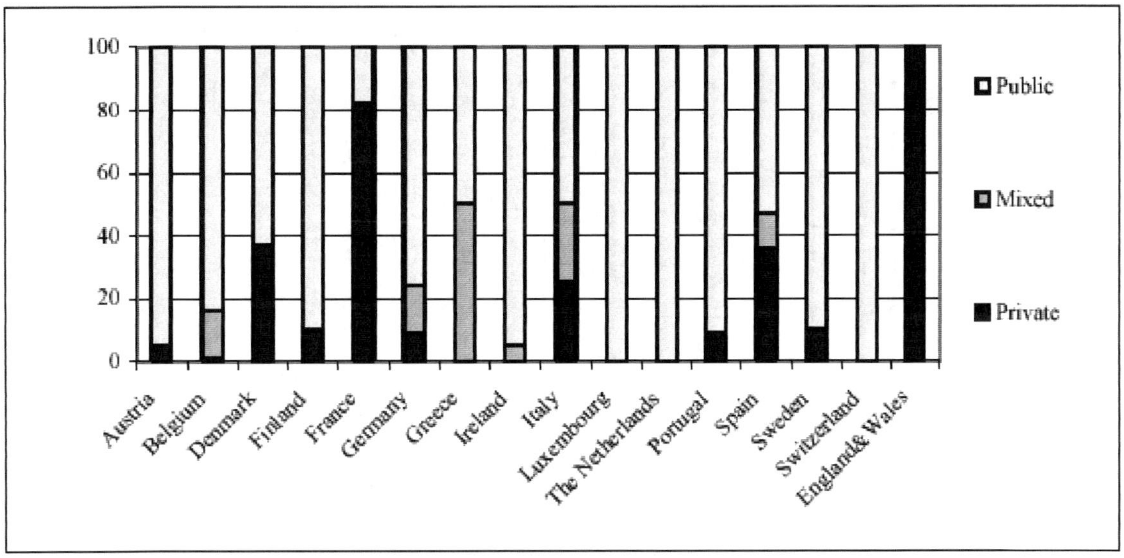

Figure1. Population served by type of operator (ownership) in Europe, 2003 [3]

The international markets for water and waste water services are dominated by the French companies Veolia, serving 108 million people, and Suez Environnement, serving 80 million people with drinking water and/or waste water services (Figure 3, Number of people served by water multinationals). The next largest company is the British Thames Water with 56 million customers, which was acquired by the German RWE Group in 2001 but sold again at the end of 2006 to Kemble Water Limited, a consortium led by the Australian Macquarie's European Infrastructure Funds [4]. With the takeover of Thames Water, RWE was the only German company that succeeded in catching up with the global players on the international water markets. Next is the American company United Utilities/US Water with 27 million customers, followed by other French and British companies. The German companies Gelsenwasser and Remondis Aqua, serving 6.5 million customers and 4.5 million customers, respectively, are lagging behind at positions nine and ten. With the takeover of Thames Water, RWE has temporarily advanced into the foremost group of global players in the water market.

3. Three Basic Models for the Regulation of Natural Monopolies in the Public Water Supply

A liberalization of the water supply can take place in different ways; the concrete basic features depend on which regulation the market for drinking water is or should be subjected to and in what way and to what extent the private sector is involved into the organization of the water supply. In the EU-15, the only country where the provision of operational services in the water supply has been totally passed to the private sector is the United Kingdom, and more specifically this is only true for England and Wales. In Scotland and Northern Ireland, independent, publicly owned companies provide the water services. Another singular case is France, where there is a mix of mainly private operating companies and municipalities that have divided the regional supply areas among themselves. In other countries like Germany, Greece, Italy, the Netherlands, Portugal and Spain either the municipalities or (majority) publicly owned companies control the water supply. In the remaining seven EU-15 countries, the water supply is organized by municipality companies only [5]. In the following, the two unique forms of privatization in France and England/Wales as well as the German method of privatization as an example of the interaction of municipalities and majority publicly owned water companies are discussed, especially with regard to the corresponding effects on competition and market structures.

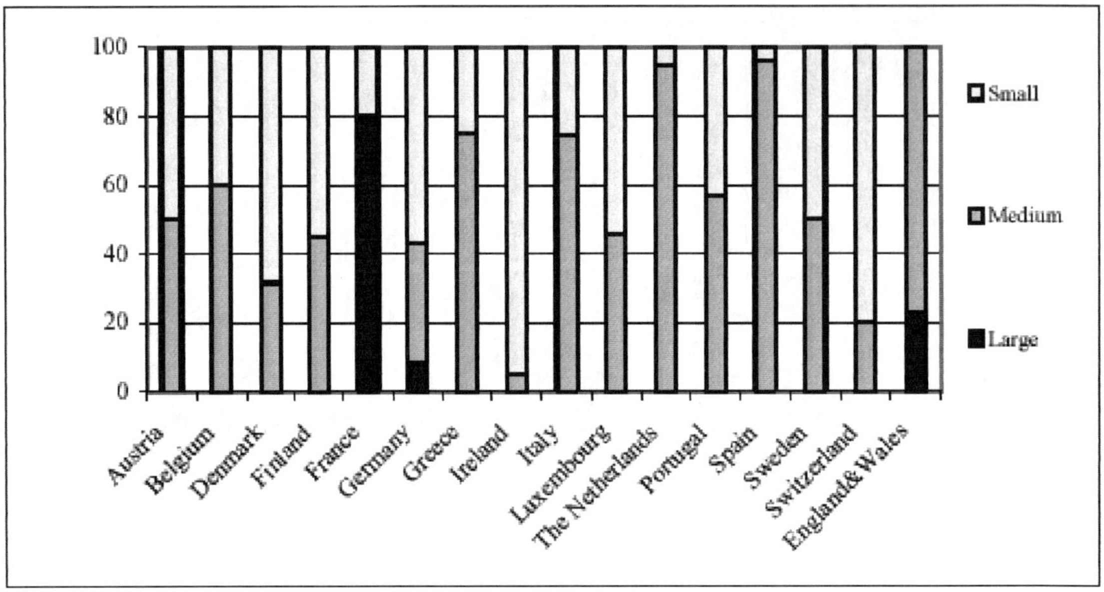

Figure 2. Population served by type of operator (size) in Europe, 2003 [3]

3.1. England and Wales: Full privatisation

With this form of privatization, which is to be found in England and Wales, publicly operated monopolies are transferred as a whole to a private enterprise oriented provider. Both ownership of the infrastructure and operation of the supply network stay in one hand, but it is no more the public sector but the private sector that is responsible. Therefore, here we speak of a "full privatization". In this case, a sale of the operator firms, including all tangible assets (such as, for example, pipelines, wastewater treatment plants and water catchment systems) to private investors takes place. In England and Wales, ten water service companies have been created in this manner, which offer the supply of drinking water and the disposal of wastewater from one hand, and whose shares have been sold publicly (in Scotland and Northern Ireland water supply and wastewater disposal on the other hand remain with the public authorities). In addition, there were some 30 water supply companies which avoided earlier nationalization and are now subject to the same regulating system as the recently privatized companies. The regulation system follows the "specialized regulation": it consists of one each of separate, independent advisory authorities for the drawing of water and discharge of wastewater, for the quality of drinking water and for water prices and supply conditions:

- the Environment Agency monitors the water quality of rivers and waters used for swimming as well as the environmental effects of the company activities.
- the Drinking Water Inspectorate is concerned with the assurance of the quality of drinking water.
- the Office of Water Services (OFWAT) sets the prices within a defined minimum and maximum whereby the performance of the individual providers is evaluated.

The price regulation allows the companies to increase their average prices per year by a factor of RPI + K, whereby RPI is the Retail Price Index and K the additional costs represented through the environmentally- and quality-related conditions. The regulating authorities set the prices such that an efficient company can expect a fair rate of return on its original business capital [6].

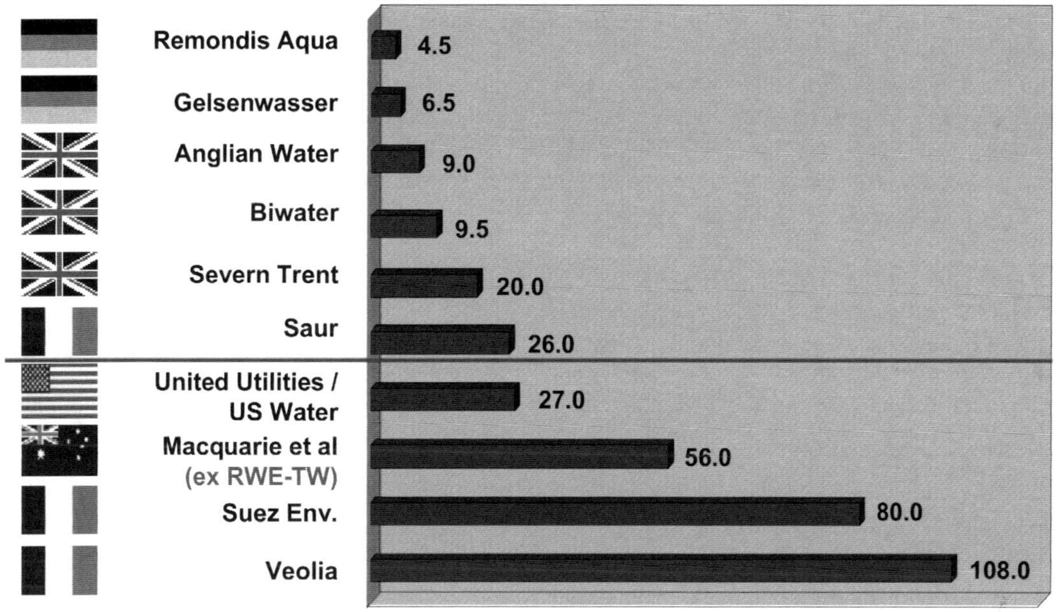

Figure 3. Number of people served by water multinationals (BMWA 2006, [7])

3.2. France: Privatisation through delegation

Under this, one understands the temporarily limited transfer of the responsibility for the operation of water networks to private operators as it is practiced in France. In this case the responsibility for the water supply (as for the disposal of wastewater) lies fundamentally with the municipalities, which however, can choose between direct management or contract management, leaving the provision of services to private companies. In the latter case, the production of goods and services are put out to tender by the municipalities; the facilities for the supply of water on the other hand, remain the property of a public body. Ownership of the infrastructure and operation of the network are separated in this case. In the contracts between the municipalities and the companies, that won the contract, it is laid down which costs the private side may include, as a maximum, in the bill. Three different forms of contracts are to be found [8]:

- the franchise agreement: here the private operator takes on the costs for new investment; the duration of the agreement is 20 to 30 years.
- the lease agreement: here the private operator does not bear the costs for new investment; the duration of the agreement is 10 to 15 years.
- the operating agreement: here only partial performances are transferred to the private side; the duration of the agreement is 6 to 10 years.

Thus the operation of existing water systems can be transferred to a private company for a relatively short period, whereby the systems as a rule remain municipal property. On the other hand, the construction and operation of new systems can be transferred to private companies, and at the end of the period the property is transferred to the municipality. Under a franchise agreement the franchise holder builds, finances and operates certain plants for the agreed period. He receives a contractual remuneration, as a rule calculated according to cubic metre of water or wastewater, respectively [6]. As opposed to the Anglo-Saxon privatization model, the tangible assets with the French privatization model remain in the ownership of the public authorities, that is the state and the municipalities or departments; only the operating responsibility is delegated to the private side if the municipality chooses contract management. In this case the operating licences are awarded by means of a bidding process.

In the six river catchment areas that were formed through the First National Water Law from 1964, two bodies in each case regulate the water supply:

- the Comité de Bassin (Committee of the Catchment Area) and
- the Agence de l'Eau (Water Agency)

The amount of the water tariffs are laid down by the Comité de Bassin. In this body, which represents a type of regional "water parliament", the state, the regions, the departments and communes, as well as the water and surface water users are represented. The setting of the objectives and priorities for the various measures are laid down there in a water management plan. Parallel to the Agence de l'Eau is a state public body that carries out the water management measures, raises charges for the water usage and water pollution as well as allocating benefits for investments and costs for treatment operations [9].

3.3. Germany: Partial privatisation with regulation by the supervisory bodies

In Germany the privatization debate is to be viewed against the background of a traditionally strong municipal administration; with the privatization of the water supply in Germany, as opposed to France and Great Britain, there is only one legal option open; in no way a national action [10]. Ownership of the infrastructure and operation of the network usually, but not necessarily in each case, remain in one hand. The German privatization model prefers a regulation of the privatized concern via its supervisory bodies. By the sending of representatives of the public authorities into this supervisory bodies, influence on the business policy of the water provider is exercised; the fixing of prices takes place according to the cost-covering principle. There are basically two different forms of this type of privatization as well as one mixed form to be differentiated:

Formal privatization or organizational privatization: In this case the task of the water supply remains with the previous administrator; only the operating agency is transformed into a business form under private law, for example, with the transformation from a municipal department or a semi-autonomous municipal agency into a municipal enterprise. Despite formal privatisation, public structures are maintained which, however, are to approximate to the management of public-law companies with regard to independence and flexibility.

Material privatization or functional privatization: Here the administrator delegates his tasks in favour of the private party. This relinquishment of the public inventory of tasks can be revocable or final [1]. A regulation of the privatised company takes place in both cases through the creation of supervisory boards and the naming of supervisory persons within the company [6].

Mixed form of privatization: Here one is concerned with a mixed form between formal and material privatisation in the shape of a participation of private companies in a municipal enterprise, which in Germany has become well-known as the so-called "Berlin model". With the partial privatisation of the Berlin Water Works (BWB) in 1998, a holding model was selected with which the Federal State Berlin received 50.1% of the shares in the strategic controlling holding Berlinwasser Holding Aktiengesellschaft. The remaining 49.9% of the shares in Berlinwasser Holding Aktiengesellschaft was acquired by an associated incorporated company that was established by an investor consortium. The business purpose of the holding is the control and further development of the competitive business and the control of the Berlin Water Works. The legal form of the Berlin Water Works as an incorporation under public law has remained unchanged, but the competitive businesses were spun off and were transferred into the Berlinwasser Holding Aktiengesellschaft [11].

The responsibility for water pollution control and the management of surface waters in most of the German Federal States is distributed over several levels. In the larger area states these are:

- the superior water authority (as a rule the Ministry of the Environment) with the responsibility for strategic decisions.
- the upper, higher or middle water authority which, as a rule, is assigned to the district committees or regional governments and is responsible for the regional water management planning.
- the lower water authority (cities, towns, urban and rural districts as well as water management offices) with monitoring, technical advice and executive functions.

The Federal State Working Group Water (LAWA), which was established for the harmonization of Federal State water laws, is formed from the superior water authorities. The Federal States have also formed working groups for the co-ordination in the management of river basins [10].

The municipal incorporations and municipal public utilities are typical in the German model for the operation of infrastructure systems for water supply, in the same way the inter-municipal agencies which were established specifically for these tasks. The German model functions essentially without formal, external regulation of the water prices, tariffs or returns on investment. As no private enterprise profit motive is present, only cost covering prices and public charges for the municipal water services are charged. The liberalization of the water supply in Germany remains rather halfhearted, and even in the case where the legal form of the water supply firms is transmitted from public to private law, the municipalities keep a substantial influence on the strategic decisions by holding at least a 50.1% majority in the new firms under private law, a construction which is summarized under the term "public-private partnership".

4. Effects on competition

The success of French water companies suggests that the clear separation of (public) infrastructure ownership and (private) operation of the water supply network encourages international competitiveness, in particular if the companies have a long experience with the practise of franchise bidding. In France private water supply companies were established already in the 19th century: in 1853 the Société Générale des Eaux and in 1880 the Lyonnaise des Eaux. In 1933 the third largest group, the Société d'Amenagement Urbain et Rural (SAUR) was founded [2]. These business groups and their successors today have the largest shares of the market in international business, which is based on the fact that under their umbrella all components for complex water management projects (plant manufacture, engineering, surface and subsurface civil engineering, development departments) can be found. They are, in essence international, competitively, vertically integrated multinational enterprises.

In Germany, privatisation of the water supply is neither linked with direct competition between municipal institutions for the market nor with an obligatory yardstick competition. The water supply companies are in a "quasi competition", as three out of four companies raise public charges in accordance with the Municipal Charges Law; these must be approved by local governments under the supervision of the federal states. Here, attention is to be paid to the principles of cost-covering and equivalence in accordance with charges law, whose consideration in the form of price-performance comparisons can be examined by the municipal supervisory authorities. The remaining quarter of the providers raises payments under private law and is subordinate to the anti-trust control of abusive practices. The anti-trust price control is oriented to the comparative market concepts and accepts price differences between providers on the strength of clearly defined criteria only [12]. Performance comparisons between the various bidders are undertaken by the municipal operators themselves by voluntary benchmarking [13]. The water supply is still organized as a regional monopoly, be it public, private or semi-private, but the unclear separation of public and private responsibilities seems to be an obstacle for success on international markets. In the German water supply the public-law forms of enterprise dominate: In 2003, 30% of the water supply companies in Germany were managed as municipal enterprise or public enterprise, 15% as semi-autonomous municipal agency, 16% as inter-municipal agency, and 6% as water and soil management associations. 29% were arrangements under

Public Private Partnership and only 3.5 % were other arrangements under private law, which in fact can be regarded as material privatization [14].

Water management competence in Germany is clearly located at the municipal level, which admittedly restricts international competitiveness. In contrast, French communes, considerably smaller than their German counterparts, are not in a position to manage the supply of drinking water supply themselves for reasons of a lack of specialist personnel and specialist knowledge [2]. An important advantage of the structures in the German water supply is that the strong communal anchoring of the German providers ensures a high degree of political involvement. With this the precautionary orientation of the management of surface waters is placed on a solid basis. This system enjoys strong acceptance amongst the population. The high level and the efficiency in the technical management are guaranteed through the close co-operation between water supply companies, industry, government agencies as well as through the activities of technical-scientific associations which set the rules. Due to the strong functional and organizational fragmentation, the influence of the German water industry on the decision processes in the European Union is, however, rather small. Through the strong division of organizational competence (water supply and wastewater disposal companies, construction firms, plant constructors, component suppliers, consulting firms, engineer offices, water laboratories and research institutes) the integrated appearance on the international market is missing in the German water industry. This is the main reason why the steadily growing market segment of the complete turnkey solutions (planning, construction, operation, maintenance, invoicing and customer service), where above all, enterprises from France and Great Britain dominate, cannot be filled by German competitors [15].

5. Conclusion

The international markets for the operation of water supply systems and complete solutions are dominated by French and British companies. The sole exception was briefly the German RWE which, however, was owing to the take over of a British company. Otherwise the typical German plant constructor either does not achieve the critical size for a global player or he lacks the necessary references as operator of water supply systems. On the other side, the water supply and wastewater disposal operators lack the financial power in order to compete with the world market leaders. In order also to be present on the market for complete solutions, a large competitive backlog compared with foreign water companies would have to be offset. For this the current, strong municipal anchoring of the water industry in Germany would have to be relaxed in favour of the build up of vertically integrated water multinationals, which can be competitive on international markets. An adjustment of the market and concentration in the German water industry would be the required prerequisite for such an internationalization, which bears the danger that the structures of German water supply, proven with regard to security of supply and drinking water quality, will have to be sacrificed for this competition with its uncertain outcome.

References

[1] Meyer-Renschhausen Martin 1996 *Die Auswirkungen der Privatisierung öffentlicher Dienstleistungen auf die Umwelt am Beispiel von Energiewirtschaft und Abwasserbeseitigung* in: Zeitschrift für öffentliche und gemeinwirtschaftliche Unternehmen.
[2] Spelthahn Sabine 1993 *Privatisierung natürlicher Monopole - Theorie und internationale Praxis am Beispiel Wasser und Abwasser* (Berlin)
[3] European Commission Community Research 2004 *Euromarket – Water Liberalisation Scenarios, Final Report Part 3* (Brussels)
[4] EUWID 2006 *Europäischer Wirtschaftsdienst Wasser und Abwasser* **43** (24.10.2006)
[5] Gordon-Walker Simon and Marr Simon 2002 *Study on the Application of the Competition Rules to the Water Sector in the European Union* prepared by WRc and Ecologic fort the European Commission - Competition Directorate General
[6] Kraemer R Andreas 1997 *Öffentliche und Private Wasserversorgung und Abwasserbeseitigung*

in Europa in: Francisco Nunes Correia und R Andreas Kraemer *Dimensionen Europäischer Wasserpolitik* (Berlin, Heidelberg)

[7] Rudolph Karl-Ulrich and Harbach Michael 2007 Private Sector Participation in Water and Sanitation for Developing Countries *CESifo DICE Report* **5 (2)** 33 - 39

[8] Der Rat von Sachverständigen für Umweltfragen [Council of Experts for Environmental Questions] 2000 *Umweltgutachten 2000 - Schritte ins nächste Jahrtausend* (Stuttgart)

[9] Langenfeld Fredy 2000 Agence de l'Eau Rhin-Meuse, Metz, Das System der flussgebietsbezogenen Wasseragenturen in Frankreich *Kongress „Wasser Berlin 2000"* (23.-27.10.2000)

[10] Kraemer R Andreas und Frank Jäger 1997 *Deutschland* in: Francisco Nunes Correia und R. Andreas Kraemer *Dimensionen Europäischer Wasserpolitik* (Berlin, Heidelberg)

[11] Mecke Thomas 2000 Das Berliner Modell - Weltweit gültig? *Kongress Wasser Berlin 2000"* (23. - 27.10.2000)

[12] Grobosch Michael 2003 *Grundwasser und Nachhaltigkeit - Zur Allokation von Wasser über Märkte (Diss.)* (Tübingen)

[13] ATT, BGW, DVGW, DWA and VKU 2005 *Branchenbild der deutschen Wasserwirtschaft*, (Bonn)

[14] Wackerbauer Johann 2007 "Struktur und Entwicklungsperspektiven der deutschen Wasserwirtschaft" *ifo Schnelldienst* **60 (01)** 14-25

[15] BMBF [Federal Ministry for Education and Research] ed 2000 *Aktionskonzept nachhaltige und wettbewerbsfähige deutsche Wasserwirtschaft* (Karlsruhe, März 2000)

[16] Auer Josef, Barbara Böttcher, Hans-Joachim Frank, Stefan Heng and Eric Heymann 2003 Traditionelle Monopole: Wachstum durch mehr Wettbewerb *Deutsche Bank Research Nr. 261* (Frankfurt am Main, 20. März 2003)

[17] Correia Francisco Nunes und R Andreas Kraemer 1997 *Institutionen der Wasserwirtschaft in Europa, Einleitung* in: Francisco Nunes Correia und R Andreas Kraemer *Dimensionen Europäischer Wasserpolitik* (Berlin, Heidelberg)

[18] Ewers Hans-Jürgen, Botzenhart Konrad, Jekel Martin, Salzwedel Jürgen, Kraemer R Andreas 2001 *Optionen, Chancen und Rahmenbedingungen einer Marktöffnung für eine nachhaltige Wasserversorgung, Endbericht*

[19] Handelsblatt dated 19.10.2004 p 5 *Verband gegen EU-Pläne für Wassermarkt*

[20] Wackerbauer Johann 2006 "*The Regulation and Privatisation of the Public Water Supply and the Resulting Competitive Effects", CESifo DICE Report* **4 (4)** 42-49

IOP Conf. Series: Earth and Environmental Science **4** (2008) 012038 doi:10.1088/1755-1307/4/1/012038

Complex hydro- and sediment dynamics survey of two critical reaches on the Hungarian part of river Danube

Sándor Baranya[1], László Goda[2], János Józsa[1] and László Rákóczi[3]

[1] Budapest University of Technology and Economics (BUTE), Budapest, Hungary
[2] Lower-Danube-Valley Environmental and Water Directorate, Baja, Hungary
[3] Water Resources Research Centre, Budapest, Hungary

E-mail: baranya@vit.bme.hu

Abstract. Detailed hydrodynamic survey of two critical river reaches has been performed from hydro- and sediment dynamics points of view, in order to explore the main features, moreover, provide calibration and verification data to related 3D flow and sediment transport modelling. Special attention has been paid to compare moving and fix boat measurement modes for estimating various flow and large-scale bed form features, resulting in recommendations e.g. on the time period needed in stationary mode operation to obtain sufficiently stabilized average velocity profiles and related parameter estimations. As to the study reaches, the first comprises a 5 km long sandy-gravel bed reach of river Danube located in Central-Hungary, presenting problems for navigation. As a conventional remedy, groyne fields have been implemented to make and maintain the reach sufficiently deep, navigable even in low flow periods. As is usually the case, these works resulted in rather complex flow characteristics and related bed topography at places. The second site is another 5 km long reach of river Danube, close to the southern border to Serbia. There the river presents navigational problems similar to the previously mentioned reach, however, having entirely sand bed conditions, abundant in a variety of dunes, especially in the shallower parts. In both study reaches ADCP measurements were done with around 2.5 Hz sampling frequency both in moving boat operation mode providing overall, though locally moderately representative picture, and in fixed boat mode at a considerable number of selected verticals with 10 minutes long measuring time.

1. Introduction

Acoustic Doppler Current Profilers (ADCPs) are extensively used in river flow measurements nowadays. The most general utilization of ADCPs is measuring river discharges from boats moving along river transects. The accuracy of these measurements can be sufficient if the application of such instruments is focusing on discharge determination, only (e.g. [1]). However, making use of the fact that the ADCP records detailed three-dimensional velocity field, besides determining the river discharge, many other flow and river morphological parameters can be derived. Recently e.g. [2] estimated shear velocity derived from fitted logarithmic velocity profiles. [3] presented a software which provides dispersion coefficients driven by either vertical or transverse velocity gradients. [4] reconstructed the apparent velocity of bed load in a gravel-bed river.

In the present study detailed hydrodynamic survey of two river reaches was performed by means of ADCP in Hungary, as an important phase of 3D flow and sediment transport modelling project. Apart from collecting calibration and verification data to the numerical modelling, the differences between the case study areas, the actual flow regime during the measurements offered the opportunity for a methodological analysis of ADCP-based hydrodynamic surveys. The main focus was on quantifying and comparing the representativeness of moving and fix boat measurement modes on various flow features, including recommendations on the time needed in stationary mode operation to obtain sufficiently stabilized average profiles and related parameters.

The mean-flow regulation of the present Hungarian reach of Danube River started in the second half of the 19. century has put an end to the natural meandering of the river channel by straightening many of its bends. Numerous pseudo-bends and irregular sections, so-called surflexions have been thus created. Figure 1 shows a regular curvature, in which the imaginary chord line, connecting the end points of the bend is leaving the river channel. In case of a pseudo bend, however, the chord remains inside the channel. The second curvature in figure 1 is, however, not a usual pseudo bend. Here two right curves follow each other while the highest velocity filament (dotted line) does not cross the centre of the channel. Instead, it turns back, towards the apex of the second curvature. At such bends, the guidance of water flow is very poor and this uncertainty generates unstable centre bars, repeatedly creating obstacles for navigation.

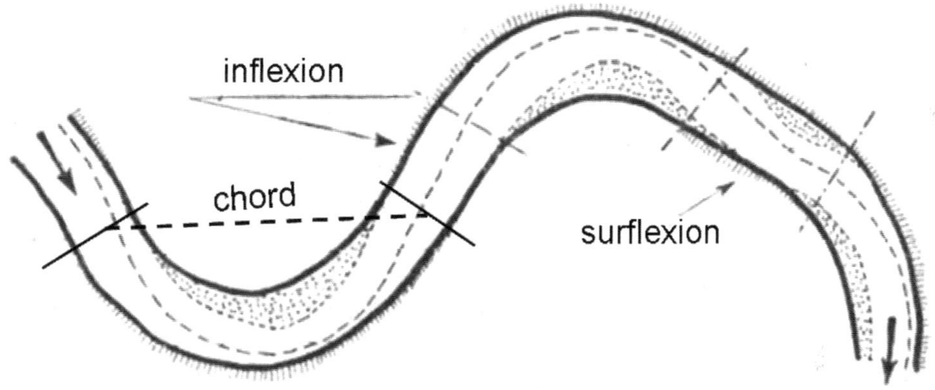

Figure 1. River bends.

Already in the first part of the last century, irregular depositions, moving bars were detected within the recently regulated mean-flow channel, and the navigation became more and more difficult. A comprehensive low-flow regulation project started erecting sets of spur dykes (groynes) alternatively at the opposite river banks (figure 2). In this way, the low flows were forced to develop low amplitude, though rather regular meanders inside the mean-flow channel. The systematic use of groyne fields has fixed the navigation routes in long straight reaches and even at pseudo bends, normalizing the occurrence of fords, thus improving the general conditions in most of the year. The handling of surflexion reaches, however, has remained a difficult task for river engineers so far. Since both banks of the regulated mean-flow channels were fixed more than 100 years ago by stone riprap embankments up to the mean flow level, their relocation, i.e. changing their platform is not feasible even at the most unfavourable reaches for economical reasons.

As to the study reaches the first one comprises a 5 km long sandy-gravel bed reach of river Danube located in Central-Hungary (figure 3), presenting problems for navigation for a long time. As a conventional remedy, groyne fields have been implemented to make and maintain the reach sufficiently deep, navigable even in low flow periods. As is usually the case, these works resulted in rather complex flow characteristics and related bed topography at places.

IOP Conf. Series: Earth and Environmental Science 4 (2008) 012038 doi:10.1088/1755-1307/4/1/012038

Figure 2. Straight reach and pseudo bend with groyne fields (River Danube – North Hungary).

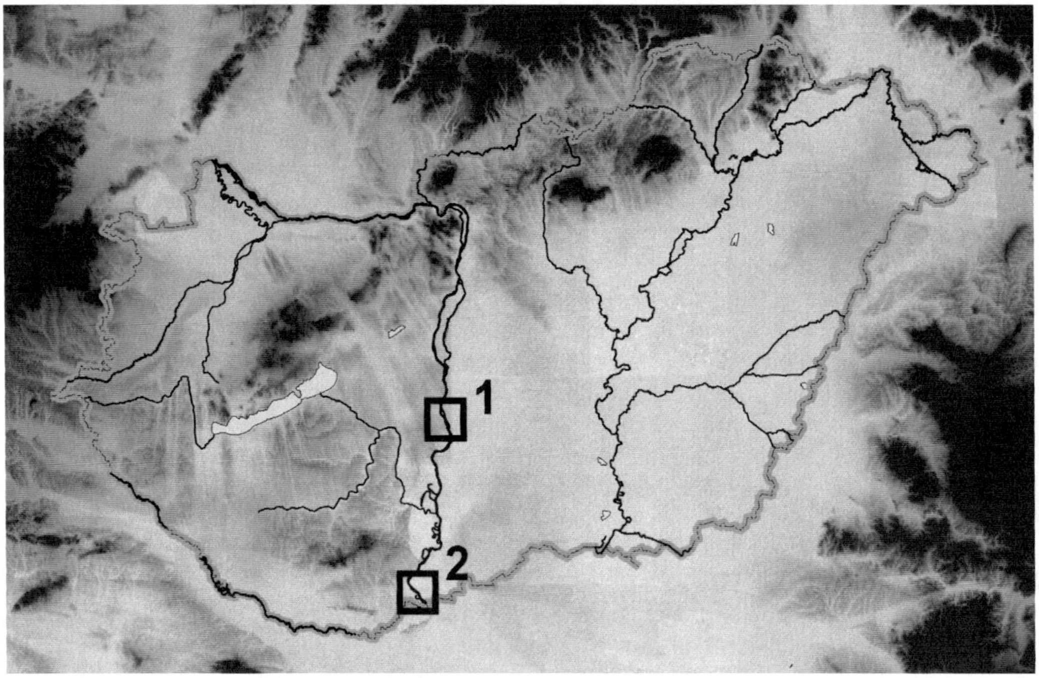

Figure 3. Location of study reaches in Hungary.

The second problem section situated in the southernmost part of Hungary (figure 3) downstream of Mohács city has got over-widened and shallow channel, also showing a surflexion character. At the upstream end the mean flow channel is about 300 m wide only, being significantly narrower than the regulated width in that reach of River Danube. Downstream of this "bottleneck", the channel widens to 600 m over a 3 km long distance. Navigation takes place along the right bank, while a 200-300 m wide shallow strip on the left side is covered by 15-30 m long and 0.5-1 m high dunes. The latter can easily be observed on low-altitude aerial photos taken in low-flow regimes.

Recently, detailed hydrodynamic, morpho-dynamic and sediment measurement campaigns have been carried out in the introduced two Danube reaches. The main goal of the surveys was to produce sufficient input data for numerical modelling. Recently a number of papers have dealt with flow and morphological analysis in river reaches with hydraulic structures (see e.g. [5], [6], [7]) and emphasized the necessity of three-dimensional investigation of flows in such regions, either by hydraulic scale models using up-to-date measurement techniques or by spatial numerical models calibrated by field measurements. Since complex bed topography, furthermore, existing as well as designed hydraulic structures are present in the study reaches, a 3D CFD turbulence code proved reasonable to be adapted in this study.

2. Measurements

In this study an RDI four beam 600 kHz Rio Grande ADCP was deployed from a vessel. On one hand, moving boat measurements were done with approximately 2.5 Hz sampling frequency with a vertical resolution of 25 cm. During the surveys DGPS position data were continuously collected which gave the opportunity to locate each measurement in absolute coordinate system. In each of the two Danube reaches seven transects were measured with moving vessel operation with an average spacing of 600 m apart, while in the confluence zone an intensive survey was performed with continuous crossings resulting in 25 transects on the roughly 600 m long zone. On the other hand, in each study area fixed vessel measurements were also done using the settings mentioned above. Five profiles per cross-section were measured over 10 min time interval. Recently, fixed vessel ADCP measurements have been done by e.g. [8] and it was shown that adequate mean and turbulence parameters of the flow can be captured, however, temporal and spatial scales of calculated flow features are strongly depend on instrument characteristics. Indeed, performing fixed boat flow measurements can have significant importance in measuring flows with strong spatiality e.g. in river bends, groyne fields, etc.

In this study the goal of measuring in fix points was to reproduce mean vertical velocity distributions from which several parameters such as bed roughness, bed shear stress, shear velocity and dispersion coefficient could be derived using fitted logarithmic profiles.

2.1. Fixed vessel measurements

In the first step the sampling time length needed to get reasonably stabilized velocity values was investigated. In order to determine the required time interval the variation of the normalized mean square error (NMSE) was used, as defined by [9]:

$$NMSE = \frac{\overline{(U_T - U)^2}}{\overline{u'^2}} \tag{1}$$

where U = instantaneous velocity, U_T = cumulated mean velocity from $t = 0$ to $t = T$ and u' = velocity fluctuation.

A characteristic plot of the error function in different depths of one profile is shown in figure 4. As can be seen a measuring interval of around 100 seconds gives adequate values fulfilling an error criteria of 5%. Despite in this plot one can see little variation of NMSE is not influenced by the vertical position of the measurement points, the relation between the required sampling time (that is to keep the error below 5%) and sampling depth was nevertheless investigated. Furthermore, the sensitivity on mean flow velocity and flow discharge was also analysed and significant correlation was not found.

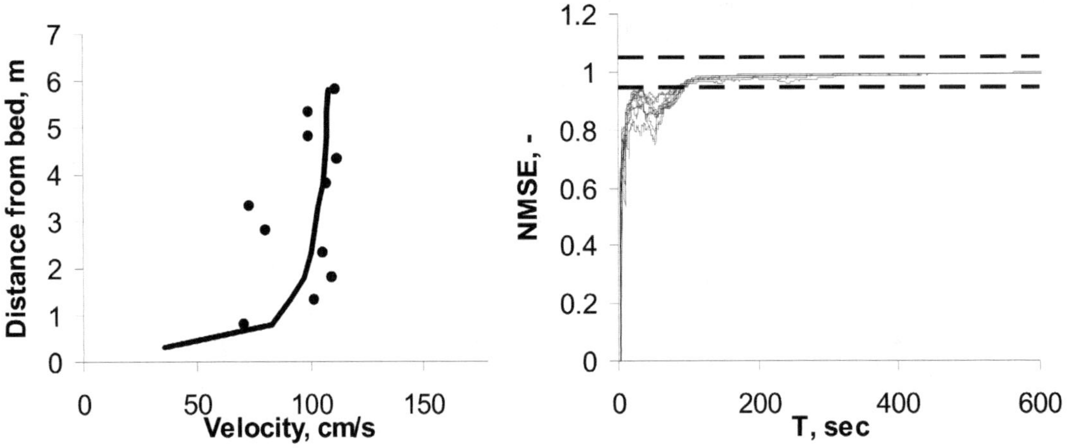

Figure 4. Instantaneous and mean velocity profiles (left) and variation of NMSE function in one measured profile (right).

A typical instantaneous and a 10 min mean velocity profile is shown in figure 4. As is expected in a flow with high Reynolds number, the instantaneous velocity distribution is significantly scattering around mean profile.

To estimate local values of bed roughness, shear velocity and bed shear stress, a logarithmic curve, typical to turbulent boundary layer type of regions, was fitted to the measured velocity profiles. To observe the accuracy of the derived characteristics, the logarithmic profile fitting (Eq. 2) was performed based both on the instantaneous and the cumulated vertical velocity distribution.

$$U = \frac{u_*}{\kappa} \ln \frac{z}{z_0} \tag{2}$$

where U = flow velocity, u_* = shear velocity, $\kappa \approx 0{,}4$ is the von Kármán constant, z = vertical position and z_0 is the height at which U is assumed to be 0.

The distributions of roughness parameter and shear velocity show significant difference between the two estimation methods. Without averaging the measured velocities the characteristics are largely overestimated, at the same time, after a few tens seconds of averaging a fairly good estimation can be obtained.

2.2. Moving vessel measurements

ADCP technology has become a standard tool for reasonable estimations of river discharges by now. In the present study besides discharge estimation a closer look has been taken to the detailed spatial velocity distribution records. Parameters presented in section 2.1 were calculated in the same way. Since the sampling time step was relatively short (~ 0.4 sec) vertical velocity distribution showed high fluctuation, thus, most of the fitted logarithmic functions provided unrealistic values of z_0 and u_*. Therefore, averaging of individual profiles was used to assess distribution of these parameters along cross-sections. To find the optimal number of profiles to average, three cases were distinguished (no averaging, and averaging 10 as well as 30 profiles). Distribution of the three derived parameters (depth-averaged velocity, roughness height and shear velocity) from a selected transect is plotted with and without moving average type of filtering in figure 5. For comparison parameters calculated from fixed vessel measurements are also displayed.

Using long-term measurements as reference values, it can be seen that parameters from instantaneous profiles are widely scattering. Averaging the raw velocity data by using a ten profiles wide moving window, a more realistic pattern develops, but still significant scattering is remains. Creating a moving average of 30 velocity data yields values showing adequate agreement with long-term measurement parameters, the latter accepted as reasonably accurate, reference values. Dealing with moving boat measurements data filtering is not only a simple time averaging, but also a space averaging. A mean boat speed of 1.4 m/s and sampling rate of 2.5 Hz represent a space averaging of a 8.5 m wide section in case of a filtering over 30 elements. This width ensures that water depth is similar when creating average, nevertheless, it points out that the boat speed has to be chosen carefully respect to changes in river bed elevation.

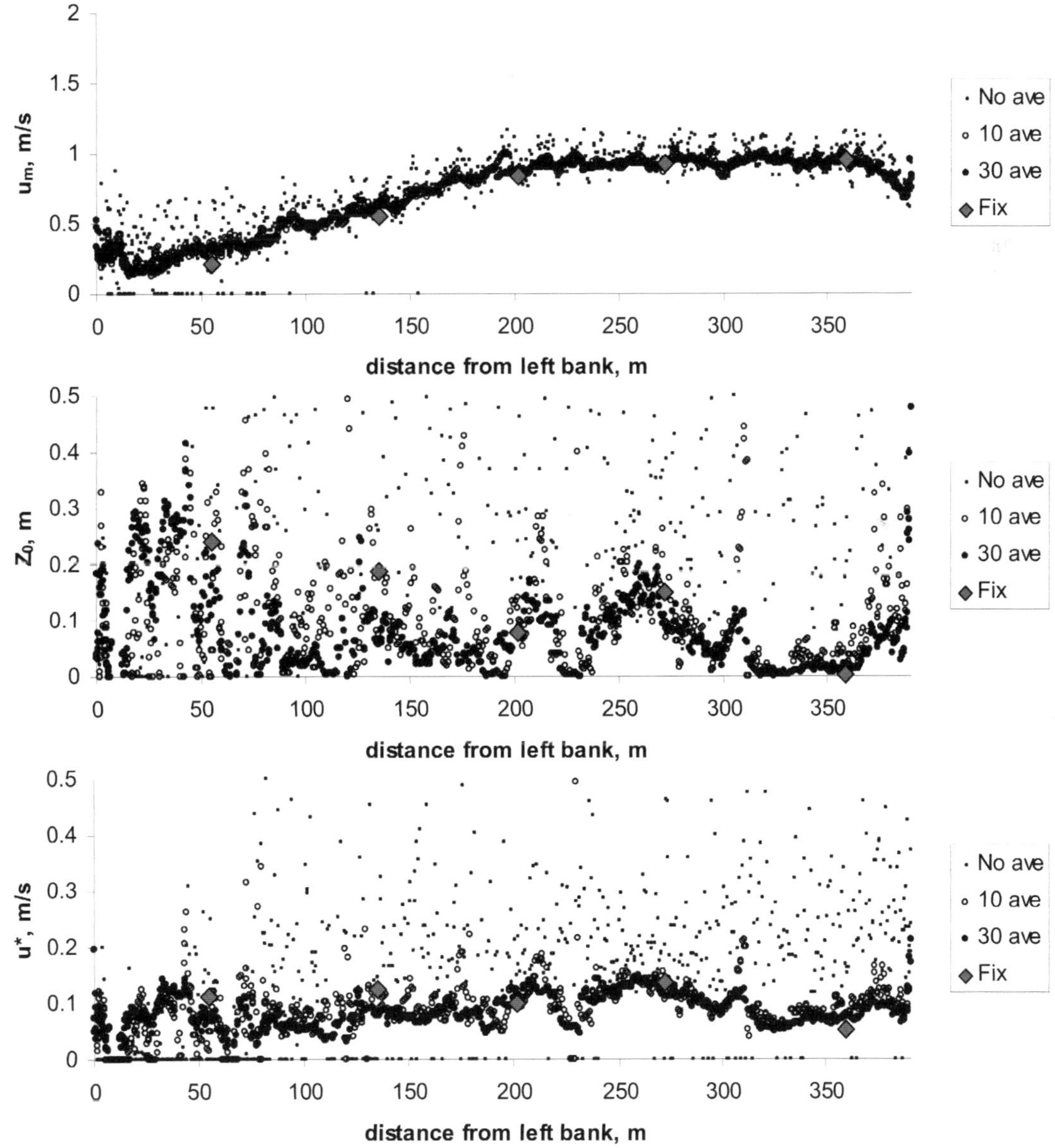

Figure 5. Cross-sectional distributions of depth-averaged flow velocity (u_m), roughness height (z_0) and shear velocity (u_*).

2.3. Sediment and bed material measurements
Suspended sediment measurements were carried out using a pump. Water samples were taken from different depths of the water column (2-5 samples per profile) in 35 profiles in all. Laboratory analysis of the samples provided concentrations related to points and grain size distributions related to profiles. Bed material samples were also collected with a van Veen grab sampler in the same 35 points. Typical suspended sediment and bed material grain size distributions can be seen on figure 6.

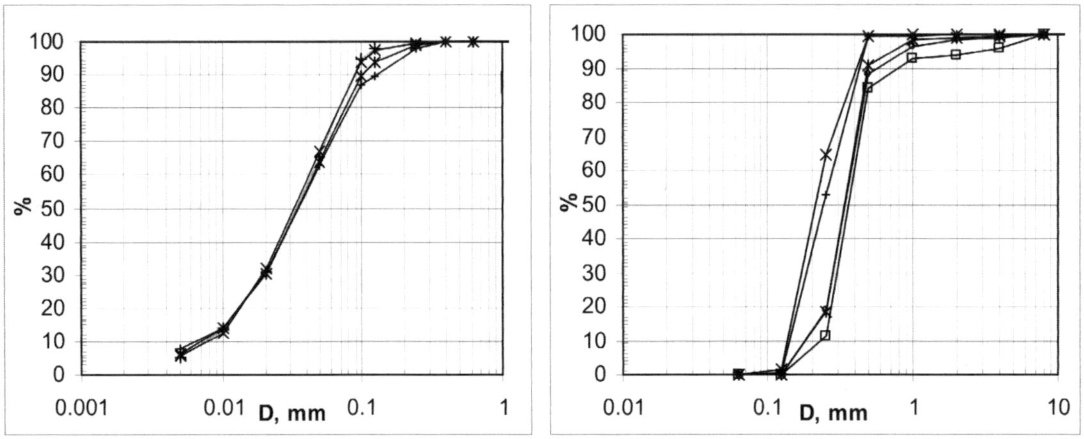

Figure 6. Typical grain size distributions for suspended sediment (left) and bed material (right).

3. Numerical modelling

3.1. Model description
The CFD code SSIIM [10] was used in this study to model flows and sediment transport. The numerical model solves the 3D RANS equations with the *k*-ε turbulence closure (see e.g. [11]) by using a finite-volume method and the SIMPLE algorithm [12] on a 3D non-orthogonal, structured grid. The momentum equations are in their complete form, without resorting to the hydrostatic assumption.

For calculating the suspended sediment transport the advection-diffusion equation of sediment concentration is solved:

$$\frac{\partial c}{\partial t} + U_j \frac{\partial c}{\partial x_j} + w \frac{\partial c}{\partial z} = \frac{\partial}{\partial x_j}\left(\Gamma_T \frac{\partial c}{\partial x_j}\right) \tag{3}$$

where c is concentration, w is settling velocity, U_1, U_2 and U_3 are velocity components, x_1, x_2 and x_3 are spatial coordinates and Γ_T is the turbulent diffusion derived from the k-ε model. Close to the bed the equilibrium sediment concentration is calculated by the formula of [13]

$$c_{bed,sl,i} = 0.015 \frac{d_i}{a} \frac{\left[\dfrac{\tau - \tau_{c,i}}{\tau_{c,i}}\right]^{1.5}}{\left\{d_i \left[\dfrac{(\rho_s - \rho_w) g}{\rho_w \nu^2}\right]^{1/3}\right\}^{0.3}} \tag{4}$$

in which d_i is the sediment particle diameter, a is a reference level set equal to the roughness height, τ is bed shear stress, $\tau_{c,i}$ is critical bed shear stress for movement of particles according to

Shield's curve, ρ_w and ρ_s are the density of water and sediment, v is viscosity of water and g is the acceleration due to gravity.

Bed load is calculated with the empirical formula of [14] given as

$$\frac{q_{b,i}}{d_i^{1.5}\sqrt{\dfrac{(\rho_s-\rho_w)g}{\rho_w}}} = 0.053\frac{\left[\dfrac{\tau-\tau_{c,i}}{\tau_{c,i}}\right]^{2.1}}{d_i^{0.3}\left[\dfrac{(\rho_s-\rho_w)g}{\rho_w v^2}\right]^{0.1}} \qquad (5)$$

where q_b is bed load.

3.2. Model calibration and validation

For model calibration field survey data from the actual project were used. For upstream boundary conditions in the hydrodynamic model the measured quasi-steady discharges were given in both study cases with the values of 1450 and 1950 m^3/s, respectively. The upstream boundary conditions were set in the sediment model, based on the measured suspended sediment concentrations from the transects closest to the inflow boundaries.

Model calibration was performed varying the value of k_s roughness height. One option could be to calculate k_s according to the empirical formulas (e.g. [15]) using a characteristic particle diameter, but in case of fine sand bed material these formulas would give too small values, furthermore, bed roughness can be extremely underestimated if ripples and dunes are situated in the river. In the second study reach low-altitude aerial photos taken at low-flow regime have shown the presence of such bed forms, moreover, ADCP also captured dunes in the shallow zones, indicating the need to take them into account. Carrying out comparisons of measured and modelled cross-sectional velocity distributions, the best agreement was observed with a homogeneous k_s of 0.06 m in the first reach and k_{s1} = 0.1 m and k_{s2} = 0.02 m in the second reach (figure 7), respectively.

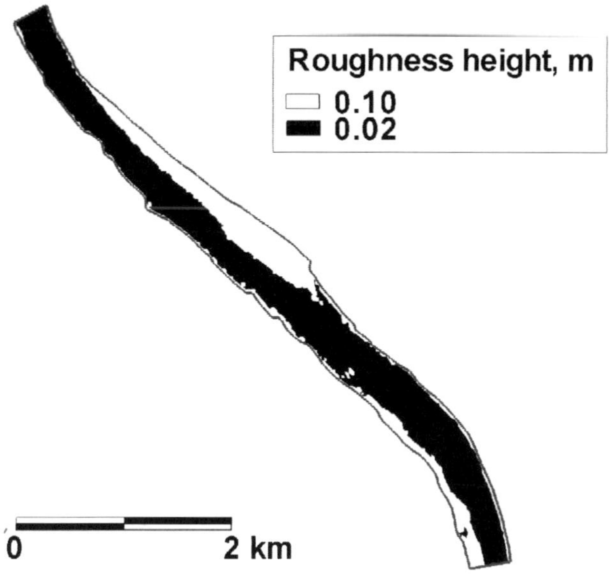

Figure 7. Applied roughness height distribution in CFD model in the second investigated reach.

The calibration and validation of the numerical model were performed assessing measured and modelled suspended sediment profiles and bed geometry changes during a two-week long moderate flood wave with a peak discharge around 3000 m^3/s. A few examples from the model validation are shown in figures 8-10, comparisons of cross-sectional velocity distributions, suspended sediment profiles and bed changes, respectively. As can be seen in figure 10, while the locations of erosion and deposition zones are fairly well estimated, however, the magnitude of sedimentation is somewhat underestimated.

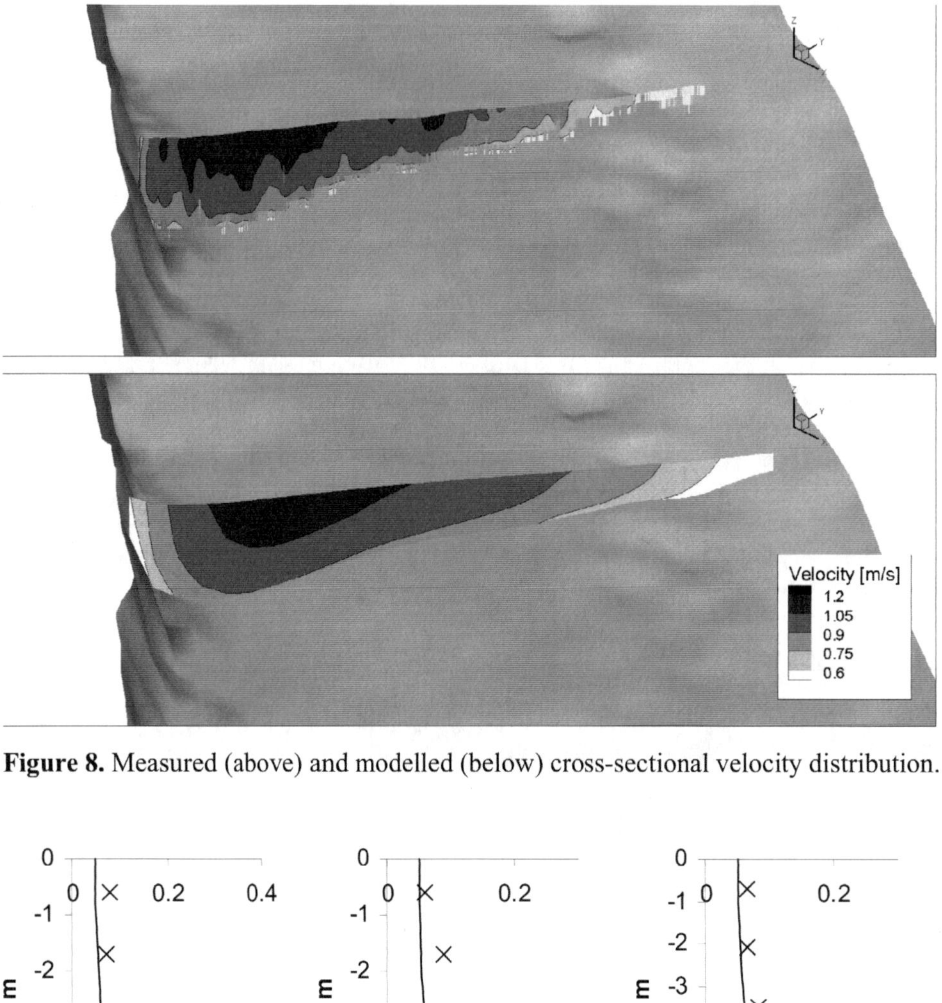

Figure 8. Measured (above) and modelled (below) cross-sectional velocity distribution.

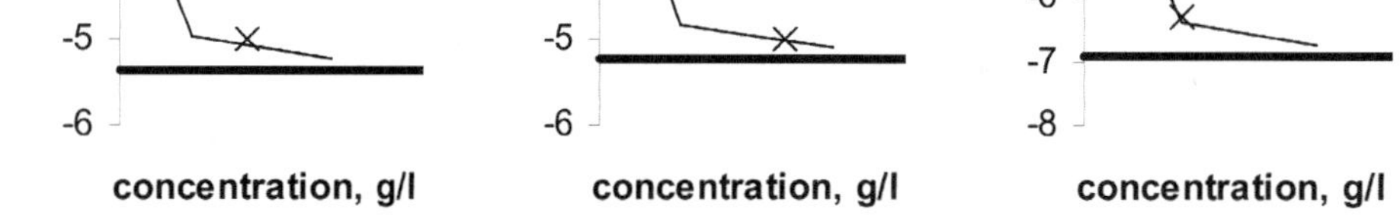

Figure 9. Measured (crosses) and modelled (continuous line) vertical concentration profiles.

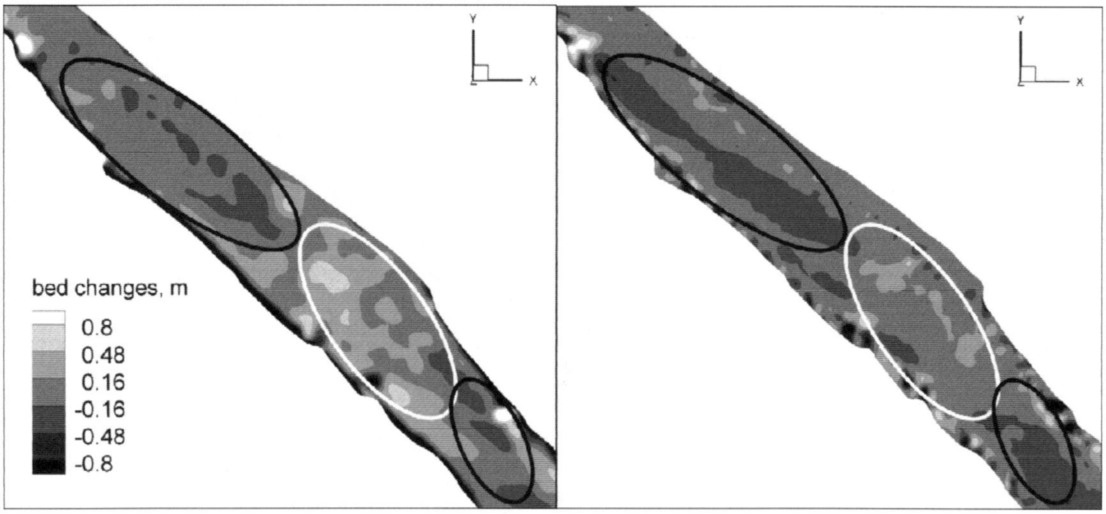

Figure 10. Measured (above) and modelled bed changes in the second study reach occurred during a moderate flood wave with a peak discharge of 3000 m^3/s (blue line – erosion, red line – deposition).

4. Results

New hydraulic structures were designed into the problematic sections of the two river reaches and the impact on river morphology has been studied by using the presented numerical model. New groyne fields are planned to be constructed in the shallow zones aiming to force sediment settling between the structures and at the same time to deepen navigation channel (figures 11-12).

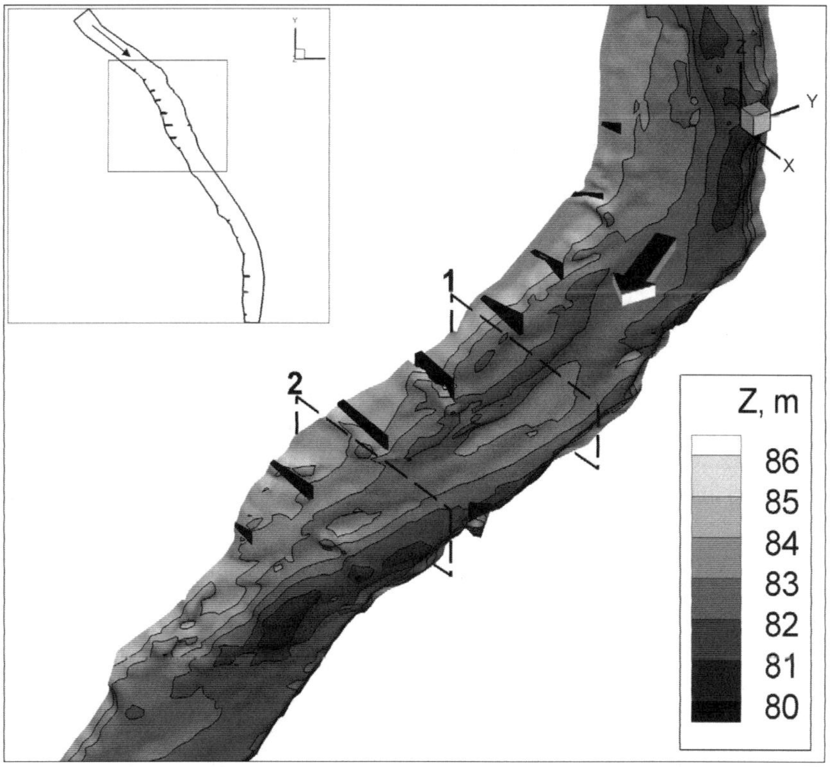

Figure 11. Layout of the designed groyne fields – first study reach.

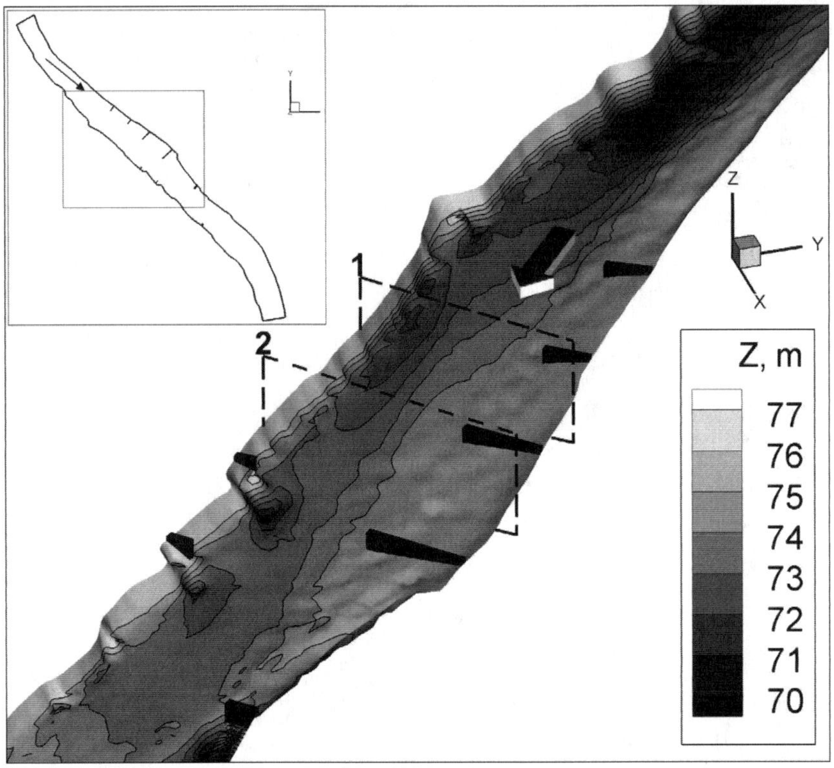

Figure 12. Layout of the designed groyne fields – second study reach.

Five month long intervals were modelled in both cases assuming constant formative discharges of 1950 m^3/s and related downstream water levels. Sediment parameters were taken from model calibration. As expected, due to narrowing, higher flow velocities are developing in the surroundings of the new structures resulting then in substantial bed erosion. In turn, between the planned groynes considerable deposition develops. Looking at marked cross-sections (figures 13-14) it can be clearly seen that, indeed, the navigational channel deepened with a magnitude of 0.5 m in the first case and 1.5-2 m in the second case indicating a positive impact of the designed structures on low-flow fluvial navigation. It has to be taken into account, however, that natural flow regimes are highly unsteady and while low and mean flows result in the above-mentioned morphological changes, high waters and even floods might produce completely different effects.

5. Summary and conclusion

Two five km long reaches of River Danube presenting problematic sections for fluvial navigation in low-flow periods have been investigated. Detailed hydrodynamic, and sediment measurements as well as bed topographic surveys have been carried out in order to get adequate data to calibrate and verify a three-dimensional CFD model. In the hydrometric analysis both fixed and moving boat measurements were performed with an ADCP providing the opportunity to compare flow parameters derived from long-term mean velocity profiles and from instantaneous data. Time-averaging long-term fixed boat ADCP data, logarithmic velocity profiles were fitted and depth-averaged flow velocity (u_m), roughness height (z_0) and shear velocity (u_*) were successfully estimated. It was shown that by horizontally space-averaging 30 neighbouring elements a fairly good estimation similar to the fixed boat ones could be obtained.

IOP Conf. Series: Earth and Environmental Science **4** (2008) 012038 doi:10.1088/1755-1307/4/1/012038

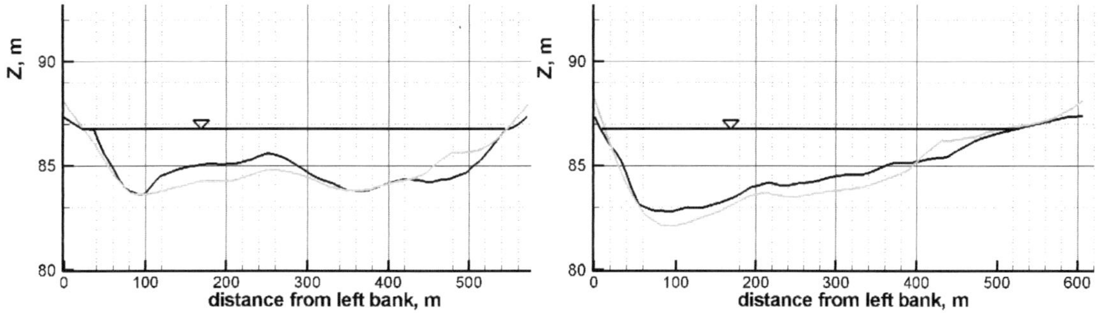

Figure 13. Changes of cross-section geometry in the marked transects in the first study reach (black – actual bed, grey – modelled bed).

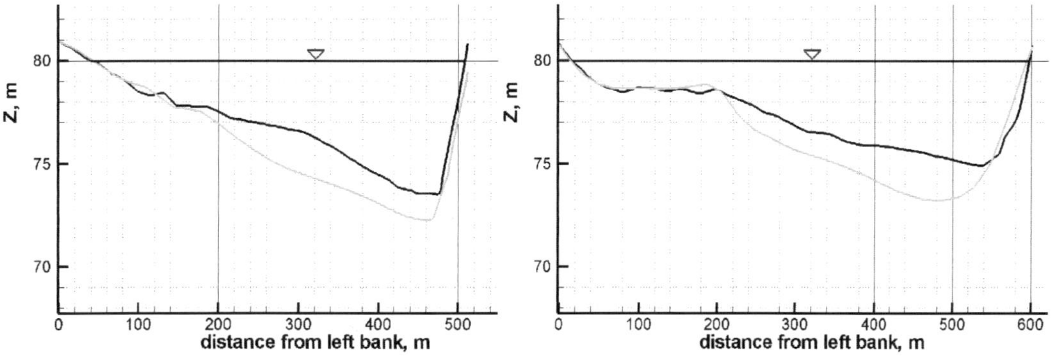

Figure 14. Changes of cross-section geometry in the marked transects in the second study reach (black – actual bed, grey – modelled bed).

Filtered velocity data, sampled suspended sediment profiles and bed elevation changes were compared with modelled results showing satisfactory agreement. Having the model validated, the impacts of designed hydraulic structures were analysed. Defining a steady bed-forming discharge at the upstream boundary, five month long time periods were modelled in both study reaches with bed topography modified by the planned groyne field. Model results showed a reasonable impact, i.e. in the contracted section of the navigation channel considerable erosion took place, whereas zones with almost zero velocity between the groynes resulted in sediment deposition. Though modelling results indicate positive effects on fluvial navigation, it has to be considered that due to local bed erosion in the study reaches, a large quantity of sediment will be transferred in downstream direction. Since the second investigated river reach is situated close to the border with Serbia, any kind of river engineering intervention has to be considered carefully.

References

[1] Yorke T H and Oberg K A 2002 Measuring river velocity and discharge with acoustic Doppler profilers *Flow Meas. and Instr.* **13** 191-5

[2] Rennie C D and Church M 2007 ADCP shear stress and bedload transport in a large wandering gravel-bed river *Proc. XXXII. IAHR Congress (Venice)*

[3] Kim D, Muste M and Weber L 2007 Software for assessment of longitudinal dispersion coefficients using acoustic Doppler current profiler measurements. *Proc. XXXII. IAHR Congress (Venice)*

[4] Rennie C D, Millar R G and Church M A 2002 Measurement of bedload velocity using an acoustic Doppler current profiler *J. Hydr. Eng.* **128** 473-83.
[5] Tan S K, Yu G, Lim S Y and Ong M C 2005 Flow structure and sediment motion around submerged vanes in open channel *J. Waterway, Port, Coastal and Ocean Eng.* **131** 132-6
[6] Jia Y, Scott S, Xu Y C, Huang S L and Wang S S Y 2005 Three-dimensional numerical simulation and analysis of flows around a submerged weir in a channel bendway *J. Hydr. Eng.* **131** 682-93
[7] Sukhodolov A, Engelhardt C, Krüger A and Bungartz H 2004 Case study: Turbulent flow and sediment distributions in a groyne field *J. Hydr. Eng.* **130** 1-9
[8] Muste M, Yu K, Pratt T and Abraham D 2004 Practical aspects of ADCP data use for quantification of mean river flow characteristics; Part II: fixed-vessel measurements *Flow Meas. and Instr.* **15** 17-28
[9] González-Castro J, Oberg K and Duncker J 2000 Effect of temporal resolution on the accuracy of ADCP measurements *Proc. ASCE Conference (Minneapolis)*
[10] Olsen N R B 2002 *A three-dimensional numerical model for simulation of sediment movements in water intakes with moving option. User's Manual*, Dept. of Hydr. and Env. Eng. NTNU Trondheim Norway
[11] Pope S B 2001 *Turbulent Flows* (Cambridge University Press)
[12] Patankar S V 1980 *Numerical heat transfer and fluid flow* (New York: McGraw-Hill Book Company)
[13] Rijn L C van 1984b Sediment transport, Part II: Suspended sediment transport *J. Hydr. Eng.* **110** 1613-41
[14] Rijn L C van 1984a Sediment transport, Part I: Bed load transport *J. Hydr. Eng.* **110** 1431-56
[15] Rijn L C van 1984c Sediment transport, Part III: Bed forms and alluvial roughness *J. Hydr. Eng.* **110** 1733-54

Torrent Classification – Base of Rational Management of Erosive Regions

Zoran Gavrilovic, Milutin Stefanovic, Irina Milovanovic, Jelena Cotric and Mileta Milojevic

Institute for the Development of Water Resources "Jaroslav Cerni", 11226 Beograd (Pinosava), Jaroslava Cernog 80, Serbia tel./fax (+381) 11-3906-461

E-mail: gavrilovicz@sbb.rs

Abstract. A complex methodology for torrents and erosion and the associated calculations was developed during the second half of the twentieth century in Serbia. It was the "Erosion Potential Method". One of the modules of that complex method was focused on torrent classification. The module enables the identification of hydro graphic, climate and erosion characteristics. The method makes it possible for each torrent, regardless of its magnitude, to be simply and recognizably described by the "Formula of torrentially". The above torrent classification is the base on which a set of optimisation calculations is developed for the required scope of erosion-control works and measures, the application of which enables the management of significantly larger erosion and torrential regions compared to the previous period. This paper will present the procedure and the method of torrent classification.
Keywords: soil erosion; torrents; torrent classification

1. Introduction

In nature, there are a great number of torrents. The most widespread are the small torrents called gullies. Their lengths are several hundreds of metres and watershed areas are of proportional sizes. Their characteristic is that water and sediment flow in their channels only during and immediately after a heavy rain. On the other end of that huge spectre of torrents are the large torrential rivers, whose lengths and watershed areas amount to several hundreds and thousands of kilometres. The prefix torrent is derived from the characteristics of flood waves which occur very quickly, carry significant amounts of sediment, but instead of flooding, they destroy whatever is in their way.

During the twentieth century, Serbia was impacted by a great number of torrents of all magnitudes. They damaged and destroyed the settlements, the roads and interrupted the railway and road transport.

The State was faced with a great problem, because it did not have even an approximate knowledge of the number of torrents and could not plan the defence against torrential floods.

For this reason, in the first half of the twentieth century, the torrents in Serbia were inventoried. More than fifty thousand torrents of various magnitudes were listed. This is on the average one torrent per square kilometre of the hilly and mountainous part of Serbia.

Soon after the first inventory of torrents in Serbia, it was determined that the collected data were not sufficient. The main deficiency of the inventory was the fact that it was designed according to the model of watercourse classification, which codes the main recipient and the tributaries of the first, the second and further orders. Such a classification is adequate for the inventory of large rivers and their

IOP Conf. Series: Earth and Environmental Science **4** (2008) 012039 doi:10.1088/1755-1307/4/1/012039

tributaries. The unsolved issue was how to classify a gully, which is a torrential tributary of the first order of a large navigable river, as well as a series of other questions.

2. Erosion Potential Method

Systematic investigations of the intensity of erosion and torrents were begun in former Yugoslavia sixty years ago ("Jaroslav Černi" Institute for the Development of Water Resources, 1947.), and they enabled the development of The Method for the Quantitative Classification of Erosion (MQCE), 1954. During the investigation work it was notified that erosion intensity could be used for computing the amount of sediment that reaches the downstream part of a river so that the investigations were extended to include the observations of the transport of the sediments to the control profiles. During last fifty years of permanent developing process results as a complex methodology for investigation erosion process, mapping, sediment calculating and torrent classification. Name of this method is "Erosion Potential Method". Since 1968. EPM is a standard method for erosion and torrent training engineering in water management [1], [2].

By the development of this original method, a complex method was created for the spheres of erosion and torrents, named "Erosion Potential Method". In short, this method has for a long time been the standard method and the tool for all engineering problems related to erosion and torrents in the field of water management and later on even for wider use. The method consists of:

Quantitative classification of erosion (1954)
Quantitative sediment regime (1955)
Torrent classification (1956)
Methods of optimizing calculations of the volume of erosion control works (1958)
The method was enhanced several times:
Erosion Potential Method I phase (1966)
Erosion Potential Method II phase (1968)
Erosion Potential Method III phase (1986)
Identification of erosive regions (1998)
Development of information and GIS procedures and applications for EPM (1985 - to date). ([3]; [4]; [6])

2.1. EPM Quantitative classification of erosion

Some modules of the "Erosion Potential"Method are well known, such as the module for erosion mapping, often under the title "Gavrilović's method or model", after the surname of the creator of the method - Slobodan Gavrilović. Under his leadership, during the period 1952-1976, the basic modules were realised, two phases were completed and the third phase of the method improvement was started.

The EPM erosion mapping procedure requires investigations and computations to determine and present on a map the surfaces with the same quantitative erosion class. The basic EPM value of the quantitative erosion intensity is the Erosion Coefficient (Z). ([3]; [4])

The quantitative value of the erosion coefficient (Z) has been used to separate erosion intensity to classes or categories. The mean value of the EPM erosion coefficient (Z) for the catchment's area is the basics value for all EPM calculations.

EPM Erosion and torrent categorization			**Table 1**
Erosion and torrent category	Qualitative name of erosion category	Range of values of coefficient (Z)	Mean value of coefficient (Z)
I	Excessive erosion - deep erosion process (gullies, rills rockslides and similar)	$Z > 1.0$	$Z=1.25$
II	Heavy erosion - milder forms of excessive erosion	$0.71 < Z < 1.0$	$Z=0.85$
III	Medium erosion	$0.41 < Z < 0.7$	$Z=0.55$

IV	Slight erosion	0.20<Z<0.4	Z=0.30
V	Very slight erosion	Z< 0.19	Z=0.10

The erosion mapping module has been presented several times to the professional audience, because it is the module which represents the base of the method. Although erosion mapping by this method and module requires a trained specialist, the apparent simplicity of the method and the similarity to other methods led to several erroneously constructed erosion maps and unprofessional modifications.

Namely, all the mistakes made during the phase of erosion mapping bring in an exponential error to all subsequent calculations. This issue was devoted much attention to in several papers.

In Serbia and in the countries of the former Yugoslavia, which make up one fourth of the Danube catchments, there are several different types of erosion mapping, for which their authors claim to have applied EPM. However, they cannot explain why their maps present the weak erosion processes in the areas with the developed gully erosion in forests, or the severe erosion in the areas of mild slopes and intensive agricultural production. For this reason, it is recommended to be cautious in using the non-authorised erosion maps.

2.2. EPM Torrent Classification

The module which was more modestly presented to the professional public so far is the module of torrent classification. This module is intended for the specialists in torrent management and planning of complex erosion control works.

Already in the introduction, it was mentioned that this is the case of a variant of hydro graphic classification adapted to the characteristics of torrents. Namely, every stream, therefore also torrents, have the essential hydro graphic characteristics that can be reduced to flow, catchments, slopes, etc. [3].

The characteristic by which torrents are distinguished from other streams is high erosion intensity, which is the source of sediment transported by the torrential flow.

Torrent characteristics also depend on climate characteristics of the region in which they occur. It is a rule of thumb that, in the optimal case, the only the data in torrential areas refer to precipitation and temperature. For this reason, this set of data is adopted for the definition of the effect of climate.

The basic idea of this module is to enable the assessment of torrential characteristics by one glance. This is made possible by the introduction of the entry called "Torrent Formula". The formula consists of three parts, i.e.:

- Torrential class
- Torrential category
- Erosion intensity.

Torrential class is determined based on the value of hydro graphic coefficient of the torrential class (Hk) calculated by the formula:

$$Hk = F \times A \times K \times \frac{Lt+1}{L+1} \tag{1}$$

Where:

F- Catchments area

L – Whole length of torrent flow

Lt - total length of the tributaries of I - II order.

K – Climatic and topographic coefficient calculated by the formula:

$$K = T \times \sqrt{Ip} \tag{2}$$

Where:

Ip - mean pondered slope of torrent bed

T is temperature coefficient calculated by the formula:

$$T = \sqrt{\frac{t^{\circ}}{} + 0.1} \quad (3)$$

Where:

t° - Mean annual temperature

A - coefficient of catchments form calculated by the formula which defines the similarity of the form of torrential catchments with the fan-like semi-circular form which characterises the most intensive torrents.

$$A = \frac{P}{L \times (\pi + 2)} \quad (4)$$

P - Catchments perimeter

The calculated value of the coefficient of hydro graphic class Hk is the base of torrent classification. They are classified into six classes of torrents, as presented in the Table 2 [3].

Torrent classification		Table 2
Torrent class	Description	Torrent hydrographic class coefficient (Hk)
A	Torrent Rivers	Hk>20
B	Small torrent Rivers	10<Hk<20
C	Torrent streams	1.0<Hk<10
D	Small temporary torrent streams	0.1<Hk<1.0
E	Landslide small torrents	0.05<Hk<0.1
F	Gullies	Hk<0.05

EPM torrent class (A-F) and erosion coefficient (Z) are basic values for antierosion and torrent training works optimization. Torrent Formula can be entered to the list in the form of alpha-numeric data of the following form:

B- III- 0587

Which means that it is a torrent

- Class **B** (Small Torrent River)
- **III** torrential category. See table 1.
- Mean value of erosion coefficient is **Z=0.587**, i.e. it is a more severe form of medium erosion.

The torrents classified in this way can also be presented in the cartographic form, because in this way the real torrential risk of the torrent can be much better evaluated. Torrent classification has improved the monitoring of the results of erosion control works and the elaboration of the optimizing calculations for the planning of the required volume of erosion control works in the function of erosion intensity and torrential class.

The inaccuracies made during the planning of this type of works, in addition to the irrationally high or low investments, can also result in a torrential disaster.

2.3. EPM Antierosion and torrent training works optimization calculation

Long duration of intensive complex antierosion and torrent training works help us to find optimal volume of the works. Long term investigation was successful. The Figures present the dependences of the optimal volume of works in the function of torrent class and erosion intensity. The dependences related to technical works and biological works are presented separately. The presented dependences are the result of fifty years of systematic work on torrent management and erosion control and the continual monitoring of the realised effects. The presented dependences relate to the region of Serbia and a greater part of the European countries. Figure 1 show relation who help us to define optimal volume of biological and biotechnical works depending on erosion intensity (Z) and torrent class. Figure 2 show similar relation for technical works [5].

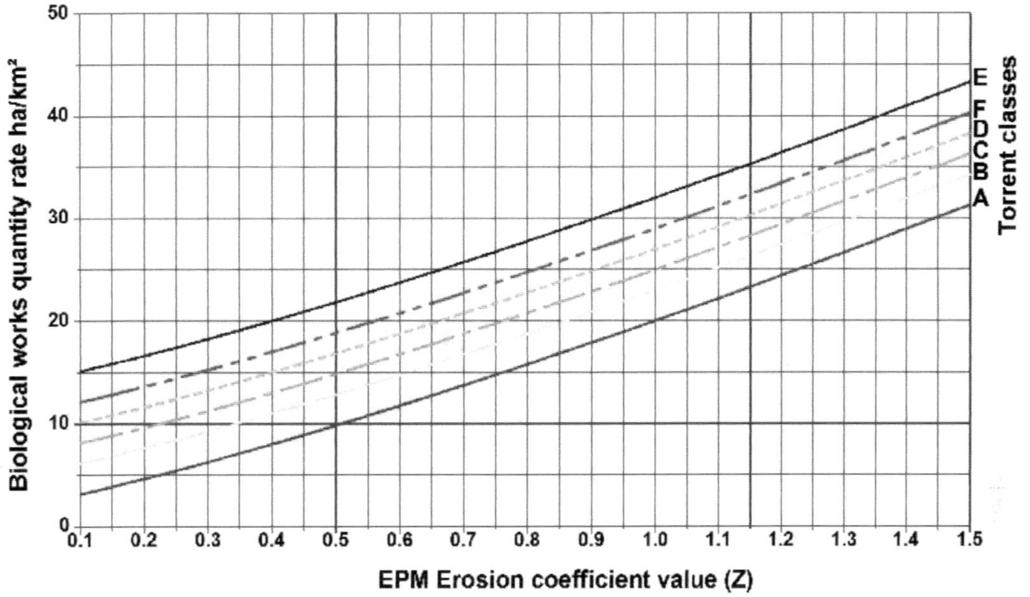

Figure 1. Relations between erosion coefficient and biological works quantity rate

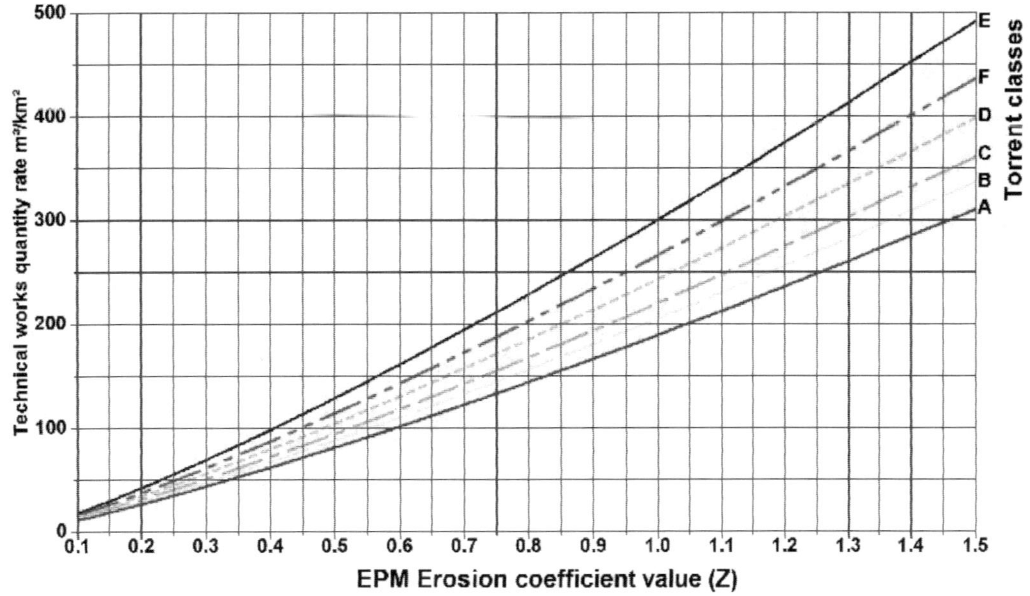

Figure 2. Relations between erosion coefficient and technical works quantity rate

Technical works are expressed in m^2/km^2, and biological works in ha/km^2. By applying the presented dependences, the required volume of technical works and biological works can be simply calculated for each classified torrent.

The calculated quantities of the necessary works are adequate solutions for the realization of reduced erosion intensity and the mitigation of torrential characteristics. Namely, it is technically impossible to realize the irresponsible promise given by many engineers, that erosion will be reduced to "zero" and that there will be no more torrential floods. By erosion control works and measures, it is only possible to mitigate the slopes and to improve the biological protective cover. However, the natural soil erodibility cannot be changed [6].

The positive effects of technical works are recorded immediately after the construction, while the period for the effects of biological works is longer than ten years. Evidently, the estimated volume of works should be realized within the period of ten to twenty years, depending on the realization of the desired effects and potential changes.

Thanks to the lack of respect of the above facts, numerous errors were made in the past both in Serbia and in other regions. Torrents were regulated exclusively by the systems of technical structures (dams and regulations), without any biological protection for the mitigation of erosion processes. Also, there were the opposite cases of the exclusive use of biological protection, which was successful by itself, but the total effect was negative, because the deep erosion processes were intensified, as it can be seen in the Figure 3.

The exclusive use of technical structures for torrent regulation was favoured in the cases of road and settlement protection, because the results were seen immediately. In such cases, there are no serious problems in the events of the usual regular torrential floods. The problem develops in the events of torrential floods of the lower probability of occurrence. In such cases, the sediment silts up or destroys the built torrent regulation structures.

Figure 3. Gully erosion downstream of the successful forestation.

Figure 4 presents a detail of the torrent regulation at the crossing with the road. Before the torrential flood, the water was 3m below the bridge.

Figure 4. The bridge destroyed during torrent flood and the channel was silted up.

Figure 5. The downstream check dam of several ones which were constructed at the same time.

IOP Conf. Series: Earth and Environmental Science **4** (2008) 012039 doi:10.1088/1755-1307/4/1/012039

In the presented case, the local technical works and biological works were performed at the higher positions of the catchments. The technical works for the protection of steep gullies in the zone of biological works were not performed.

A special problem occurs in the regulation of small torrents, for which the total volume of required works is proportionally low. In such cases, the investors try to perform all the necessary works in the shortest possible term.

In Serbia, the usually applied torrent classification used to be carried out based on the torrent cadastres, which helped the engineers and the Ministries in bringing the decisions and the road planners in selecting the optimal traces. Also, the above classification was applied for the improvement of the eroded catchments of the storage reservoirs. However, because of the economic problems in Serbia, torrent cadastres had to be discontinued. The consequence is the use of the old data from the time when the catchments were eroded, and the volume of the performed erosion control works was several times lower.

Those optimizations relations make fundamental changes during antoerosion and torrent training works. Optimized quantity of applied works directly mean greater territory covered and protected.

3. Conclusion

Torrent classification based on EPM ensures a set of short and clear information for the assessment of the torrent state. Taking into account that the classification of torrentiality depends directly on the quality and scale of erosion mapping, the mapping scale of 1 : 25,000 provides the required data set also for the gullies, whereas the smaller scales are applicable for the large torrential rivers and rivulets.

Torrent management works and erosion control works belong to the group of long-term activities that have to be planned and synchronized in the realization.

References

[1] Gavrilovic S 1957 *Classification of Torrents in Grdelica Canyon and Quantitative Regime of Torrent-Borne Sediments* (Beograd)

[2] Gavrilovic S 1966 *Methodology for Classification of Erosion Processes and Mapping of Eroded Areas. Stage I* (Beograd, Institute "Jaroslav Černi")

[3] Gavrilovic S 1972 *Engineering of Torrents and Erosion special* issue (Beograd: Izgradnja)

[4] Gavrilovic Z 1998 The Use of an Empirical Method (Erosion Potential Method for Calculating Sediment Production and Transportation in Unstudied or Torrential Streams" *International Conference on RIVER REGIME* (Wallingford, England)

[5] Gavrilovic Z, Stefanovic M, Milojevic M and Cotric J 2006 "Erosion Potential Method" An Important Support for Integrated Water Resource Management *Balwois Conference 2006.* (Ohrid. Macedonia)

[6] Gavrilovic Z and Stefanovic M 1998 *Methodology for identification and proclamation erosion zones* (Belgrade: Institute for development of Water Resources "Jaroslav Černi")

[7] Gavrilovic S 1968 *Proposed Classification of Torrents for the Water Management Plan of Yugoslavia* (Beograd: Institute "Jaroslav Černi")

[8] Gavrilovic S 1970 Contribution to the Methodology for Research in Erosion Plots *International Seminar: Erosion, Torrents, and Alluvial Deposits* (Beograd)

IOP Conf. Series: Earth and Environmental Science **4** (2008) 012040 doi:10.1088/1755-1307/4/1/012040

Erosion control works and the intensity of soil erosion in the upper part of the river Toplica drainage basin

S Kostadinov, N Dragović, M Zlatić and M Todosijević

Faculty of Forestry, Belgrade University, Kneza Višeslava 1, 11030 Belgarde, Serbia

E-mail: kost@EUnet.yu

Abstract. Aiming at the protection of the future storage "Selova" against erosion and sediment, and also to protect the settlements and roads in the drainage basin against torrential floods, erosion control works in the upper part of the river Toplica basin, upstream of the storage "Selova", started in 1947. The works included building-technical works (check dams) and biological works (afforestation and grassing of bare lands and other erosion risk areas). Within the period 1947–2006, the following erosion control works were executed: afforestation of bare lands on the slopes 2,257.00 ha, grassing of bare lands 1,520.00 ha, and altogether 54 dams were constructed in the river Toplica tributaries. This caused the decrease of sediment transport in the main flow of the river Toplica. This paper, based on the field research conducted in two time periods: 1988 and in the period 2004–2007, presents the state of erosion in the basin before erosion control works; type and scope of erosion control works and their effect on the intensity of erosion in the river Toplica basin upstream of the future storage "Selova".

1. Introduction

Properly designed and performed erosion control works cause the decrease of soil erosion intensity (Kostadinov, S., Marković, S., 1996; Kostadinov *et al.,* 2000). In this respect, biological and technical works have different activities and effects. Biological works (afforestation, grassing, orchard establishment on terraces) are carried out on catchment slopes. They cause the lowering of erosion intensity and can reduce the intensity to normal geological erosion. Technical works can be longitudinal, which protect stream banks against erosion and sloughing, and transversal, which prevent bed erosion and retain the sediment (first of all bedload).

Of all water management structures, water storages are the most endangered structures by erosion and sediment.

In the storages, in addition to the loss of the available storage capacity, sediment siltation also has chemical and biological impacts. By topsoil washing from the catchment slopes, the streams bring various chemical elements to the storage. The input elements can cause chemical and biological reactions, which affect water quality directly and significantly. The degree of risk imposed by the storage sediment depends on the quantity of sediment input, i.e. on the size of the drainage basin and erosion intensity in it.

Aiming at the protection of the future storage "Selova" against erosion and sediment, and the defence of settlements and roads against torrential floods, erosion control works in the river Toplica drainage basin, upstream of the storage "Selova", started in 1947. The works consisted of structural

IOP Conf. Series: Earth and Environmental Science **4** (2008) 012040 doi:10.1088/1755-1307/4/1/012040

works (check dams) and biological works (afforestation and grassing of bare lands and other areas endangered by erosion).

Based on field researches in two time periods: 1988 and 2004–2006, this paper presents the state of erosion in the drainage basin before the erosion control works; the type and scope of erosion control works, and their effect on the intensity of erosion in the river Toplica drainage basin upstream of the future storage "Selova".

2. Study area and method

2.1. Study area

The River Toplica (South Serbia) is the largest left tributary of the Južna Morava, both by the water flow and by the basin area which is altogether 2,180 km^2. The storage drainage basin encompasses the east slopes of Mt. Kopaonik. The storage "Selova" will be the source of water supply for the population of the settlements in the districts of Niš and Toplica. Other functions are: protection against flood, sediment storage, generation of hydro-energy, fish farming, irrigation, tourism, etc. The structural works (rock filling) of the dam "Selova", on the river Toplica 18 km upstream of Kuršumlija, were completed in the spring 2007. The other accessory works were still to be completed, after which the filling of the storage could begin.

The storage controls 346.05 km^2 of headwaters of the Toplica drainage basin. The main characteristics of the dam and of the storage are:

- Height of earth dam .... 70.6 m
- Total storage volume ....70.5 mil. m^3
- Volume of dead storage..................................7.3 mil. m^3
- Elevation of the crest.....................................527 m a.s.l.

The dam is made of tipped rockfill, with slant clay core, with watertight diaphragm (curtain) of variable depth.

The drainage basin of the storage "Selova" is characterised by intensive erosion, which produces high quantities of sediment. The sediment transported by the Toplica and its tributaries will be delivered to the future lake, where it will be deposited, thus reducing the available storage capacity. It will also cause mechanical and chemical contamination of the water in the storage.

2.2. Method

The method consisted of the following phases:

1. Collection of documents and data on the state of erosion in the river Toplica drainage basin before the beginning of erosion control works;
2. Collection of documents and data on the type and scope of erosion control works;
3. Investigation of the drainage basin natural characteristics;
4. Assessment of erosion intensity and distribution in the drainage basin in 1988;
5. Assessment of the state of erosion-control means: biological and technical.
6. Assessment of erosion intensity and distribution in the drainage basin in 2006;
7. Calculation of sediment yield and transport in different time periods;
8. Discussion of the results;
9. Conclusion.

The geological map of the drainage basin is based on map scale 1:100,000. The soil characteristics were determined based on the soil map of Serbia, issued by the Institute of Soil Science in Belgrade, scale 1:50,000, as well as by direct field tests (opening of the soil profiles) and laboratory study. Land use maps of the drainage basin for two time periods, 1988 and 2006, were based on satellite images and direct reconnaissance.

The assessment of erosion intensity and distribution in the drainage basin in 1956 was adopted from the existing technical documents. The intensities of soil erosion in 1988 and 2006 were defined based on erosion mapping. Erosion map was constructed by Gavrilović's methodology with the addition of satellite images and direct reconnaissance. Erosion map scale is 1:25,000.

Sediment yield and transport were calculated by Gavrilović's method, aiming at the comparison of the state of erosion in different time periods.

3. Results

3.1. The main physico-geographical characteristics of the drainage basin

Table 1 presents the main parameters of the drainage basin, significant for the genesis of soil erosion processes and sediment transport. The drainage basin parameters point to a significant predisposition of the terrain for the development of intensive erosion processes and sediment transport.

Table 1. The main parameters of the drainage basin

Parameters	Symbol	Value
Basin area	$A - km^2$	346.05
Basin length	$L - km$	33.2
Basin perimetre	$O - km$	94.06
Mean altitude of the basin area	N_{sr} – m.n.m	971.10
Mean altitudinal difference	$D - m$	514.7
Mean slope of the basin area	$I_{sr} - \%$	37.44
Stream bed slope	$I_t - \%$	4.72
Local erosion basis	$H - m$	1 ,560.6
Drainage density	$G - km \cdot km^2$	2.23

3.1.1. *Bedrock composition and hydro-geological features of the terrain.*The lithologic composition, especially the development of the Tertiary clastic sediments, affects the development of instability and erodibility. Sericite-chlorite schists consist of fine-grained quartz, sericite and chlorite. They occur as the marginal, contact zone of the granodiorite massif Kopaonik. Crystalline limestones and marbles occur as alternating layers of crystalline limestones, dolomites and marbles. They also belong to the series of low-metamorphised rocks of the margin of Mt. Kopaonik granodiorite massif. Serpentinites are found in the east periphery of Ibarski ultrabasic massif. They occur in tectonised zones, often schistose and hydro-thermally modified. Gabbros do not occur as large masses in the diabases of the Kuršumlija tectonic zone. Diabase-hornstone formation consists of sedimentary rocks and igneous rocks: sandstones, limestones, hornstone, argillites and diabases. It is in the tectonic relation with the surrounding rocks. The oldest sedimentary rocks are the Lower Cretaceous basal breccias, siltstones and sandstones. The most represented rocks are micro-conglomerates, arenytes, siltstones and marls. The calcareous sediments of the Senonian occur as layered slaty and massive reef limestones which occasionally end with marmorised limestones. The facies diversity of the Senonian flysch is characterised by alternating sandstones, siltstones and marls. Sedimentary-volcanogenic series is represented by the rhythmical alteration of sandstones, marls, argillites, limestones and tuffs. Granitoid rocks consist of porphyroid quartz-moconites and granular granodiorites.

IOP Conf. Series: Earth and Environmental Science **4** (2008) 012040 doi:10.1088/1755-1307/4/1/012040

The products of volcanic activity are quartz-latite rocks and minor masses of dacite-andesite. The youngest formations are the unbound sediments of alluvion, diluvial-proluvial screens and screens of the Quaternary period.

3.1.2. *Soil in the drainage basin of the storage "Selova".* The following soil types are present in the drainage basin:

- Humus-siliceous soil with two subtypes:
  - eutric on flysch, serpentinite, dacite-andesite, schists, gabbro-diabase and sandstone, and
  - dystric on dacite-andesite, schists, flysch and sandstone.
- Brown soil, also in two subtypes
  - eutric soil on flysch, andesite and gabbro-diabase and
  - dystric soil on flysch.
- Chernozem on compact limestones
- Alluvial sediment
- Colluvium
- Lithosols, formed at places and creating a mosaic with humus siliceous soils and brown soils.

Of all the above soils, the most represented are eutric humus-siliceous soils on flysch on the area of about 133 km^2 (52%), and then eutric humus-siliceous soils and lithosols on other parent rocks 66.87 km^2 (19.2 %).

Eutric brown soils cover the area of 9.25 km^2 (2.65%), on about 38 km^2 (10.9%). They are distributed in a conglomerate with humus-siliceous soils and lithosols. Dystric brown soils cover the area of 0.25 km^2 and they are mosaically distributed with humus-siliceous and lithosols on about 1.0 km^2 (2.9%). Chernozems occupy about 4 km^2 (1.15%), and the colluviums cover 0.25 km^2. Alluvial sediments, calcareous and non-calcareous loamy, deep, occur in the river valleys on the area of altogether 7.5 km^2 (2.15%).

3.1.3. Climate characteristics. Climate, the most important physico-geographical factor, has an essential effect on hydrological features of the Toplica drainage basin, which is situated in the region of temperate continental climate (Central European climate). The headwater part of the drainage basin belongs to the mountain zone characterised by abundant atmospheric precipitation. Winters are severe and cold, and summers are fresh. Springs are changeable, with frequent alternation of warm and cold and windy days and frequent rains. Autumn is relatively warmer than spring.

Mean air temperature in a multiannual period, measured at the weather station Kuršumlija, as the nearest to the drainage basin, is 10.3°C. January is the coldest month and the only month in the year with negative mean monthly temperature (–0.49°C). The warmest month is July, with 19.8°C. Based on the isohyet map of the drainage basin, the mean multiannual precipitation sum amounts to 891.9 mm. In addition to precipitation and temperature, the regime of water and sediment in the drainage basin is also affected by wind. An especially high effect is that of the spring south wind "jugovina" which melts the snow which remains even in April in the higher parts of the drainage basin (above 1,200 m). Snow melting becomes fast and leads to sudden increase of water level, and great quantities of mud are delivered to the watercourses.

Table 2. Mean monthly and annual precipitations for the river Toplica drainage basin, upstream of the storage "Selova"

Station	Annual values	Period of the measurement
Kuršumlija	639.4	1946–2003
Velika Šatra	828.5	1955–1999
Selova	669.8	1955–1999
Lukova	931.3	1951–1999
Štave	908.4	1955–1999
Kopaonik	959.4	1950–2003
Trećak	714.5	1954–1999
Blaževo	812.1	1946–2003

3.1.4. *State of vegetation in the drainage basin.* The greatest part of the drainage basin area is occupied by forests (more than 40 %) and thin - degraded forests (about 14%)

Forest associations in the drainage basin are as follows:

1. Forest of Hungarian oak and Turkey oak (Quercetum farnetto cerris) is climateogenic forest of this part of Serbia and it covers the greatest part of the area under forests.
2. Forest of sessile oak (Quercus petraeae s.lat.) is a belt above the forests of Hungarian oak and Turkey oak.
3. Mountain beech forest (Fagetum montanum Rud.) is for the most part orographically conditioned, i.e. it occurs as a permanent stage at the cooler and fresher sites of oak region.
4. Forest of beech and fir (Abieto-Fagetum serbicum Jov.) grows above montane beech forests and occupies a significantly smaller area.

Table 3 presents land use in the drainage basin in two time periods: 1988 and 2006.

Table 3. Land use in the drainage basin in 1988 and 2006

No	Culture	1988		2006	
		km^2	%	km^2	%
1.	Forests	140.29	40.54	149.03	43.07
2.	Degraded forests	49.26	14.23	48.05	13.88
3.	Meadows and pastures	57.61	16.65	72.81	21.04
4.	Degraded pastures	12.55	3.63	11.47	3.31
5.	Ploughed land	25.89	7.48	13.21	3.82
6.	Mixed cultures	29.72	8.59	29.19	8.44
7.	Bare land	23.87	6.90	15.43	4.46
8.	Setlements	6.86	1.98	6.86	1.98
Total		346.050	100.00	346.05	100.00

IOP Conf. Series: Earth and Environmental Science **4** (2008) 012040 doi:10.1088/1755-1307/4/1/012040

Agricultural areas include farmlands, pastures, orchards and meadows. A special group consists of: farmsteads, meadows, thin forests, pastures and orchards, as the areas where, all these land uses occur on a small space, on somewhat gentler slopes, usually with medium intensity of erosion, so they have not been dealt with separately for rationality purposes.

3.2. Soil erosion before erosion control works
At the beginning of the fifties of the last century, it was decided to construct the dam and the storage "Selova" in the river Toplica drainage basin, for the purposes of water supply of Prokuplje, Blace, Merošine and Niš. In 1953, the river Toplica drainage basin became an experimental drainage basin for hydrological research. Based on the field reconnaissance, it was concluded that the prevailing water erosion processes in the drainage basin were very intensive, i.e. between intensive and, excessive according to Gavrilović's classification, with the coefficient of erosion Z = 1.0.

The most intensive erosion processes occurred in the drainage basins of the following tributaries of the river Toplica: Mala Reka, Kačaruša, Đerekarska Reka, Pavaštička Reka and in the upper part of the Lukovska Reka drainage basin. The observed erosion processes were: sheet, rill and gully erosion, as well as landslides.

The state of erosion in the drainage basin can be illustrated also by the transport of sediment from the drainage basin. Table 7 presents the calculation of sediment yield and transport by Gavrilović's method in three different periods: 1956, 1988 and 2006.

3.3. State of erosion in 1988
There was a considerable decrease of erosion intensity compared to the state in 1956. This was, by all means, the consequence of the performed erosion control works. Based on erosion map in 1988, the coefficient of erosion was calculated after Gavrilović, for the entire drainage basin upstream of the storage "Selova" and it amounted to Z = 0.49, which means that medium erosion processes prevailed in the drainage basin, but sediment yield and transport were high.

3.4. Erosion control works and their effect
Based on the data mainly obtained from enterprise "Erozija" Niš, Tables 4 and 5 present erosion control works in the Toplica drainage basin upstream of the storage "Selova". The works started in 1947, but very few, i.e. only five check dams were constructed till 1968.

Table 4. Technical erosion control works in the river Toplica drainage basin upstream of the dam "Selova"

Period	Check dams	
	Number	Masonry [m^3]
1947–1968	5	1,580.0
1968–1980	34	2,881.3
1980–1987	5	1,980.0
Total 1947–1987	44	6,441.3
1988–2006	10	4,182.0
Total 1947–2006	54	10,623.3
Average annual	0,90	177.06

Table 5. Biological works in the river Toplica drainage basin upstream of the dam "Selova"

Period	Afforestation [ha]	Grassing [ha]
1955-1987	737.0	20.3
1978-1987	1,280.0	930.7
1988-2006	240.0	380.0
Total 1955-2006	2,257.0	1,520.0
Average annual	37.6	25.3

A much higher intensity of works was carried out in the period 1968–1987. The following works were intensified: building of check dams for bedload storage in the Toplica tributaries, and biological works (afforestation and grassing) on the drainage basin slopes. Afforestation was carried out on bare lands attacked by erosion. Most often, it was dense planting in pits, and the most frequently applied species were Austrian pine and Scotch pine. Grassing was performed on ploughlands on steep slopes and on bare lands, with the application of a special grass mixture for erosion control. The beginning of dam construction for the storage "Selova" attracted public attention on the significance of erosion control works.

3.5. Analysis of the effects of erosion control works

The total area treated by biological works is 3777.0 ha which is 10.91% of the total area of the drainage basin.

The effect biological works is expressed by the decrease of erosion intensity, i.e. the reduction of sediment yield in the treated areas. This is reflected on the reduction of sediment transport through the hydrographic network, which also reduces the intensity of siltation in the storage. Based on the detailed Erosion map produced in 1988, it was concluded that erosion intensity in the drainage basin was significantly reduced compared to the state in 1956, when the drainage basin was attacked by severe erosion close to very severe erosion. Based on the mapping in 1988, the prevalent erosion process upstream of the storage "Selova" was medium erosion (after S Gavrilović), with coefficient of erosion $Z = 0.49$ (mean value for the whole drainage basin). It should be noted that there were some areas where afforestation was not completely successful, mostly in the headwater area of the Blaževska Reka (Boranačka Reka) drainage basin.

Another problem regarding the biological works is the fact that afforestation was most often performed by dense planting of conifers (more than 10,000 plants per 1 ha), which was not followed by tending measures, i.e. thinnings, etc. Such plantations do not have a good economic value now, and they are at risk caused by wildfires, decline, and due to too dense canopy, they are not resistant to snow-breaks and wind-breaks.

The building and technical works consisted of classical torrential dams. So far, altogether 54 check dams of different heights have been constructed, but the effective height of the majority of dams is between 3 and 5 m. They are mainly made of stone masonry, mortarless stone and, somewhat less, of concrete. The effect of check dams is primarily reflected on bedload checking, stream stabilisation, reduction of torrent bed slope, by which, thanks to the lower transport capacity of the stream, bedload delivery to the storage is indirectly reduced.

The greatest number of check dams is built of stone (dry dams, or dams in stone masonry). Only three dams are made of concrete, two in the Rankovica catchment, and one in Magovski Potok. A part of the check dams have a complete construction, with apron and subsidiary dam. As for the check dam dimensions, the check dam average height is about 3 m, and the width of the dam mouth is 4–6 m.

3.6. State of erosion in 2006

Erosion map of the drainage basin is based on the research and field reconnaissance of the drainage basin in the period 2005–2006. Table 6 presents the distribution and intensity of erosion in the river Toplica drainage basin upstream of the dam "Selova" in 2006.

The coefficient of erosion for the whole drainage basin now is 0.404, which means the prevalence of medium erosion processes, at the lower level, close to the weak erosion, according to Gavrilović's classification. Although the intensity and the scope of erosion control works in the period 1988–2006 was reduced compared to the previous period, the trend of reducing the intensity of erosion in the drainage basin continued, on the average. In addition to erosion control works, this was by all means also the result of demographic changes in the drainage basin in that period. Namely, the population from the mountainous villages which prevail in the drainage basin (except Lukovska Banja Spa) migrates to the towns, looking for jobs. In future research, the problem of migrations and reducing the number of active population should be studied in more detail.

Table 6. Erosion categories in the river Toplica drainage basin, 2006

No	Erosion category	f [km^2]	%	Z	fZ
1.	Weak	149.029	43.07	0.30	44.7087
2.	Weak	4.050	13.88	0.35	16.8175
3.	Medium	72.807	21.04	0.50	36.4035
4.	Medium	11.467	3.31	0.55	6.3069
5.	Medium	13.210	3.82	0.65	8.5865
6.	Medium	29.199	8.44	0.45	9.53955
7.	Intensive	15.430	4.46	0.85	13.1155
8.	Medium	6.858	1.98	0.65	4.4577
Total		346.050	100	O.404	139.93585

The calculations of sediment yield and transport in 2006 indicate that in future considerable quantities of sediment will be delivered annually to the storage "Selova". In addition to the reduction of storage volume, great damage is caused by mechanical and chemical contamination of the storage water. Namely, together with soil particles, fertilisers (mineral and organic), are also washed down from the slopes, then pesticides applied in plant production, and also some heavy metals present in the soil in the upper parts of the drainage basin (Đerekarska Reka, Blaževska Reka, Mala Reka), which can have a serious impact of the water quality and cause additional costs for its purification.

Based on the so far performed erosion control works in the drainage basin upstream of the storage "Selova", and based on the actual state of erosion, it can be concluded that the system of erosion control protection is not complete. For this reason, the storage "Selova" is at risk, both by mechanical means (siltation), and by ecological means, because erosion sediment is the carrier of all types of chemical contamination.

Table 7. Intensity of erosion and mean annual sediment transport in the drainage basin before and after the erosion control works

Period	Erosion intensity	Erosion category	Erosion coefficient – Z	W_{year} $[m^3y^{-1}]$	W_{year}/sp $[m^3y^{-1}km^{-2}]$	G_{year} $[m^3y^{-1}]$	G_{year}/sp $[m^3y^{-1}km^{-2}]$
1956	Intensive, excessive	I	1.0	969,694.40	2,802.2	624,483.8	1,804.6
1988	Medium	III	0.49	332,605.2	961.15	214,197.7	619.0
2006	Medium	III	0.404	249,004.3	719.60	160,358.8	463.4

Figure 1. Check dam for bedload checking

3.7. Forecast of siltation dynamics of the storage "Selova"

The process of suspended sediment deposition in the conditions of the established dam and storage "Selova" cannot be precisely analysed without the complex mathematical model. This model should take into account the regime of the storage, the regime of suspended sediment in the stream, as well as the coincidence of the torrential waves and the levels in the storage (the elevations from 468 to 526 m). Taking into account the relatively great length of the zone of wedging out of the "Selova" backwater (about 7 km) and hydraulic-morphological conditions in this zone, the greatest part of the coarser fractions of suspended sediment would be deposited in this sector. Without the exact calculation, it is difficult to estimate which part of the total transport of suspended sediment would be deposited in the domain of the available storage capacity, but this percentage could attain even 60%. This means that about 80,000 m^3 of suspended sediment from the river Toplica would be deposited in the available volume of storage. In other words, the available capacity of the storage "Selova" would be reduced annually on the average by about 80,000 m^3.

As for the bedload, it is certain that, in the conditions of the formed storage "Selova", this sediment would not reach the dead storage (below the elevation of 468). The main hydraulic reason of sediment retention is the insufficient transport capacity of the slowed down river Toplica for the sediment of certain sizes. Taking into account the estimated average annual bedload transport of the Toplica and Lukovska Reka, annual volume of bedload deposits would amount to about 20,000 m^3.

Based on the above, it can be concluded that, in the silting up process of the storage "Selova", annual deposition in the available storage capacity amounts to about 80,000 m^3 of suspended sediment, and about 20,000 m^3 of bedload. The loss of the available storage capacity of about 100,000 m^3/year is relatively small compared to its total size ($70 \cdot 10^6$ m^3). However, this does not mean that no measures should be taken for the reduction of sediment input in the storage. The erosion control management of the drainage basin would surely have very beneficial effects, both for the reducing of the lake silting up with sediment, and from the aspect of ecological and sanitation protection of the storage "Selova", aiming at the preservation of water quality.

4. Conclusion

The results of the research in the river Toplica drainage basin upstream of the dam "Selova", show a very significant effect of erosion control works on the decrease of intensity of erosion, the consequence of which is a considerable decrease of sediment yield and transport.

The greatest part of erosion control works was realized in the period 1967 - 1987. That was the period of the greatest decrease of erosion process intensity, sediment yield and sediment transport in the drainage basin.

In the period before 1956, there were very intensive water erosion processes in the upper part of the river Toplica drainage basin. The value of erosion coefficient was Z = 1.0 (severe to very severe erosion), the volume of sediment yield was W_{year} = 2,802 $m^3year^{-1}km^{-2}$, and sediment transport G_{year} = 1,804.6 $m^3year^{-1}km^{-2}$.

After large-scale erosion control works, the research in 1988 showed a significant decrease of the values of erosion intensity, sediment yield and transport: Z = 0.49 (medium erosion), W_{year} = 961.15 $m^3year^{-1}km^{-2}$ and G_{year} = 619.0 $m^3god^{-1}km^{-2}$ respectively.

At the end of the study period, in 2006, the intensity of erosion, sediment yield and transport were lower than in 1988: Z = 0.404 (medium erosion at the lower level, close to weak erosion); W_{year} = 719.60 $m^3year^{-1}km^{-2}$, G_{year} = 463. 4 $m^3year^{-1}km^{-2}$ respectively.

Although in the period 1988–2006 there was a reduction of erosion control works, the trend of decreasing intensity of erosion, sediment yield and transport continued, but with a lower rate of decrease. This can be explained by the fact that biological works performed in the period 1980–1987, attained their full functionality by the mid nineties of the 20th century, when the canopy of the new forest plantations was established, also the result of demographic changes in the drainage basin in that period.

In the last 60 years, sediment yield and transport in the river Toplica drainage basin, upstream of the dam and the storage "Selova", decreased by almost four times, however every year, considerable volumes of sediment will still reach the future storage "Selova". The reason is the fact that the erosion control system has not yet been completed. The problem is not the quantity of sediment, i.e. the storage siltation, but a much more significant problem is the mechanical and chemical contamination of storage water by the sediment. Namely, together with suspended sediment, many chemically detrimental substances will be delivered to the storage: organic and mineral fertilisers, pesticides, and other substances found in the soil, and among them some heavy metals. For this reason, it is necessary to go on with erosion control works in the drainage basin, especially in its headwater part., in order to protect the storage "Selova" against sediment silting and to protect the water quality from contamination.

References

[1] Gavrilović S 1972 Inženjering o bujičnim tokovima i eroziji *Časopis "Izgradnja"* (Beograd: Specijalno izdanje, DGA) pp 1-296

[2] Kostadinov S 1996 Soil Erosion and Sediment Transport Depending on Land Use in the Watershed *In the book: "Hydrological Problems and Environmental Management in Highlands and Headwaters"* ed Josef Krecek, Rajwar G.S. and Martin J. Haigh (New Delhi, Calcuta: Oxford and IBH Publishing CO. PVT. LTD) pp 43-51

[3] Kostadinov S and Marković S 1996 Soil Erosion and Effects of Erosion Control Works in the Torrential Drainage Basins In South - East Serbia *IAHS Publication no 236 ISSN 0144-7815 "Erosion and Sediment Yield: Global and Regional Perspectives"* ed Walling D E and. Webb B W (England: IAHS Press Walingford) pp 321-332

[4] Kostadinov S, Zlatić M and Dragović N 2000 Effects of Erosion Control Works in the Grdelička Klisura Gorge *Zbornik radova: "75 godini visše lesotehničesko obrazovanie v Balgaria"; sekcija: Gorsko stopanstvo* (Sofia: Lesotehničeski universitet) pp 115-124

[5] Kostadinov S 2004 Soil Erosion and Sediment Transport Within Small Torrential Drainage Basins of Serbia *Poster Report Booklet; Symposium AHS: Sediment transport through the fluvial system* (Moscow, August 2004) pp 16-21

[6] Petković S, Kostadinov S et al. 1989 *Studija antierozione zaštite sliva akumulacije "Selova"* (Beograd: Šumarski fakultet)

IOP Conf. Series: Earth and Environmental Science **4** (2008) 012041 doi:10.1088/1755-1307/4/1/012041

Response of conservation measures from small cultivated watersheds, concerning runoff and erosion, under the impact of extreme rainfall events

N Popa

Research and Development Centre for Soil Erosion Control Perieni Barlad, Romania

E-mail: nelu_c_popa@yahoo.com

Abstract. The study has been made in a representative small watershed with gently to hilly slopes from Tutova Rolling Hills, Romania. The system of conservation measures is represented by stripcroping, bufferstrips, bench terraces, a grassed waterway and a drainage network. The monitoring of hydrological response of agricultural units has been made in two cross sections corresponding to each of the land use type by means of two concrete triangular weirs. The most important soil losses were caused by three extreme rainfall events from August 2004, May 2005 and September 2007. At the date of the first rainfall event, the soil was generally very well protected against erosion by the vegetative cover, excepting parcels that were just ploughed after the mash crop. In that case, it was estimated that the value of soil losses ranged between 20.0 and 24.5 t/ha while for the other crops like corn and soybean, soil losses they were 1.0-1.5 t/ha and 0.5-0.8 t/ha respectively. Damages caused by the rainfall from September 2007 were much more important because at that time about 30% from the entire surface was just prepared for rape seeding. Maximum value of erosion was 95 t/ha on a parcel with 16% slope and 50m length along the slope.

1. Introduction

Romania has a variety of landscapes resulting from its wide range of major relief forms which are very well proportioned: 36% Carpathian Mountains and Subcarpathians, 34% hills and tablelands, and 30% plains. Within its boundaries live 23.5 millions people.

Hot summers and cold winters, variability in the distribution of rainfall, fluctuating length of growing season typify Romanian continental-temperate climate. There is a crossing of three main climatic influences as follow: Mediterranean (subtropical) climate, central European climate (under Atlantic air masses) and east European climate by excessive continental nuance.

Under these natural conditions, erosion processes are numerous and the following forms represent them: sheet erosion, gully erosion and landslides.

Traditional conservation measures on the agricultural fields are still maintaining in Romania. Among them, are mentioned the following: contouring (the practice of tillage and planting on the contour), stripcropping, buffer stripcropping, agroterracing on cropland, wine and fruit plantation terraces, gully control structures, perennial grasses in the crop rotation system and artificial revegetation of the denuded fields by shrubs and trees.

Unfortunately, these methods have not been generalized. Consequently, in some areas were practiced on large scale methods that enabled increasing of erosion as fallow: up-and-down hill plot layout and farming, land clearing for agricultural purposes on hillsides with high risk of erosion. Many indigenous forests were removed by ax and fire, while pastures were overgrazed or the sod turned by plow into cropland.

During the last decades erosion experiments have been developed on small watersheds with predominant agricultural or forestry land use. There have been elaborated methods of runoff and erosion estimation. The concept of an integrated system has been elaborated for rational treatment and use of the eroded soils. There have been elaborated treatment drafts, which take in hypotheses to choose the solutions depending on the climatic zone, soil conditions and relief.

The present trend of scientific research is to find indicators easy to measure and to define the behaviour of the environment in relation to runoff and erosion risks.

Effectiveness of soil conservation measures is proved mainly during the extreme rainfall events when the amount of soil losses may be very important.

2. Materials and methods

2.1. Study site

The study area is located in a representative small watershed named Gheltag (46^{0}15'40"N, 27^{0}35'30"E) with gently to hilly slopes from Tutova Rolling Hills, Eastern Romania (figure 1) where the soil is represented by grey wood soil, highly eroded. The terrain that has almost 100 ha arable land and 18 ha pasture was fit out by different conservation measures in the period of 1975-1977 when it belonged to the Stat [3].

After Romanian revolution from 1989, according to the new laws on land property, the entire surface of Gheltag basin was realloted to 83 old landowners, on the original locations that means that the plots consisted in long and narrow strips, disposed up-and-down the hill. This situation made impossible applying of the contour tillage system and leads to abandon of the majority of conservation practices.

Since 2002 the terrain was leased by Perieni Research Centre and was completely rehabilitated. The following soil conservation measures were used: intercropping, contour planting, strip cropping, bench terraces, a grassed waterway and a drainage network (figure 2).

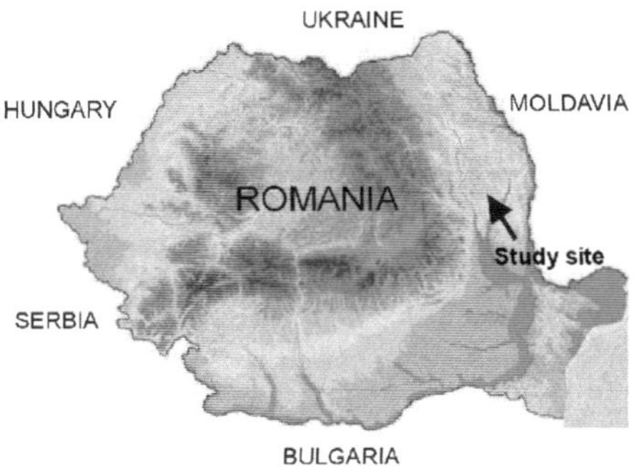

Figure 1. Location of the study site

Figure 2. Bird's-eye view of Gheltag basin, Romania

Two stations provided meteorological data. One of them is located 5 km S-E, in Barlad town and it registered data since 1941. The other one is an automatical station and is positioned in the neighbourhood of the site, on the territory of Perieni Research Centre.

Historical registrations indicated that yearly average of precipitation and temperature at Barlad station are 494 mm and 9.8 °C respectively. A maximum monthly average of precipitation is placed in June with 76.8 mm while the minimum is 24.1 mm in March. The critical season of erosion is between

May and August but in the last decade data showed that these limits are about to change and they trend to enlarge until the middle of September.

The last extreme rainfall events that were associated with the highest values of runoff and erosion were registered in August 29, 2004 (76.3 mm), May 07, 2005 (53.5 mm) and September 05, 2007 (88.5 mm).

2.2. Experimental methods

The monitoring of hydrological response of agricultural units was performed in two cross section corresponding to each of the land use type (cultivated land and pasture) by means of two short-crested triangular weirs. This weir type was developed by Natural Resources Conservation Service – US Department of Agriculture [1] and its characteristics are presented in figure 3 and equation (1).

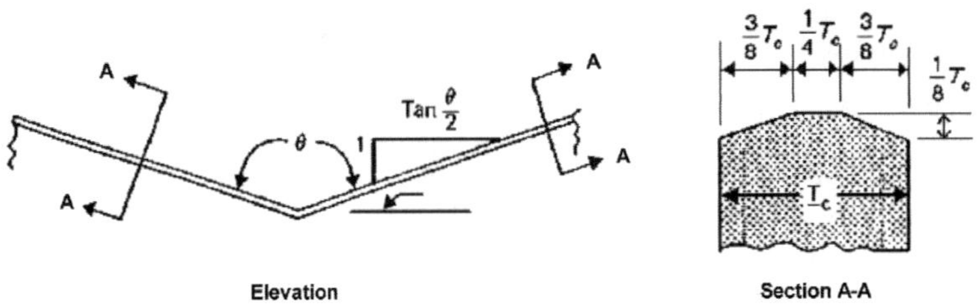

Figure 3. Short – crested triangular weir

The discharge equation for the short – crested triangular weir is the following:

$$Q = \left(C_0 + C_1 Log_{10} \frac{h}{T_c} \right) \sqrt{g} \tan \frac{\theta}{2} \left(h + \alpha \frac{v^2}{2g} \right)^{2.5} \qquad (1)$$

$\theta = 150^{\circ}$

$\alpha = 1.33$

v = average velocity in cross section 3 m upstream from center of crest;

h = head 3 m upstream from center of crest;

T_c = crest thickness (0.4 m);

C_o, C_1 = correction factors.

Assessing soil losses on every parcel was performed mainly by conventional methods (topographical measurements) after every most important rainfall events. However, on some plots with steep slopes, the Cs-137 technique was adopted to evaluate the effectiveness of conservation measures on medium term. Thus, 7 sampling points disposed on a representative transect were selected to characterize the effectiveness of the narrow terraces in diminishing of soil erosion processes on hilly slopes. In soil losses calculating, Cs-137 inventory was determined and some conversional models like Mass Balance Models developed by Walling DE [4] were used. The main advantage of this technique is the possibility to analyze erosion and to have a diagnostic of the soil status even for a single visit in the field.

3. Results and discussions

On August 29, 2004, when the rainfall reached 76.3 mm in 90 minutes, the stage of the crops was the following: 24 ha from the total of 100 ha were just ploughed after the mash crop; 52 ha were covered by winter wheat stubble; the height of corn plants (12 ha) had the maximum value; 12 ha were occupied by soybean.

On areas with gently to rolling slopes, the most important soil losses were caused by the pattern of relief formed many years ago when the tillage up-and-down the hill was usually executed. Even today, where contour cropping and strip cropping systems are applied, the former microchannels still exist.

Measured data were compared to those obtained from the parcels situated in the neighborhood which, approximately are in the same conditions of soil and relief but they are disposed up-and-down the hill.

On the surface covered by corn it was found, for example, that values of erosion ranged between 6.5 and 14 t/ha while in Gheltag basin soil losses were 1.0-1.5 t/ha for corn and 0.5-0.8 t/ha for soybean.

The worst situation was met on the plots just ploughed after the mash crop where it was estimated that the values of soil losses varied between 20.0 and 24.5 t/ha. On the similar conditions, the parcels disposed up-and-down were eroded by 35-45 t/ha.

On the plots with winter wheat stubble, the remainders of vegetation constituted a very good protection of soil against erosion and soil losses were insignificantly.

The grassed waterway worked good enough, thus excess of water was rapidly evacuated without trigging rill erosion. Maximum values of runoff registered on short – crested triangular weirs were 4.14 m^3/s for the cultivated land and 2.81 m^3/s for the grazing land. It was calculated that through cross section corresponding to the cultivated land run about 9100 m^3 of water that represent cca. 12% from the total precipitation.

At the time of rainfall from May 07, 2005 (53.5 mm in 80 minutes) 48 from 100 ha were covered by winter wheat and the rest of 52 ha were just seeded by pea (11 ha), soybean (19 ha), mustard (16 ha) and corn (6 ha). Therefore, not merely the terrain was uncovered by vegetation but also the surface of soil was finely hoed, creating ideal conditions for surface erosion. It was estimated that erosion varied locally between 15 and 35 t/ha, depending on the hill slope. For the whole basin the average of erosion was about 10-12 t/ha, having in sight the influence of stripcroping system where the parcels with bare soil alternated with those covered by winter wheat.

In the same manner like in previous described case, the most important soil losses were registered on the bare soil, situation that also owed to the pattern of soil surface characterized by small microchannels formed during the former agricultural system.

Measurements on above described weirs indicated that maximum values of runoff were 4.63 and 3.17 m^3/s. They showed that even the level of precipitation was lower than the first analyzed situation, conditions concerning soil status were much more favorable for developing runoff and erosion.

The last analyzed is the rainfall from September 05, 2007 which reached 88.5 mm in more than 200 minutes. Practically, 4 rainfall events in a single day succeeded with short intervals between them.

Damages caused by these rainfalls were significantly higher than the previous described rain events because, at that time, about 30% from the entire surface was just prepared for rape seeding. Maximum value of erosion was 95 t/ha on a parcel with 16% slope and 50m lengths along the slope. Net erosion for the entire watershed was estimated to 22 t/ha. Runoff through the main cross sections have maximum values of 6.25 and 4.05 m^3/s and the surface drainage network discharged about 13000 m^3 of water. They were the highest values registered in Gheltag basin since it was monitored.

On the bare soil from small private plots disposed up-and-down, estimated soil losses varied between 47 and 110 t/ha depending on relief parameters. On the other hand, an insignificant percentage of private lands were ploughed and hoed at that date for the reason that majority of small landowners do not plant rape which is usually seeded in September but winter wheat that is seeded one month later. Thus, more than 95% of terrains were covered by vegetation and a low level of erosion process was noticed.

Finally, to have an image about behavior of conservation practices on the steep slopes, some measurements using Cs-137 technique for a representative part of the basin were performed. Methodology took into account the effects of all factors that influenced erosion processes till the end of 2007.

On narrow terraces, redistribution of soil was mainly assigned to tillage erosion. Thus, measurements in the upper part of bench terraces, from which values of ^{137}Cs inventory were below the reference, indicated that significant erosion in the last decades occurred. Year by year, soil was translocated to the lower edge of terraces, process reflected by high values of ^{137}Cs activity in the lowest sampling points of terraces.

Values of soil losses calculated by Mass Balance 3 Model [4] ranged between 8.3 and 18.6 t/ha/y while values of deposition were situated between 22.1 and 57.8 t/ha/y. However, net erosion for the entire transect was 1.8 t/ha/y that is an acceptable value if compared to soil loss tolerance, which is 2-6 t/ha/y in Gheltag basin.

The above-presented situations are representatives for the natural conditions of the investigated site and showed that arable land has to be much more protected by vegetation, especially in the spring and in the autumn, when the crops are seeded. That means that alternating crops, contouring system, stripcrops system, and generally, all conservation practices have to be more carefully managed having in sight that the rainfalls are already much more aggressive due to changing climate.

4. Conclusions

- Much as professionally realized, soil conservation measures can be affected in different degrees by erosion if the rainfalls go beyond the limits took into account in designing activity. However, on places where conservation practices were applied, soil losses caused by the extreme rainfall events were about three times more reduced than those from the parcels disposed up-and-down the hill.
- Alternating crops, contouring system, stripcrops system, and generally, all conservation practices from agricultural land play more and more important roll in controlling runoff and erosion because rainfalls are already much more aggressive due to changing climate.

References

[1] Brakensiek, DL, Osborn HB and Rawls WR 1979 *Field Manual for Research in Agricultural Hydrology,* Agricultural Handbook 224, U.S. Department of Agriculture, U.S. Government Printing Office, Washington, DC.

[2] Popa N, Filiche E and Purnavel G 2002 *Stages of soil degradation by erosion*, Scientific papers", Agronomy Series of University of Agricultural Sciences and Veterinary Medicine Iasi, Romania.

[3] Popa N, Nistor D, Năstasă V, Purnavel G, Filiche E and Petrovici G 2005 *Gheltag basin from Tutova Rolling Hills, Romania, model of land management for the private agricultural land.* „Scientific papers" Vol. 48, Agronomy Series of University of Agricultural Sciences and Veterinary Medicine Iasi, Romania.

[4] Walling DE, Zhang Y, He Q 2007 *Models for Converting Radionuclide (^{137}Cs, Excess ^{210}Pb , and ^{7}Be) Measurements to Estimates of Soil Erosion and Deposition Rates* CD Including Software for Model Implementation.

Model based estimation of sediment erosion in groyne fields along the River Elbe

Sandra Prohaska, Thomas Jancke and Bernhard Westrich

University of Stuttgart, Institute of Hydraulic Engineering, Pfaffenwaldring 61, 70550 Stuttgart, Germany

E-mail: sandra.prohaska@iws.uni-stuttgart.de

Abstract. River water quality is still a vital environmental issue, even though ongoing emissions of contaminants are being reduced in several European rivers. The mobility of historically contaminated deposits is key issue in sediment management strategy and remediation planning. Resuspension of contaminated sediments impacts the water quality and thus, it is important for river engineering and ecological rehabilitation. The erodibility of the sediments and associated contaminants is difficult to predict due to complex time depended physical, chemical, and biological processes, as well as due to the lack of information. Therefore, in engineering practice the values for erosion parameters are usually assumed to be constant despite their high spatial and temporal variability, which leads to a large uncertainty of the erosion parameters. The goal of presented study is to compare the deterministic approach assuming constant critical erosion shear stress and an innovative approach which takes the critical erosion shear stress as a random variable. Furthermore, quantification of the effective value of the critical erosion shear stress, its applicability in numerical models, and erosion probability will be estimated. The results presented here are based on field measurements and numerical modelling of the River Elbe groyne fields.

1. Introduction

The prediction of flow, sediment and contaminant transport in natural rivers is very difficult due to river morphology varying in time, complex sediment transport phenomena, and uncertainties related to chemical properties of contaminants. Beside physical factors, biological characteristics of deposited sediments have also a great influence on resuspension. For example, sediment resistance is increased by extracellular polymeric substances (EPS) produced by macrofauna, algae, and bacteria [1, 2]. All mentioned factors (physical and biological) are lumped in numerical modelling to the critical erosion shear stress ($\tau_{cr,E}$). The spatial and temporal variability of the $\tau_{cr,E}$ is not yet included in the constitutive description of cohesive sediment erosion [3]. This emphasizes the difficulty to determine the effective value of the $\tau_{cr,E}$, to be used in transport modelling for engineering practice. Therefore, managing cohesive sediment erosion and resuspension is still a challenging task. In order to avoid inappropriate application of the $\tau_{cr,E}$, natural sediments are sampled and subsequent laboratory measurements are performed to quantify numerical model parameters.

The measuring results presented here are from three different sites of the groyne fields of the River Elbe, sampled during different seasons of the year. Based on these measured data and their statistics, various spatial distributions of the $\tau_{cr,E}$ are generated and applied in the numerical model. The applied numerical model is 1D multi-strip model, which predicts the suspended sediment transport in rivers trained by groynes [4]. The aim of the study is to compare the deterministic approach assuming constant critical erosion shear stress and an innovative approach which takes the critical erosion shear stress as a random variable. Thus, the uncertainty aspect is taken into account. Furthermore, quantification of the effective value of the critical erosion shear stress, its applicability in numerical models, and erosion probability are estimated.

1.1. Investigated area
The current research is focused on a representative Middle Elbe reach, for which flow and suspended particulate matter (SPM) regime is influenced by two tributaries (Mulde and Saale), see figure 1. The groyne structures are built on both river sides. The $\tau_{cr,E}$ is measured in three groyne fields (GF): Coswig (km 235.0), Steckby (km 280.0), and Fahlberg (km 318.0). Coswig is situated on the right river bank in the strong bend of the Elbe (GF width of about 60 m). Steckby is situated on the left river bank with the groyne field width of about 80 m. The biggest investigated groyne field is Fahlberg, situated on the left river bank (GF width of about 90 m).

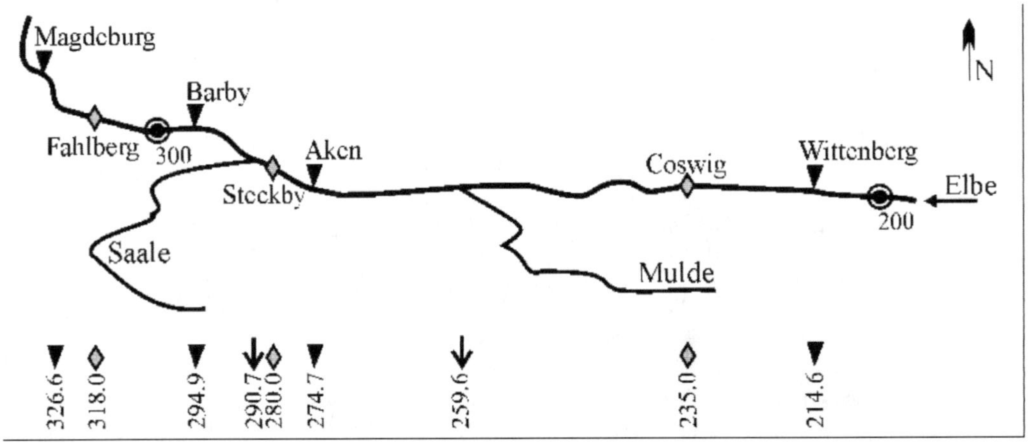

Figure 1. Middle Elbe: modelled area from Wittenberg to Magdeburg and sampled sites (Coswig, Steckby and Fahlberg).

The simulated domain extends from Wittenberg (km 214.6) to Magdeburg (km 326.6). Cross sections are determined by ultra sonic profiling at each 100 m, together with information on groyne field dimensions.

2. Sediment sampling campaigns
As mentioned in section 1, the samples are taken from three different sites. At Coswig the undisturbed sediment cores of diameter 13.5 cm are taken in July 2005, at Steckby in July 2005, and at Fahlberg in July, August and November 2005 in zones with approximately highest deposition. Additional sediment sampling at various spots within the groyne field at Fahlberg is performed in October 2006 in order to obtain the spatial variability of the measured parameter. The $\tau_{cr,E}$ is measured in a laboratory channel, so called SETEG system [5]. SETEG consists of a rectangular channel, where the shear stress is a function of the controlled flow rate based on calibration, see figure 2. The sediment cores are inserted from below and sediment is slowly moved upwards. The flow rate is increased until the critical shear stress is reached, i.e., when significant particle erosion can be observed. Using this setup, the critical erosion shear stress can be determined for different sediment layers at every 1 cm down to 50 cm.

In figure 3 measured depth profiles of the $\tau_{cr,E}$ are shown for three groyne fields at different time and location. Measurements show strong variation of the $\tau_{cr,E}$ depending on the type of the groyne field and the sampling location, see figure 3a. Nevertheless, the values measured at Steckby are considerably higher in the first 12 cm compared to the other locations because of their age and consolidation. The results of the seasonal measurements are shown in figure 3b where surprisingly higher values are observed for summer period compared to the autumn one. It is however, expected that in summer period biomass and macrofauna are higher and thus, the $\tau_{cr,E}$ to be lower [1].

In figure 3c a large spatial heterogeneity of the bottom sediments within the groyne field Fahlberg is shown. The sediment stability testes at Fahlberg show that $\tau_{cr,E}$ strongly differs depending on the exact location of the sampling point. The difference can be very high, sometimes up to one order of magnitude. Fahlberg groyne field and the exact sampling points are shown in figure 4.

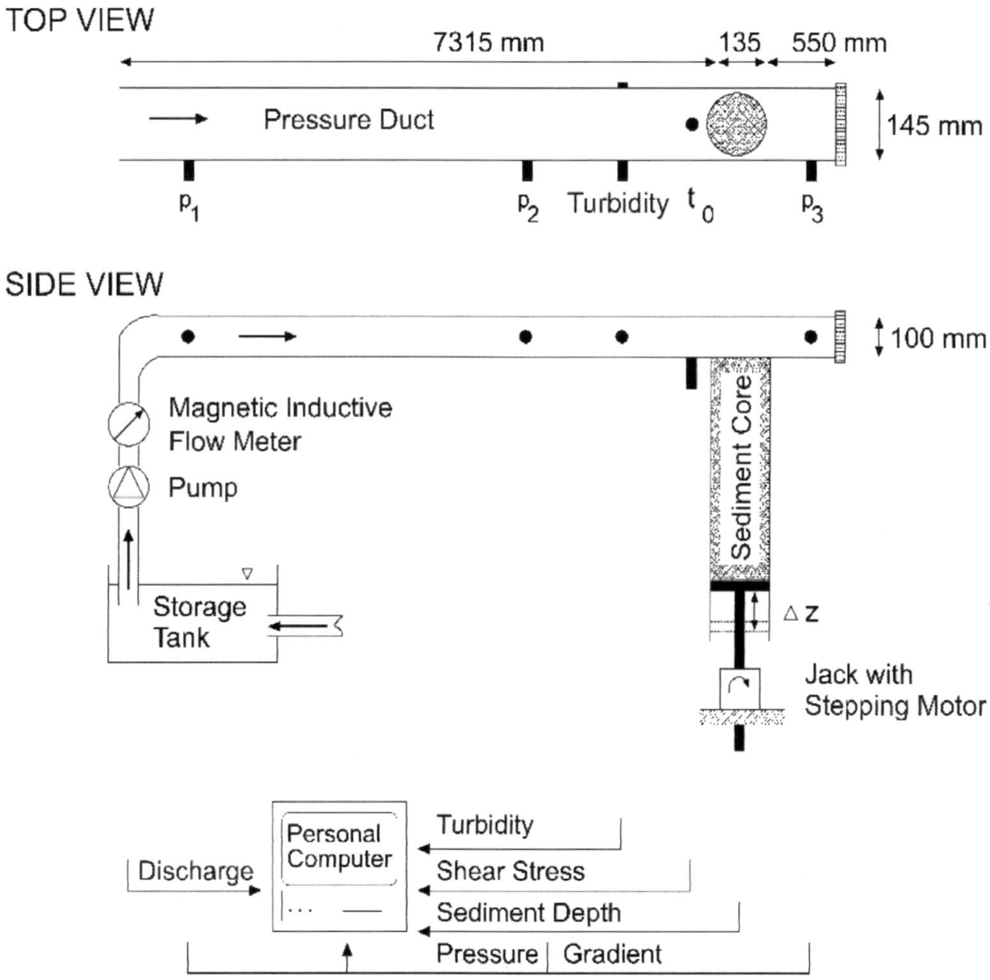

Figure 2. Schematic configuration of the SETEG-System [5]. Top and side views are shown.

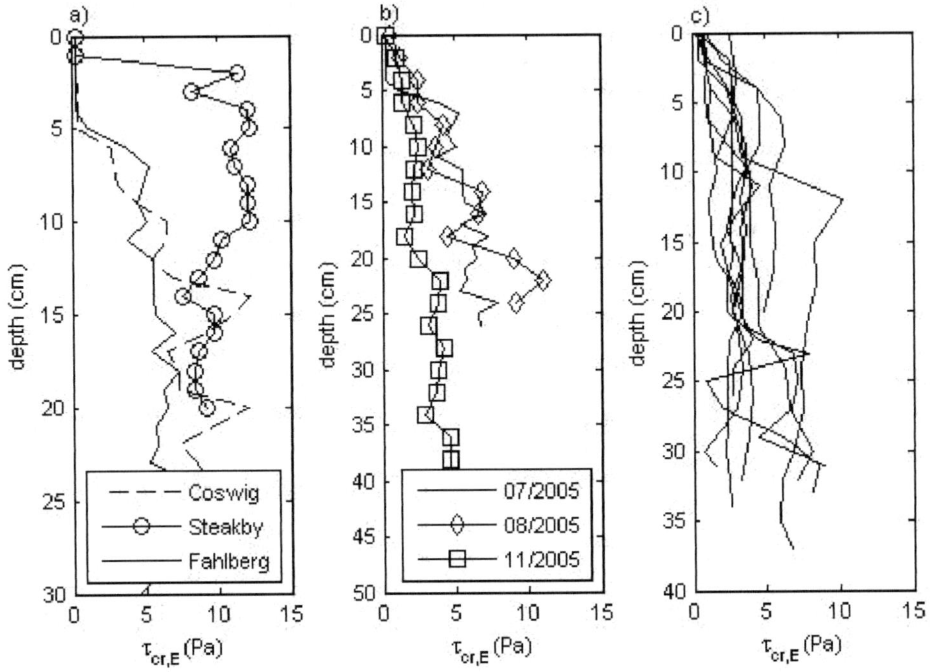

Figure 3. Measured values of the critical erosion shear stress in depth a) for three groyne fields (July 2002); b) at Fahlberg (July, August, and November 2005); and c) at Fahlberg at different locations within the groyne field (October 2006).

Figure 4. Left panel: Fahlberg GF. Right panel: exact location of the sampling within the GF (black circles) and depth of deposited sediments.

2.1. Statistics of measured data

In order to analyze the measured values of the $\tau_{cr,E}$, basic statistical parameters are defined and summarized in table 1. All measured data are taken for the statistical analysis, regardless their spatial and temporal variability. A relatively large data range of 12 Pa and high variance confirm the large spectrum of the $\tau_{cr,E}$, even though the measurements are restricted to the top layer of 10 cm of the sampling cores. For further analysis, the measured data are considered as independent random variables.

Table 1. Basic statistical parameters of the measured critical erosion shear stress (94 measurements): minimum, maximum, arithmetic mean, expected value, standard deviation, and variance.

$\tau^{min}_{cr,E}$ [Pa]	$\tau^{max}_{cr,E}$ [Pa]	$\tau^{mean}_{cr,E}$ [Pa]	$E(\tau_{cr,E})$ [Pa]	σ [Pa]	var [Pa^2]
0.28	12.27	3.04	3.13	3.20	10.24

The histogram of relative and cumulative frequencies is presented in figure 5. The obtained distribution of the measured data is asymmetric with skewness and high variance. The standard deviation is the same order of magnitude as the arithmetic mean value, which means that an effective value of the $\tau_{cr,E}$ can vary highly. The distribution of the measured data is fitted to an exponential function:

$$F(\tau_{cr,E}) = 1 - 1.13 \cdot \exp(-0.38 \cdot \tau_{cr,E}), \quad (1)$$

which is also shown in figure 5. This theoretical exponential distribution is further used for generating spatial distribution of the $\tau_{cr,E}$.

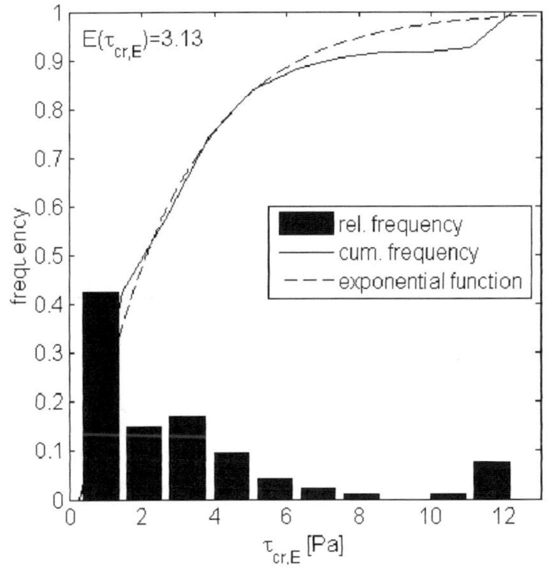

Figure 5. Relative and cumulative frequency of the measured critical erosion shear stress with fitted exponential distribution function. Expected value is shown as well.

After analyzing the measured data it can be concluded that high variability of the $\tau_{cr,E}$ is observed depending on: (i) location of a groyne field itself along the river course, (ii) the sampling spot within a groyne field, (iii) sediment depth, and (iv) time, i.e., season when the samples are taken. Additionally, deeper layers with high values of the $\tau_{cr,E}$ are not likely to be eroded and they protect layers below with lower stability from erosion. Therefore, it is very difficult to assign the effective value of the $\tau_{cr,E}$, which represents each groyne field in a river section. In a conservative practice, a constant value of $\tau_{cr,E}$ is often used - irrespective of spatial or temporal variability. The aim of the numerical investigations is to explore the applicability of the traditional approach and to investigate the sensitivity of the $\tau_{cr,E}$ on eroded mass if a different approach is used.

3. Model simulations and results

3.1. Setup of the Elbe model
Numerical simulations are performed with the 1D multi-strip model for a river reach from Wittenberg to Magdeburg as shown in figure 1. The multi-strip model is developed aiming to describe the erosion, dispersion and deposition dynamics of fine suspended sediments for regulated rivers with typical training works such as near bank groyne fields. The total river cross section is subdivided into three compartments: main channel, adjacent groyne fields and flood plains on the left and right bank, respectively. The set of 1D transport equations for the strips are coupled by dispersive and advective exchange terms [4].

For all simulations, boundary conditions and model parameters are fixed, except the critical erosion shear stress. Simulations are performed for steady state flow and transport with constant inflow conditions for a time period of one day. Since the aim is to investigate the influence of $\tau_{cr,E}$ on erosion, flood discharge conditions are applied. The simulated Elbe discharge of 1790 m^3/s corresponds to a five years return period flood. The Mulde and Saale discharges are 144 m^3/s and 503 m^3/s, respectively. Inflow of suspended sediments from the Elbe, Mulde and Saale is 58 g/l, 43 g/l, and 43 g/l, respectively.

3.2. Basic ideas of the statistical approach
The statistical approach to determine the spatial distribution of the parameter needed for numerical models is obtained by means of stochastic analysis. The spatial distribution of the $\tau_{cr,E}$ is generated based on a theoretical distribution of the measured data. Adjustment of the measured data to a theoretical distribution is one of the key factors indicating the quality of obtained results. Therefore, it is indispensable to have a large number of measurements in order to do statistical analysis and determine their distribution, which is used as distribution for generating the spatial distribution of the $\tau_{cr,E}$.

3.3. Spatial distribution of critical erosion shear stress
In order to generate random numbers according to the known exponential distribution, equation (1), the following procedure is used: (i) random numbers ($u[0,1]$) are generated using the MATLAB generator [6]; (ii) generated random numbers are sorted in increasing order saving the original order; (iii) for each generated random number corresponding value of $\tau_{cr,E}$ is calculated as inverse value of the exponential distribution, see equation (1) (i.e., $F(\tau_{cr,E})=u[0,1]$); and (iv) the obtained values are returned in the saved original order. Generated values of the $\tau_{cr,E}$ are determined for each groyne field in the simulated domain of the Elbe.

3.4. Comparison between deterministic and statistical approach
In order to compare the deterministic conservative (cases 1, 2, and 3) and statistical (case 4) approaches, different values of the $\tau_{cr,E}$ are applied:

- Case 1: constant mean measured value ($\tau_{cr,E}$ =3.04 Pa),
- Case 2: constant mean value plus standard deviation ($\tau_{cr,E}$ =6.24 Pa),
- Case 3: constant minimum measured value ($\tau_{cr,E}$ =0.28 Pa), and
- Case 4: randomly spatially distributed values comprising the whole data range.

The changes in bed elevation for all four cases for the left strip (groyne fields on the left river side) are shown in figure 6. Erosion and deposition heights are presented in terms of total sedimentation and erosion rates ($\Delta z = f(\dot{S} - \dot{E})$). For the case 2, lower erosion and deposition occur in comparison to the case 1. Case 3 gives the extreme case of very high erosion that would rarely occur in nature. Case 4 results in higher erosion and deposition in comparison to the case 1.

Using a constant value of the $\tau_{cr,E}$ (cases 1 to 3), the bed elevation differences occur only due to varying the $\tau_{cr,E}$, whereas it is independent on the size of a groyne field. This means that larger groyne fields contribute more to the total mass balance than smaller ones, because of bigger area that can erode. Therefore, the erosion/deposition patterns remain the same for the cases 1, 2, and 3, see figure 6. The differences in patterns are only evident in the stretches where deposition is altering to erosion, depending on the $\tau_{cr,E}$, for example in the upstream part of the simulated domain. In this case, by decreasing the $\tau_{cr,E}$, erosion increases so that $\dot{S} - \dot{E} < 0$. Contrarily to the cases 1-3, for randomly distributed $\tau_{cr,E}$ (case 4) the differences in bed elevation are caused by both the varying value of the $\tau_{cr,E}$ and the groyne field size. It might happen in a large groyne field that the random value of the $\tau_{cr,E}$ is smaller than the mean, which would result in increased erosion. To the small size groyne field this would result in almost no change in erosion, due to small area that can be eroded.

Nevertheless, randomly distributed $\tau_{cr,E}$, due to mass balance dependency on groyne field size as well as on the parameter value, requires further detail statistical investigation by performing a large enough number of numerical simulations.

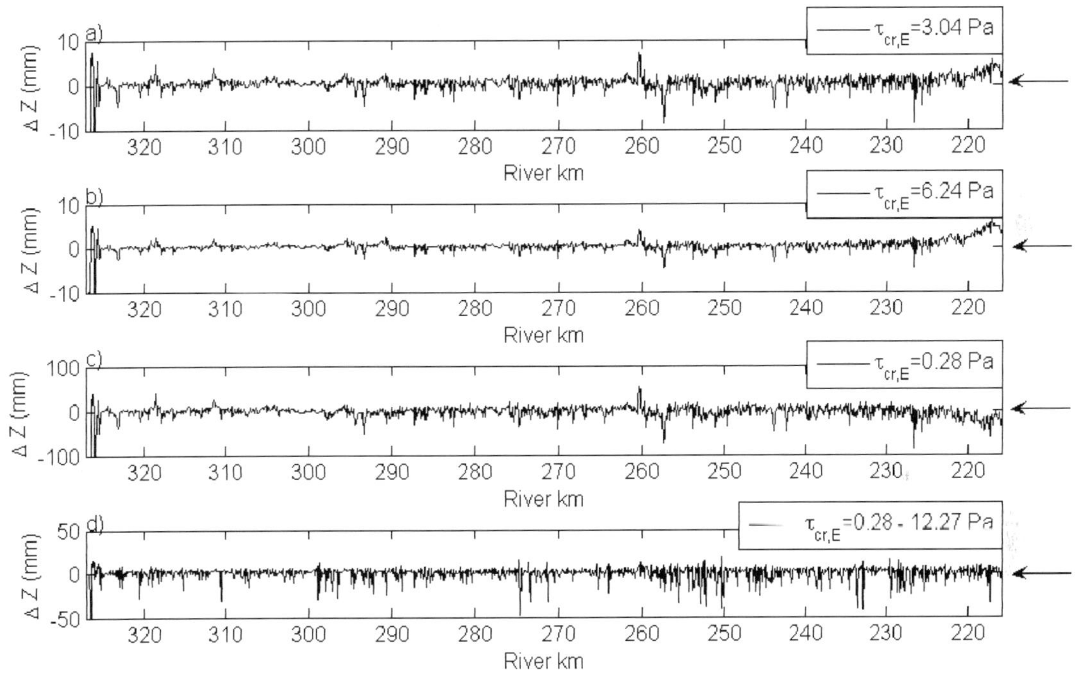

Figure 6. Longitudinal change in bottom elevation of groyne fields on the left side (note different scales): a) case 1; b) case 2; c) case 3; and d) case 4.

3.5. Estimating the effective critical erosion shear stress

Since this study aims at estimating the effective value of the $\tau_{cr,E}$ and probability that a certain amount of groyne field sediments will erode, a Monte Carlo procedure is applied [7]. Many randomly distributed values of $\tau_{cr,E}$ are generated and for each realisation total net erosion is calculated by the multi-strip model. Since the result of each numerical simulation is different, they are analysed statistically. The number of realisations is gradually increased, from 20, 50, to 200.

Considering only one chosen groyne field, in figures 7a and b it can be seen that generated random numbers have the same distribution as measured data. Calculated erosion rates for each generated $\tau_{cr,E}$ at the chosen groyne field give also the same distribution with the expected value of erosion of 1.02e-4 kg/m^{2}s, see figure 7c. Summarizing the erosion for the whole river reach (i.e., summarizing independent and identical distributions shown in figure 7c), the mean total net erosion rate of 0.22 kg/m^{2}s is obtained. Furthermore, with increasing the number of realizations the exponential distribution tends to Gaussian distribution according to the central limit theorem [7], see figure 8. This implies that an effective value of the $\tau_{cr,E}$ (constant for the whole domain) which would give the erosion rate of 0.22 kg/m^{2}s is 1.42 Pa.

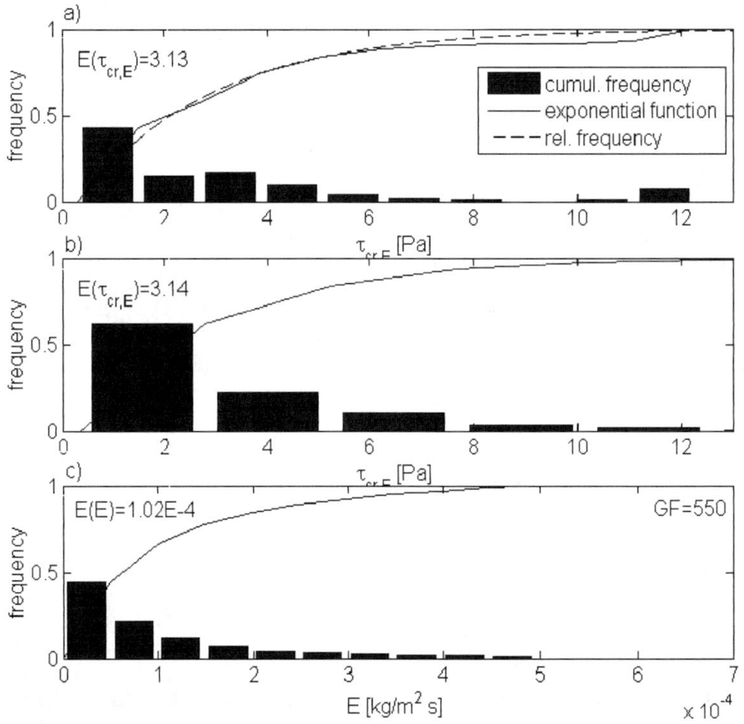

Figure 7. Statistics of a) the measured data; b) generated values at chosen GF; and c) the calculated erosion rates by the multi-strip model for the chosen GF.

The effective value based on statistics is smaller than the mean measured value of 3.13 Pa. This underlines the importance of statistical analysis in determining the effective value of the parameter due to considering the whole range of measured values, not only one constant value. The suggested statistical method applies, in an indirect way, the depth profiles of the critical erosion shear stress, which is important for long lasting floods where deeper layers with different values of the $\tau_{cr,E}$ can erode as well.

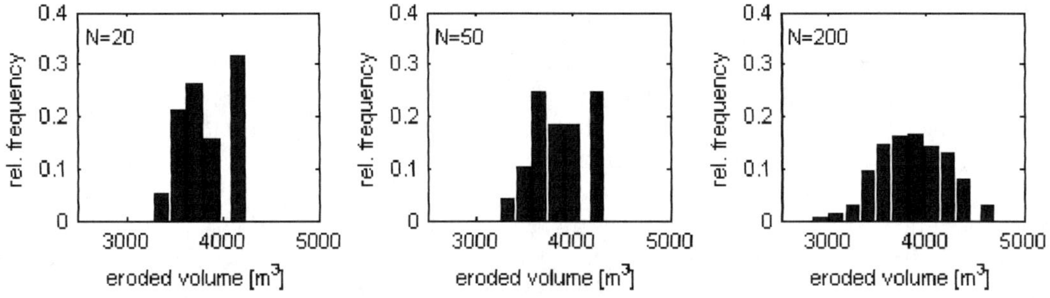

Figure 8. Histograms of relative frequencies of eroded volumes for different number of realisations, showing the approach to Gaussian distribution.

4. Erosion uncertainty estimation

The erosion uncertainty estimation provides a summary of all parameters and variables involved in erosion calculation, estimating their distributions. This includes a summary of: (i) side strips' areas distribution, (ii) side strips' bed shear stresses distribution, (iii) calculated eroded volume distribution, (iv) analysis of uncertainties and likelihood that erosion will occur. In order to perform the erosion probability characterization, the following discharges are applied in the Middle Elbe: Q_{100}=3500 m^3/s, Q_5=1790 m^3/s, and $Q_{1.2}$=1100 m^3/s. For all three discharges the $\tau_{cr,E}$ values are generated as explained before and 200 numerical simulations are performed by the multi-strip model.

Flow variables, calculated for each discharge, are directly influenced by a discharge probability and river geometry. In the left panel of figure 9, distributions of the side strips' areas along the whole river stretch of 112 km are shown, calculated for all three discharges. It can be seen that the distributions tend to normal, with higher variance during higher discharge. Unlikely the side strip's area, the side strip's bed shear stress distribution is asymmetric and shifted to skewness (see right panel of figure 9), as the critical shear stress distribution. A higher discharge causes higher variance, as for the areal distributions. All these spatial variations, areas and bed shear stresses, together with a variation of the $\tau_{cr,E}$ are incorporated into the erosion calculation.

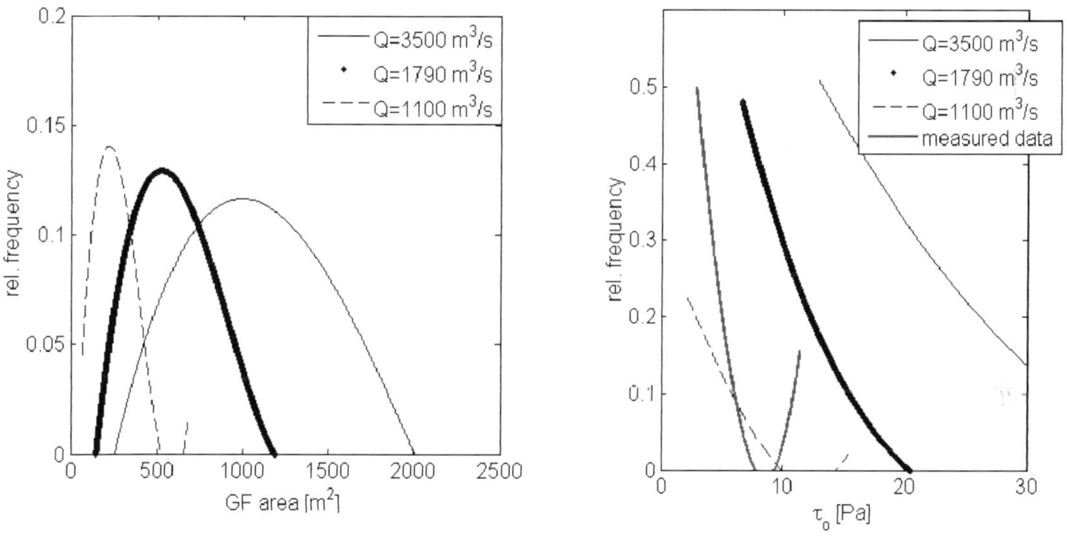

Figure 9. Distributions calculated for different flood discharges: Left panel: areas of side strips; Right panel: bed shear stresses of side strips.

Figure 10 exhibit distributions of eroded and deposited volumes during the simulated discharges. Applying a discharge of $Q_{1.2}$=1100 m^3/s both deposition and erosion occur depending on generated $\tau_{cr,E}$ values. This means that a spatial distribution of the $\tau_{cr,E}$ represents a significant part of uncertainty and therefore, it is very important for transport characterization. A discharge of Q_5=1790 m^3/s results in higher total net erosion, whereas distribution tend to normal with higher variance. A high flood discharge of Q_{100}=3500 m^3/s results in even higher total net erosion with higher variance and thus, higher relative uncertainty.

A large number of realizations allows erosion probability estimation. The probability that a certain volume of sediments will erode from groyne fields is shown, for 200 realizations in figure 11. The results present erosion probability depending on the applied discharge, which also has a certain probability of occurrence. Furthermore, uncertainty of erosion calculation increases with increasing discharge, whereas variance increases as well. Therefore, all these *joint distributions* represent significant parts of the uncertainty inherent to groyne field erosion. The randomness of both the discharge and critical erosion shear stress confirms the application of the probabilistic approach.

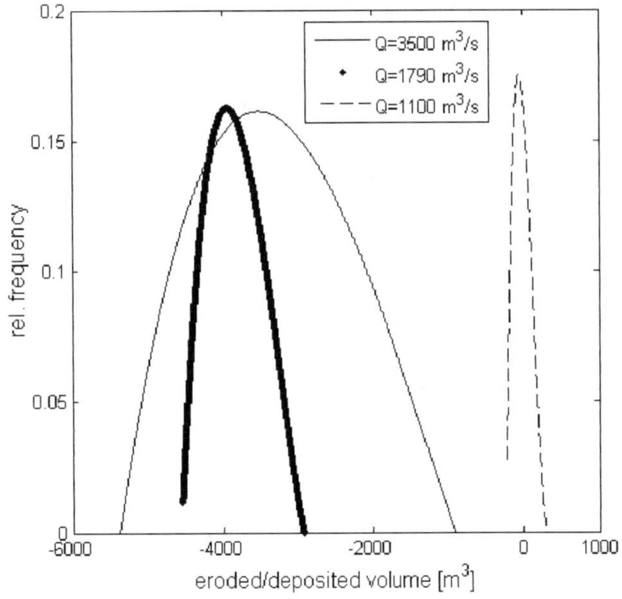

Figure 10. Distributions of side strips' net eroded/deposited volumes calculated for different flood discharges.

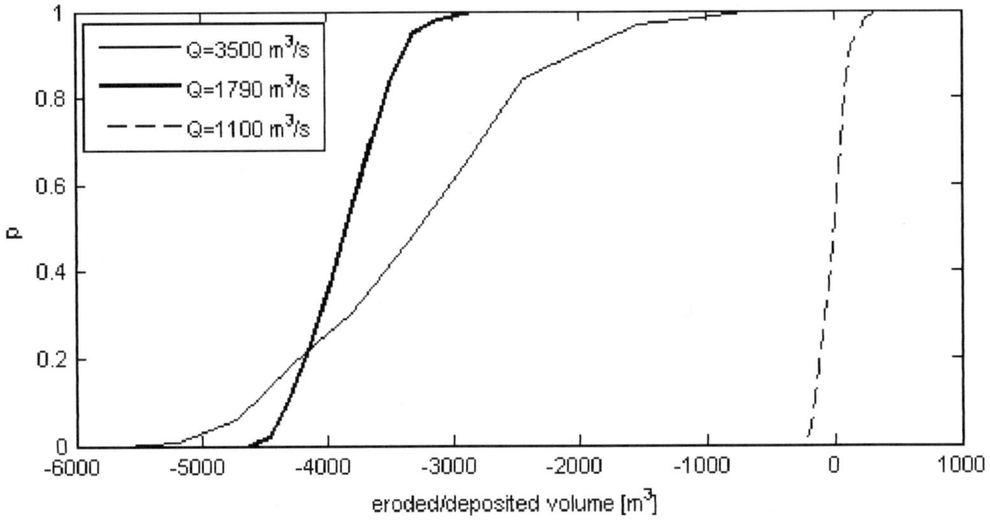

Figure 11. Probability of groyne field erosion calculated for different flood discharges.

5. Summary and Conclusions

The new methodology is implemented in a study of a critical erosion shear stress and its application in numerical modelling. Stochastic methods, which are widely used in geostatistics and prediction of pier scour [8, 9], are coupled with a numerical modelling of suspended sediment transport. In order to estimate an effective value of the parameter and erosion probability a spatial distribution of the critical erosion shear stress is generated and a large number of realisations are calculated by the multi-strip model.

The results of the stochastic analysis, suggested in this paper by treating critical erosion shear stress values as random variables, are compared with a traditional approach, which evaluates erosion by referring to a mean constant value of the parameter applied to the whole river reach. This approach assumes that erosion can be estimated by using a single value of the critical erosion shear stress, ignoring the fact that with on going erosion deeper layers with different sediment stability will be exposed to flow. The analysis gives the effective value of the critical erosion shear stress to be smaller compared to the mean measured value. The results showed that the deterministic method for determining erosion lead to underestimates of groyne field erosion. Therefore, the new statistical methodology coupled with a numerical modelling of suspended sediment transport presents an innovative and valuable framework for assessing erosion processes and associated contaminant mobilization.

Acknowledgements
The authors are grateful to Jing Li for useful comments and suggestions. This work was supported by European Commission; project MODELKEY (Contract No 511237-GOCE).

References

[1] Gerbersdorf S U, Jancke T and Westrich B 2005 Physico-chemical and biological sediment properties determining erosion resistance of contaminated riverine sediments – Temporal and vertical pattern at the Lauffen reservoir/River Neckar, Germany *Limnologica* **35** 132-144

[2] Debnath K, Nikora V, Aberle J, Westrich B and Muste M 2007 Erosion of cohesive sediments: resuspension, bed load, and erosion patterns from field experiments. *Journal of hydraulic engineering* **133** (5) 508-520

[3] Mehta A J 1988 Laboratory studies on cohesive sediment deposition and erosion *Physical Processes in Estuaries* ed J Dronkers and W van Leussen (Berlin: Springer) pp 427-445

[4] Prohaska S and Westrich B 2006 One-dimensional multi-strip transport model for long distance sediment transport in navigation channels, case study the River Elbe *Proc. 23rd Conference of the Danubian Countries on the Hydrological Forecasting and Hydrological Bases of Water Management* (Belgrade, Serbia, 28-31 August 2006) ed S Bruk and T Petkovic p 78

[5] Kern U, Schuerlein V, Holzwarth M, Haag I and Westrich B 1999 Ein Strömungskanal zur Ermittlung der tiefenabhängigen Erosionsstabilität von Gewässersedimenten: das SETEG-System *Wasserwirtschaft* **89** 72-77

[6] Hahn B and Valentine D 2002 *Essential MATLAB for Engineers and Scientists* (Oxford: Elsevier)

[7] Montgomery D C and Runger G C 2003 *Applied statistics and probability for engineers* (USA: John Willey & Sons)

[8] Kitanidis R K 1997 *Introduction to goestatistics, application in hydrogeology* (UK: Cambridge University Press)

[9] Brandimarte L, Montanari A, Braud J L and D'Odorico P 2006 Stochastic Flow Analysis for Predicting River Scour of Cohesive Soils *Journal of Hydraulic Engineering* **132** 493-500

IOP Conf. Series: Earth and Environmental Science **4** (2008) 012043 doi:10.1088/1755-1307/4/1/012043

Flood Risk and Flood Hazard Maps – Visualisation of Hydrological Risks

Karl Spachinger[1], Wolfgang Dorner[1], Rudolf Metzka[1], Kamal Serrhini[2], Sven Fuchs[3]

[1] University of Applied Sciences Deggendorf, Germany
[2] Université de Technologie de Compiègne, Génie des Systèmes Urbains, France, and Université François Rabelais, Unité Mixte de Recherche, Tours, France
[3] Institute of Mountain Risk Engineering, University of Natural Resources and Applied Life Sciences, Vienna, Austria

E-mail: karl.spachinger@fhd.edu

Abstract. Hydrological models are an important basis of flood forecasting and early warning systems. They provide significant data on hydrological risks. In combination with other modelling techniques, such as hydrodynamic models, they can be used to assess the extent and impact of hydrological events. The new European Flood Directive forces all member states to evaluate flood risk on a catchment scale, to compile maps of flood hazard and flood risk for prone areas, and to inform on a local level about these risks. Flood hazard and flood risk maps are important tools to communicate flood risk to different target groups. They provide compiled information to relevant public bodies such as water management authorities, municipalities, or civil protection agencies, but also to the broader public. For almost each section of a river basin, run-off and water levels can be defined based on the likelihood of annual recurrence, using a combination of hydrological and hydrodynamic models, supplemented by an analysis of historical records and mappings. In combination with data related to the vulnerability of a region risk maps can be derived. The project RISKCATCH addressed these issues of hydrological risk and vulnerability assessment focusing on the flood risk management process. Flood hazard maps and flood risk maps were compiled for Austrian and German test sites taking into account existing national and international guidelines. These maps were evaluated by eye-tracking using experimental graphic semiology. Sets of small-scale as well as large-scale risk maps were presented to test persons in order to (1) study reading behaviour as well as understanding and (2) deduce the most attractive components that are essential for target-oriented risk communication. A cognitive survey asking for negative and positive aspects and complexity of each single map complemented the experimental graphic semiology. The results indicate how risk maps can be improved to fit the needs of different user groups. Recommendations were developed of how to provide stakeholder-oriented information on hydrological risks.

1. Introduction

On the European level, a lot of research institutes work in the field of flood risk assessment in different projects, such as EUROFlood, EUROTAS, FLOODAWARE or FLOODsite. In the field of interregional cooperation (Interreg), risk maps have been developed and tested as a first element of risk management in different projects, e.g. IRMA and TIMIS for the Rhine catchment. Complex considerations and assessments of the flood process chain, linking different processes leading to flood disasters have been carried out, from precipitation to runoff generation and concentration in the catchment towards flood routing in the river network. Possible failures of flood protection measures, inundation within certain areas and the economic damage were evaluated through different methods[1]. A number of methods were integrated to repeat random sampling and to compute reliable results, such as Monte Carlo approaches. Such approaches are most suitable for computational use because of their reliance on repeated computation and random or pseudo-random numbers, and tend to be applied when it is infeasible or impossible to calculate an exact result by a deterministic algorithm – e.g. if probabilities have to be assessed in the field of risk analyses. With respect to the application of Monte Carlo simulations, a considerable amount of research results is currently available – on a local, regional, national and international level, and also related to practitioners' needs. However, with respect to flood risk mapping, harmonisation is still outstanding. In 2006 the European Union published the drafts of the Directive on the assessment and management of flood risk to harmonise the research activities and the final results in the different European countries, which were put into force in 2007 [2].

The project RISKCATCH, funded under the CRUE ERA-Net initiative by the German Federal Ministry of Education and Research and the French Ministry of Ecology and Sustainable Development, addressed the issue of harmonisation in the assessment of hydrological risks and vulnerability assessments focusing on the flood risk management process. The RISKCATCH project aimed to deliver new, practical and viable solutions for an integrated risk-assessment-based management of natural hazards in Alpine environments and related forelands with a special focus on the interaction of technical and non-technical measures during recent flood events in catchment areas. By evaluating already available methods and results from other projects and experiences from other countries, the status quo and historical development of risk was compared and possible future developments were deduced based on scenarios. Necessary action was derived and efficiency of non-technical measures of flood defence was evaluated. Based on the assessment of historical and possible future development of the hazard, the values at risk and the associated vulnerability, non-technical measures were evaluated from an economic and technical point of view. Regarding different scenarios of the temporal development of risk, maps were generated and assessed. These maps provided the basis for the study on risk perception using the method of experimental graphic semiology.

2. Methodology

The main focus of the project, also presented in this paper, was a user-oriented transfer of the results to different stakeholder groups, e.g. outputs of hydrological and hydrodynamic models, necessary modifications and final output in exemplary maps, which were finally evaluated by test persons from target groups. Focusing on medium-scale information on risk, test sites in the Alpine foreland were chosen. The project area in Bavaria is situated at the two rivers Vils and Rott. The Vils river has a catchment size of about 1450 km^2 and is flowing into the Danube river after 120 km. The catchment of the Rott river is 1200 km^2, and its length is about 100 km. Both catchments are part of the tertiary hilly landscape, an intensively used agricultural landscape. Rural structures, settlements with scattered buildings, the intensive agricultural use are determining factors. However, during the last two decades there was a remarkable increase of industrial and commercial use of land in potential inundated areas.

Figure 1. Project areas situated in Bavaria, Germany

2.1. Hydrology

There is a consensus that hazards resulting from hydrological extremes are on the increase (e.g., [3]). This fact is confirmed by evidence both for recent changes in the frequency and severity of floods as well as droughts and for outputs from climate models which predict increases in hydrological variability[4]. Because of major flood events in Europe in 2002 and 2005 there is much concern about the levels of protection provided by existing flood defences and their consequences.

During the last decades the characteristics of floods and land use patterns in the flood plains changed fundamentally. These changes are affected by climatic developments, land use changes, river development and training and also by measures of flood defence, enabling building in former flood plains and reducing the natural detention and profile for flood runoff. The main question is how many upcoming impacts and actual changes will influence the future runoff scenario and change the characteristics of floods and extreme discharges.

In combination with land use maps, development plans and historical maps describing the changes of land use and damage potential in the flood plain, hydraulic and hydrological data sets were used to analyse the developments of flood-related problems for the status quo (ex-post perspective), i.e. flood risk. In combination with data for existing technical facilities of flood defence the effects (levels of protection and restrictions) of classical flood defence were evaluated and compared with non-technical measures. Future developments and underlying scenarios were derived and assessed with respect to risk management.

2.2. Hydrodynamic Modelling

Hydrodynamic models delivered the basic data for the compilation of hazard information. These data were derived from models, pre-processed and compiled in maps. HydroAS 2D in combination with SMS as a standard software model was used to simulate flood events; consequently, data were derived from this software. In a further step, these data were pre-processed in SMS, evaluated and visualised using ArcGIS 9 to derive the relevant maps (inundated area, water depth and flow velocity).

IOP Conf. Series: Earth and Environmental Science **4** (2008) 012043 doi:10.1088/1755-1307/4/1/012043

2.3. Hazard and Risk

As shown by Dorner et al. [5], maps indicating different information about floods and related risks have considerable relevance to visualise hydrological data in a spatial context (see Fig. 2). Regarding hydrological aspects especially the visualisation of probabilities is bound to restrictions e.g. if run-off is influenced by controlled detention structures.

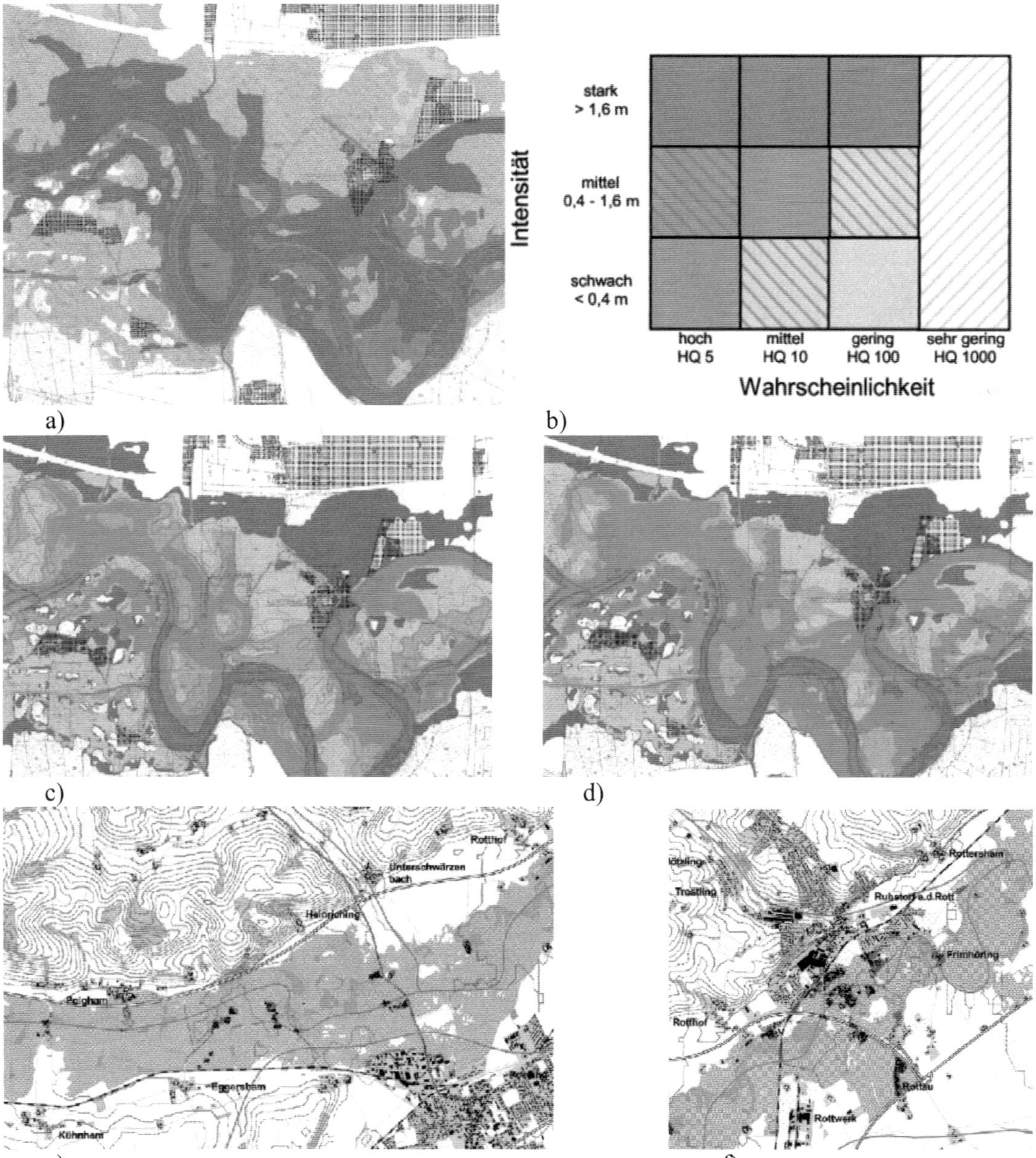

Figure 2. a) an intensity-probability diagram or risk index enables the combination of risk components in one map; b) explains the derivation of the different risk levels; c) and d) show the extent of flood with different recurrence probabilities; e) and f) indicate the extent of extreme events (HQ100) on a large scale to provide an overview e.g. for the broad public.

IOP Conf. Series: Earth and Environmental Science **4** (2008) 012043 doi:10.1088/1755-1307/4/1/012043

In the area of natural hazards, risk is defined as a function of probability of occurrence and extent of damage. Extent of damage is constituted by the two factors damage potential and vulnerability. In general, this function has gained acceptance in accordance with the definition of the United Nations (e.g., [6],[7]) and is indicated in the form of a product as follows:

$$R = p_{Si} \cdot A_{Oj} \cdot p_{Oj,Si} \cdot v_{Oj,Si} \qquad (1)$$

where	R	=	risk
	p_{Si}	=	probability of scenario i
	A_{Oj}	=	value at risk of object j
	$p_{Oj,\,Si}$	=	probability of exposure of object j to scenario i
	$v_{Oj,\,Si}$	=	vulnerability of object j, dependent on scenario i

In Alpine countries, mainly in Austria and Switzerland, the procedure of hazard assessment is methodologically reliable in determining the hazard potential and the related probability of occurrence (p_{Si}) by mapping, examining, modelling, and assessing individual processes and defined design events (see e.g. [8], [9], [10], [11]; eq. 1). So far, little attention has been given to the damage potential (A_{Oj}) affected by hazardous processes, particularly concerning spatial patterns and temporal shifts [12, 13]. Studies related to the probability of exposure of an object ($p_{Oj,\,Si}$) to a defined scenario and the appropriate vulnerability of the object ($v_{Oj,\,Si}$) have predominantly been carried out so far in terms of expert's reports. Only few approaches and conceptual proposals determine the risk of property and human life (e.g.,[14], [15], [16]).

The different approaches of several German federal states concerning hazard maps showed some similarities. The approach of Switzerland (e.g. [17]; [18]) to indicate hazard uses the two parameters intensity and probability. It combines them in a matrix (probability-intensity-matrix) to indicate different hazard levels. This approach was extended by a third dimension to include also the vulnerability and receive a scale of color indicators to visualise risk [18].

In a further step the hazard maps were developed towards a risk map by taking further geographical data but also statistical data on elements at risk into account. Modifying the approach of Switzerland, a 3-dimensional cube was developed representing three input values, (1) probability, (2) intensity, and (3) vulnerability (see Fig. 3). Depending on the intended use of the map, different additional parameters were added.

The European Flood Directive recommends the creation of flood risk maps with different hazard criteria, (1) flood events with a high probability (HQ 10), (2) flood events with a medium probability (HQ 100), and (3) flood events with a low probability (extreme event). The content should include information on water depth and velocity as well as areas with embankment erosion and sediment deposit. Furthermore, the number of potentially affected inhabitants and the economic as well as environmental damage should be depicted. Apart from these considerably complex content required, it is necessary to keep the basis of these maps very flexible in order to allow for a computer-aided automatised creational processes due to the necessary update of these risk maps with a six-year frequency.

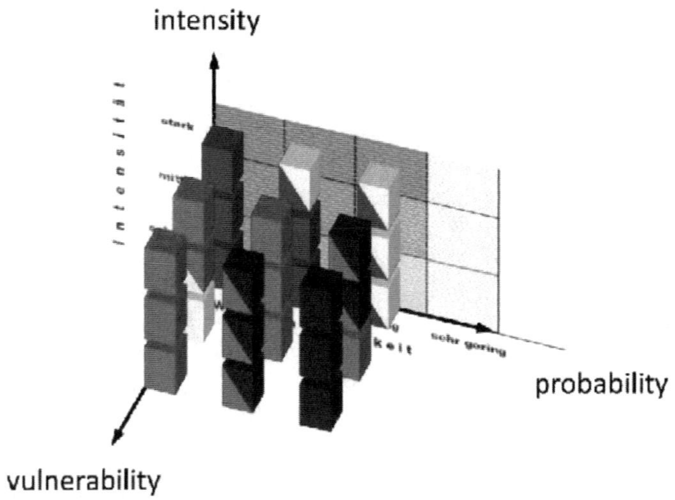

Figure 3. Risk cube including 3 dimensions: probability, intensity and vulnerability

The hazard intensity is reflecting two different factors: water depth and velocity of the water body. Critical water depths and velocities had been explored and were derived from hydrodynamic models. The values evaluated for the hazard maps were in an additional step used for the creation of risk maps. Therefore, hazard intensities were grouped into three different levels in order to fulfil the criteria of the risk cube, (1) 0-0.5 m, (2) 0.5-1.2 m, and (3) >1,2m. According to the European Flood Directive three relevant flood events were defined: (1) HQ10, (2) HQ100, and (3) HQ1000 (extreme event).

Regarding the extent of flood plains in the test sites, it was necessary to use cumulated data to efficiently consider vulnerability in the risk equation. In order to meet the requirements of a small-scale analysis, an economic approach was chosen to assess vulnerability. In doing so, vulnerability was calculated on an object basis for different types of buildings using damage functions depending on the water depth. Since for the test sites no record of historical or previous flood damage was available, design values and design loss functions were applied according to suggestions made by Meyer & Mai [19]; this procedure was possible due to comparable economic settings in the test sites. Due to the large amount of buildings situated in flood-prone areas, these damage functions were not applied to individual buildings but to groups of buildings and homogenous settled areas by aggregation. The spatially representation of vulnerability turned out to be an appropriate method for a small-scale analysis.

This procedure is analogously equivalent to the small-scale approach described in the Swiss guidelines [16] to provide an area-wide overview on susceptibility to hazard processes. Hence, vulnerability values varied between low (such as agricultural areas and individual farm estates), medium (such as dispersed settlements and small villages), and high values (city centres and industrial zones).

2.4. Experimental Graphic Semiology

Since quality is an important aspect when looking at the European Directive on the Assessment and Management of Flood Risks [2], two major questions have to be addressed during the compilation of flood risk maps, (1) How to evaluate the quality of a map, and (2) How to produce a map adapted to the issue of communication and especially with respect to the public recipient concerned?

To quantify risk perception, the output maps were presented to a group of stakeholders from different European countries. The method used is based on the approach of experimental graphic semiology, reversing the traditional communication pattern from transmitter to receiver by a feedback loop allowing to integrate different perception patterns of multiple end-users (Fig. 4). Starting from receiver, the maps were presented to the test persons using an ophthalmic device for the record of the eye movements during picture reading. The test was accompanied by a specific survey; hence, the cognitive perception of risk maps was evaluated. All maps were presented to the test persons for a relatively short time period to identify the most attractive components of each map. The eye movements were subsequently statistically analysed in order to assess patterns of visual perception for each map and to study the reading behaviour for text elements included in the maps. The visual strategies of each test person were quantified.

The approach of experimental graphic semiology followed a three-stage procedure:

1. Compilation of the maps and the experimental protocol: Exchanges between the RISKCATCH partners allowed the development of reference maps while varying various elements of graphic semiology. Examples of significant maps in terms of information of the public and risk management of flood were considered. The RISKCATCH partners carried out a first proposal of maps to be used as support for the ocular pre-tests. Based on these pre-tests, draft amendments were made. These amendments were the origin for the choice which variables to be tested (position of the title and the legend, level of detail of the legend, merges of maps, etc) and for the subsequent development of 17 maps to be used as reference for the ocular tests and the cognitive study. The creation of a form for the cognitive investigation (in order to specify the results of the ocular recordings) and the identification of the various candidates also belonged to this preliminary stage.
2. Realisation of experimental graphic semiology approach: Recording the ocular movements of the selected candidates and undertaking the cognitive survey. Furthermore, the data were extracted and prepared for the statistic, static, and dynamic analyses.
3. Analysis of the recordings and interpretation of the results obtained: The results resulting from the statistical processing (calculation of the averages by map and by group of candidates), from the recording of the ocular movements (static, spatial and dynamic) and from the cognitive investigation were analysed and interpreted. Furthermore, recommendations for the compilations of risk maps were derived.

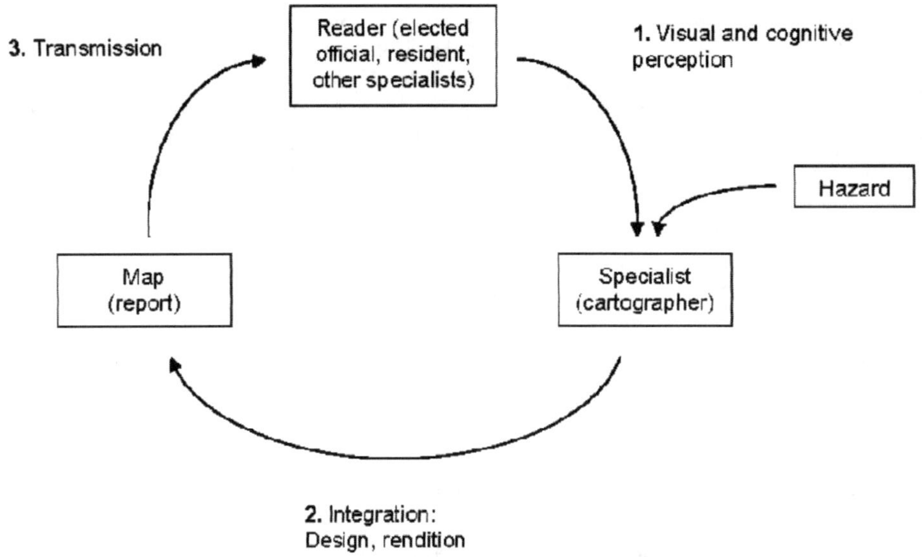

Figure 4. Principle of the cyclic model of experimental graphic semiology

3. Results

3.1. Historical Development

Analysing the temporal development of risk by using historical and actual risk maps resulted in remarkable changes in land use patterns, especially during the last 20-50 years. In Figure 5, the development of residential areas affected by the HQ100 and HQ1000 is presented for the Lower Vils region, based on the analysis of a multi-temporal series of land register maps. It was shown that land use increased considerably in recent decades, an associated increase in risk was proven.

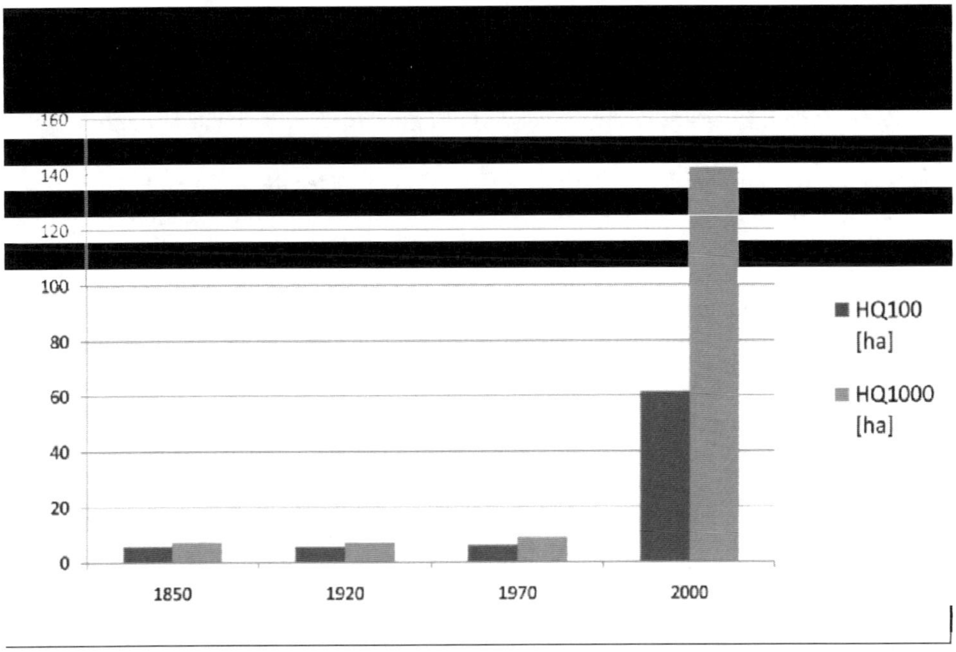

Figure 5. Concerned developed areas of the Lower Vils region within the time period of 1850-2000

When comparing the different time periods, a tenfold increase of the developed area affected by the inundated area of a one hundred year flood resulted. Considering the thousand year flood, the affected area in the year 2000 was twenty times larger as it had been in 1850. Results from the GIS-based analysis of the entire study area supported these findings (Fig. 6).

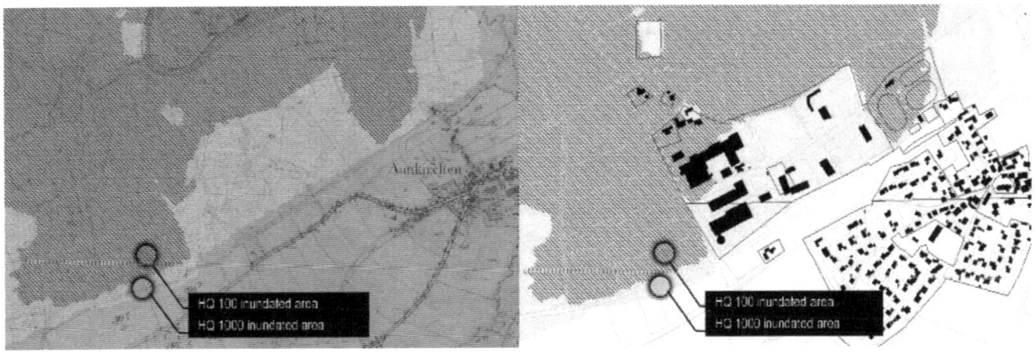

Figure 6. Comparison of affected areas (left: year 1850, right: year 2000)

IOP Conf. Series: Earth and Environmental Science **4** (2008) 012043 doi:10.1088/1755-1307/4/1/012043

3.2. Hydrological Issues

Within the catchments of Vils and Rott two major storage reservoirs are available for flood control. The storage reservoir situated in the middle of the Rott catchment was built between 1968 and 1971 and has a total capacity of 13.9 Mio m³ and a flood retention capacity of 12.75 Mio m³. The main purposes for the erection were (1) the protection of fertile agricultural areas, (2) flood defence for settlements in the downstream riparian areas, as well as (3) recreational purpose. The hydrological calculations undertaken within the RISKCATCH project showed that the estimation for the design floods used for the construction of the reservoir were underestimated. The evaluation of the control strategy of the reservoir based on actual data and a new hydrological model for the river Rott proved that the expectations about the detention effect were considerably optimistic. For the storage reservoir an innovative control strategy was suggested, including (1) an increase of the flood discharge without any detention and (2) a transition to a flexible control strategy for higher discharges.

At the Rott already minor events result in the inundation of large areas of the intensively used flood plain. Modelling a discharge of 40 m³/s at the outlet of the reservoir, the first bridge and agricultural sites were affected, and during a discharge of 120 m³/s, first buildings and settlements were concerned. Thus, at the moment the basic discharge is set to a maximum of 40 m³/s. An increase up to 80 m³/s would be possible without considerable harm to buildings or infrastructure. However, the results had shown that even with an improved steering concept the reservoir, full protection for all assets in the underflow cannot be guaranteed due to the limited capacity (Figure 7, [4]).

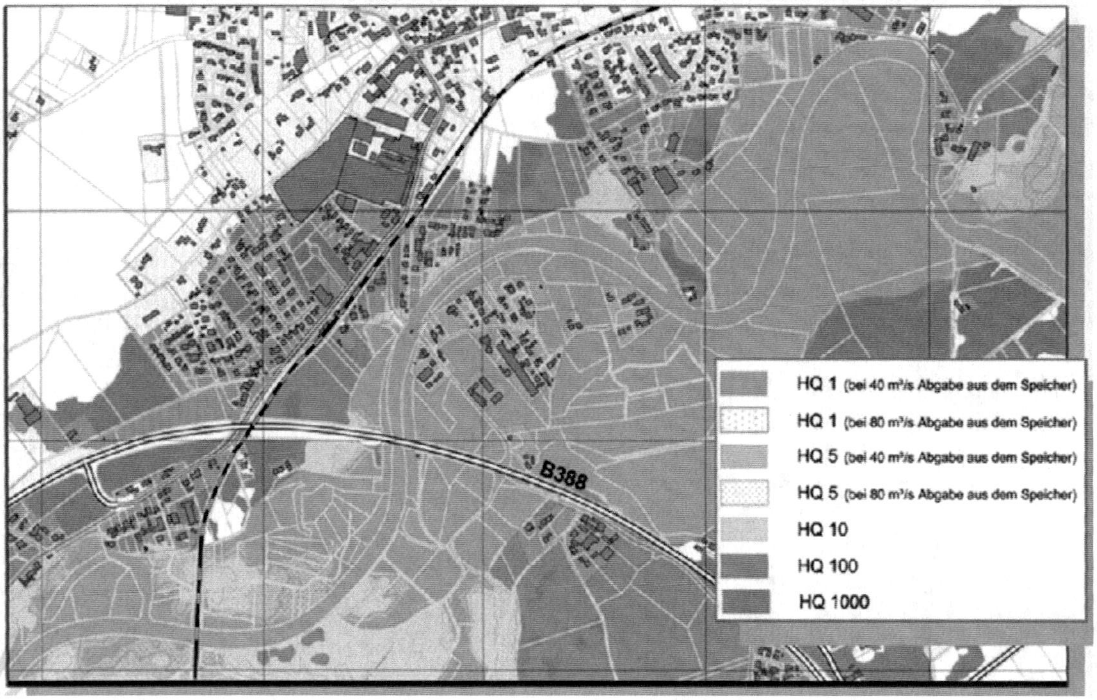

Figure 7. Probability map with different detention pond steering scenarios (HQ1 and HQ5)

3.3. Risk Maps

The initial approach to use the population density for risk mapping is presented in Figure 8 for a section of the Lower Rott test site. In this Figure, the amount of affected inhabitants is shown for different flood events (HQ10, HQ100, HQ1000). This initial and quite simple approach turned out to be effective, in particular considering precision and accuracy of data available and level of detail versus costs for data acquisition and compilation.

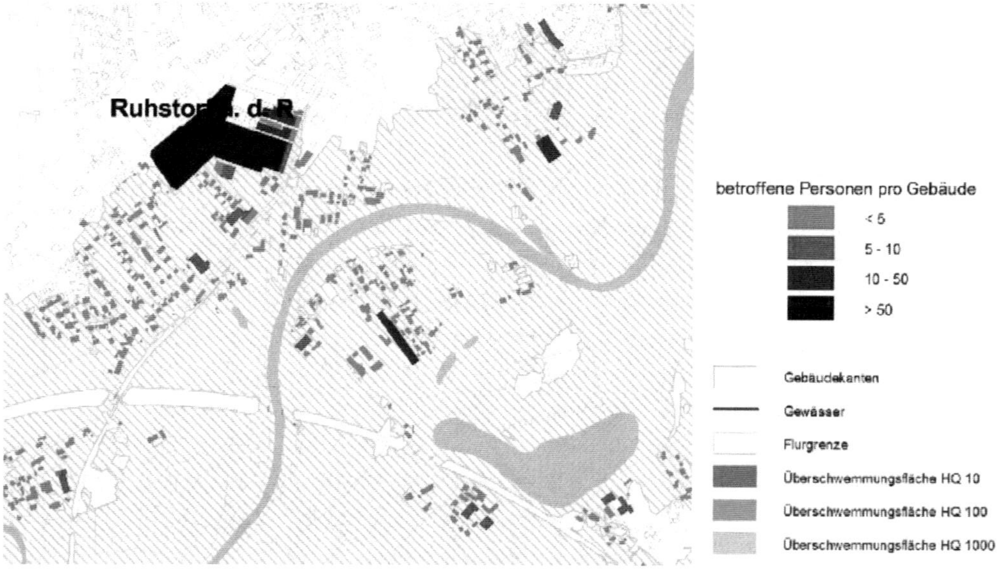

Figure 8. Detail of a risk map (variant 1: inhabitants)

In a second set of calculations, this approach was modified in order to achieve another type of risk information to be tested for communication purpose. Hence, the distribution of inhabitants provided the basis for risk information, and a separation of housing structures and industrial and commercial structures was performed on the basis of land use data. Simultaneously the number of employees of each single community was attributed to the commercial areas within the test site (Fig. 9).

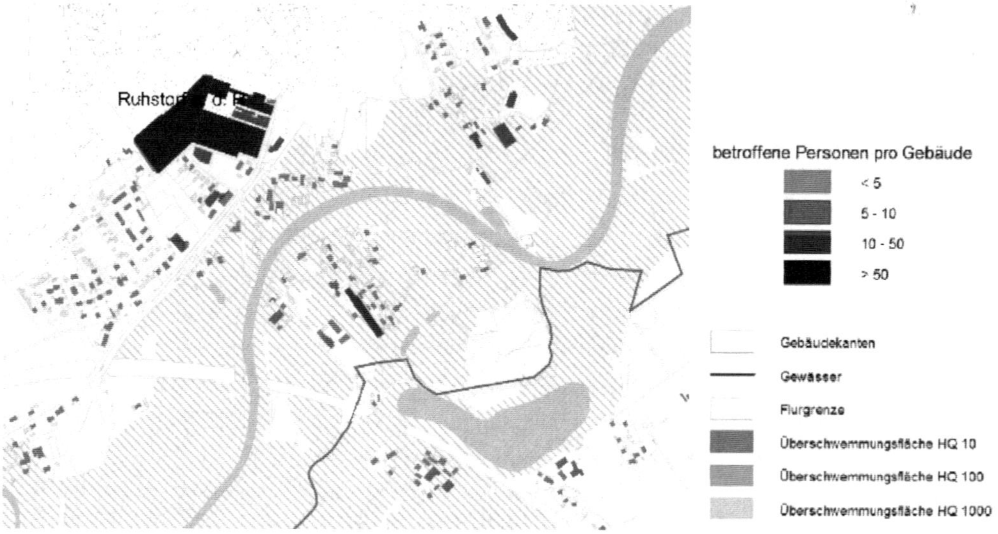

Figure 9. Extraction of a risk map (Variant 2.1: inhabitants and employees)

In a further step the principle of the risk cube was applied to all risk maps (Fig. 10) in order to provide aggregated risk values to the group of test persons.

IOP Conf. Series: Earth and Environmental Science 4 (2008) 012043 doi:10.1088/1755-1307/4/1/012043

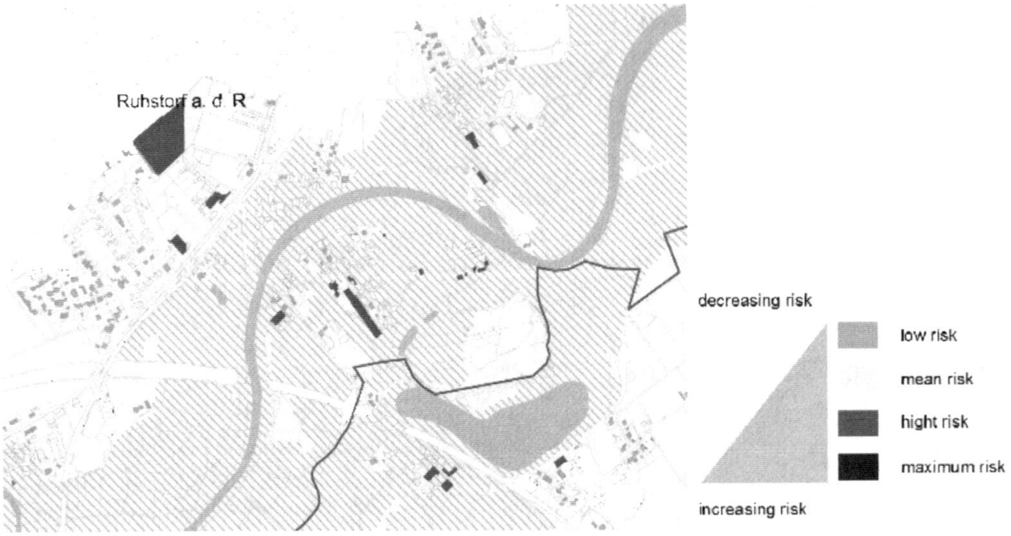

Figure 10. Extraction of a risk map based on the risk cube (Variant 2.2: inhabitants and employees)

To indicate the spatial distribution of risk, three to seven classes of risk were used based on either individual objects or the aggregation of values, e.g., settlement areas, industrial areas, agricultural areas, and forestry.

Linking process intensities, vulnerabilities and values at risk, small-scale risk maps were compiled for the test sites (Fig. 11). In doing so the focus was not only on different parameters of how to express risk, but also on the different communicative purposes with respect to the presentation of these maps to different stakeholder groups. All maps were compiled using a landscape layout in DIN-A1 format. The sets of maps were modified according to several variables in order to test the different accessibility and readability, and thus understanding by different stakeholders.

Considering that scale and consequently level of detail varied due to different size of the test sites, the concept of visualising risk was remarkable successful using spatial signatures already at scales ≤ 1:10000. This required the recalculation of values at risk for individual objects in order to adapt the values to larger areas.

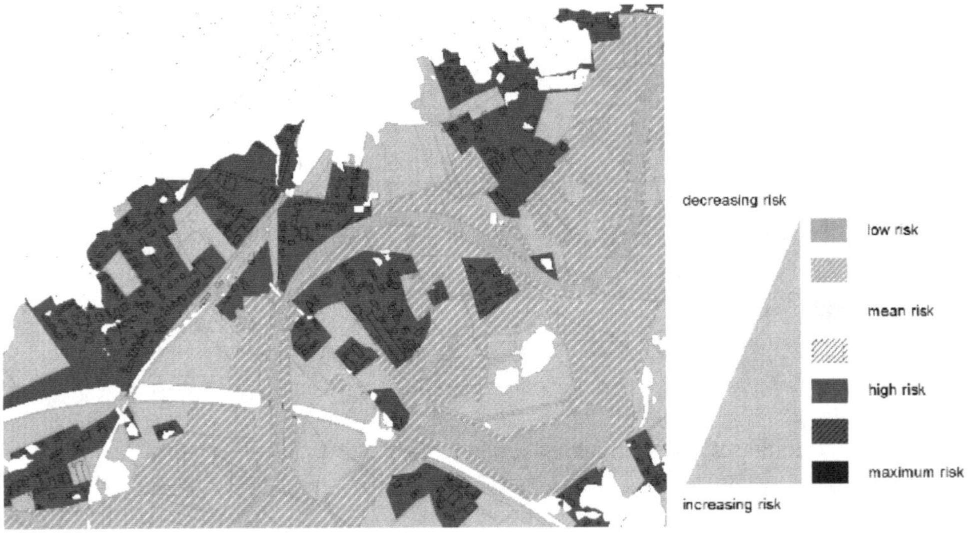

Figure 11. Extraction of a risk map (economical damage)

3.4. Experimental Graphic Semiology

The results of the eye tracking using experimental graphic semiology resulted in several findings related to the preferred layout of risk maps.

3.4.1. *Statistic Analysis.* The average number of fixations per map provided information concerning the extent of visual exploration; however, the 17 maps did not induce the same visual impact. Maps that included more graphic information, e.g., an infrared or coloured orthophoto representing the overall setting of the situation depicted, resulted in significantly more fixations than those maps that contained only a black and white land register plan for background information instead. Hence, the visual accessibility of maps is highly dependent on minimising information where possible and highlighting only necessary information since the eye visually fixes the most outstanding elements successively.

However, the average number of fixations per map is not only dependent on the map content, but also on (1) the effect of habituation, in particular the control group of laypersons, that gradually get accustomed to map reading, (2) the effect of tiredness, since the recording process spanned a period of 25 minutes per person, and (3) the repetitiveness character of certain maps with only marginal differences in depiction and thus visual distinction.

Those risk maps that followed general rules of design with respect to natural hazards, i.e., red and yellow hazard zones and only sketchy depiction of the overall setting, resulted in a considerable number of fixations within the group of specialists. On the other hand, a strongly coloured map background seemed to induce a noticeable attractiveness to the group of laypersons; the visual exploration was stimulated in particular for those maps that were based on infrared and coloured orthophotos. The group of people sensitized for map production or flood risk showed rather variable visual behaviour.

As a result of the statistic analysis, specialists in risk mapping do not scatter their attention over the map (less fixations per map) but do more thoroughly access the map content (more time for each fixation), while laypersons spend most of the time in a random access of the map content with no particular focus except from a certain interest in coloured information. A correlation between the number of fixations (attractive zones visually fixed) and the number of saccades (ocular movements that enable the passage from one zone of the picture to another) was observable.

3.4.2. *Static Analysis.* The static analysis was used to identify systematic and regular spatial patterns in map exploration. First, based on the analysis of video recordings, certain repetitive elements were identified that attracted the gaze of the entire sample of test persons. Second, certain specific elements of the maps could particularly be assigned to individual groups of test persons. In general, the focus of visual perception was on textual and coloured information, and areas with clear contrast; 90 % of the fixations were concentrated in these areas. There was a clear tendency that the gaze of all test persons followed along those information that was arranged in a vertical or diagonal order, e.g. along the river that was depicted on some of the maps.

The map title provided the first major information concerning the displayed content. If the title was located at the upper part of the legend section, the whole group of test persons looked at it during their first ten fixations. If the tile was located at the bottom, this pattern was not as clearly observable, in particular if the title was additionally written inverse in bright colours with a dark coloured box. Therefore, the title seems to fulfil its informative function best when placed above the legend and emphasized by good contrast (preferably black coloured text on a bright coloured background). This was also confirmed by the cognitive survey that accompanied the study.

The map legend is fundamental information for the understanding of the map content and enables the comprehension of the graphic symbolic system used. The test persons devoted between three and seven seconds out of 15 to the reading of the legend, which equals to approximately 20-50 % of the total exposure time. Placed either on the right or on the left side of the map, the legend attracted the eye; and vertical or quasi-vertical saccades clearly represented the process of vertical reading. However, the most coloured elements of the legend were the areas of major concentration of fixations. This tendency had an explicitly strong appearance if the legend was highly complex and contained a large amount of different information. A simple legend, containing up to five classes and two sets of information resulted in one set of ocular movements, while the visual strategy necessary for a detailed and complex legend generated two or even three sets of ocular movements.

The central element of every map is the cartographic information depicted. This element in general drew the most attention and the test persons spent 65-80 % of the exposure time of 15 seconds for analysing information. In doing so, the ocular movements were concentrated systematically on the most coloured areas, which represented the 'colour effect' [18]. In addition, the more a specific information was visually distinguishable – namely through the contrast created by the overlapping of various colours used and the map background – the more fixations focused on this information. Conversely, the less specific information appeared in contrast to the map background the more fixations seemed to be dispersed.

3.4.3. *Dynamic Analysis*. The dynamic analysis was based on the assessment of eye movements for all 21 test persons on individual sectors of the maps. It could be shown that the gazes were drawn to the main elements of the map, the legend and the figure part.

Maps with various centres of strong visual contrast resulting from a considerable spread of cartographic information lead to an accessible delivering of information. Conversely, for low-contrast maps it was shown by the dynamic analysis that the location and focus of the gaze scattered considerably. By the end of the time intervals during the test, these maps were almost entirely covered by the gaze due to little visual contrast between the main elements and the background. During statistical analysis, this was proven by a large number of short fixations.

Moreover, the sectoral analysis for the whole sample had shown that approximately two third of the observation time is devoted to less than one fourth of the map surface. While maps with dark background and only little visual contrast between the various elements depicted were entirely covered by the ocular movement of the test persons, maps with a clear distinction between central element and background showed less disperse fixations.

4. Conclusion

Risk management for natural hazards is based on risk assessment techniques, including methods to determine the hazard potential and procedures to analyse and evaluate the damage potential exposed. For these management issues, risk maps provide the basis (1) for any planning and implementation of mitigation measures by public authorities as well as for the prioritisation of these measures, and (2) for any activities concerning regional development, land-use and construction engineering. Since maps are changing their function from final products to value-added geoservices, and the development of geoservices is more user-driven than purpose-driven, the overall aim of risk mapping includes (1) the delineation of areas endangered by defined risk thresholds, (2) the assessment of exposure levels in such areas, and (3) the communication of risk to various stakeholders, e.g. politicians, residents and other people concerned. It is axiomatic that, for communication to be successful, the message must be received, understood, accepted and acted upon. Therefore, the impact of information has to be

IOP Conf. Series: Earth and Environmental Science **4** (2008) 012043 doi:10.1088/1755-1307/4/1/012043

assessed in order to understand the attitudes of the target groups and to provide these issues appropriately.

Using the method of experimental graphic semiology had shown that the structure of maps influences the visual strategy of the readers; therefore, map perception is iconographic. The more accessible visual information is the more effective it will be in terms of visual transmission of information. Moreover, particular reading behaviour of specialists, sensitised people and laypersons led to the conclusion that perception is anthropic. Hence, risk maps should be compiled according to these different needs, in particular bearing in mind that approximately 65 % of the observation time of subjects is devoted to less than 25 % of the map surface.

To summarise, the spatial and dynamic analysis highlighted certain aspects that were identified as being important for an efficient design of risk maps, and will contribute to professional risk communication:

- Coloured zones and written information concentrated approximately 90 % of the fixations.
- The concentration of information in the legend needs to be visible (contrast and colour used) and accessible (limited number of information), to attract the eye and deliver information.
- The spatial localisation of information considerably influences the perception by the reader.

Even if the study is based on some restrictions and constraints, above all the limited number of 21 participating test persons, a number of general conclusions originating from visual strategies resulted (Fig. 12). Specific elements of semiology for a cartographic representation of risk include

- a map background in bright colour to increase the contrast to informative elements, and to avoid an overload of information;
- a sufficiently large legend, preferably on the right side of the central element of the map, with a limited number of information (five classes of discretisation) comprised from one range in colour only and arranged in decreasing values; and
- a sufficiently large scale that the elements of the map are recognisable sufficiently rapid.

A risk map compiled according to these conclusions would result in a visual strategy that is composed from three clear sets of ocular movements (Fig. 13). Starting from the centre, the eye moves to the title of the map (1), following a vertical axis downwards the legend section (2) and returning back to the central element of the map (3). If there is sufficient time, the additional peripheral elements of the figurative part are explored subsequently.

It had clearly been shown that if risk maps will be designed according to certain guidelines, the information could be delivered in a visually efficient manner. Specific elements of semiology that have to be taken into account when designing risk maps include the contrast, the level of discretisation and the colour range and hue. Consequently, if risk maps are adjusted to these findings, risk communication will be enhanced, and awareness-building of the public will be increased. In particular concerning the European Flood Risk Directive, but also with respect to the overall aim of building hazard-resilient communities, future studies might include the applicability of risk maps within flood risk management plans.

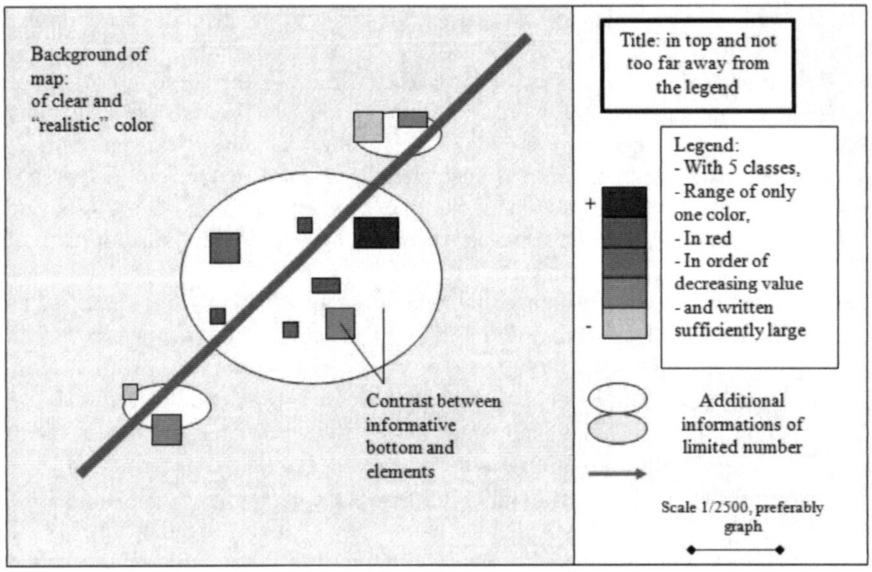

Figure 12. Conceptual map suitable for efficient risk communication to different stakeholder groups

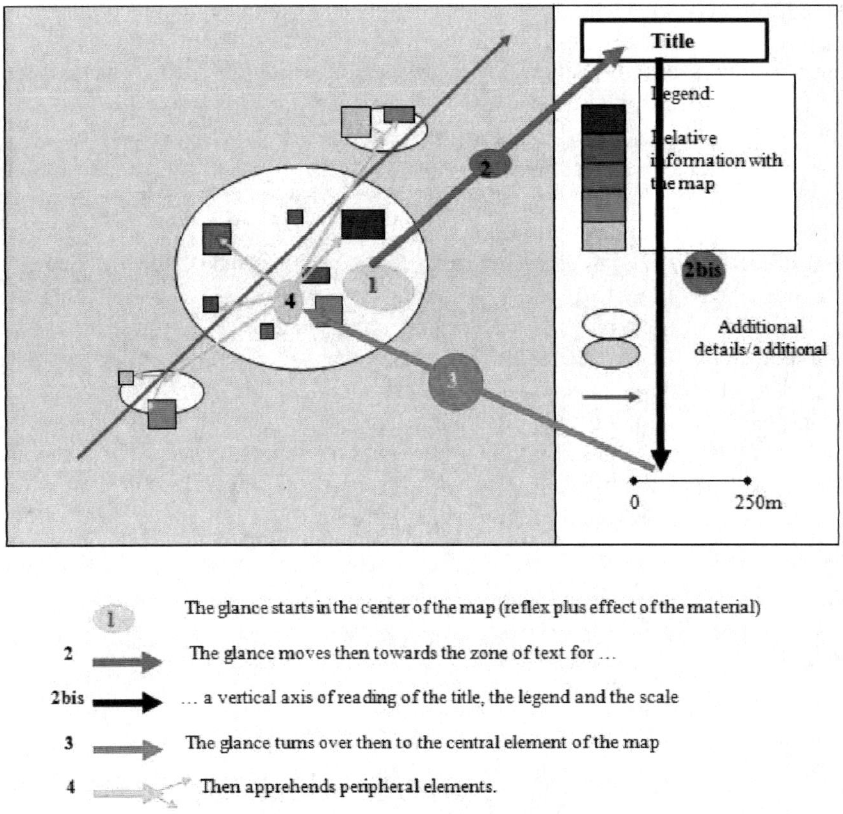

Figure 13. Representation of the visual strategy generated by the application of Fig. 12

References

[1] Merz B and Apel H 2004 Risiken durch Naturgefahren in Deutschland. *Abschlussbericht des BMBF-Verbundprojektes Deutsches Forschungsnetz Naturkatastrophen (DFNK). GeoForschungsZentrum Potsdam* Scientific Technical Report 04/01 (Potsdam)

[2] Commission of the European Communities 2007 *Directive 2007/60/EC of the European Parliament and of the Council of 23 October 2007 on the assessment and management of flood risksDirective 2007/60/EC on the assessment and management of flood risks.* http://eur-lex.europa.eu/LexUriServ/LexUriServ.do?uri=OJ:L:2007:288:0027:0034:EN :PDF (14 August 2008)

[3] Munich Re 2008 *Topics Geo. Natural catastrophes 2007* (München: Munich Re.)

[4] IPCC 2007 Climate Change 2007 *Synthesis report – An assessment of the Intergovernmental Panel on Climate Change* (Cambridge: Cambridge University Press)

[5] Dorner W, Schrenk C, Spachinger K and Metzka R 2006 Flood risk management plan - Relevance of the hydrological behaviour of a catchment. In ed Bruck S and Petkovic T *XXIII Conference of the Danube Countries on the hydrological forecasting and hydrological bases of water management* CD-ROM

[6] Varnes D 1984 *Landslide hazard zonation: A review of principles and practice* (Paris: UNESCO)

[7] UN/ISDR 2004 *Living with risk. A global review of disaster reduction initiatives* (Geneva: United Nations)

[8] Kienholz H 1977 *Kombinierte geomorphologische Gefahrenkarte 1:10.000* von Grindelwald (Bern: Geographica Bernensia G4)

[9] Grunder M 1984 *Ein Beitrag zur Beurteilung von Naturgefahren im Hinblick auf die Erstellung von mittelmassstäbigen Gefahrenhinweiskarten (mit Beispielen aus dem Berner Oberland und der Landschaft Davos)* (Bern: Geographica Bernensia G23)

[10] Bollinger D, Noverraz F and Tripet J P 1992 Die kartographische Darstellung der Bodenbewegungsgefahren als Planungsgrundlage: Ein Beitrag zur Vereinheitlichung der Gefahrenaufnahmen in der Schweiz. Proc. Int. Symp. Interpraevent 1992 (3): 191-200

[11] Kienholz H and Krummenacher B 1995 *Symbolbaukasten zur Kartierung der Phänomene. Bundesamt für Wasserwirtschaft (BWW)* (Bern: Bundesamt für Umwelt, Wald und Landschaft (BUWAL))

[12] Fuchs S, Keiler M, Zischg A and Bründl M 2005 The long-term development of avalanche risk in settlements considering the temporal variability of damage potential *Natural Hazards and Earth System Sciences* **5 (6)** 893-901

[13] Keiler M, Sailer R, Jörg P, Weber C, Fuchs S, Zischg A and Sauermoser S 2006 Avalanche risk assessment – A multi-temporal approach, results from Galtür, Austria *Natural Hazards and Earth System Sciences* **6 (4)** 637-51 Borter 1999 Risikoanalyse bei gravitiativen Naturgefahren *BUWAL Umweltmaterialien* **107/I,II** Bern

[14] Wilhelm C 1997 *Wirtschaftlichkeit im Lawinenschutz* Mitt. Eidgenössisches Institut für Schnee- und Lawinenforschung 54 Davos

[15] Heinimann H, Hollenstein K, Kienholz H, Krummenacher B and Mani P 1998 Methoden zur Analyse und Bewertung von Naturgefahren. *BUWAL Umweltmaterialien* **85** Bern

[16] Borter 1999 Risikoanalyse bei gravitiativen Naturgefahren *BUWAL Umweltmaterialien* **107/I,II** Bern

[17] BWW, BRP and BUWAL 1997 *Berücksichtigung der Hochwassergefahren bei raumwirksamen Tätigkeiten. Bundesamt für Wasserwirtschaft (BWW), Bundesamt für Raumplanung (BRP), Bundesamt für Umwelt, Wald und Landschaft (BUWAL)* Bern und Biel

[18] Serrhini K, Rochman J, Fuchs S, Dorner W and Spachinger K 2008 Sémiologie graphique expérimentale et cartographie du risque d'inondation. In Pinet, F & Miralles, A. (eds.) Actes de l'Atelier «Systèmes d'Information et de Décision pour l'Environnement». *Proc. XXVIème*

IOP Conf. Series: Earth and Environmental Science **4** (2008) 012043 doi:10.1088/1755-1307/4/1/012043

Congrès INFORSID – Fontainebleau (27.-30. Mai). Clermont Ferrand and Montpellier: Cemagref : 31-40 Casale R and Samuels P 1998 *Hydrological risks: Analysis of recent results from EU research and technological development actions* (Brussels: European Commission, Directorate-General Science, Research and Development)

[19] Meyer V and Mai S 2004 Überflutungsschäden im Küstenhinterland nach Deichbruch. *Wasserwirtschaft* **11** 23-24

[20] IPCC 2001 Climate Change 2001: Impacts, Adaptation and Vulnerability *Contribution of Working Group II to the Third Assessment Report of the Intergovernmental Panel on Climate Change (IPCC)* (Cambridge, Cambridge University Press)

[21] Kienholz H, Krummenacher B, Kipfer A, Perret S 2004 Aspects of integral risk management in practice – Considerations with respect to mountain hazards in Switzerland *Österreichische Wasser- und Abfallwirtschaft* **56** 43-50

[22] Casale R. and Samuels P 1998 *Hydrological risks: Analysis of recent results from EU research and technological development actions* (Brussels: European Commission, Directorate-General Science, Research and Development)

IOP Conf. Series: Earth and Environmental Science **4** (2008) 012044 doi:10.1088/1755-1307/4/1/012044

Variability of bed sediments in channel network of Rye Island (Slovakia)

Y Velísková and R Dulovičová

Institute of Hydrology, Slovak Academy of Sciences, Račianska 75, 831 02 Bratislava, Slovakia

E-mail: Yvetta.Veliskova@savba.sk

Abstract. Aim of the paper was to evaluate the changes of bed silt state of the Rye Island channel network in the period from 1993 to 2004.
The Rye Island is part of the Danube Lowland, it is area between the two branches of the Danube river - the Danube and the Small Danube. During previous centuries (up to the 19th century), the Danube river branched into multiple streams and frequently changed course within its own alluvial sediments. At present, the process of river migration has stopped or is being controlled.
Measurements of bed silt thickness in Rye Island channel network were done in 1993 at the channels: Aszód, Gabčíkovo-Topoľníky, Aszód-Čergov, Čergov-Komárno, Čalovo-Holiare and Holiare-Kosihy. Then in 2004 were done measurements at selected profiles of Aszód, Gabčíkovo-Topoľníky and Komárňanský channel for checking of the silting up variability. According to results of measurements at the three selected channels we can say that the channel silting up has not been changed considerably in monitored period.
The observed facts could be useful for channels maintenance, as a base for expectation of silting up process in the channels. Also the knowledge about silt thickness in the channels is important from the view of interaction between channel network and groundwater, especially when this information would be supplemented by characteristics of silt permeability. So the other aim of field measurements was determination of hydraulic conductivity. Their values were calculated by empirical formulas. These characteristics will be used for simulation of interaction between channel network and groundwater at the Rye Island area in the oncoming period.

1. Introduction

The Danube River and its nearby area Rye Island has brought in many benefits, but sometimes also problems. Water resource management have to solve problems during dry periods or floods, which are connected with irrigation or drainage function of the Rye Island channel network. The regulation of water regime at the Rye Island area needs information about mutual interaction between channel network and groundwater at this area. The aim of the built channel network has been to drain wet area and to irrigate fields during dry period. Groundwater at this area is strongly related to water regime of channel network. Various factors have changed their mutual interaction (e.g. design and operation of Gabčíkovo water work). One of them is also bed aggradation of channels with suspended load silts, which affects the channel permeability.

2. Description of locality

The Danube River flows through a relatively narrow valley along the Austrian-Slovak border. The river crosses a strip of mountains connecting the Alps and the Karpatians, passes Bratislava, and divides into two branches just below Slovak capital, i.e., the Danube and the Small Danube – see figure 1. Two branches of the Danube flow separately approximately 100 km and then join each other again near the town Komárno. The area between the two branches of the Danube River is called Rye Island, it is part of the Danube Lowland. The average width of Rye Island is 20 km; its area is approximately 2000 km^2 , which represents about 4 % of the Slovak territory.

The Rye Island area forms a flat plain with only small differences in altitude. The surface of Rye Island decreases in the south-east direction. Its average slope is about 0,25 ‰, and it was one of the reasons for building channel network at this area (as a drainage system). Rye Island is the result of sedimentation of the Danube River, with sediments from upstream mountains (Alps) being spread over its territory. Sedimentation of transported materials has created thick layers of gravel, with a tendency to elevate the river bed. The creation of river branches, which flow according to the Danube River water level, is another aspect of the sedimentation process. During previous centuries (up to the 19th century), the river branched into multiple streams and frequently changed course within its own alluvial sediments. At present, the process of river migration has stopped or is being controlled, because irregular sedimentation provides changes in the river bottom and changing conditions for shipping.

Water in alluvial deposits of the Danube River serve as a source of water for such agglomeration as Vienna, Bratislava, Budapest, Belgrade, and others. The body of sediments between Bratislava and Gabčíkovo is of great importance in terms of supplying water to surrounding areas. The area is often designated as an inland Danube delta, made up of a gravel layer of up to 400m thick and containing an underground water reservoir of about 20 millions cubic meters of high-quality water.

The general direction of groundwater flow during periods of low discharges in the Danube River along the length of Rye Island is in a south-eastern direction, more or less parallel to the Danube flow direction. When the water level in the Danube is relatively high, large amounts of water infiltrate into the Rye Island aquifer, with velocity vectors of the seepage water being close to 90 degree to the Danube River flow direction [4].

As it was mentioned above, the slope of the Rye Island area is very low, so natural draining has not developed. For this reason the network of channels has been built at this area since the end of 19th century. Today the channel network length is nearly 1000 km. Its density is approximately 1 km/1,25 km^2. System of channels can be utilized for drainage and irrigation.

The problem of water level regulation at this area is complex. Therefore, many specialists were interested in solution of it ([1], [3], [4], [6], [7], [9]). Channels, manipulating objects and pumping stations as basic elements of this channel network allow controlling water level in channels to achieve optimal position of groundwater table during vegetation period. The thickness and structure of bed silts are factors influencing interaction, therefore it has been necessary to assess also the impact of channel network silting up on it. In this paper we dealt with the changes of silt thickness in the period from 1992 to 2004.

3. Field measurements of silt thickness and results

The measurements of silt thickness were performed in 1993 at following channels: Aszód, Gabčíkovo-Topoľníky, Komárňanský channel, channel Aszód-Čergov, Čergov-Komárno, Čalovo-Holiare-Kosihy. The scheme of the area is in figure 2. Measurements were done at the distance 1,0-2,0 m in cross-section direction and distances of evaluated cross-sections were in the range 2,0-5,0 km. The measured data included distance of a measured point from the right channel bank, water depth to the bottom of channel, water depth to the silt and silt thickness (see table 1). It is important to remind that all measurements were done from surface water level, so only the differences between silt top and channel bottom levels were estimated. The average silt thicknesses were determined as a ratio of a cross-section profile silt area and width of the channel.

The volume of suspended load sediments was calculated from average silt thickness and from its distribution along a channel. The total volumes of suspended load sediments are summarized in table 2.

Table 1. Example of measured data record (Komárňanský channel)

Stream log (of profile) (km)	Distance from R bank (m)	Water depth (to bottom of channel) (m)]	Water depth (to silt) (m)	Thickness of silt (m)
	0,00	0,00	0,00	0,00
	1,00	-0,66	-0,54	0,12
	2,00	-0,81	-0,58	0,23
	4,00	-1,09	-0,94	0,15
27,00	7,50	-1,28	-0,83	0,45
	11,00	-0,85	-0,65	0,20
	13,00	-0,69	-0,59	0,10
	14,00	-0,45	-0,45	0,00
	15,00	0,00	0,00	0,00

Table 2. Total volume of suspended load sediments in the Rye Island channel network(1993)

Channel	Volume (m^3)
Aszód	111 436
Gabčíkovo-Topoľníky	93 793
Komárňanský channel	197 824
Aszód – Čergov	38 419
Čalovo – Holiare - Kosihy	113 059
Čergov - Komárno	58 350

3.1. Control field measurements

Control measurements for checking of the silting up changes were performed at selected profiles of Aszód, Gabčíkovo-Topoľníky and Komárňanský channels in 2004. The control profiles were selected at the beginning, middle and ending part of channels and the measurements were carried out approximately at the same profiles along channels like in 1993. The results of some measurements are shown for illustration in figure 3a, figure 3b and figure 4.

3.1.1. *Aszód channel.* Measurements were done at three profiles of the Aszód channel: Aszód (km 1,2), Veľký Meder-Dolný Štál (km 12,3), Ňarád (km 25,3). The silt thickness did not change significantly at the beginning part of channel (Aszód). We also calculated the average silt thickness: the value was 0,3 m in 1993 and in 2004 it was 0,32 m. In the middle part (Veľký Meder-Dolný Štál) we did not evaluate the changes because there was noticed occurrence of water vegetation, thus it was not possible to measure silt thickness along whole cross-section. By comparison of calculated average values of silt thickness in 1993 (0,48 m) and 2004 (0,76 m) in profile Ňarád the silt thickness increased.

3.1.2. *Komárňanský channel.* The changes of silt thickness are variable along the channel and the shape of cross-sections along this channel has been changed, too. At the beginning part it is without change (Hadovce, km 2,2), but in the middle part (Okoličná na Ostrove, km 14,8) it increased from 0,4 m to 0,8 m. This part of channel was rather wide and deep with dense water vegetation. In the last measured profile Nový Golyáš (km 26,0) distribution of silt through this cross-section in 2004 showed relatively uniform trend. In this and previous parts there were observed moderate increase of silt. The total volumes of suspended load sediments at the Komárňanský channel during evaluated period increased by calculations from 161 344 m^3 to 234 729 m^3 .

3.1.3. *Gabčíkovo-Topoľníky channel.* The measurements were done at three profiles, as well (Trhová Hradská-Topoľníky, km 1,6; Kútniky-Dolný Bar, km 10,0; and Baka, km 26,0). The comparison of silt thickness in the beginning part of the channel was not made because in 2004 measurements was not performed at the same location. In the middle part the measurement was not completed again, but partial results show increase of silting up. The comparison of measurement outputs shows that silting up at the ending part of this channel (Baka) was increased. Additionally, the silt included large content of organic mass.

4. Assessment of silt permeability

The observed facts could be useful not only for channels maintenance, but the knowledge about silt thickness in the channels is also important from the view of interaction between channel network and groundwater because the Rye Island is also known as the biggest source of groundwater in Slovakia. The water level in the channel network system has effect to groundwater level and in reverse. From view of mutual interaction between channel network and groundwater it has been necessary to know not only the silt thickness in the channels, but also permeability of silts. So the other aim of field measurements was determination of the silt permeability, which is characterized by hydraulic conductivity.

For that reason we took at the same time as we measured the silt thicknesses in 2004 also the samples of silts for determination of its granularity and hydraulic conductivity. The samples were taken from top and also from bottom of silt layer. However, in some cases there was not possible to take them. For example, the banks of the profile Hadovce at the Komárňanský channel were overgrown with vegetation; at the profile Topoľníky of the channel Gabčíkovo-Topoľníky the alluvium had liquid consistence and thin thickness, so we did not manage to take samples of it; or at the profile Kútniky, there was flagrant silt, but as we already mentioned, the silt included large content of organic mass. Therefore it was not possible to establish its granularity.

It is possible to use several empirical relationships for determination of hydraulic conductivity, but we can apply them very carefully for their limited validity. As we could not take of undisturbed samples, we could use only the relationships by Beyer-Schweiger (cited in [5]) and by Špaček [8]. These relationships are functions of d_{10} – particle diameter in 10% of soil mass (m) and d_{60} – particle diameter in 60% of soil mass (m). Both of them were determined from granularity curves of the silts – some examples are in figure 8.

Values of grain and filtration characteristics of silts are in table 3.

5. Summary and conclusions

Aim of this paper was to evaluate the variability of bed sediments in channel network of the Rye Island in the period from 1993 to 2004. Measurements of silt thickness were done in 1993 more detailed at the channels: Aszód, Gabčíkovo-Topoľníky, Komárňanský channel, channel Aszód-Čergov, Čergov-Komárno and Čalovo-Holiare-Kosihy. Then in 2004 were done measurements at selected profiles of Aszód channel, Gabčíkovo-Topoľníky channel and Komárňanský channel for checking of the silting up changes. According to results of measurements at the three selected channels we can say that at the beginning parts of channels the silt thickness are nearly without changes. In the middle parts we observed luxuriant water vegetation, so it was not possible in each channel to measure silt thickness along whole cross-section. The silt thickness increased in Komárňanský channel and slightly in Gabčíkovo-Topoľníky, too. In the ending parts of channels Aszód and Gabčíkovo-

Topoľníky silts has been increased in this monitored period, in the case of Komárňanský channel (which flows into the Váh River) there was observed only moderate increase of silts. The silts from middle and ending part of Gabčíkovo-Topoľníky channel included large content of organic mass.

Table 3. Values of grain and filtration characteristics of silts

Channel	Sampling site	d_{10} (mm)	d_{60} (mm)	d_{60}/d_{10}	Hydraulic conductivity K (m.s^{-1})
Aszod ^{T}	Meder/Štál	0,074	0,116	1,57	^{Sp} 2,36E-05
Aszod ^{B}	Meder/Štál	0,075	0,293	3,91	$^{B\text{-}Sch}$ 5,10E-05
Aszod ^{T}	Aszód	0,006	0,080	13,32	^{Sp} 1,50E-06
Aszod ^{B}	Aszód	0,0085	0,078	9,18	^{Sp} 2,14E-06
Aszod ^{T}	Ňárád	0,038	0,213	5,61	^{Sp} 2,09E-05
Aszod ^{B}	Ňárád	0,0076	0,086	11,32	^{Sp} 1,89E-06
Kom. ^{T}	Golyáš	0,008	0,164	20,5	^{Sp} 1,92E-06
Kom. ^{B}	Golyáš	0,0073	0,144	19,73	^{Sp} 1,76E-06
Kom. ^{T}	Okoličné	0,082	0,321	3,92	$^{B\text{-}Sch}$ 6,09E-05
Kom. ^{B}	Okoličné	0,007	0,154	21,89	^{Sp} 1,69E-06
Gab.Topol.^{T}	Baka	0,008	0,088	11,05	^{Sp} 1,99E-06
Gab.Topol.^{B}	Baka	0,0093	0,291	31,29	^{Sp} 7,64E-06

^{T} top layer of silt sample;
^{B} bottom layer of silt sample;
$^{B\text{-}Sch}$ by Beyer-Schweiger;
^{Sp} by Špaček

Generally, we can say that the channel silting up has not been changed considerably in this monitored period.

The silt thickness measurements along the whole cross-sections were managed to do in 2004 only at Komárňanský channel, for that reason the numerical evaluation of changes of silt volume in period from 1993 to 2004 were done for this channel only. The values of silt volume in km 4,0 – 26,0 of Komárňanský channel document that silts have been increased from 161 344 m^3 to 234 729 m^3 in monitored period.

The observed facts could be useful for channels maintenance, but the knowledge about silt thickness in the channels is also important from the view of interaction between channel network and groundwater, especially when this information is supplemented by the hydraulic conductivity. The values of this characteristic were calculated by empirical formulas mentioned above and results are shown in table 3. These characteristics will be used for simulation of interaction between channel network and groundwater at the Rye Island area in the oncoming period.

Acknowledgments
This work was supported by the Science Granting Agency under the contract No. VEGA-2/0101/08 and APVV 51-017804.

IOP Conf. Series: Earth and Environmental Science **4** (2008) 012044 doi:10.1088/1755-1307/4/1/012044

References

[1] Burger F 2003 Subsurface hydrological system responses to changes in waterworks influence zones *Acta Hydrologica Slovaca* **4** No.1 203-9 (in Slovak)

[2] Kališ J and Bačík M 1983 Modelling of reservoir aggradation *Water Research Institute Report* Bratislava (in Slovak)

[3] Kamenský J 1993 Suspended loads in some channels at Rye Island *Faculty of Civil Engineering STU Report* Bratislava (in Slovak)

[4] Kosorin K 1997 Spatial groundwater dynamics of the Rye Island aquifer. *J. Hydrol. Hydromech.* **45** 348 – 64

[5] Mucha I and Šestakov V M 1987 *Fluid Mechanics of Ground Waters* (Bratislava: Alfa) (in Slovak).

[6] Novák V Šútor J Majerčák J Šimunek J and Genuchten V M 1998 *Modelling of water and solute movement in the unsaturated zone of the Rye Island Region, South Slovakia* (Bratislava: Institute of Hydrology SAS)

[7] Šoltész A Gramblička M and Baroková D 1999 *Proc. Int. Watermanagement Colloquy XIth Int. Scientific Conf.- Occassion of the 100th anniversary of Establishing the Czech Technical College in Brno and Initiating in the Civvil Engineering Branch of Study (Brno, Czech Republik, 20th October 1999)* p 70 –79 (in Slovak)

[8] Špaček J 1987 Determination of filtration coefficient from total grain-size curves *J. Meliorace* **23** No.1 1-13 (in Czech)

[9] Šútor J and Štekauerová V 2000 *Hydrophysical characteristics of soils at Rye Island*IH SAS (Bratislava, Slovakia: Institute of Hydrology SAS) (in Slovak)

IOP Conf. Series: Earth and Environmental Science **4** (2008) 012045 doi:10.1088/1755-1307/4/1/012045

Interrill soil erosion on flysch soil under different types of land use in Slovenian Istria

M Zorn[1], **S Petan**[2]

[1] Anton Melik Geographical Institute, Scientific Research Centre of the Slovenian Academy of Sciences and Arts. Gosposka ulica 13, 1000 Ljubljana, Slovenia

[2] Faculty of Civil and Geodetic Engineering, University of Ljubljana, Jamova cesta 2, 1000 Ljubljana, Slovenia

E-mail: matija.zorn@zrc-sazu.si, spetan@fgg.uni-lj.si

Abstract. In this paper the results of interrill soil erosion measurements from recent years in Slovenian Istria are presented. Eight 1-m2 erosion plots were set up on locations with different land use types: on bare soil in a young olive grove (2), in an overgrown meadow (2) and in a forest (4). Surface runoff from each of the erosion plots was collected in reservoirs. As a rule, samples from the reservoirs were taken weekly. The samples were dried in the laboratory, where the concentration of undissolved particles was determined. A tipping bucket rain gauge was located next to the erosion plots for monitoring of precipitation and intensity of erosive events. The results show that only a few major erosive events are responsible for the greater part of the eroded soil. Interrill soil erosion in the first year (May 2005–April 2006) was estimated at 90.1 t/ha on bare soil with a slope of 5.5°, and 118.2 t/ha in the second year (August 2006–July 2007), despite the lower cumulative rainfall amount.

1. Introduction

"*Little research has been done on soil erosion in Slovenia*" says the evaluation of the implementation of the United Nations Convention to Combat Desertification in Slovenia [1]. In April 2005 researchers at the Anton Melik Geographical Institute of the Scientific Research Centre of the Slovenian Academy of Sciences and Arts began measurements of interrill soil erosion in cooperation with the Faculty of Civil Engineering of the University of Ljubljana. The measurements took place near the village of Marezige in the Dragonja river basin.

The Dragonja river basin is situated in SW Slovenia; more precisely in the Northern part of the Istrian Peninsula, on the border between Slovenia and Croatia (Figure 1). The river flows from west to east, into the North Adriatic Sea. The basin area is mainly hilly and covered by forest. The surface area amounts to 91 km2. More than half the river basin is situated at an elevation of between 200 and 300 m above sea level and has a slope of between 5° and 15°. The majority of settlements have developed on the long flat ridges (cl. up to 400 m) above the deep and narrow river valleys [2].

The geological composition of the river basin is relatively simple and homogenous. Eocene flysch sediments predominate, mostly consisting of marl and sandstones. The bedrock is relatively soft and erodible. Soil is mostly carbonate rendzina, while some hill ridges have brown eutric soil. In the valley, fluvisol on carbonate alluvium developed. The alluvium consists mainly of clay with small amounts of gravel and sand. There is hardly any groundwater flow. The alluvial aquifer lies on the impermeable flysch that covers most of the basin area [3].

The Dragonja river basin belongs to the sub-Mediterranean climate region with an average annual temperature of 14°C on the sea coast and 10°C on the easternmost side. The total annual precipitation by the coast has an average of 950 mm. This value rises as we move further inland and it reaches 1200 mm at the eastern end of the river basin. Historical data show that the maximum measured daily sum of precipitation reached 151.4 mm in October 1980. The average 15-minute shower typically contributes 20 mm of precipitation (the measured maximum is 49.1 mm) and the average one-hour shower contributes 34 mm (the measured maximum is 66 mm) [2].

Figure 1. The Dragonja river basin with monitoring sites.

2. Soil erosion

Soil erosion is mechanical soil degradation as a result of the activities of natural forces, often supported by human activities. Soil particles are detached from the ground and moved distances of varying lengths, and thus lose their natural role. The soil is well protected in the natural environment by dense vegetation cover, such as forests or meadows. As a rule, on cultivated areas, vegetation is removed and crops harvested, and so the land surface stays bare and exposed. The natural soil erosion changes into an accelerated anthropogenic erosion, which directly affects the physical, chemical and biological soil characteristics [4].

The soil erosion processes consist of soil detachment, transport and deposition. Soil particles are separated from the ground due to the raindrop impact force (splash erosion) and the shearing force of the surface water flow, which is also responsible for the down slope transport of the detached soil. The transport capacity decreases with infiltration rate and increases with water flow velocity. Because the slope tends to level due to the soil movement down slope, the flow velocity decreases and deposition occurs [5].

Soil erosion can occur as rill erosion, in the small channels or rills induced by the surface water flow, or interrill erosion, in the area between the rills. Most of the interrill soil erosion is caused by the impact of the raindrops upon soil. The raindrop splashes also contribute to the soil transport but only at sloping surfaces. The majority of the soil transport is caused by the thin surface water flow supported by the raindrop impact, which increases the turbulence of the surface flow and makes it capable of transporting the detached soil particles. The detached soil quantity depends on the rainfall intensity, the kinetic energy of the raindrops and the clay content of the soil [6].

According to the Revised Universal Soil Loss Equation (RUSLE), the soil loss quantity can be calculated as the product of the active (hydrologic and topographic) factors and reactive factors (erodibility, land cover and land use) [5]. The crucial factor is the hydrological one, the rainfall erosivity factor R, which is defined as the product of the erosive event kinetic energy and the maximal 30-minute rainfall intensity (I_{30max}).

2.1. Interrill soil erosion

Interrill soil erosion is the consequence of splash erosion (erosion impact of raindrops) and surface wash, before the surface water flow is combined into jets of water and starts to erode downwards. Although it is considered as denudation, it is classified as an erosion process. Without measurements interrill erosion is difficult to assess and quantify, which is why its consequences are often underestimated [7; 8; 9; 10]. Such erosion is high especially on agricultural land; accumulation can be seen along the lower boundaries of fields or along the fences at the end of fields [11; 12; 13]. Soil erosion from the fields and the heavy work of the peasants who carried the eroded soil back to the fields found its way into Slovenian literature, for example in the story "Passion above the precipice" written by Prežihov Voranc (1893–1950) [14]. Sources about this can also be found elsewhere: for example, in the hills of Slovenian Istria water erosion was always present despite the terracing, which is why "*the peasants carried each year the eroded soil back upslope*" [15]. For this purpose they had by "*modern terraces [...] a ditch to be made on the top of each terrace*" [15]. Similar tales also come from the lithologically and climatically similar Goriška Brda hills (W Slovenia) [16]: "*There are no rare circumstances that the locals dug ditches on the lower parts of the fields into which the rainwater washed the soil; the soil was afterwards in baskets carried back to the fields*".

Figure 2. Carrying the soil near the Brezovica village (Brkini hills, SW Slovenia) just after the Second World War (photographer: Andrej Pagon Ogarev; © The Regional Museum in Koper – Ethnological collection).

3. Measurement methodology

Near the village of Marezige eight 1-m^2 erosion plots were set up on locations with different land use types: on bare soil in a young olive grove (2), in an overgrown meadow (2) and in the forest (4) (Figures 3 and 4). The erosion plots in the forest were placed on soil with two different slopes, 8° and 21°. Following the plot size classification made by Poesen et al. [17], our plots are useful for micro or meso scale measurements. Measurements on erosion plots of similar size were for example made in Spain [18; 19; 20; 21].

Figures 3 and 4. Erosion plots in the olive grove on bare soil and the tipping bucket rain gauge (left), and the erosion plot in the forest (right).

For the construction of erosion plots we consulted the papers of Vacca et al. [22] and Ollesch and Vacca [23], and for the installation into the ground we used the works of Lal and Elliot [24] and Dunjó et al. [19]. A visit to the erosion plots near the Abrami village in Croatian Istria was also of great assistance.

The erosion plots are composed of three metal plates 1 m long and 30 cm wide (one in the back and two on the sides), which were dug into the soil approximately 10 cm deep (the same depth is also mentioned in the paper by Ollesch and Vacca [23]), and a metal sink in front. From the sink the outflow goes through a plastic tube into a plastic reservoir (30 l), dug into the soil. All metal parts of the erosion plots are placed so that the plots are separated from the surroundings – they belong to the so-called 'closed erosion plots' [9].

As a rule, the measurements were performed weekly. Firstly, the volume of the collected surface runoff was determined, and then the samples of water and detached soil mixture from the reservoirs were taken. The samples were dried at 105°C in the laboratory, where the concentration of undissolved particles was determined. A tipping bucket rain gauge equipped with a data logger is located next to the experimental plots for monitoring of precipitation and intensity of erosive events [9]. It has a resolution of 0.2 mm and an accuracy of 1% up to 20 mm/h.

4. Results

Interrill soil erosion in the first year of measurements (May 2005–April 2006) was estimated at 9013 g/m^2 (90 t/ha) on bare soil in an olive grove with an inclination of 5.5°; while on an overgrown meadow with an inclination of 9.4° erosion it was estimated to be 168 g/m^2 (1.68 t/ha), and in the forest with an inclination of 7.8° it was 391 g/m^2 (3.91 t/ha) and in the forest with an inclination of 21.4° it was 415 g/m^2 (4.15 t/ha). The total amount of precipitation in this reference year was 1044 mm and was slightly lower than a long year average, which in the surroundings of the Marezige village amounts to between 1100 and 1200 mm [2].

In the second year (August 2006–July 2007) interrill soil erosion on bare soil in the olive grove was 11.822 g/m^2 (118 t/ha), on the overgrown meadow 90 g/m^2 (0.90 t/ha), in the forest with lower inclination 421 g/m^2 (4.21 t/ha) and in the forest with bigger inclination 115 g/m^2 (1.15 t/ha). This was despite the fact that the total cumulative rainfall amount was lower and amounted to only 930 mm.

Although our measurements are not long-term, we could still notice that "bigger" erosive events contributed a great portion to the total yearly soil loss. In the week between August 5th and August 12th 2005 the greatest erosive rainfall event occurred in the first reference period (weekly erosivity was 1235.91 MJ/ha mm/h; on August 11th the maximum 30-minute precipitation was 42.8 mm, daily erosivity was 1110.5 MJ/ha mm/h). On bare soil in the olive grove during this week 30% of all yearly soil loss in the first reference year was recorded. In an overgrown meadow 24% of yearly soil loss was recorded. Because of the rainfall interception on treetops the amount of soil loss in the forest was lower. In the forest with lower inclination during this week 15% of yearly soil loss was recorded; in the forest with greater inclination the amount of soil loss did not stand out from other extremes [9]. From these data it is seen that the vegetation period is of considerable importance for soil erosion.

The total erosivity of precipitation in the second reference year (August 2006–July 2007) was 3216 MJ/ha mm/h. Almost 93% of yearly erosivity was the consequence of 26 rainfall events, which contributed 71% to total yearly precipitation. The greatest soil loss was recorded in August 2006, when during five major erosive rainfall events 262 mm of rain fell. These rains were co-responsible for 53% of the total yearly soil loss on bare soil in the young olive grove, for 78% of total yearly soil loss in the overgrown meadow, for 68% of total yearly soil loss in the forest with lower inclination and 51% of total yearly soil loss in the forest with greater inclination. The biggest erosive rainfall event was on August 25th 2006 (erosivity was 650 MJ/ha mm/h and maximum 30 minute precipitation 25.2 mm), during which 21% of total yearly soil loss on bare soil was recorded and almost 52% of total yearly soil loss in the overgrown meadow [25].

4.1. A closer look at the results from 2006 in 2007

In 2006 33 erosive events were registered, contributing 875 mm of rainfall, while 43 events in 2007 contributed 939 mm of rainfall. The erosive events were chosen as rainfall events of at least 6 mm; two erosive events were distinguished if there was a 6-hour period between the two events that contributed not more than 1 mm of rainfall. Table 1 presents the measured and calculated annual values for rainfall erosivity (R factor), soil loss and average run-off coefficients.

Table 1. Annual values for R factor, soil loss and average run-off coefficient for the measurements performed in 2006 and 2007.

quantity	unit	year	bare soil (olive grove)	overgrown meadow	forest (slope of 8°)	forest (slope of 21°)
R factor	[MJ/ha mm/h]	2006	2090.4			
		2007	3551.3			
Soil loss	[kg/m^2]	2006	8.434	0.101	0.136	0.192
		2007	10.129	0.064	0.049	0.027
Run-off coef.	[-]	2007	0.32	0.19	0.14	0.14

The erosion measurements in 2006 unexpectedly stopped for a period of three months, from 26 April to 1 August. This is the reason why Table 1 gives the average run-off coefficients for 2007 only. We were able to estimate the missing soil loss quantities from the R factor of the registered rainfall events that occurred in the period. These 6 events contributed approximately 11% of the annual R factor and 10% of the annual soil loss on the bare soil.

The total R factor for 2007 was approximately 75% higher than in 2006, while the annual soil loss on bare soil was higher by only 25%. On the other hand, the erosion plots in the meadow and in the forest gave much lower values for the soil loss in 2007. The relation between the soil loss and the R factor is not linear. In 2006 the 3 most intensive erosive events represented 55% of the annual R factor, while for the same portion of the annual R factor in 2007 5 erosive events were needed. This can be seen in Figure 5, where the cumulative relative contribution of the erosive events, arranged in decreasing order by their intensity, to the total annual R factor and kinetic energy is shown.

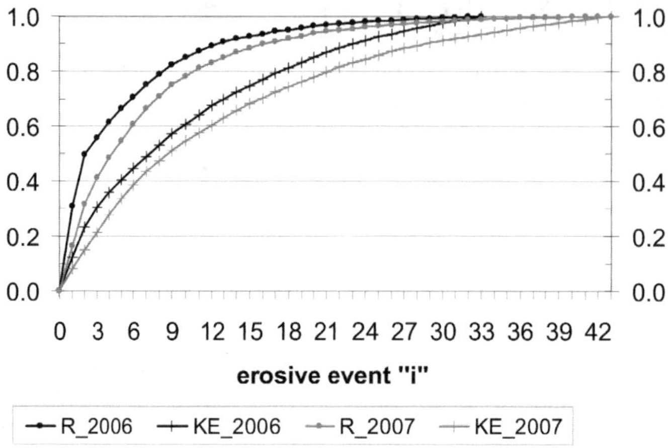

Figure 5. Cumulative relative contribution of the "i" most intensive erosive events to the total annual R factor and kinetic energy (KE). The erosive events were arranged in decreasing order by their intensity.

The most intensive erosive event in 2006 occurred on 25 August with 49.4 mm of rainfall, I_{30max} of 50.8 mm/h and R factor of 650 MJ/ha mm/h, producing approximately 24% of the annual soil loss. The most intensive event in 2007 occurred on 18 September, bringing 43.6 mm of rainfall and 584 MJ/ha mm/h, with I_{30max} of 50 mm/h, producing nearly 12% of the annual soil loss. On 11 August 2005, during the period of preliminary measurements, an extreme event occurred. The measured event rainfall was 45.6 mm with maximum 30-minute rainfall intensity of 84.8 mm/h, creating an erosive event with R factor of 1071 MJ/ha mm/h and 2.67 kg/m^2 of soil loss on the bare soil.

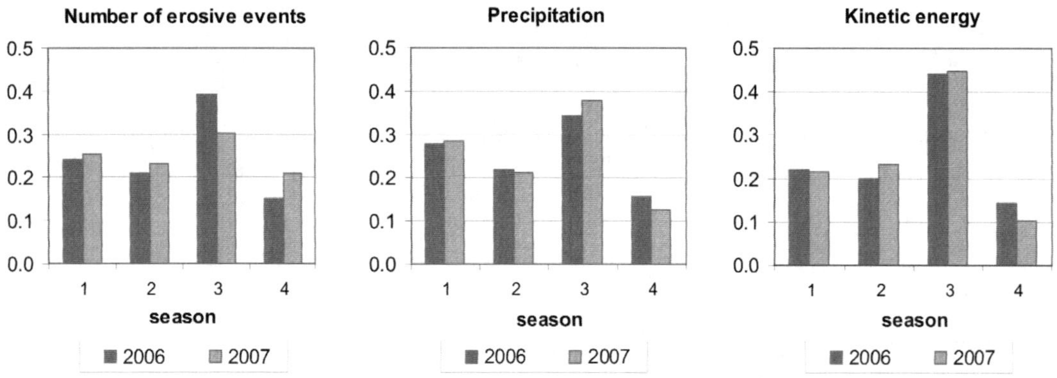

Figure 6. Seasonal distribution of the annual portions of the number of erosive events, precipitation quantity and the kinetic energy. Seasons are marked with numbers: 1 – Jan-Mar; 2 – Apr-Jun; 3 – Jul-Sep; 4 – Oct-Dec.

We also investigated the seasonal patterns of the measured and calculated quantities. Figure 6 shows the seasonal distribution of the annual portions of the erosive events, precipitation and the kinetic energy. The high similarity between these histograms is obvious. This is also clearly seen in Figure 7, where the annual portions of the R factor and the soil loss on bare soil are shown. The effect of the developed canopies on the soil erosion protection in the forest can be seen from Figure 8 (the values for January and December are not shown in the diagram due to their high amounts; January – 51, December – 15). The value for April is missing because there was not any rainfall during the month. It is obvious that the portion of the forest soil loss in the summer period, when the vegetation cover is nicely developed, is much lower than the soil loss on bare soil.

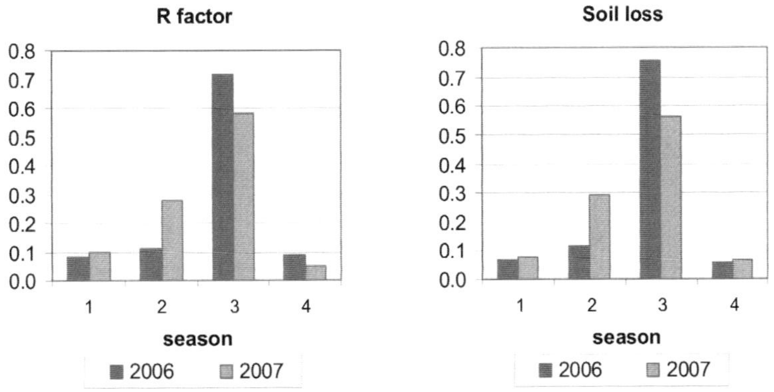

Figure 7. Seasonal distribution of the annual portions of R factor and the soil loss on bare soil. Seasons are marked with numbers: 1 – Jan-Mar; 2 – Apr-Jun; 3 – Jul-Sep; 4 – Oct-Dec.

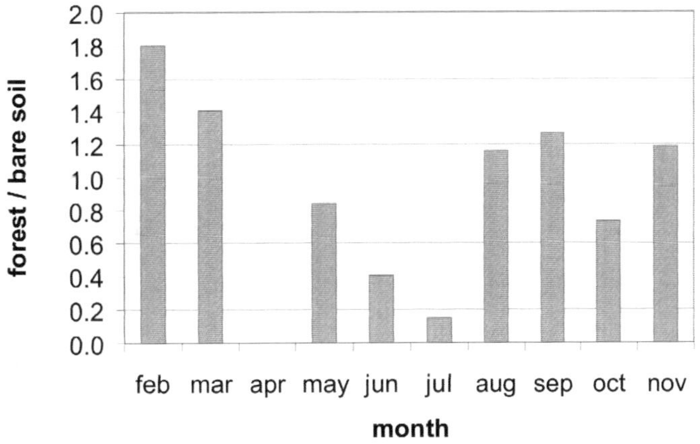

Figure 8. Ratio between the monthly relative contributions to the annual soil loss in the forest and on the bare soil. January and December are omitted due to the high values (January – 51, December – 15).

5. Conclusion

An annual denudation rate of 1 cm due to soil erosion was in Slovenia rarely mentioned in the past [26]. Mikoš and Zupanc [27] stated that on average 5–10 mm of "fertile soil, or even more in severe conditions" are lost in Slovenia every year due to soil erosion. These values were considered high in the past. However, our measurement data show that these values are quite common: the bare soil surface in an olive grove is lowered by 9–11 mm per year. It should be noted that only interrill soil erosion was measured and not total (rill and interrill) soil erosion. The estimation of lower erosion rates in the past could be assigned to the fact that measurements were rare; the data were based on empirical equations, which strongly underestimate the actual soil erosion, as is also shown by our research. Thus we hope our measurement data will be helpful for the calibration of foreign erosion models in Slovenian conditions.

The research also highlighted the importance of vegetation in the battle against soil erosion (for example, leafy canopies reduce the rainfall intensity). The surface denudation on the overgrown meadow amounted to between 0.09–0.16 mm. Since the difference compared to the bare soil denudation is very high, it is not necessary to stress the importance of grass cover on cultivable land as one of the soil erosion control measures.

References

[1] Ocena izvajanja Konvencije Združenih narodov o degradaciji tal v Sloveniji (Evaluation of the implementation of the United Nation Convention of Soil Degradation in Slovenia) 2005. http://www.rec-lj.si/projekti/NCSA/Dokumenti/NCSA_DT.pdf (10. 11. 2007).

[2] Ogrin D 1995 *Podnebje slovenske Istre* (The Climate in the Slovene Istria) (Koper: Annales).

[3] Globevnik L 2001 *Celosten pristop k urejanju voda v povodjih* (Integrated Approach to Water Management). (Ljubljana: Unpublished Ph. D. Theses)

[4] Kisić I, Bašić F, Butorac A, Mesić M, Nestroy O and Sabolić M 2005 *Erozija tla vodom pri različitim načinima obrade* (Water Soil Erosion under Different Land Use) (Zagreb: Agronomski fakultet)

[5] Hahn C T, Barfield B J and Hayes J C 1994 *Design hydrology and sedimentology for small catchments* (San Diego: Academic Press)

[6] Petkovšek G 2000 Process based soil erosion modelling *Acta hydrotechnica* **18(28)** 41–60

[7] Komac B and Zorn M 2005 Soil erosion on agricultural land in Slovenia - measurements of rill erosion in the Besnica valley *Acta geographica Slovenica* **45(1)** 53–86

[8] Zorn M and Komac B 2005 Erozija prsti na kmetijskih zemljiščih v Sloveniji (Soil erosion on agricultural land in Slovenia) *Ujma* **19** 163–174

[9] Zorn M 2007a *Recentni geomorfni procesi na rečno-denudacijskem reliefu na primeru porečja Dragonje* (Recent geomorphic processes on fluvial-denudational relief on the example of Dragonja river basin) (Ljubljana: Unpublished Ph. D. Theses)

[10] Zorn M 2007b Ali se zavedamo hitrosti erozijskih procesov – primer iz slovenske Istre (Are we aware of the speed of erosion processes – an example from Slovene Istria) *Dela* **28** 183–194

[11] Lazarević R and Miličević S 1983 Erozija u slivu Botonege (Erosion in the Botonega catchment) *Erozija – stručno-informativni bilten* **11** 51–65

[12] Natek K 1989 Vloga usadov pri geomorfološkem preučevanju Voglajnskega gričevja (The role of landslides in the processes of landform transformation in the hills east from Celje) *Geografski zbornik* **29** 37–77

[13] Hrvatin M, Komac B, Perko D and& Zorn M 2006 Slovenia *Soil erosion in Europe* ed J Boardman and J Poesen (Chichester: Wiley) pp 297–310

[14] Voranc P 1994 The self-sown and Passion above the precipice *Litterae Slovenicae* **32(2)**

[15] Titl J 1965 *Socialnogeografski problemi na koprskem podeželju* (Social-Geographical Problems in Koper Countryside) (Koper: Lipa)

[16] Vrišer I 1954 Goriška brda – gospodarska geografija (Goriška Brda hills – economical geography) *Geografski zbornik* **2** 51–113

[17] Poesen J W, Torri D and Bunte K 1994 Effects of rock fragments on soil erosion by water at different spatial scales: a review *Catena* **23(1-2)** 141–166

[18] Dunjó G, Pardini G nad Gispert M 2003 Land use change effects on abandoned terraced soils in a Mediterranean catchment, NE Spain *Catena* **52(1)** 23–37

[19] Dunjó G, Pardini G and Gispert M 2004 The role of land use-land cover on ronoff generation and sediment yield at microplot scale, in a small Mediterranean catchment *Journal of Arid Environments* 57(2) 239–256

[20] Usón A and Ramos M C 2001 An improved rainfall erosivity index obtained from experimental interrill soil losses in soils with a Mediterranean climate *Catena* **43(4)** 293–305

[21] Boix-Fayos C, Martínez-Mena M, Calvo-Cases A, Arnau-Rosalén E, Albaladejo J and Castillo

IOP Conf. Series: Earth and Environmental Science **4** (2008) 012045 doi:10.1088/1755-1307/4/1/012045

V 2007 Causes and underlying processes of measurement variability in field erosion plots in Mediterranean conditions *Earth Surface Processes and Landforms* **32(1)** 85–101
[22] Vacca A, Loddo S, Ollesch G, Puddu R, Serra G, Tomasi D and Aru A 2000 Measurement of runoff and soil erosion in tree areas under different land use in Sardinia (Italy) *Catena* **40(1)** 69–92
[23] Ollesch G and Vacca A 2002 Influence of time on measurement results of erosion plot studies. *Soil and Tillage Research* **67** 23–39
[24] Lal R and Elliot W 1994 Erodibility and erosivity *Soil Erosion Research Methods* ed R Lal (Delray Beach: St. Lucie Press) pp 181–208
[25] Zorn M and Petan S 2007 Meritve medžlebične erozije na različnih rabah zemljišč v slovenski Istri (Measurement of interrill soil erosion under different land use in Slovene Istria) *Strategija varovanja tal v Sloveniji* ed M Knapič (Ljubljana: Pedološko društvo Slovenije) pp 51–61
[26] Hrovat A 1953 *Kraška ilovica: njene značilnosti in vpliv na zgradbe* (Karstic clay: its characteristic and influence on buildings) (Ljubljana: DZS)
[27] Mikoš M and Zupanc V 2000 Erozija tal na kmetijskih površinah (Soil erosion on agricultural land) *Sodobno kmetijstvo* **33(10)** 419–423

IOP Conf. Series: Earth and Environmental Science **4** (2008) 012046 doi:10.1088/1755-1307/4/1/012046

Automatic calibration of the Watbal model

A Cotar[1] and M Brilly

Chair of Hydrology and Hydraulic Engineering, Faculty for Civil Engineering, Jamova 2, 1000 Ljubljana, SI

E-mail: acotar@fgg.uni-lj.si; mbrilly@fgg.uni-lj.si

Abstract. Slovenia is very rich with water. Sava river basin is the biggest one in Slovenia and it covers 53% of Slovenian area, which means approximately 10760 km^2. Because of its importance in year 2006 water balance analysis of Sava river basin under the name SARIB was done [10]. In the SARIB project model Watbal was used [6]. We had a lot of problems with calibrating it via standard way using Excel's Goal Seek function in combination with method of trials and errors. Because of that we developed a new algorithm PETTON in Matlab for automatic, fast and reliable calibration of model Watbal. The new calibration results were better and model Watbal was more successful in prediction. In this article we will briefly present the main ideas of optimisation and calibration, model Watbal and algorithm PETTON. At the end we will present comparison between SARIB results and new results, which we get by using PETTON.

1. Introduction

Hydrology is a discipline, which studies water cycle in nature [4]. It is a part of geophysical sciences, which studies circulation and distribution of water on Earth, its chemical and physical properties and its connection with nature [2]. Elements of water circle in nature are precipitation, evapotranspiration, surface runoff, infiltration and underground flow [8]. Studying water cycle is important because water is an important medium for our lives and our economy. In Slovenia an analysis of water balance of Sava river basin under the name SARIB was done in 2006 [10]. It was done in connection with water balance analysis of Danube river basin [7]. Both analyses were done with Watbal model.

In SARIB project we were experiencing problems with calibration of Watbal. This process of calibration was an important and time demanding task. Dietrich claims that there is not enough attention paid to the calibration process in general [1]. In hydrology this is a very serious problem, because calibration is usually very hard because of data lack and resolution problems. In this article we will present an algorithm PETTON for automatic calibration of Watbal model. What we were trying to make is a better, more reliable, faster and automatic algorithm for calibration of model Watbal in comparison with standard way of calibrating model Watbal. Standard way of calibrating model Watbal is by using Excel's Goal Seek function in combination with method of trials and errors [6].

[1] To whom any correspondence should be addressed.

2. Background

2.1. Models and calibration

Model is a system, mathematical or physical, that simulates reality and is more or less successful in doing that. Model is not right or wrong, it is only better or worse approximation of reality. We can divide models, which simulate some system, in four different categories [5]:

- Models, which are symplified teories of that system.
- Models, which are analogies of some others systems, that can be real (atom-solar system).
- Models generated from data of that system (generalisation, statistic).
- Models, that are examples of simple teories of that system.

Every model must be calibrated [5]. Inspection of many articles has shown that there is little attention paid to calibration, which is usually very hard, long lasting and important task [1].

2.2. Optimization

In general optimization problem we have a function of one or many variables. Our goal is to find such values of variables (usually called optimal value of parametric vector), that our function takes the extreme value [9]. We call such function a criterial or objective function *f*. We usually also have some constraints over variables in forms of functions g_i and h_j. Optimization process is given by the set of equations, where x* represents optimal parametric vector:

$$\frac{df(x^*)}{dx} = 0 \qquad (1)$$

$$g_i(x^*) \leq 0, i \in I = \{1,2,...,m\} \qquad (2)$$

$$h_j(x^*) = 0, j \in J = \{1,2,...,n\} \qquad (3)$$

General ideas about optimization are [9]:

- Optimization should be fast, cheep and not memory demanding.
- Method should compute the value of objective function as few times as possible.
- Optimal method for optimisation does not exist.
- Avoiding local minimums is impossible.

2.3. Matlab

Matlab is a powerful interactive program environment for numerical computation. It is especially useful in natural and technical sciences [3]. Computation in Matlab is based on matrixes [11]. Matlab is short for Matrix Laboratory.

3. Watbal

Watbal is a model for water balance analysis and it is developed on equation (4) for water balance [6], where *P* stands for precipitation, *Q* for runoff, *E* for evapotranspiration and ΔS for water changes in underground. Schematic presentation of model Watbal is given on figure 1.

$$P = Q + E + \Delta S \tag{4}$$

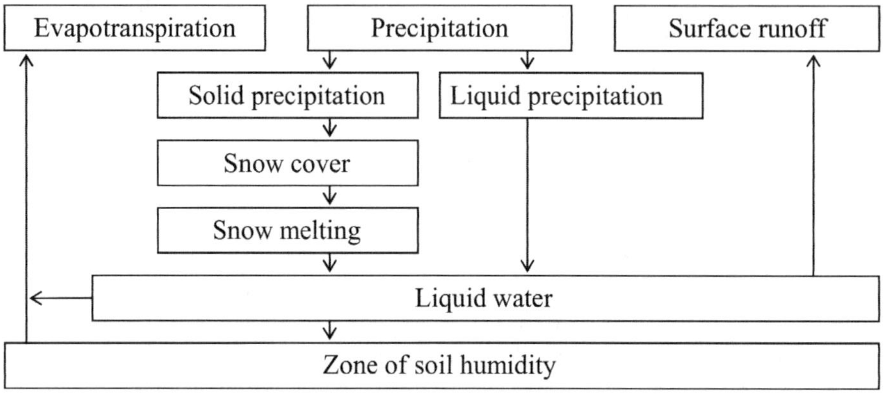

Figure 1. Schematic presentation of model Watbal.

Model Watbal is written in Microsoft Excel. The whole study basin is divided into subbasins. It is very strongly recommended, that every subbasin has at least three stations for all meteorological data. Every subbasin is defined by its elevation median. Every input data are linearly interpolated on that median. Watbal takes the input data and returns the output data in the form of time series.

First we prepare data about mean monthly river runoff, total monthly precipitation, mean monthly temperature and mean monthly air humidity. Secondly we prepare data on potential evapotranspiration. Then we calibrate Watbal. We must calibrate six parameters WCRIT, WSFFC, ATSNOW, ATRAIN, MDGFAC and PRIESK [6]. Calibration of model Watbal is done with Excel's function Goal Seek in combination with method of trials and errors. The problem is that Goal Seek optimizes in one direction in parametric space at the time. So many parametric subspaces are left uncheck as the figure 2 presents. Optimization criteria are WDELTAMm, BDELTA, WDELTAMd.

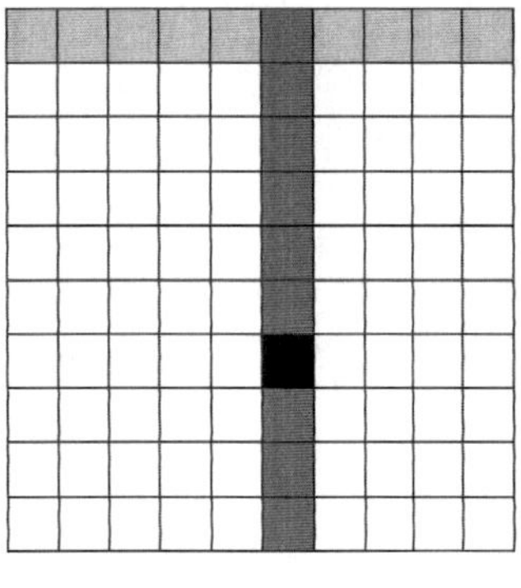

Figure 2. Excel's Goal Seek function leaves all white cells of parametric space uncheck and find only one extreme (black cell). First it finds an extreme in a first row and secondly it finds an extreme in a column under the first extreme. So we must use a method of trials and errors to start from somewhere else. This is usually not a good approach, because it is not systematic.

So, in summary we must calibrate six parameters using three objective functions. All that is done with Excel's Goal Seek function in combination with method of trials and errors.

4. Algorithm PETTON

We wrote an algorithm PETTON in Matlab, which automatically finds the optimal values for all six parameters to calibrate model Watbal. Schematic presentation of algorithm is given by figure 3.

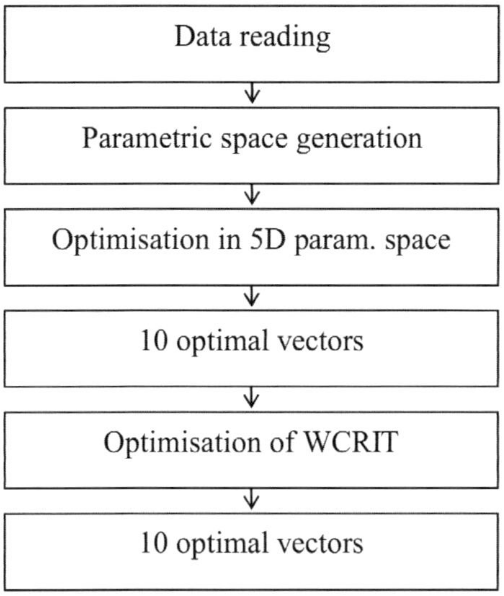

Figure 3. Schematic presentation of new algorithm for automatic calibration of model Watbal. We must first put in data time series and then through process of optimization algorithm finds ten most optimal parameter vectors.

Objective function of which minimum we look for, is given by

$$\sum_{i=1}^{N} \left(Qm_i - Qc_i\right)^2 = \min \tag{5}$$

where Qm_i stands for measured river runoff and Qc_i for computed river runoff at month i. So we were looking for the minimum value of objective function over whole time series of river runoff.

Algorithm starts in Matlab with input in Command line:

- [ming, time] = petton(std,stu,sts,rtd,rtu,rts,mfd,mfu,mfs,wcd,wcu,wcs,pd,pu,ps),

where ming is computed optimal matrix of ten best optimal parametric vectors and time is elapsed time for computation. The variables on right side are put in by user and they have a following meaning:

- std is lower value for parameter ATSNOW, stu is upper value for parameter ATSNOW, sts is changing step for parameter ATSNOW. Other variables are analogies for other parameters.

This possibility of user input is appropriate because a user can define physically sensible ranges of parameters.

The result is given by matrix ming:

- ming = [WDELTAMm, BDELTA, WDELTAMd, ATSNOW, ATRAIN, MDGFAC, WSFFC, PRIESK, WCRIT, SM2, WSURF]

In this matrix we get the optimal values for all six parameters and the values for all three standard objective functions and also for some other two criteria, which meaning can be seen in technical documentation of Watbal [6].

5. Sava river basin

Sava river basin covers 53% of Slovenia, which means 10760 km^2. Sava river basin was divided into twelve subbasins (table 1).

Table 1. Area cover and geometric characteristics of Sava river basin. Sava river basin is mostly cover by forest. Its lowest point is 125 meters about see level and its highest point is mount Triglav, which is 2864 meters about see level.

Subbasin name	Procent [%]				Elevation [m. a. s. l.]		
	Urban	Agri.	Forest	Water	Min	Max	Mediana
Dolinka	2.45	9.84	87.62	0.12	410	2864	1160
Bohinjka	0.80	9.04	89.00	1.17	410	2800	1187
Kranj	4.26	23.03	72.47	0.23	306	2520	700
Sora	2.43	27.09	70.48	0.00	311	1674	665
Kamniska Bistrica	6.08	35.38	58.31	0.23	266	2558	425
Ljubljanica	5.12	27.31	66.28	1.28	265	1780	593
Litija	3.10	27.15	68.82	0.93	187	1210	463
Celje	1.20	30.83	67.77	0.20	230	2440	595
Veliko Sirje	2.36	48.37	48.89	0.37	187	1455	407
Catez	1.56	43.68	54.05	0.71	125	1200	353
Dvor	2.92	30.73	65.99	0.36	168	1250	484

6. Results

With our algorithm we got better results than with standard way of calibrating. The prediction was better with new calibration. Also the values for six parameters were inside physically sensible range. The figure 4 shows the result and comparison between measured and computed runoff (from both cases, namely PETTON and SARIB) in a graphical way for all subbasins. We used next equations to get statistical evaluation of prediction quality, which is summarized in table 2. In equations (8) and (9) the last parts represents mean values. Also a correlation coefficient R was computed. It should be realized that *R* and *d* should be as close to one as possible and *S* should be close to zero. From table 2 we can see that in general the PETTON was more successful achieving that than SARIB.

$$S = \left(\frac{1}{\mathrm{N}} \sum_{\mathrm{i}=1}^{\mathrm{N}} \left(Qm_{\mathrm{i}} - Qc_{\mathrm{i}} \right)^2 \right)^{\frac{1}{2}} \tag{6}$$

$$d = 1 - \frac{\sum_{\mathrm{i}=1}^{\mathrm{N}} \left(Qm_{\mathrm{i}} - Qc_{\mathrm{i}} \right)}{\sum_{\mathrm{i}=1}^{\mathrm{N}} \left(\left| Qm_{\mathrm{i}}^{*} \right| - \left| Qc_{\mathrm{i}}^{*} \right| \right)} \tag{7}$$

$$Qm_{\mathrm{i}}^{*} = Qm_{\mathrm{i}} - \overline{Qm} \tag{8}$$

$$Qc_{\mathrm{i}}^{*} = Qc_{\mathrm{i}} - \overline{Qc} \tag{9}$$

Table 2. Statistical comparison between measured and computed runoff in two cases. First case is with new algorithm calibration. Second case is with standard calibration used in SARIB project [10]. From the numbers it is seen, that PETTON was more successful.

Subbasin name	PETTON			SARIB project	
	R	d	S	R	S
Dolinka	0.83	0.80	5.8	0.78	5.4
Bohinjka	0.92	0.94	8.5	0.91	7.5
Kranj	0.75	0.84	3.5	0.59	4.8
Sora	0.84	0.84	6.3	0.85	6.2
Kamniska Bistrica	0.11	0.37	7.6	0.54	8.0
Ljubljanica	0.76	0.86	4.6	0.60	7.4
Litija	0.40	0.63	4.6	0.48	3.9
Celje	0.21	0.48	5.1	0.06	5.5
Veliko Sirje	0.33	0.58	3.7	0.38	3.6
Catez	0.03	0.39	4.1	0.13	4.0
Dvor	0.71	0.64	7.1	0.72	7.5

Dolinka

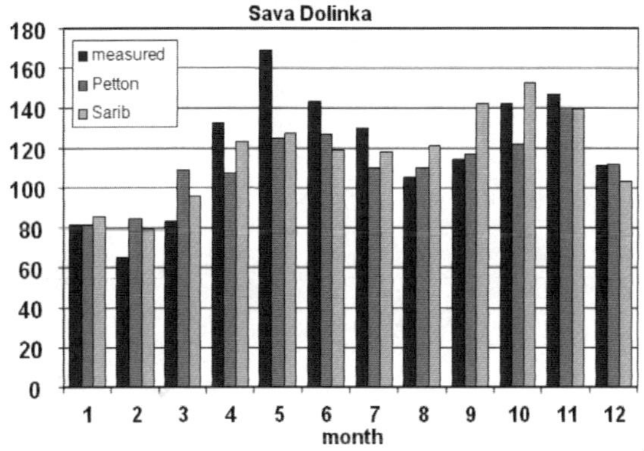

Bohinjka

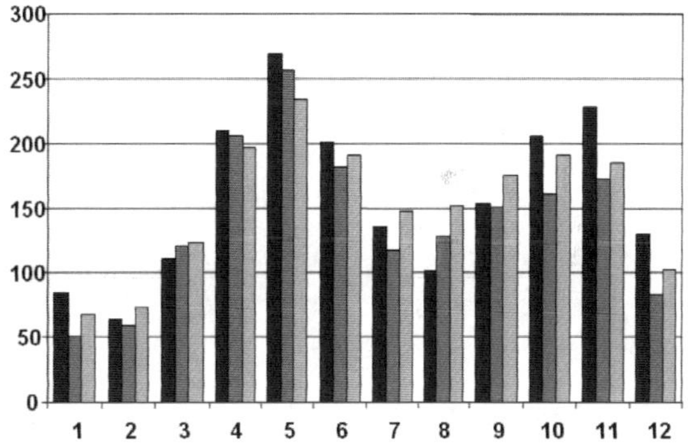

Sora

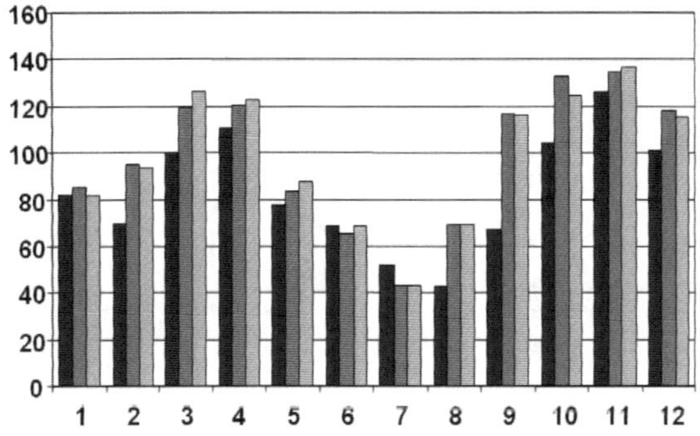

Kamniska Bistrica

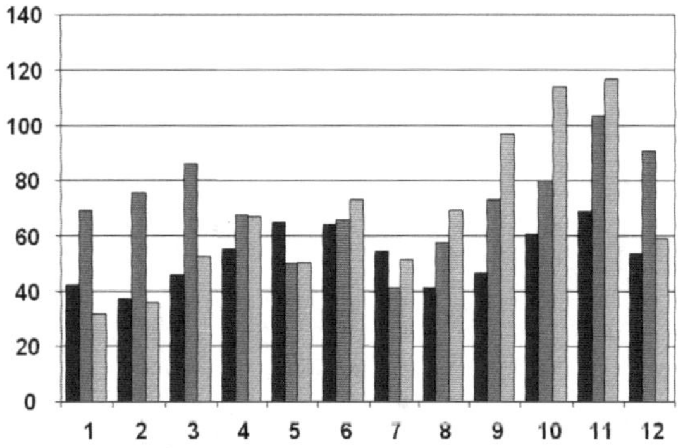

IOP Conf. Series: Earth and Environmental Science **4** (2008) 012046 doi:10.1088/1755-1307/4/1/012046

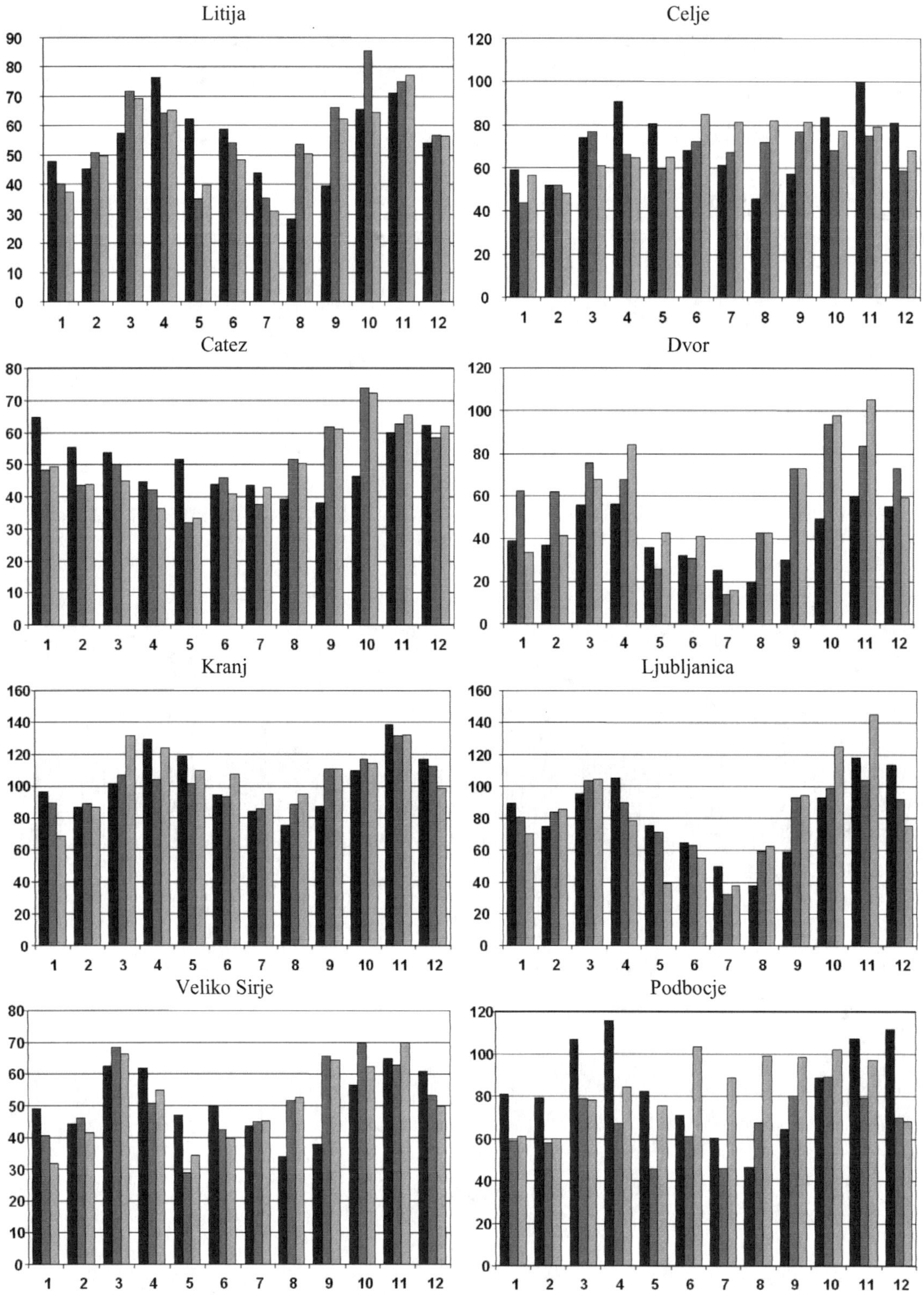

Figure 4. Figure 4 stand for all 12 top graphs. The graphs represents comparison of measured and computed runoff from using standard calibration in SARIB and new

calibration with PETTON. In all graphs first colomn represents measured runoff, second one computed runoff with PETTON, third one computed runoff with SARIB. The abscissa units are months and the ordinate unit is runoff in cubic meters per second.

7. Conclusions

The predicted runoff was better when using Watbal calibrated via PETTON than with standard calibration process. The process of calibration of model Watbal using PETTON was much faster, measured in minutes in comparison to days in the SARIB project. The calibration was also derived inside physically sensible parametric space, again in contradiction with SARIB, which was not completely able to achieve that. So we can say, that calibration should be a systematic task, because it is very complicated and trouble making problem.

References

[1] Dietrich C R 2001 On Simulation, Calibration and Ill-conditioning with Application to Environmental System Modelling *Model Validation – Perspectives in Hydrological Science* vol 2, ed Anderson M G and Bates P D (John Wiley & Sons) pp 77-116

[2] Gray M D 1973 Introduction to Hydrology *Handbook of the Principles of Hydrology* vol 2, ed Gray M D (New York, Water Information Center) pp 1.1-1.5

[3] Kermit S 1992 *Matlab Primer* vol 2 (University of Florida, Department of Mathematics) p 29

[4] Maidment D R 1993 Hydrology *Handbook of Hydrology* ed Maidment D R (McGRAW-HILL) pp 1.1-1.15

[5] Morton A and Suarez M 2001 Kinds of Models *Model Validation – Perspectives in Hydrological Science* vol 2, ed Anderson M G and Bates P D (John Wiley & Sons) pp 11-22

[6] Petrovič P Water Balance Model and Tuning Parameters *22nd Conf. on Danube Countries on Hydrological Forecasting and Hydrological bases of Water Management* 30. 8.-2. 9. 2004. (Brno, Czech Republik, Czech National Commitee for the IHP UNESCO) p 10

[7] Petrovič P, Nachtnebel H P, Zimmerman L et all 2006 *Basin – Wide Water Balance in the Danube River Basin. The Danube and its Basin – Hydrology Monograph* vol 3 (IHP UNESCO, Bratislava, Slovakia) p 161

[8] Ponce V M 1989 *Engineering Hydrology, Principlies and Practices* (New Jersey, Prentice-Hall) pp 1-94

[9] Press W H, Flannery B P, Teukolsky S A and Vetterling W T 1989 *Numerical Recipes – The Art of Scientific Computing* vol 5 (Cambridge University Press) pp 274-334

[10] Sraj M 2006 *SARIB Sava river basin water balance analysis* (Univerza v Ljubljani, Fakulteta za gradbeništvo in geodezijo)

[11] Zupan D 2005 *Programski jezik Matlab* (Univerza v Ljubljani, Fakulteta za gradbeništvo in geodezijo) p 99

IOP Conf. Series: Earth and Environmental Science **4** (2008) 012047 doi:10.1088/1755-1307/4/1/012047

Flood warning level forecasting for ungauged catchments by means of a combined API storage concept

T Lehmann and H Holzmann

Inst. f. Water Management, Hydrology and Hydraulic Engineering, Univ. of Natural Resources and Applied Life Sciences Vienna (BOKU), Muthgasse 18, 1190 Vienna, Austria

E-mail: thilo.lehmann@boku.ac.at

Abstract. The knowledge of the expected dimension of the flood peak is of major importance for security and warning services to take preventive measures. In this paper the authors want to introduce the concept of the Antecedent Precipitation Index (API) as a possible variable to estimate runoff warning classes. The aim was (a) to define API warning classes which correspond to runoff warning classes at a certain runoff gauge and (b) apply the method to ungauged basins. To consider time and state dependant rainfall losses a spatially distributed linear storage concept was applied to intercept the actual rainfall. The 3-parameter API function was fitted to several flood events at observed gauges within the district of lower Austria and lead to a set of optimized parameters. Through extreme value statistics the 1, 5 and 30 years API extremes were derived and set into correlation to the corresponding flood events. These API extremes together with the optimized API parameters were spatially interpolated and thus transferred to ungauged basins. The calculated flood events had the tendency to underestimate the smaller flood frequencies whereas the extreme flood classes could be reliably performed.

1. Introduction

Flood forecasting with sparse data availability is still a great challenge to hydrologists. This is especially the case when a forecast should be made at an ungauged catchment. There already exist different methods to transfer model parameters from gauged to ungauged catchments by linking them to catchment characteristics like slope, geology, landuse etc. [1] [2]. But still the uncertainty is high when trying to forecast real hydrographs.

Most frequently disaster prevention services deal with warning levels corresponding with flood frequencies. These warning levels are linked to certain measures. In this case the knowledge of the flood dimension and consequently the warning class would be very helpful to take preventive measures [3]. Additionally through forecasting warning classes the forecast uncertainty in comparison to forecasting hydrographs is diminished through the range in which the flood peak can be located. By using the API model, warning classes could be forecasted by only using precipitation as input data. It is an attempt to directly transform irregular rainfall into runoff warning classes.

IOP Conf. Series: Earth and Environmental Science **4** (2008) 012047 doi:10.1088/1755-1307/4/1/012047

2. Study region and data

Lower Austria is the largest province in Austria and is located in the Northeast of the country. In the North it borders on the Czech Republic and in the East on the Slovak Republic. The City of Vienna is enclosed by Lower Austria (see figure 1).

The landscape reaches from mountainous areas in the South with mountains up to 2000 m a.s.l. over soft hills in the North to rather flat areas in the East. South and North are divided by the Danube River. The geology is dominated by limestone in the mountainous Southwest whereas in the Northwest mostly granite is found. Forrest and agriculture are the dominant landuse.

For analysis, precipitation and discharge data from 1990 until 2005 were used. For calibration 31 gauges have been chosen. Precipitation data in hourly resolution from about 84 rain gauges were used for interpolating catchment mean values.

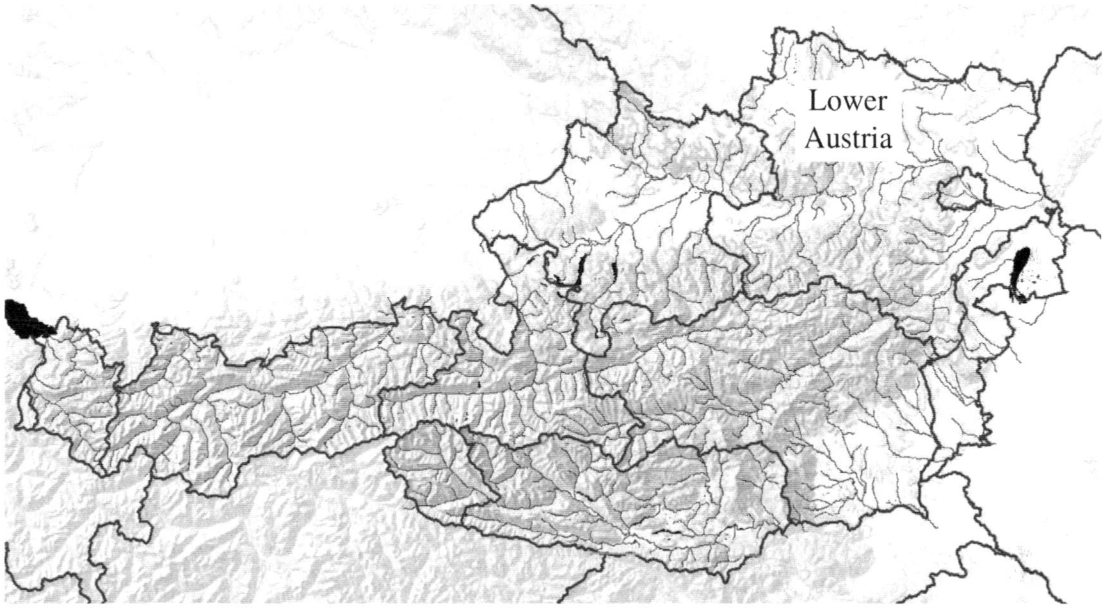

Figure 1. Austria and Lower Austria

3. Methodology

The traditional concept of the API refers to precipitation on daily basis. It describes the declining impact of past precipitation in time [4]. The API can be interpreted as a kind of runoff disposition in the catchment [5] [6]. With a higher temporal resolution of available precipitation data also the API has to be adapted to a higher resolution as the runoff disposition changes with every precipitation element. In this study data with an hourly resolution were used.

Runoff losses were considered by a linear storage reservoir. The depletion of the reservoir, which corresponds to factors like basin size, hydrogeology or soil moisture redistribution, was described by a retention factor. Only the precipitation excess of the storage (P_{eff}) contributes to the API computation (see figure 2). This leads to a better correlation between discharge and API, which is of importance for deriving API warning classes that should correspond to flood warning classes [7].

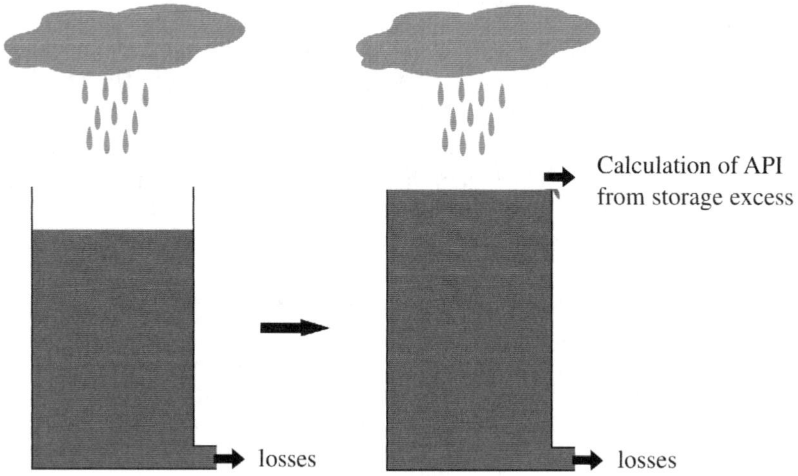

Figure 2. Linear storage.

$$P_{eff} = P - (S_{\max} - S_{act})$$

$$S_{act,i} = S_{act,i-1} * e^{\frac{-1}{Sval}}$$

$$API_i = API_{i-1} * e^{\frac{-1}{24*dt*akvri}}$$

P	precipitation
P_{eff}	effective precipitation
S_{max}	maximum storage capacity
S_{act}	actual storage content
S_{val}	retention factor
akvri	API coefficient
dt	factor for temporal resolution (1 for hourly calculations)
API	Antecedent Precipitation Index
i	timestep

4. Results

The storage parameters like maximum capacity (S_{max}) and retention constant (S_{val}) as well as the API coefficient (akvri) were optimized until the best statistical accordance between the API function and the observed hydrograph was achieved. As the evaluation criteria the correlation coefficient has been used. This process was repeated for several flood events at all gauged catchments and led to a set of optimized parameters for each catchment. Thereby the correlation of the peak values was of major importance. This means a good correspondence between the temporal occurrence of runoff peaks and API peaks was aimed. Through the implementation of the linear reservoir this task could be satisfactory performed (see figures below).

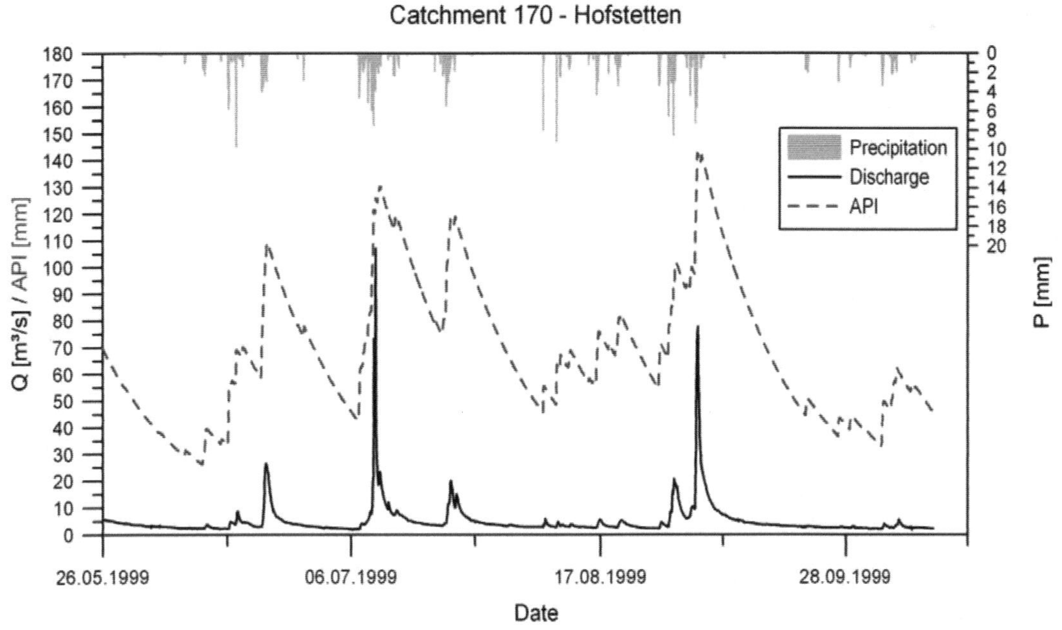

Figure 3. Comparison between discharge (solid) and API without storage (dashed)

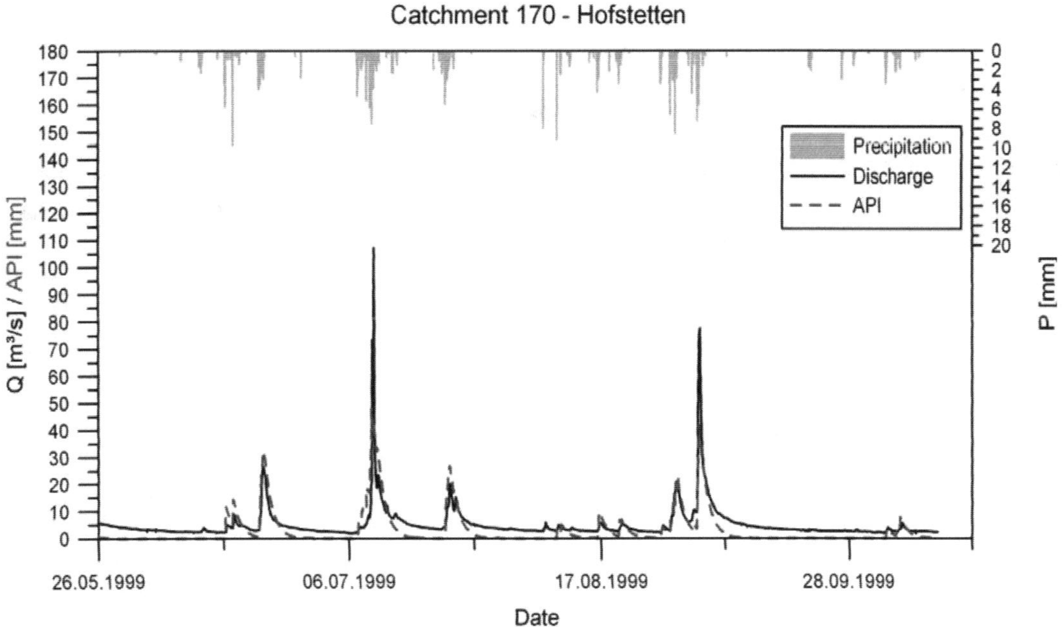

Figure 4. Comparison between discharge (solid) and API with storage (dashed)

In above figures the discharge (solid line) is shown in comparison to the API (dashed line) at gauge Hofstetten. It is apparent that through the use of the storage (Figure 4) the correlation between discharge and API is much better. The occurrence time of the peaks is well expressed and higher discharge peaks meet with higher API peaks. This is an important fact for the definition of API warning classes that should correspond to the HQ warning classes.

4.1. Derivation of warning classes
From the optimized parameter sets of each catchment, the mean values were chosen as the characteristical catchment values. Eventually API timeseries for each catchment have been created from a 15 year precipitation period (1990 – 2005). From these timeseries the 1, 5 and 30 years frequencies events have been deduced as the three different warning levels. For this purpose the Gumbel distribution has been used. Then the API warning levels have been related to the 1, 5 and 30 years flood frequencies at the correspondent catchment. At this point the API extremes did not meet the flood extremes in a good manner. Especially the smaller events like the 1 years flood frequencies were underestimated in all catchments. Therefore the API extremes have been reduced by 20% until they reached a good accordance with the flood extremes. At this stage the flood events could be reliably forecasted by the API values, although there still was an underestimation of smaller flood events, especially the 1 years frequencies events. As an example, Table 1shows the forecasting results at several catchments.

Table 1. Estimation of warning classes

ID	Number of events	Too low	Correct	Too high
170	19	5	12	2
171	18	5	11	2
172	18	10	8	0
173	18	5	12	1
174	18	9	9	0
177	19	7	12	0
180	20	8	11	1
188	19	9	8	2
193	14	4	10	0
194	20	7	13	0

At e.g. catchment 170, from a total of 19 flood events, 12 were classified in the correct warning class, two events were overestimated (too high) and 5 were underestimated (too low). Most frequently the underestimated events have been 1 years flood events. This outlines the difficulties with that kind of events. A further decreasing of the 1 years warning level led to a higher overestimation rate. So it is better to underestimate small events than making false estimation at higher events.

4.2. Transfer warning levels and API parameters to ungauged catchments
After the process of calibration at gauged sites, the API parameters and warning levels have been transferred to ungauged catchments by using Ordinary Kriging.

In the following the quality of the interpolation for the 1, 5 and 30 years extremes that refer to the three warning levels is presented. For validation purposes some of the known gauges were assumed to be unknown and be used as interpolation targets. The interpolated values obtained through Kriging were then compared to the values obtained through calibration. One example of the many possible constellations is presented subsequently.

Example:
In figure 5 an excerpt of the available gauges is shown. The gauges marked with a square have been pretended to be known and used for Kriging. The ones marked with a circle were intended to be forecasted. The following tables show the comparison between the kriged values and the values achieved through calibration for these basins. Table 2 shows the interpolation results for the 1, 5 and 30 years flood frequencies that are correspondent to the three warning levels (API_WL_1,

API_WL_2, API_WL_3). Table 3 refers to the storage parameters S_{max}, S_{val} and akvri and table 4 provides the classification results of different flood events from the period 1990 – 2005. API values are in [mm].

Figure 5. Example interpolation area

Table 2. Comparison API warning level calibration – kriging

		Calibration			Interpolation		
Id	Name	API_WL_ 1	API_WL_ 2	API_WL_3	API_WL_ 1	API_WL_ 2	API_WL_ 3
172	Atzenbrugg	22,54	53,62	89,09	26,68	61,89	103,68
173	Siegersdorf	24,96	62,41	105,53	26,68	60,33	97,93
177	Fahrafeld	25,65	66,33	113,44	26,68	66,56	115,42

Table 3. Comparison API storage parameters calibration – kriging

		Calibration			Interpolation		
Id	Name	Smax	Sval	akvri	Smax	Sval	akvri
172	Atzenbrugg	15,00	50,50	1,45	13,17	47,75	2,22
173	Siegersdorf	13,30	49,25	1,90	12,86	48,35	2,22
177	Fahrafeld	12,90	46,75	1,30	14,12	44,37	2,22

Table 4. Classified flood events when using interpolated values

ID	Number of events	Too low	Correct	Too high
172	18	6	12	0
173	18	4	13	1
177	19	9	10	0

The classification results provided in Table 4 are really promising. In comparison to Table 1, the results shown in Table 4 indicate that in this case the Kriging interpolation is an acceptable method to use.

In the following four figures, the Kriging interpolation results are demonstrated graphically. The first and second figure correspond to the API storage parameters S_{val} and S_{max}. The third and fourth figure refer to the warning levels. Warning level 1 (1 years flood frequency) and the storage parameter akvri have been left out since spatial difference was very low in this case. The graduation of the parameter bands is shown in different colours. Light grey means low values and dark black refers to high values of the accordant parameter. Every figure shows the same detail of Lower Austria and hast the same extent. In the northern part the landscape is rather flat whereas in the South it is hillier. Here, most of the flood events are caused by intensive rainfall of short duration. In the North the flood levels are lower and floods are rather caused by long duration rainfall. This leads to lower storage capacities but higher retention constants than in the South. The hydrographs in the southern part are steeper and the flood events are of shorter duration. Consequently the API storage has to drain faster which results in a lower retention constant.

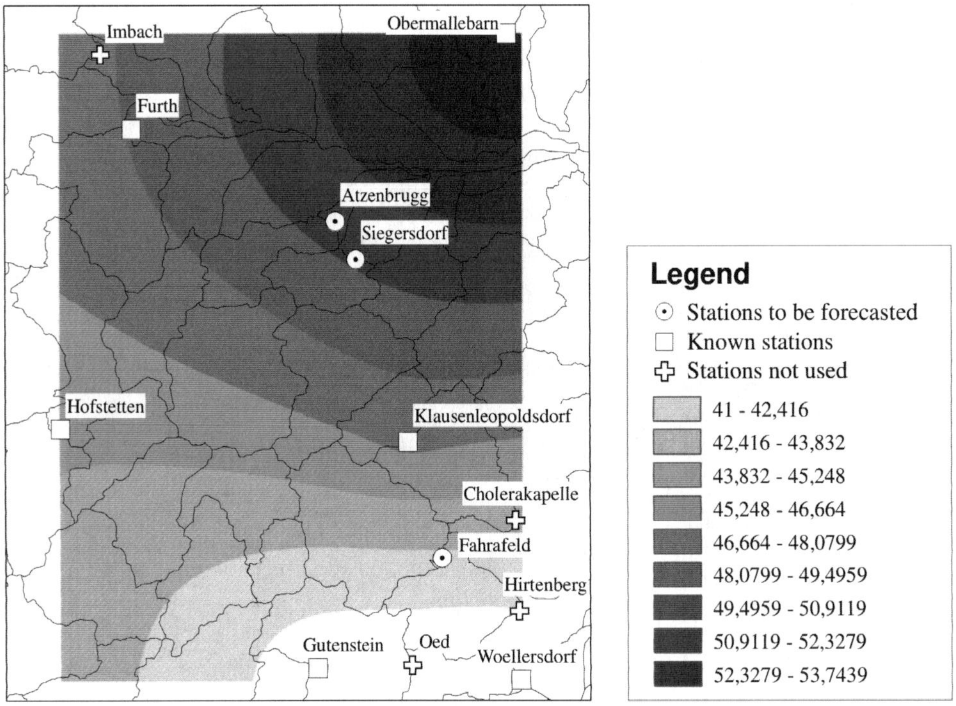

Figure 6. Spatial distribution of retention constant (Sval)

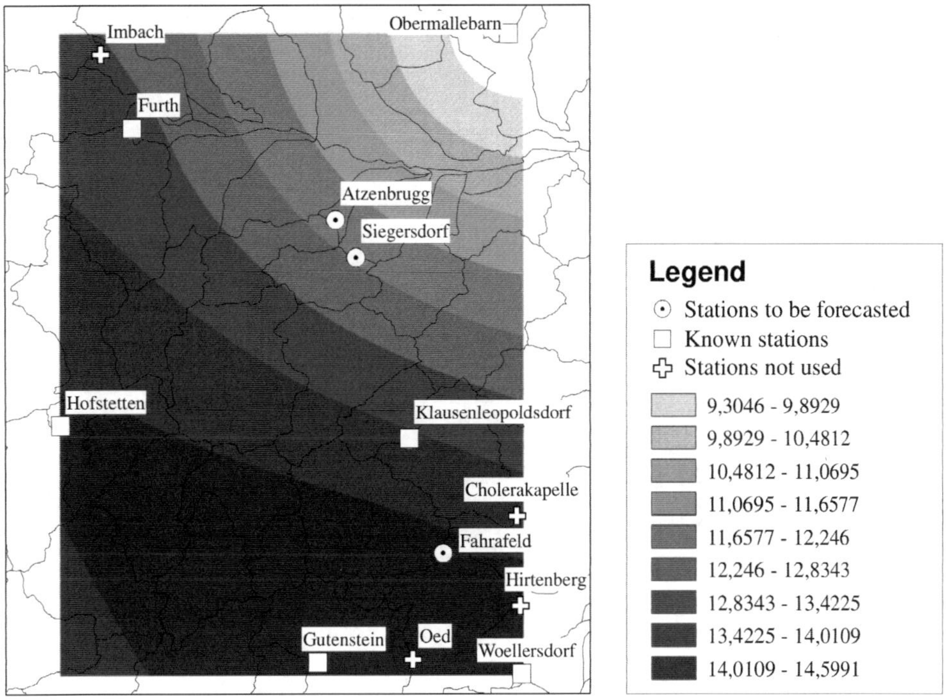

Figure 7. Spatial distribution of maximum storage capacity (Smax)

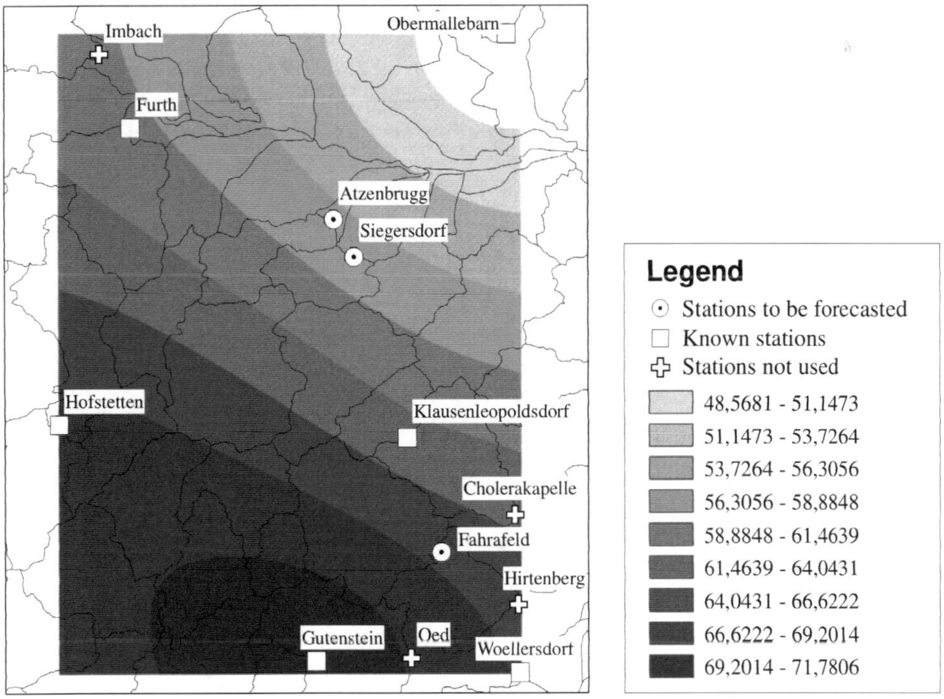

Figure 8. Spatial distribution of 5 years flood frequency (warning level 2)

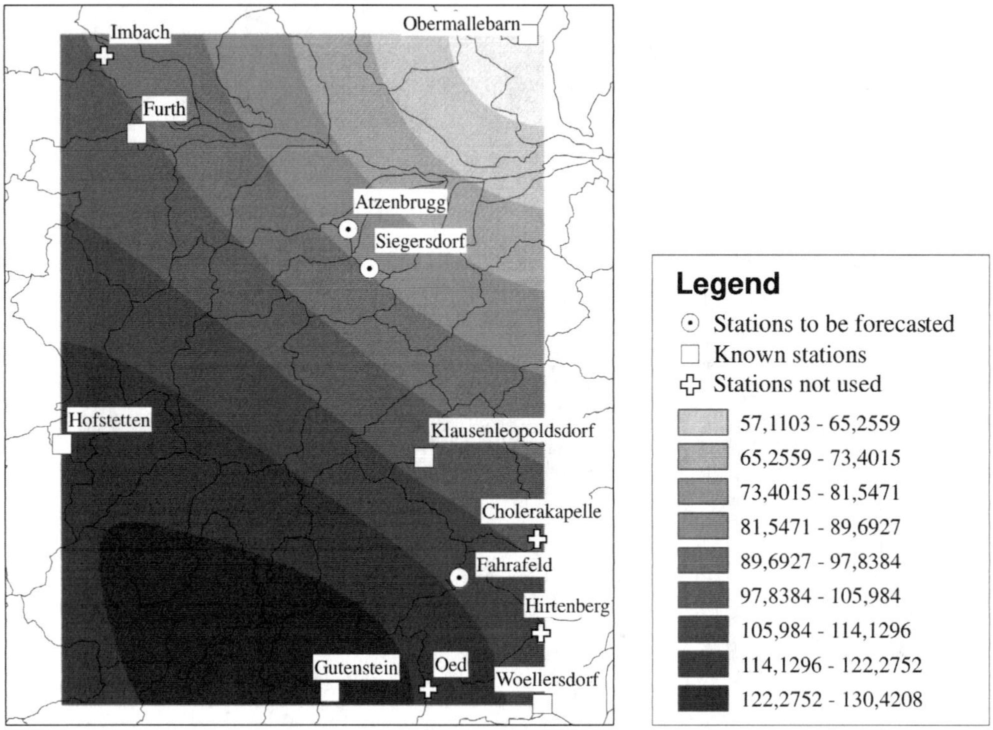

Figure 9. Spatial distribution of 30 years flood frequency (warning level 3)

5. Conclusions

The API in combination with the linear storage is a possible concept for forecasting warning levels in ungauged catchments. It is an easy to use method to classify and forecast flood frequencies (warning levels). It only requires precipitation as input data. Higher flood events such as events around the 10 or 30 years floods could be reliably forecasted. Difficulties exist at forecasting small events like 1 years flood frequencies. Here the threshold between precipitation that causes a flood and precipitation that causes no flood is hard to define. But after all the method provided promising results that need further investigation in terms of storage parameter estimation and linking the storage parameters to catchment characteristics.

References

[1] Xu C-Y 1999 Estimation of Parameters of a Conceptual Water Balance Model for Ungauged Catchments *Water Resources Management* **13** 353-368

[2] Lowe L and Nathan R 2006 *Transposing gauged streamflows to ungauged catchments* (SKM Technical Papers)

[3] Harum T, Saccon P, Hofrichter J and Fuchs K 2005 Estimation of the flood probability for ungauged catchments of the region of Styria (Austria) *Abstract Proc. of the Workshop II Dissesto Idrogeologico Venezia* (Venezia, 19.-20. April 2005)

[4] Linsley R, Kohler M and Paulhus J 1958 *Hydrology for Engineers*, Mcgraw-Hill Book Co., New York

[5] Holzmann H and Nachtnebel HP 2002 Abflussprognose für mittelgroße Einzugsgebiete – Methodik und Anwendungen, *Österr. Wasser- und Abfallwirtschaft* **54, 9-10** 142-153

[6] Holzmann H and Nachtnebel HP 2002 Sequential development of a conceptual hydrological model considering alpine basin processes. In: Rizzoli, A., Jakeman, A. (Eds.) *Integrated assessment and decision support* (Lugano, Juni 2002) Proc. of the 1st biennial meeting of the Int. Environmental modelling software society Lugano. Vol. 1 pp 416-421 Lugano

[7] Holzmann H and Lehmann T 2007 *Hochwasservorhersage für Nebeneinzugsgebiete*, final report, part 3 – Hydrology, Vienna

WaterNet: The NASA water cycle solutions network - Danubian regional applications

Dave Matthews[1], Mitja Brilly[2], Mira Kobold[3], Mark Zagar[3] and Paul Houser[4]

[1]Hydromet DSS, LLC, Silverthorne, CO 80498-1848
[2]FGG University of Ljubljana, Slovenia
[3]Environmental Agency of the Republic of Slovenia, Ljubljana, Slovenia
[4] Center for Research on Environment and Water & George Mason University, Calverton, MD 20705

E-mail: hydrometdss@comcast.net

Abstract. *WaterNet* is a new international network of researchers, stakeholders, and end-users of remote sensing tools that will benefit the water resources management community. This paper provides an overview and it discusses the concept of solutions networks focusing on the *WaterNet.* It invites Danubian research and applications teams to join our *WaterNet* network. The NASA Water cycle Solutions Network's goal is to *improve and optimize the sustained ability of water cycle researchers, stakeholders, organizations and networks to interact, identify, harness, and extend NASA research results to augment decision support tools and meet national needs.* Our team will develop *WaterNet* by engaging relevant NASA water cycle research resources and community-of-practice organizations, to develop what we term an "actionable database" that can be used to communicate and connect NASA Water cycle research Results (NWRs) towards the improvement of water-related Decision Support Tools (DSTs). Recognizing that the European Commission and European Space Agency have also developed many related Water Research products (EWRs), we seek to learn about these and network with the EU teams to include their information in the *WaterNet* actionable data base and Community of Practice. *WaterNet* will then develop strategies to connect researchers and decision-makers via innovative communication strategies, improved user access to NASA and EU - Danubian resources, improved water cycle research community appreciation for user requirements, improved policymaker, management and stakeholder knowledge of research and application products, and improved identification of pathways for progress. Finally, *WaterNet* will develop relevant benchmarking and metrics, to understand the network's characteristics, to optimize its performance, and to establish sustainability. This paper provides examples of several NASA products based on remote sensing and land data assimilation systems that integrate remotely sensed and in situ observed forcing data that drive a Land Surface Model to simulate surface hydrologic variables of value to water and emergency managers. It clearly shows the limitations of coarse resolution observations and models especially during flash flood events. Snow water equivalent and soil moisture estimates for water and drought management purposes appear to have some value at the coarser resolutions of ¼ degree; however, higher resolution information at the 1 to 5 km scale is desirable. In conclusion, this paper invites interested water-cycle researchers and decision-makers to join our WaterNet Community of Practice to exchange information and share the latest emerging technologies of value to decision-makers.

Keywords: networking, land surface models, remote sensing, hydrologic modelling, mesoscale modelling, water cycle community of practice, *WaterNet* , NASA, ESA, QPE, QPF, SWE, soil moisture, runoff, floods, droughts, water managers, emergency managers

1. Introduction

WaterNet is a new international network of researchers, stakeholders, and end-users of remote sensing tools that will benefit the water resources management community. It was established in 2007 to bring together researchers and decision-makers involved in water cycle related work. This paper provides an overview and it discusses the concept of solutions networks focusing on the *WaterNet.* It invites Danubian research and applications teams to join our *WaterNet* network. The NASA Water cycle Solutions Network's goal is to "*improve and optimize the sustained ability of water cycle researchers, stakeholders, organizations and networks to interact, identify, harness, and extend NASA, ESA, and related research results to augment decision support tools and meet national and international needs".* Research results are the products of research that have been put in a form that end-users can readily apply in their decision-making processes, tools, models, etc. This paper also provides examples of research results that are readily available to users in the Danubian region with real-time web links to observations, analyses and forecasts of value to various decision-makers. It concludes with a request for Danubian researchers and decision-makers to join our WaterNet Community of Practice (CoP).

1.1. Key disciplines of interest

WaterNet addresses the full range of water cycle related disciplines and end-users from hydrology and meteorology to water resource, environmental systems and disaster managers. Figure 1 shows the conceptual flow of information across research disciplines to end-user decisions using the "Integrated System Solutions Approach". Note that this approach integrates new observations from remote sensing through water cycle models that produce water cycle predictions ranging from soil moisture to ocean currents (column 2). These water cycle research results are then input to water cycle decision support tools (DSTs) that use their information to produce decisions of value and benefit to the society and nation. DSTs range from models of agricultural efficiency to disaster management systems and river management daily and monthly time-step models.

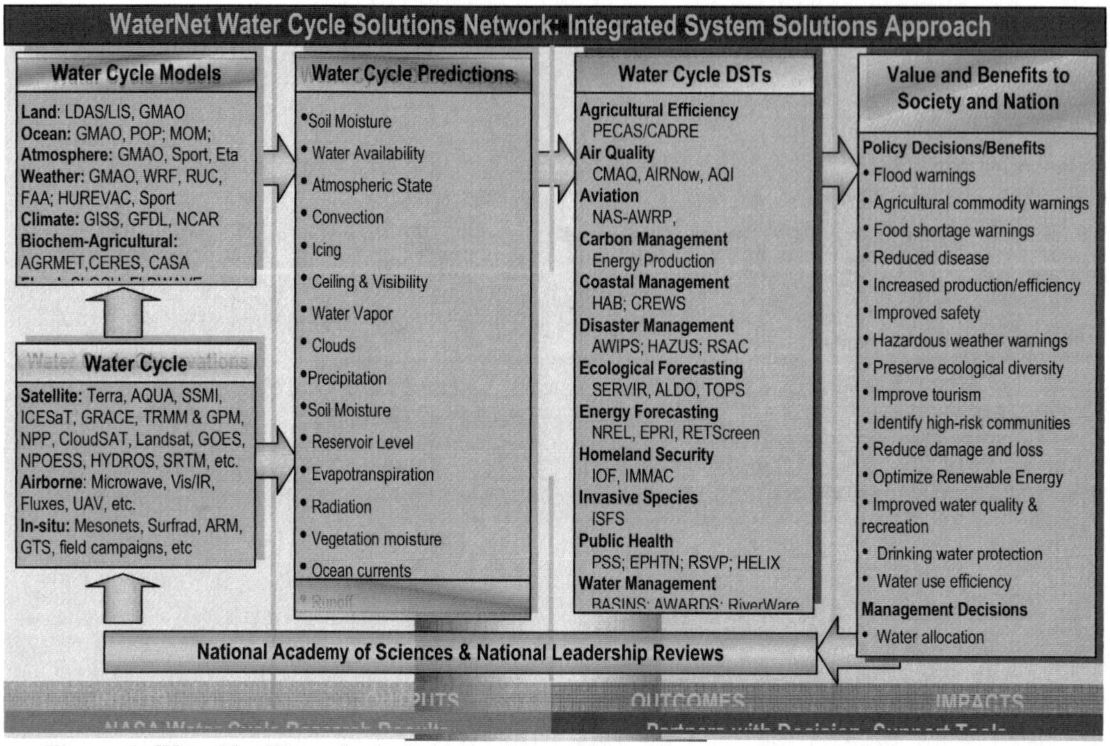

Figure 1. WaterNet Water Cycle Solutions Network: An Integrated System Solution Approach showing the multi-disciplinary nature of WaterNet and anticipated benefits to society.

1.2. Remote sensing observational and Land Surface Modelling capabilities

Remote sensing systems provide much information that may be used to initialize models with observations of precipitation, snow cover, soil moisture, ice shelf sizes, density, etc. Figure 2 shows the range of satellites available and planned from NASA and GEOSS related to the water and energy cycles.

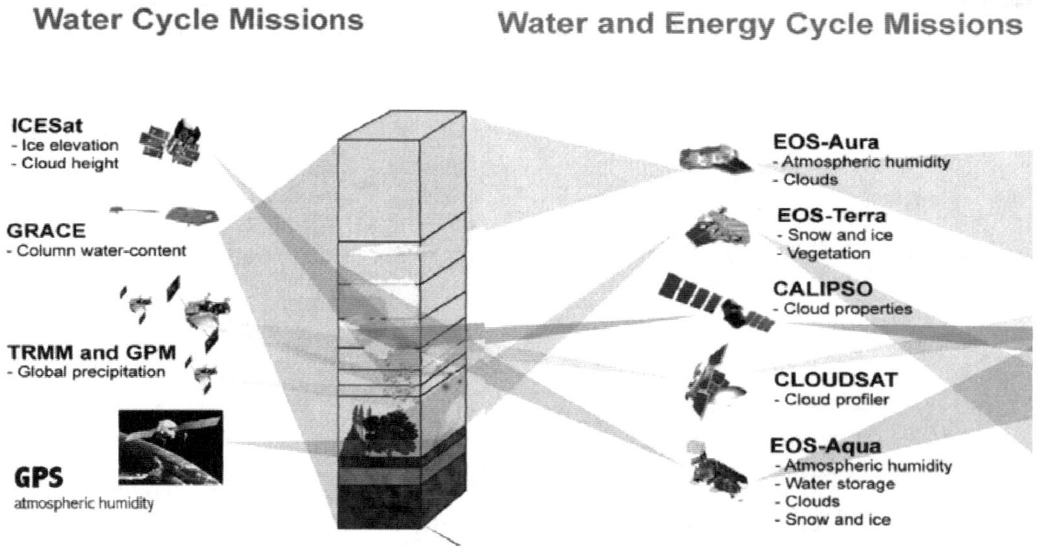

Fig. 2a **Fig 2b**

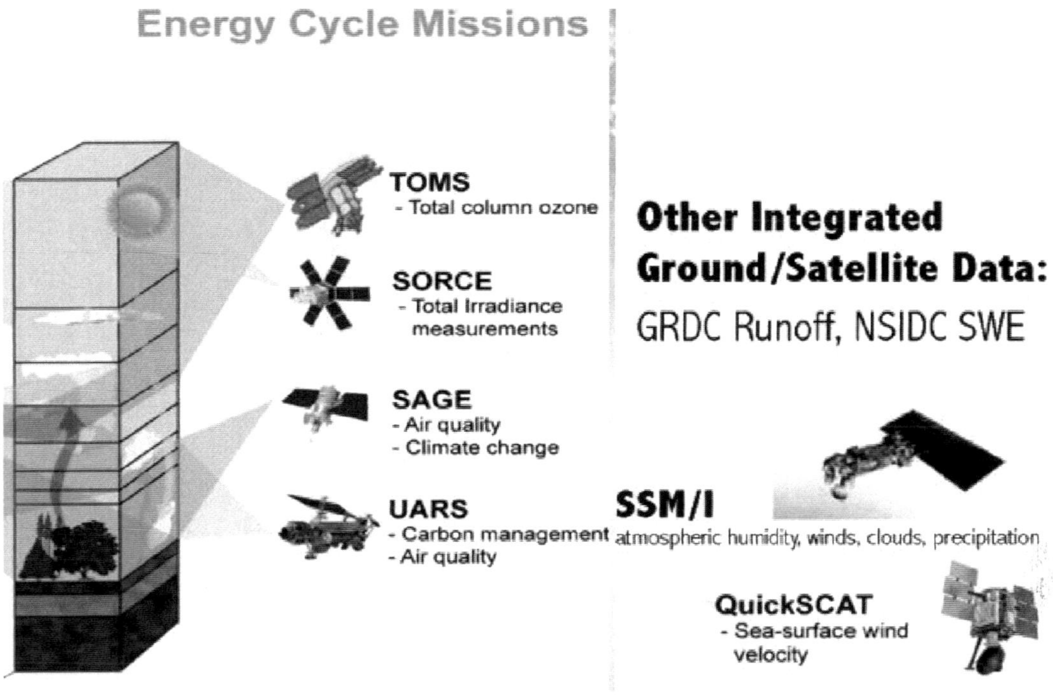

Fig. 2c **Fig. 2d**

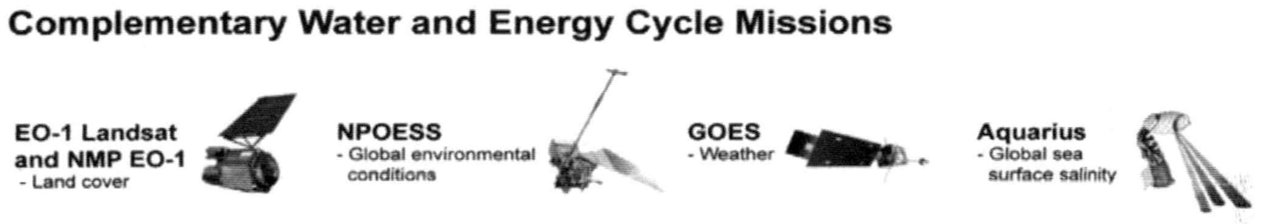

Fig. 2e

Figure 2. Water and Energy Cycle observation capabilities showing the satellites for the Water Cycle (2a), Water and Energy Cycle Missions (2b), and the Energy Cycle Missions (2c), Other integrated satellite and ground systems, (2d), and complemantary Water and energy cycle missions (2e) with the parameters observed and concept of integrative modelling to benefit society.

Data assimilation and modelling are the central focus of NASA's Land Information System which is designed to integrate remote sensing and in situ observations through advanced data assimilation techniques to drive land surface models (LSMs). These LSMs are able to use numerical physical and statistical models of surface hydrology to provide outputs of direct value to decision-makers from drought and water management teams to farmers and flood control emergency managers. Figure 3 shows the primary components of the Land Data Assimilation System (**LDAS**) and the outputs from Land Surface Models. The LDAS is operational over the United States and runs simulations at a spatial resolution of 1/8 degree (~12 km) to 1 km. It is run routinely to initialize the surface boundary conditions (energy and moisture fluxes) for the numerical weather forecast models of NOAA's National Centers for Environmental Prediction (NCEP). It is part of NASA's Land Information System **(LIS)** used for a variety of end-users and decision-makers. The Global Land Data Assimilation System (GLDAS) provides information over the Danubian region. Examples are provided in the next section.

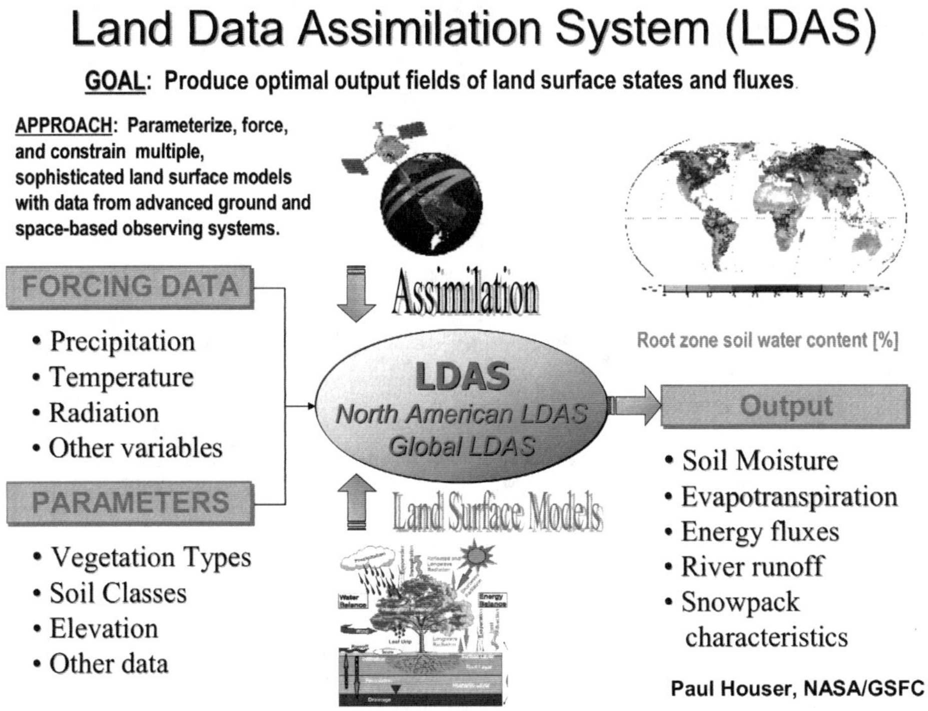

Figure 3. Land Data Assimilation System schematically showing the approach of integrating key forcing data, and parameters to drive land surface models that produce outputs of value to decision-makers.

2. Danubian Examples of Research Results

The Global Land Data Assimilation System (**GLDAS**) integrates data much like LDAS but covers the globe. It's results are routinely available at 1 and 1/4 degree spacing. GLDAS results in March 2006 and June 2007 over the Danube provide an indication of the type of information available for decision-makers. These research results include Snow Water Equivalent (SWE), Quantitative Precipitation Estimates (QPE), surface soil moisture (Qs) and root zone soil moisture (0 to 10 cm depth average – Qsb). GLDAS analyses were performed using the **GRaDS** analysis tools.

The area examined by these analyses covers the headwaters to the mouth of the Danube. Figure 4 shows the region using Google Earth and locates points where time series were obtained

Figure 4. Google Earth view of Danube analysis region used in GLDAS from the headwaters in Germany to the mouth in the Black Sea. Yellow Pins locate the time series inter-comparison points used to demonstrate the time and spatial sensitivity of the analysis.

2.1. Snow Water Equivalent

Snow water equivalent (SWE) analyses provide important information for water and emergency managers who are interested in the water supply and potential for floods in rivers fed by snow melt like the Danube. Figure 5(a) shows the GLDAS 0.25 degree grid spacing analysis on 15 March 2006. Note the maximum over the Alps (200-300 kg/m^2) and in Montenegro (300-400 kg/m2). Maximum values in the mountains of Montenegro reached 800 cm and remained at that level until 31 March 2006. However the area of the 200 cm contour decreased in the Alps and the region to the south. Figure 5(b) shows the melt of SWE by 31 March 2006 smaller areas and magnitude of 200 kg/m^2 over Montenegro. Figure 6(a) Shows the time series of spatial sensitivity across the Alps at 46.9 N from Tyrol to South of Salzburg.

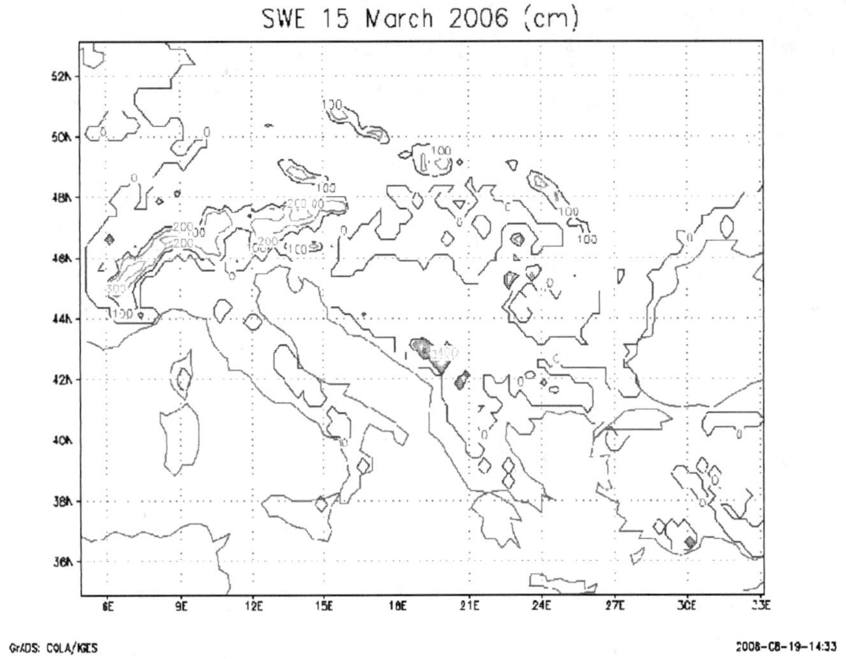

Figure 5(a) GLDAS estimate of Snow Water Equivalent (SWE) over the Danubian Basin on 15 March 2006. Note the large area of 200-300 cm values over the Alps, and the local maximum of 300-400 cm over Montenegro (maximum value of 800 cm).

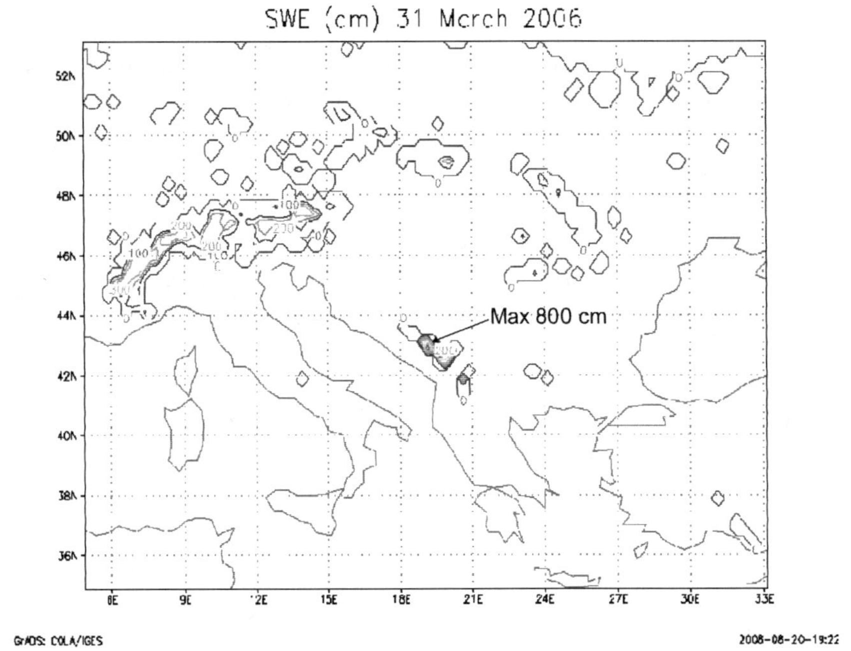

Figure 5. Fig. 5(a) and 5(b) showing the area extent and magnitude of snow water equivalent over the Danube on 15 and 31 March 2006 respectively. By 31 March the area of the 300 cm SWE had increased over the Alps, while the size of the local maxima over Montenegro decreased somewhat.

The large spatial variations in the Alps reflect the local elevation, aspect and azimuth orientation of the slopes which change dramatically in this area. The land surface models must properly simulate these very local effects on the evolution of snow water equivalent. Hence, they must have very high spatial resolutions often down to less than 1 km. Figure 6 clearly shows these dramatic changes in space and time as the March 2006 snow pack evolved. Note that the central area at 11.9 E lost all of the SWE by the end of the month; yet the other sites had from 195 to 198 cm remaining. This suggests the need for high spatial resolution.

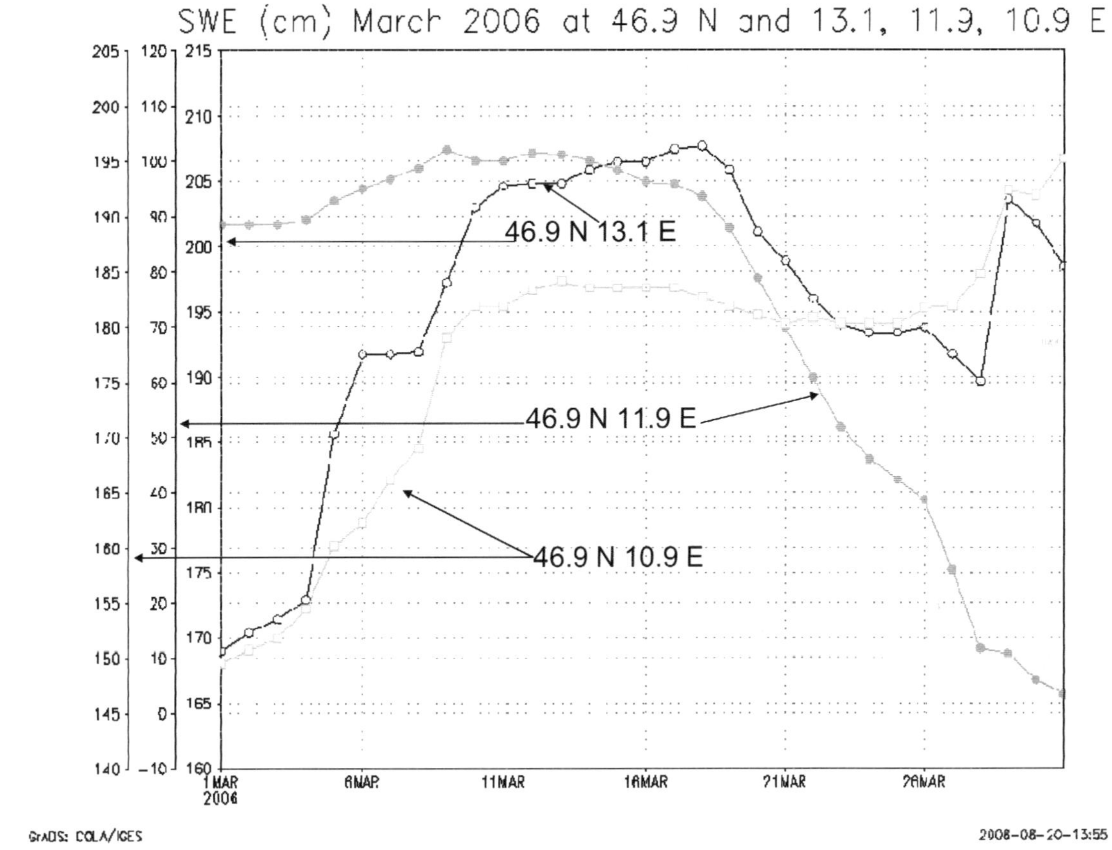

Figure 6. Time series of SWE at 46.9 N from 13.1 to 10.9 E showing the spatial and temporal sensitivity in GLDAS 0.25 degree analyses.

GLDAS appears to perform well in distinguishing between the alpine high elevation regions and lower broad river basins as shown in the SWE time series shown in Figure 7. Note the dramatic change from the Alps to the Panonian Plain and Sava River Valley at 14.8 and 15.8 E longitude.

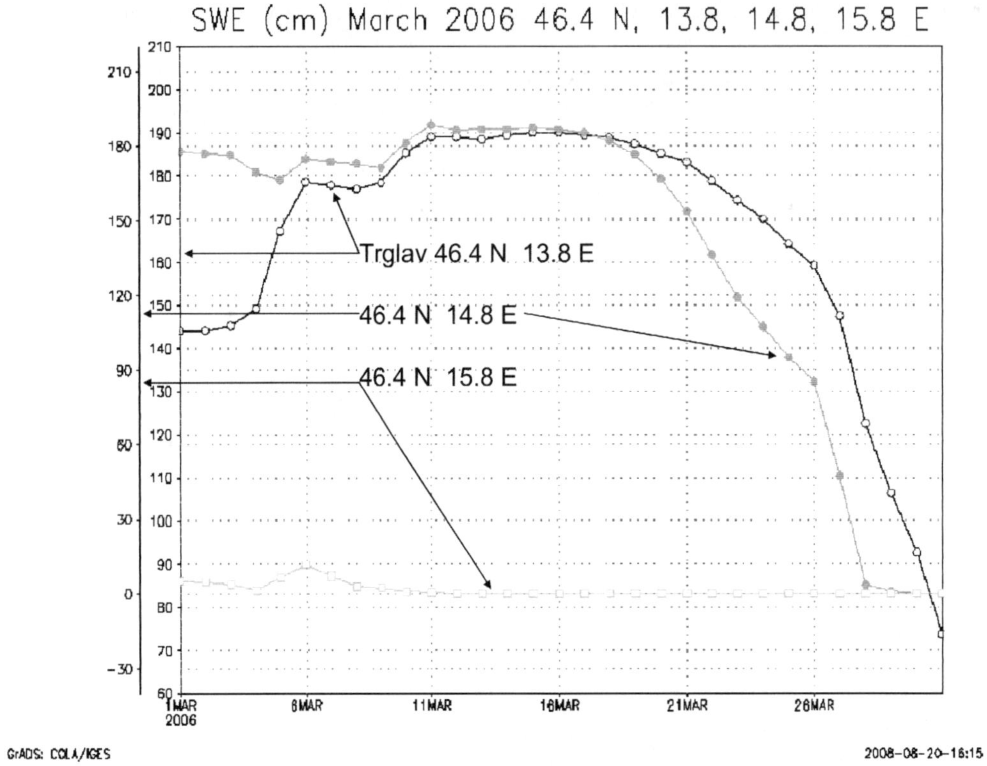

Figure 7. SWE time series at 46.4 N from Triglav in the Julian Alps to Ptuj on the Drava in the Panonian Plain dropping from 190 cm to 12 cm respectively, this is a plausible result.

Surface and subsurface runoff from GLDAS at Budapest show typical patterns of sensitivity in figure 8. The surface runoff is a direct response to precipitation and snow melt; however, the subsurface runoff appears as an aggregate effect of the surface and ground water flows.

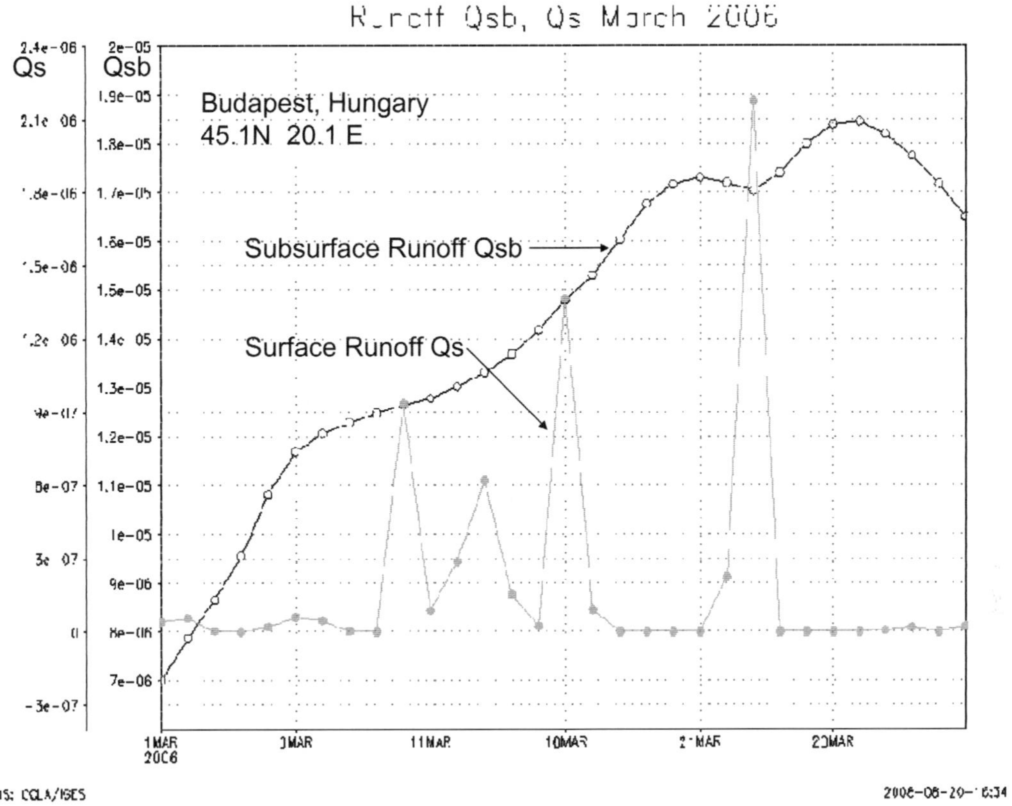

Figure 8. March 2008 time series of runoff at Budapest for surface and sub-surface runoff show the immediate response and cumulative responses to precipitation.

2.2. Precipitation analyses from remote sensing satellite data – TRMM-GPM

Analyses of the June 2007 drought conditions are shown for surface and 0-10 cm deep averaged soil moisture. These clearly show the convective spatial variability and areas of very dry conditions over the region.

IOP Conf. Series: Earth and Environmental Science 4 (2008) 012048 doi:10.1088/1755-1307/4/1/012048

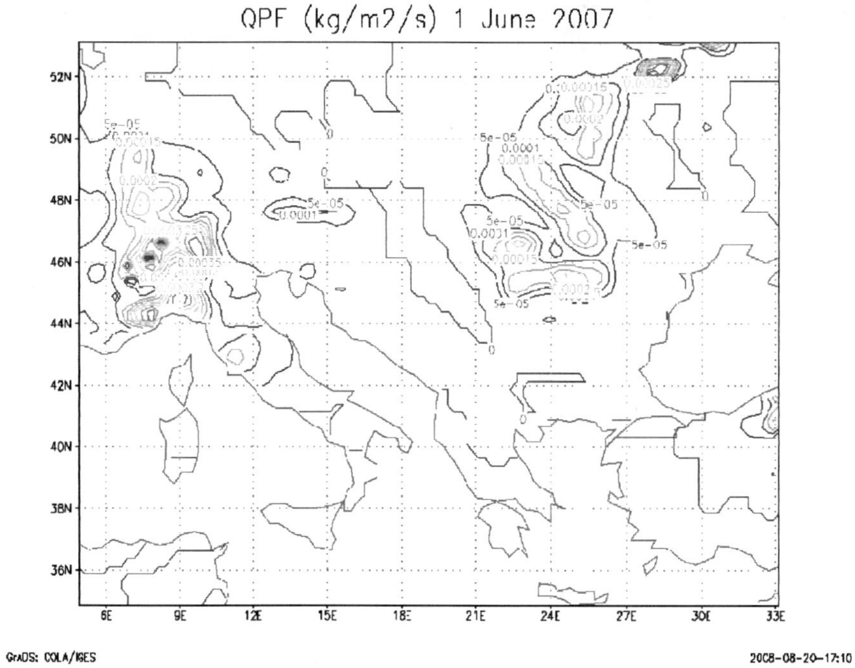

Figure 9. Precipitation pattern on 1 June 2007, showing a maximum over the Julian Alps and dry conditions in Greece, where forest and brush fires were a clear danger in 2007.

By 30 June the conditions through out the area had dried considerably as shown in the 30 June 2007 precipitation (fig. 10) and soil moisture (fig. 11) analyses.

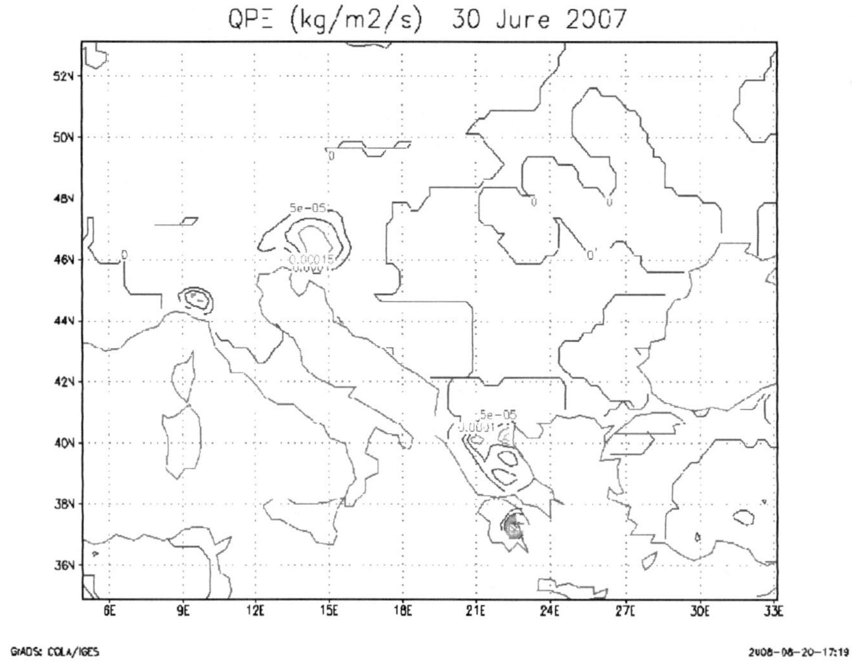

Figure 10. Precipitation from GLDAS on 30 June 2007 shows the very dry conditions in the Danube mouth region.

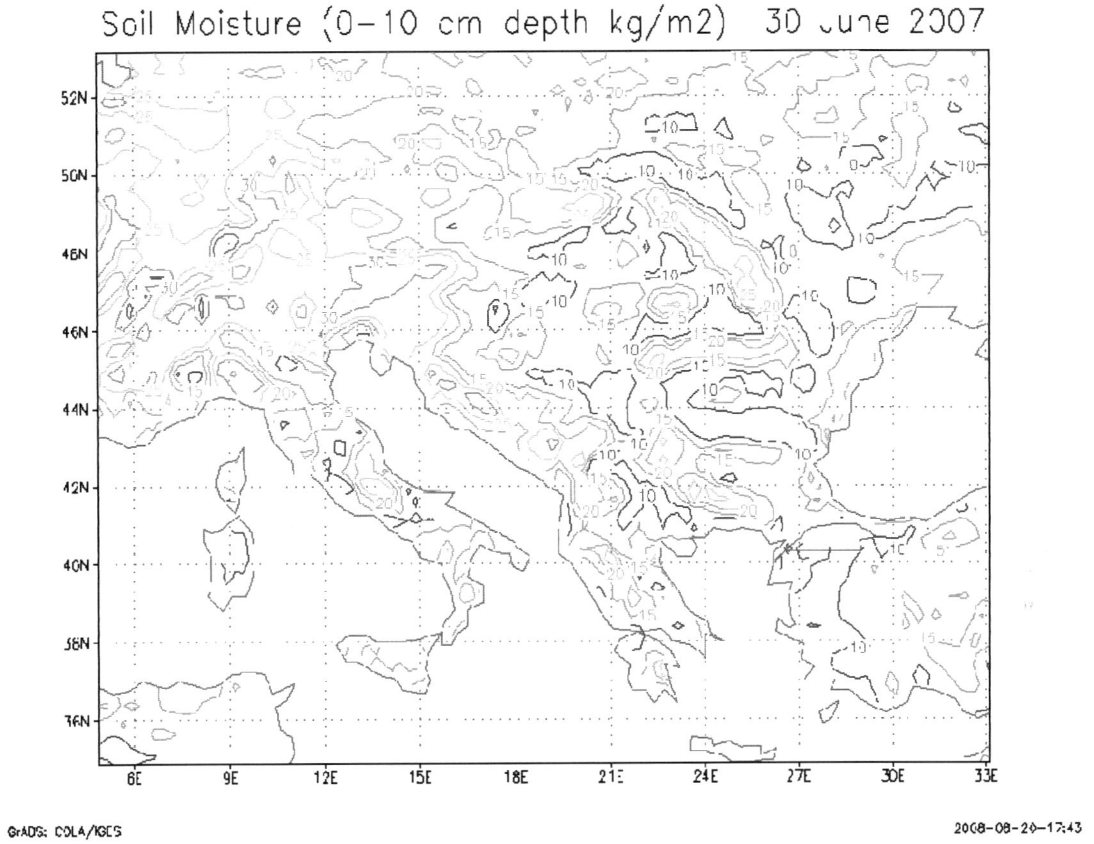

Figure 11. Soil moisture average from 0-10 cm depth on 30 June 2007 shows wet conditions in the Alps 25-30 kg/m2 vs dry 10-15 kg/m2 over Romania, Bulgaria and Greece.

Time series of precipitation and soil moisture clearly show the convective nature of precipitation and the integrating factors of soil moisture and the sensitivity of GLDAS to summer variations in soil moisture. Figure 12 shows the highest values over the Julian and Italian Alps with very dry conditions in Albania from 15-30 June. Differences between Budapest and Ljubljana also properly reflect the drier conditions in Budapest shown in figure 13.

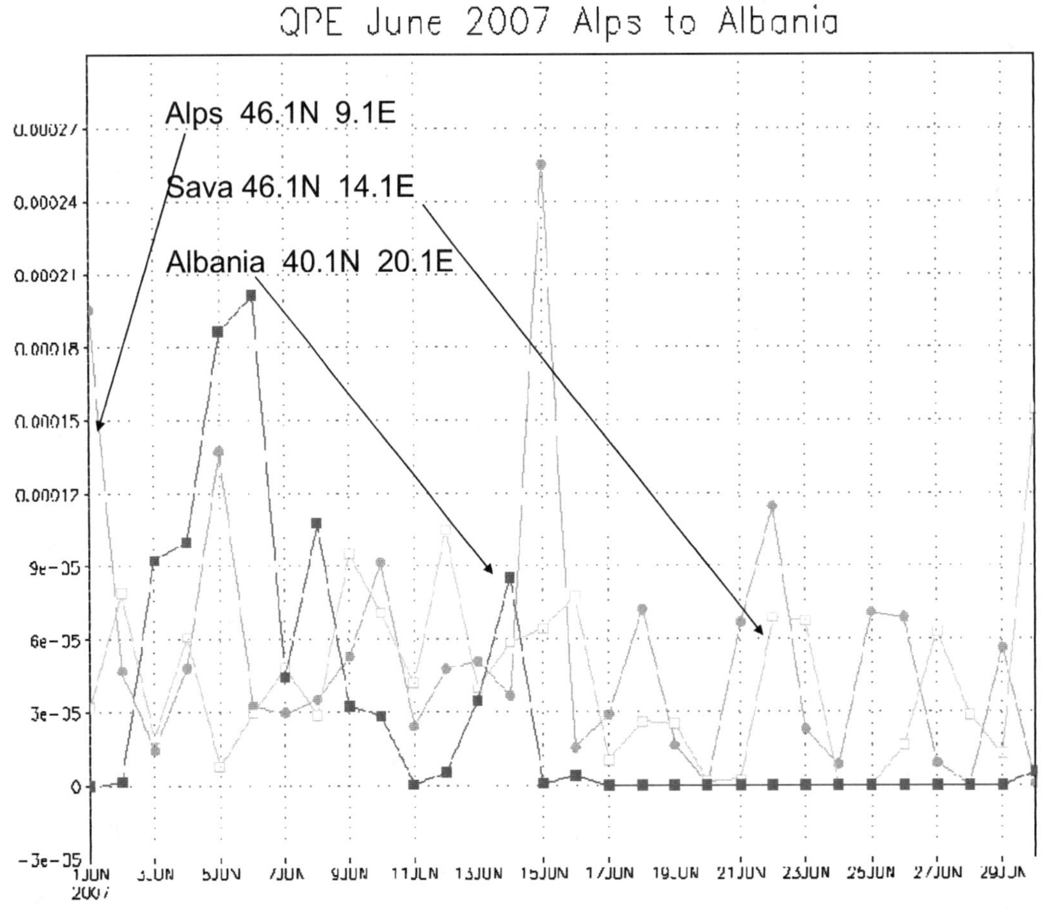

Figure 12. Time series of precipitation showing spatial and temporal variability from the Italian Alps to the Julian Alps and Albania from the GLDAS in June 2007 drought conditions.

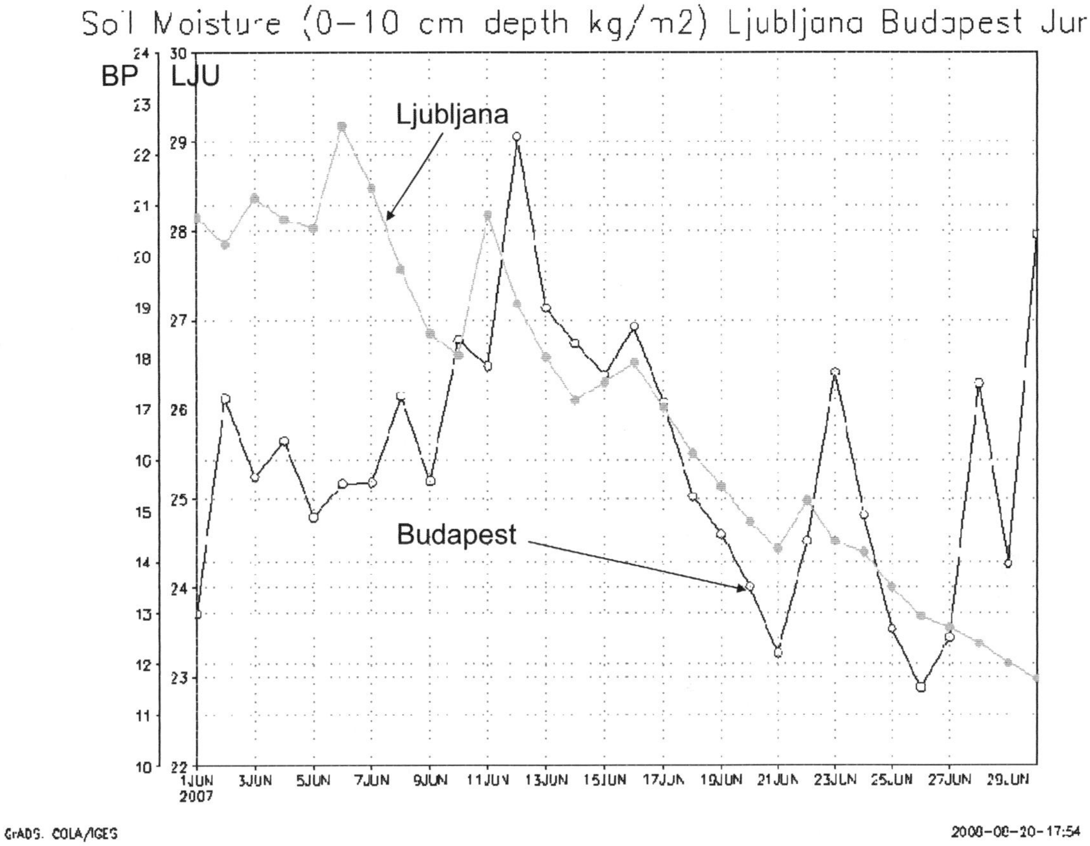

Figure 13. Time-series of root zone soil moisture (0-10 cm depth) showing differences predicted for Ljubljana and Budapest.

2.3. Extreme Flash Flood Event on 18 September 2007

These examples from the extreme flash flood event of 18 September 2007 which caused over 250 M Euro in damages to roads, homes, and other infrastructure in the mountains west of Ljubljana indicate capabilities of mesoscale models and land surface models. Kobold et al (2008) describe the detailed hydrology of this flash flood event. Heavy and abundant precipitation which captured the western, north-western and northern parts of Slovenia on 18 September 2007, caused quick rise of river discharges especially in the region of Baška grapa, Davča, the Cerkljansko and Škofja Loka hills. In that area the streams Selška Sora, Davča and Kroparica caused real destruction. The Savinja was high in the middle and lower stream. The discharges of the tributaries Hudinja and Ložnica exceeded 100-year return period. The Dravinja flooded in the middle and lower stream. The return period of flood was between 50 and 100 years. On the most affected area the peak discharges exceeded the periodical maximum discharges measured at the water stations.

The operational Numerical Weather Prediction system at Environmental Agency of the Republic of Slovenia (EARS) consists of several potential configurations of the ALADIN limited area model. The configurations can be setup and run with optimal options regarding model physics and dynamics for a certain spatial resolution. At the present the spatial resolution is 9 km, which will be increased to 4 km in 2008. The initial and boundary conditions for the operational model come from either French Meteo, or the European Center for Medium Range Weather Forecasting (ECMWF). The 1-km Weather Research Forecasting (WRF) model Quantitative Precipitation Forecast (QPF) was reasonably good as shown in Figure 14. Zagar et al (2008) describe the WRF model configuration over Slovenia and the methods used to initialize the model during this extreme event.

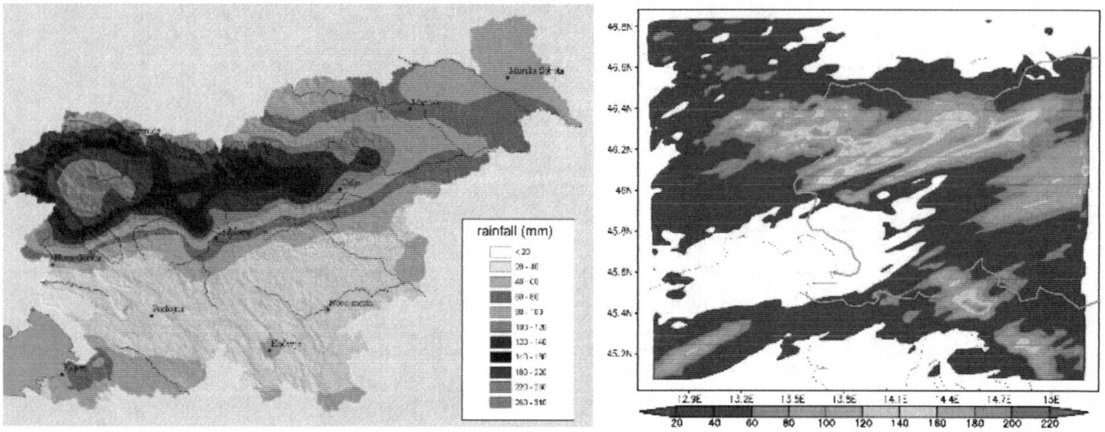

Figure 14. Observed precipitation on 18 September 2007 (left) and WRF model QPF for this event.

GLDAS results shown in figure 15a show the capability of the ¼ degree simulations for this extreme event. Clearly there is value in this simulation; however, a 1 km simulation would provide significant improvement, hence our recommendation to NASA to fund improvement of these model resolutions. The GLDAS simulations of QPE show a maximum in the area south-central Slovenia and south of the observed flash flood event. This shows the limitations of coarse resolution satellite and model estimates of precipitation.

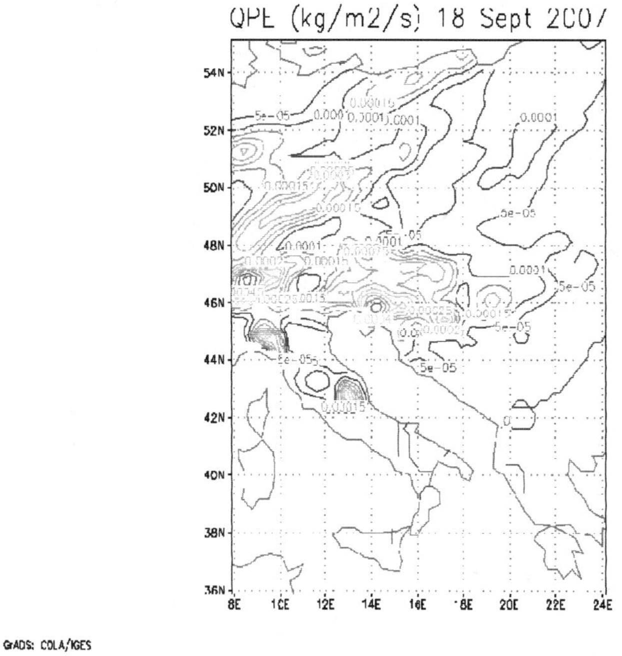

Figure 15(a). QPE 24-hour total estimate from GLDAS showing a maxima over the Italian Alps north of Milano, and in Slovenia in the vicinity of the flash flood. While this indicates the general location of the flood event, it fails to provide the detailed structure required by hydrologic models as shown in figure 14 observations and WRF model QPF simulations. Clearly higher resolution spatial information is needed.

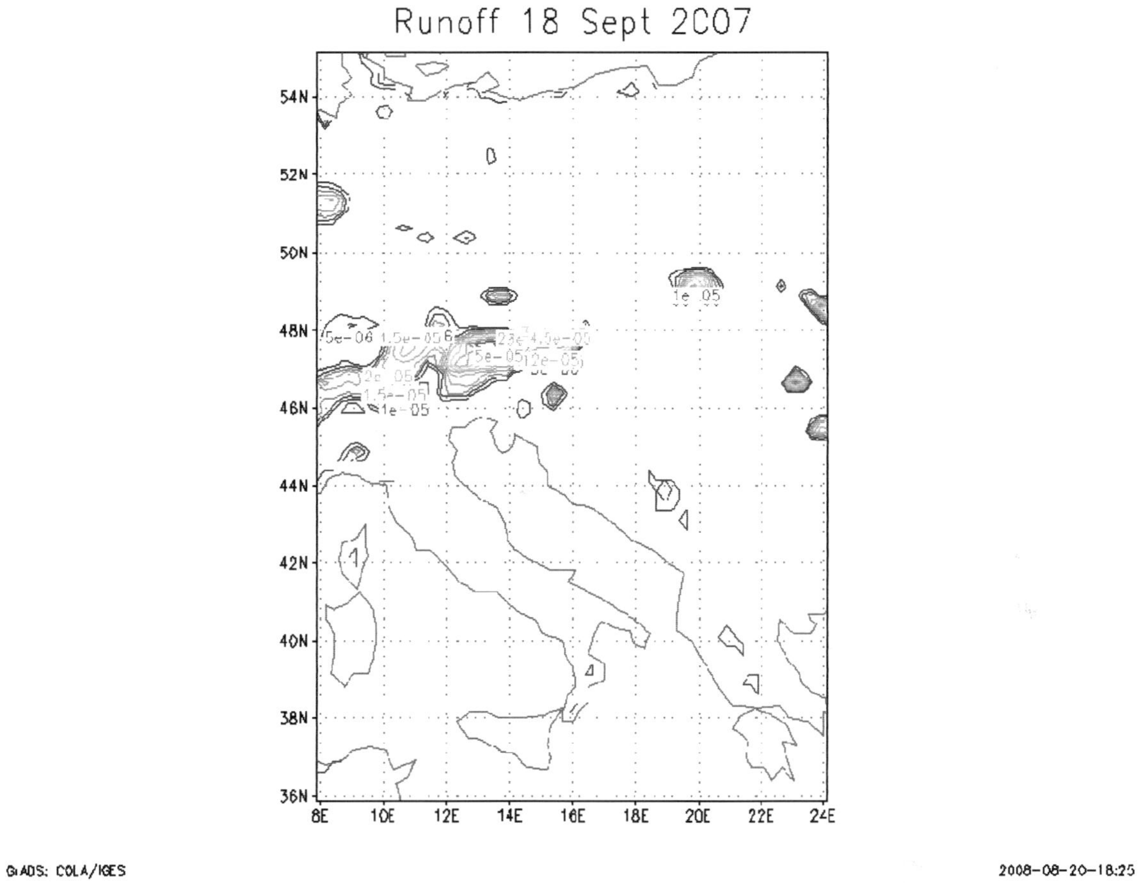

Figure 15(b) Surface runoff estimated by GLDAS showing maximum values in the north of Italy in the Alps of Austria, Slovenia and Italy on 18 September 2007. This area is north of the flash flood event again showing the limitations of coarse resolution Land Surface Models.

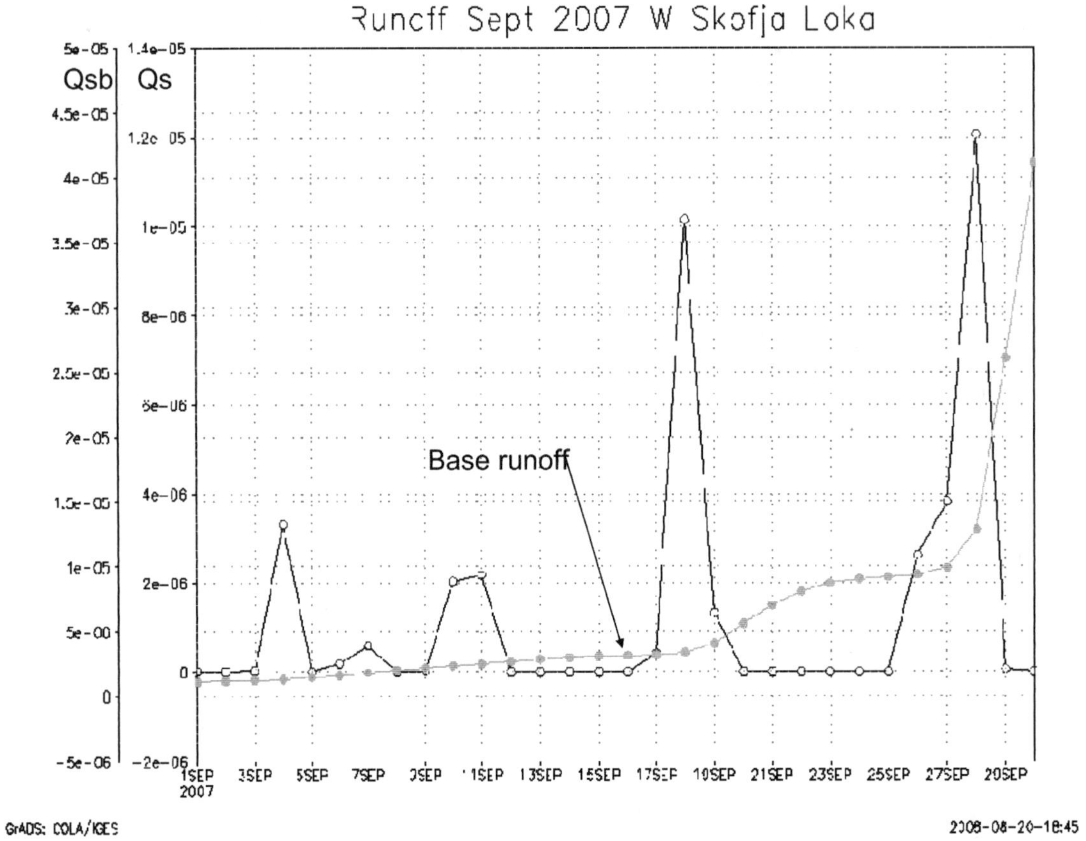

Figure 15. Examples from GLDAS on 18 September 2007 flood event-Rainfall rate (kg/m2/s) –Fig. 15(a), (runoff (kg/m2/s), Fig. 15(b); and September time-series of precipitation (daily white, total surface and groundwater (green) – Fig 15(c). Note the heavy precipitation maxima in the area of the flood, and time series maximum on that date.

Extreme precipitation events like the 18 September 2007 flood often develop on the mesoscale driven by locally enhanced air flows over complex terrain. The complex terrain largely controls the location and intensity of the dynamic and thermodynamic forcing within the storm system at a spatial scale of 1 to 5 km resolution. These GLDAS results reinforce our understanding that high resolution remote sensing systems and numerical models are required for accurate estimation of precipitation and surface hydrologic characteristics, and for nowcasting and short-range predictions. The *WaterNet* project is designed to apply existing technologies and identify gaps so that new technologies may be developed to help forecasters and decision-makers provide the best information to the public. It encourages the direct interaction among researchers, engineers, operational forecasters, and decision-makers – water and emergency managers.

3. Real-time information over the Danubian Region

WaterNet seeks to facilitate the rapid access of information of immediate value to decision-makers by linking them to the research teams and their products. In this section we provide web links (highlighted) to products that are currently available on-line for direct access. These links are available at the hydrometdss.org web page under the top menu for *WaterNet*.

Forecasts of **precipitation**, **soil moisture**, **temperature**, and **current analyses of weather** for Europe from the GFS modelling system and GLDAS are updated each day. Model data is available for a suite of models using a variety of interactive tools at the **CREW modelling** web page. Satellite estimates of precipitation are available at the **TRMM** web site on a global basis. The European Commission's Joint Research Centre provides real-time analyses of **soil moisture and drought** conditions over Europe.

4. Summary and Conclusions

The *WaterNet* - Water Cycle Solutions Network was described with examples of potential products. One key goal of *WaterNet* is to identify gaps in current technology and encourage NASA's and GEO funding to solve these problems to meet practical needs of decision-makers. WaterNet examples of various land surface model products over the Danubian Region were presented that show the type of information available for snow water equivalent, surface and sub-surface soil moisture, precipitation, and satellite estimates of precipitation from TRMM.

Established partnerships represent a cross-section of individual and networked NWRs and DSTs from government, private, and academic domains, that will enable us to quickly complete an operational solutions network, entrain more partner nodes and networks, and move *WaterNet* toward self-sustainability. EU projects like AWARE, and the flood and drought forecasting research efforts, and GMES projects are potential projects that may directly benefit from this *WaterNet* networking.

Specific goals and objectives, methods of communication, and invitation to join the *WaterNet* were discussed. Examples of NASA products from the MODIS, TERRA AQUA, and other satellites and Land Surface Model results over the Danubian countries show potential value added to water resource management in the region. Examples of snow water equivalent, soil moisture, surface temperature, runoff estimates showed how Global Land Data Assimilation System (GLDAS) and its Land Surface Models (LSM) are valuable for decision-makers.

Danube scientists, engineers, decision-makers and stakeholders are encouraged to join the WaterNet Community of Practice and express their interests and needs for research products that can meet their decision-making needs. Please contact us at hydrometdss@comcast.net or phouser@gmu.edu

References

[1] ARSO 2007 *Izjemen padavinski dogodek 18. septembra 2007.* http://www.arso.gov.si/vreme/poro%c4%8dila%20in%20projekti/padavine_18sep07.pdf

[2] Kobold M and Brilly M 2006 The use of HBV model for flash flood forecasting. *Nat. Hazards Earth Syst. Sci.* **6** 407–417 www.nat-hazards-earth-syst-sci.net/6/407/2006/

[3] Kobold M, Sušnik M, Robic M, Ulaga F and Lalic B 2008 Hydrological Analysis of High Waters and Flash Floods that Occurred in September 2007 in Slovenia. *Proceedings of Conference of the Danubian countries* ed Brilly M and Šraj M (Ljubljana: Slovenian National Committee for the IHP UNESCO)

[4] Kumar S V, Peters-Lidard C D, Tian Y, Houser P R, Geiger J, Olden S, Lighty L, Eastman J L; Doty B, Dirmeyer P et al 2006 Land Information System - An Interoperable Framework for High Resolution Land Surface Modeling *Environmental Modelling & Software,* Vol. 21, 1402-15

[5] Matthews D, Brilly M, Gregoric G, Polajnar J, Houser P, Cosgrove B and Rodell M 2008 Extreme Event Analyses from WaterNet *The NASA Water Cycle Solutions* Network, 2008 EGU Annual Meeting Proceedings (Wein, Austria)

[6] Matthews D, Brilly M, Houser P 2006 WaterNet: The NASA Water Cycle Solutions Network-Opportunities for EU Scientists, 6th Annual Meeting of the European Meteorological Society *6th European Conference on Applied Climatology* (Ljubljana, Slovenia: 4 - 8 September

IOP Conf. Series: Earth and Environmental Science **4** (2008) 012048 doi:10.1088/1755-1307/4/1/012048

2006)

[7] Matthews D A, Brilly M, Gregoric G, Polajnar J, Zagar M, Sulejl K, Kobold M, Houser P. 2005 Hydrological Analysis and Short-range Forecasting Needs of Water Management Decision-makers in Slovenia *WSN05-World Weather Research Programme Symposium on Nowcasting and Very Short Range Forecasting, Proceedings* (Meteo, Toulouse, France)

[8] NASA Applied Sciences Program 2005 *Extending NASA Earth-Sun System Research Results Through a Systems Engineering Capacity.*

[9] Zagar M, Strajnar B and Pristov N 2008 Limited Area Modeling Activities in Slovenia. *Newsletter of the 29th EWGLAM and 14th SRNWP meetings (Dubrovnik, Croatia: 8th-11th October 2007)* (Croatian Meteorological and Hydrological Service) pp 66-69

IOP Conf. Series: Earth and Environmental Science **4** (2008) 012049 doi:10.1088/1755-1307/4/1/012049

Hydrological and seasonal export mechanisms for nitrate transport from a forested catchment

S Rusjan, M Mikoš and M Brilly

University of Ljubljana, Faculty of Civil and Geodetic Engineering, Jamova cesta 2, 1000 Ljubljana, Slovenia

E-mail: srusjan@fgg.uni-lj.si

Abstract. Understanding of interactions between hydrological and biogeochemical responses of catchments on rainfall events is usually unclear from periodic measurements and requires tracing of the temporal dynamics of the processes. Smaller streams reflect strong connections between hydrological processes of the rainfall runoff formation and biogeochemical processes in the catchment; consequently, the responsiveness of the streamwater chemistry to changed hydrological states is very high. The study was carried out in 2007, within the 42 km^2 forested Padež catchment in the southwestern part of Slovenia, which is characterized by distinctive flushing, an almost torrential hydrological regime influenced by impermeable flysch geological settings. Recorded hydrographs which, in the hydrological and biogeochemical sense, differed substantially, disclosed a highly variable, but at the same time a strong linkage between hydrological, biogeochemical and particular topographic controls of nitrate exports from the spatial perspective of a studied catchment. The role of specific hydrological events on the nitrate mobilization proved to be important as the size of the accumulated nitrate pool available for mobilization was large throughout the observed hydrographs. The biogeochemical environment of the forest soils presumably significantly affects the size of the available nitrate pool in the studied catchment.

1. Introduction

Forest hydrology studies show a great complexity of interacting influences on the water cycle. Forest systems notably affect the hydrological responses of the basins. In order to understand the formation of forest hydrologic flowpaths and the way how these flowpaths affect stream chemistry, we need chemical measurements at time scales that correspond to hydrological dynamics. Conversely, the stream chemistry tracing enables us to decipher the hydrological processes [1, 2, 3]. High frequency chemical measurements are also essential for testing hydrological models. Hydrological data alone are rarely sufficient to test the process assumptions embedded into a typical hydrological model. For analysis, future prediction and modeling of biogeochemical fluxes across land-lentic and land-lotic interfaces within the landscape, catchments become an important studying spatial unit. Both, hydrological and geochemical or other water chemistry time-series data are needed to identify appropriate model structures and constrain their parameters [4].

The understanding of how hydrological conditions trigger flushing of labile nutrients on a catchment scale is still rather poor, especially if we move from the timescale of seasonal variability towards the timescale of a single hydrological event. But it is just during hydrologically versatile periods that important interactions between hydrologic biogeochemical flowpaths can be observed.

Furthermore, the ecological zones of transition constitute soil-water interfaces where the biogeochemical environments controlling the fate of nutrient transformations change rapidly during hydrological events. Factors that affect the hydrological routing and transport of different chemical forms of nutrients in these zones include [5, 6]: antecedent wetness conditions, temperature conditions, pedologic factors, vegetation composition and topographic position (i.e. proximity to the stream channel). In addition, near-stream variable source areas typically control the stormflow hydrologic responses of upland forested catchment and are sites of rapid water-table responses [7]. Consequently, these areas express great magnitude of dynamics, both in its hydrological response and biogeochemical environment.

We believe that for progress to be made in understanding the role of integral hydrological controls on nitrate flushing, a shift away from the complexity of individual microscale processes observed at low time frequency is needed towards making improvements of observations on a time step of hydrological events. Hydrological routing through variable biogeochemical environments is undoubtedly an area of great uncertainty in regards of process understanding; however a spatially holistic view at a high-frequency temporal basis can offer additional opportunities to study the hydrologically induced biogeochemical routing of different forms of nutrients.

This paper presents a study that examines the seasonality of the nitrate dynamics in a mesoscale catchment in Slovenia covered mainly by deciduous forest with the emphasis on high-frequency measurements of streamwater nitrate concentration during rainfall events of contrasting magnitudes occurring in different seasons.

2. Study area

The Padež stream catchment comprises 42.1 km^2 and is situated in the Southwestern part of Slovenia (Figure 1).

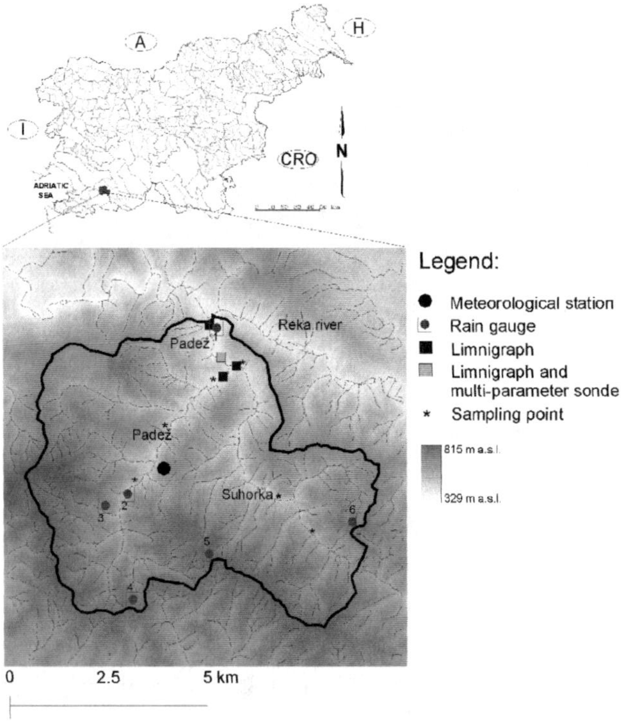

Figure 1. The study area and the monitoring system.

IOP Conf. Series: Earth and Environmental Science **4** (2008) 012049 doi:10.1088/1755-1307/4/1/012049

The Padež stream is a tributary of the Reka river, one of the widest known sinking streams of the Classic Karst area in Slovenia. The Padež catchment reaches deeply into the hilly area of Brkini in the south (altitude up to 811 m a.s.l.), the outflow to the Reka river is at 368 m a.s.l. The Brkini hilly area, and the Padež catchment as its part, consist of Eocene flysch (mainly marl and sandstone layers) underlain by deep cretaceous carbonate bedrocks which also surround the wider area of the Brkini flysch pool. From the hydrogeological point of view, the Padež catchment has a uniform structure characterized by low permeability of erodible flysch layers and a consequent well developed, dense and highly incised stream channel network with a drainage density of 1.94 km/km^2. The lowest parts of the main valleys (the Padež and Suhorka stream valleys) are covered by up to 4-m thick alluvial deposits. The hydraulic conductivity of flysch is low (in the range 10^{-6} m/s to 10^{-5} m/s), the hillslopes are steep (average slope derived from the digital elevation model amounts to 33%), the average slope of the Padež stream channel being almost 3%. The long-term mean annual discharge of the Padež stream amounts to 1.12 m^3/s. The hydrologic response of the catchment is very fast, which is reflected in the flushing, almost torrential regime of the Padež stream. Most of the year, water is present only in the Padež stream and its major tributary, the Suhorka stream, other smaller streams in the catchment being intermittent.

In the climatic sense, the Brkini hilly area is a transitional area between the mediterranean and continental climate with mean annual temperature of 9.6 °C. Mean annual precipitation is approximately 1440 mm [8]. The prevailing movement of the wet air masses is in the southwest–northeast direction. The majority of the precipitation falls during the October–March period with periodical snowfall on the highest parts of the Brkini hills, which does not have substantial influence on the catchment hydrology.

The Padež stream catchment is minimally disturbed by human activity; it has already been used for drinking water supply and, as such, it is also foreseen as an additional source of drinking water for the water-deficient area of the Slovenian coastal region. According to the CORINE 2000 land cover data, 82% of the catchment is covered by forest (79% by broad-leaved forest), and 18% of the catchment comprises complex cultivation patterns (mainly meadows with significant areas of natural vegetation) which are all in the state of successive afforestation. Lower parts of the catchment are almost completely covered by deciduous forest. Main tree species that can be found in the Padež catchment are Sessile oak (*Quercus petraea*), Black alder (*Alnus glutinosa*), Beech (*Fagus sylvatica L.*), and Hornbeam (*Carpinus betulus*).

3. Methods

The monitoring system at the Padež catchment which was established in 2005 is shown in figure 1. Precipitation data were obtained from tipping bucket rain gauges located within the Padež catchment; the meteorological data were gathered with the automatic meteorological station positioned in the middle of the catchment as shown in figure 1. The Thiessen polygons were defined to consider the spatial distribution of the rainfall which proved to be variable especially during summer storms. Water level was recorded continuously with a 5 minute time step on four locations using 1-D Doppler instrument with an integrated logger. The instrument enables recording of current velocity, water stage and water temperature. Flow was gauged on stream sections equipped with limnigraphs using two instruments. During low flow conditions, a salt-dilution flowmeter was used whereas during middle to high flows, 2D/3D handheld Doppler velocimeter was used. The resulting water-level records were converted to volumetric discharges by empirical ratings that were validated by gauging at different flow levels [8].

In 2007, the stream chemistry was periodically measured continuously on an hourly time step using water quality multi-parameter data-sonde. The multi-parameter sonde is designed for on-site and flow-through applications and measures water chemistry parameters simultaneously. The multiple parameters include: nitrate, temperature, electric conductivity, depth, dissolved oxygen, Total Dissolved Solids (TDS), Oxidation Reduction Potential (ORP) and pH. In order to achieve accurate water chemistry readings of the multi-parameter sonde, the maintenance and calibration procedures including regular cleaning of the sensors and replacement of ion-selective nitrate sensor was carefully followed. Additionally, biweekly grab water samples mainly during baseflow conditions were taken form January to November 2006 at the site where multi-parameter sonde was installed for laboratory analysis in order to control the multi-parameter sonde readings.

Measurements of stream water and rainfall electric conductivity were used to perform a two-component hydrograph separation and define the proportion of event and pre-event water in the hydrographs. The electrical conductivity of stream water was measured continuously during hydrograph episodes whereas the electrical conductivity of rainfall was measured in a bulk sample. Measurements of the nitrate concentration in bulk rainfall samples during different episodes showed small concentrations of nitrate in rainfall (i.e. bellow 0.2 mg/l). Thus the atmospheric input of nitrogen with rainfall was not considered as an important source of stream water nitrate during hydrological events.

The field measurements of water chemistry using multi-parameter data-sonde are undoubtedly less accurate than laboratory analysis and will probably never match the dependability of laboratory analysis on bottled samples, but the multi-parameter sondes provide temporal flexibility of measurements that cannot be captured by sporadic grab sampling [3, 9].

4. Results and discussion

4.1 Continuous streamwater nitrate concentration measurements

In 2007, the continuous nitrate concentration measurements were carried out mainly in late spring summer and early autumn. Some hydrological characteristics of the observed rainfall events are summarized in table 1.

Table 1: Characteristics of the observed rainfall events.

Rainfall event	Hydro. No.	Rainfall [mm]	$^aAPI_3$ [mm]	$^aAPI_7$ [mm]	$^aAPI_{14}$ [mm]	$^bT_{14}$ [°C]	^cQ [m^3/s]	^dEW [%]	$^eNO_3$ [mg/l-N]
Maj I	1	7.0	0	0	0.4	14.2	0.27	13	2.5
	2	27.0	5.6	5.6	5.9	13.9	1.39	25	4.4
	3	8.6	26.1	26.1	26.3	13.8	0.79	21	3.5
Maj II	1	17.6	1.3	1.3	3.4	17.0	0.27	20	1.3
	2	33.6	12.1	12.9	14.2	16.9	2.76	36	2.9
	3	21.2	33.2	41.0	41.8	16.3	2.01	29	2.6
Avgust	1	12.8	17.0	17.0	23.6	18.2	0.27	15	1.1
	2	23.2	18.1	18.1	23.3	18.6	1.29	24	3.5
September	1	29.6	0	0	13.3	14.0	2.12	32	4.5
	2	42.8	55.0	55.0	63.5	13.6	5.78	56	3.8
	3	19.2	55.0	55.0	63.5	13.5	4.98	55	3.9
	4	19.6	0.5	1.1	14.4	13.7	0.88	27	3.5

^{a}APIx: Antecedent precipitation index determined for x preceding days.
^{b}Tx: Average hourly temperature for x preceding days.
^{c}Q: Hydrograph peak discharge.
^{d}EW: Event water contribution to total discharge.
$^eNO_3$: Streamwater nitrate concentration.

The time series of nitrate concentration shown in figures 2 and 3 point out a variety of hydrologically induced nitrate concentration oscillations during the late spring hydrological events. During the sequence of the May I hydrographs (the peak discharge during the second hydrograph reached 1.39 m^3/s) the streamwater nitrate concentration varied substantially and the maximum nitrate concentration reached 4.4 mg/l-N (Figure 2).

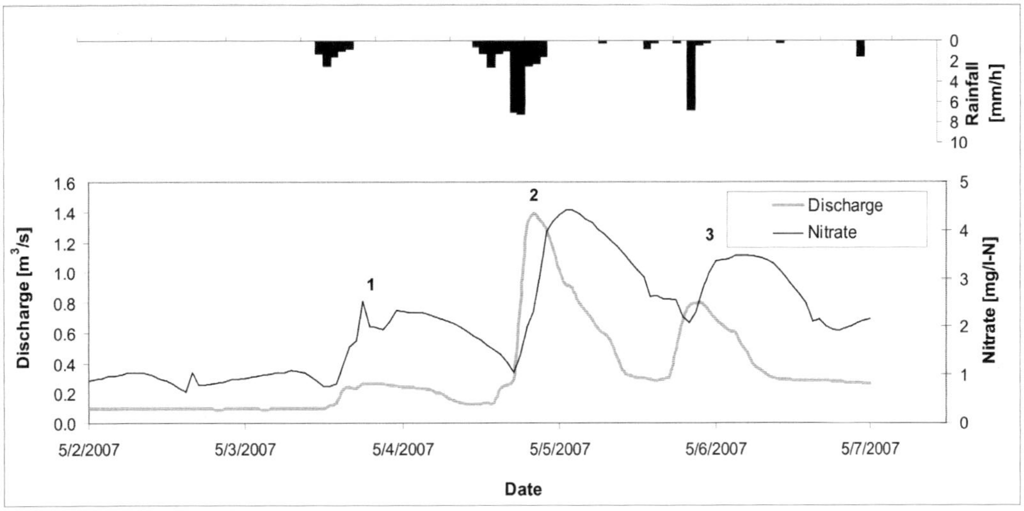

Figure 2. Streamwater nitrate concentrations during first May rainfall event (May I).

During the second rainfall event in the end of May (Figure 3), the streamwater nitrate concentration fluxes were smaller than during the first rainfall event (Figure 2). The hydrograph peaks during the second rainfall event were higher (up to 2.76 m^3/s) than during the rainfall event May I, however the streamwater nitrate concentration was less responsive to changed hydrological conditions which could indicate a minor depletion of the nitrate pool in forest soils during the rainfall event May II.

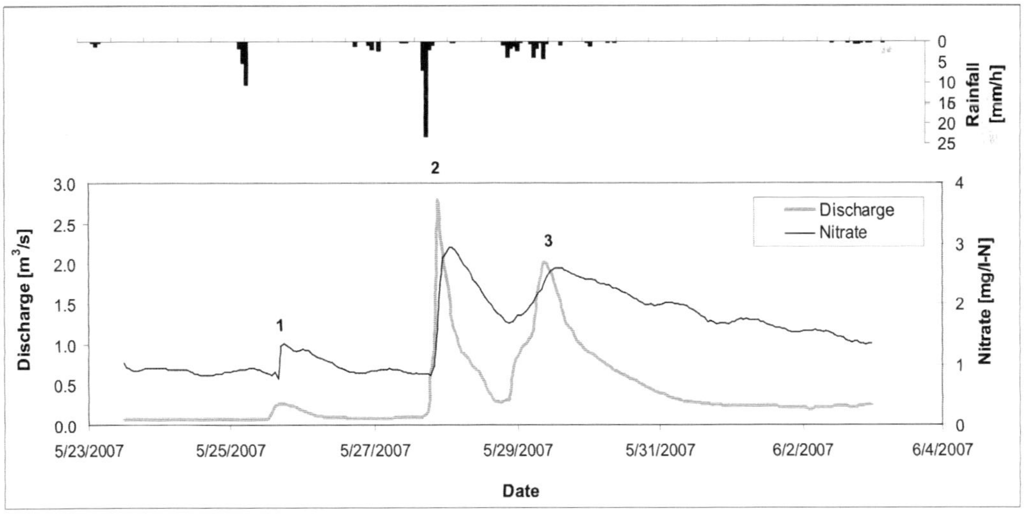

Figure 3. Streamwater nitrate concentrations during second May rainfall event (May II).

A situation similar to the rainfall events in May was also observed during the rainfall event in August (Figure 4). The period preceding the August rainfall event was considerably wet (values of API_7 and API_{14} were 17 and 23.6 mm of rainfall, respectively) which might, similarly as it was the case in the May II rainfall event, indicate the washout of nitrate in the antecedent relatively moist period.

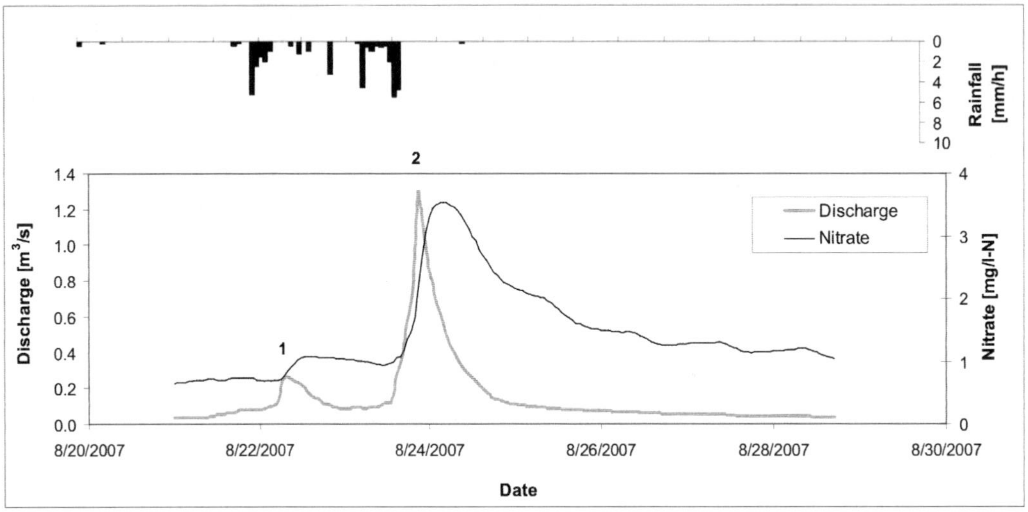

Figure 4. Streamwater nitrate concentrations during the August rainfall event.

The most changeable situation in terms of streamwater nitrate concentrations was observed during the rainfall event in September (Figure 5). The period preceding the September rainfall event was relatively dry (values of API_7 and API_{14} were 0 and 13.3 mm of rainfall, respectively). During the first hydrograph the streamwater nitrate concentration raised very sharply from 1 mg/l-N to more than 4 mg/l-N. During the consecutive hydrographs (No. 2 and 3), the streamwater nitrate concentrations decreased and increased again after the occurrence of hydrograph peaks. Signs of minor nitrate concentration dilution during the periods of hydrograph peaks No. 2 and 3 can be associated to high portions of event water in the total discharge which contained lower nitrate concentration. The event water portion was defined from hydrograph separation based on electrical conductivity measurements. In both hydrographs the event water portion exceeded 50 % (Table 1).

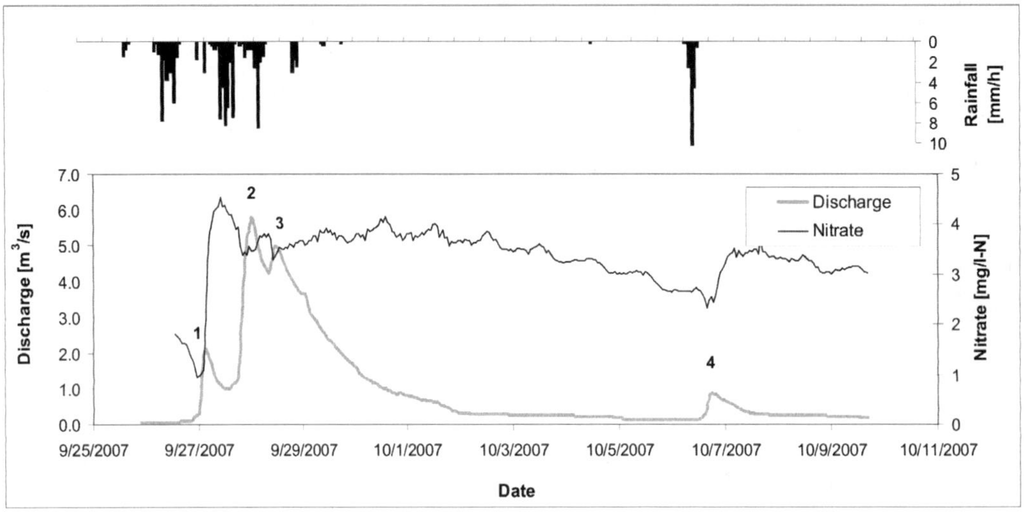

Figure 5. Streamwater nitrate concentrations during the September rainfall event.

The overall decreases of the nitrate concentrations in the falling limbs of the hydrographs were gradual, with different time intervals necessary for the nitrate concentrations to return to the pre-event baseflow concentrations. In the case of September rainfall the time span necessary to return the streamwater nitrate concentration to pre-event values extended to more than two weeks. This presumably indicates that in autumn the size of the movable nitrate pool in forest soils is large an actually exceeds the instant flushing capacity of single rainfall event.

4.2 Temporal distribution of hydrograph peak and peak streamwater nitrate concentration formation
During all observed rainfall events in year 2007 (Figures 2 to 5) an interesting temporal distribution of hydrograph peaks and peak streamwater nitrate concentration was evident. Peak streamwater nitrate concentration emerged with a certain time delay after the hydrograph peak formation. Figure 6 shows peak discharges (Q_{max}) vs. discharges in the time of peak nitrate concentrations ($Q_{max\ nitrate}$) for all recorded hydrographs. A relatively good linear relationship between the two discharges was derived although the observed events differ substantially, both in hydrological and seasonal biogeochemical sense.

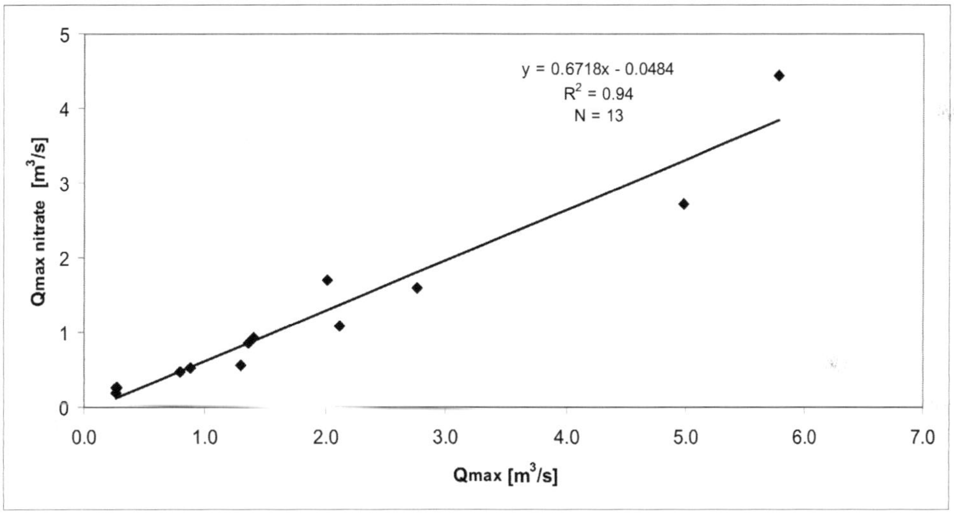

Figure 6. Relationship between peak hydrograph discharges (Q_{max}) and discharges at the time of maximum streamwater nitrate concentration ($Q_{max\ nitrate}$).

In accordance to the hypothesis proposed by [10] and [11], the resulting connection between the two discharges, which is evident beyond changeable hydrological and seasonal biogeochemical settings, could be associated to particular topographical settings present at the studied catchment. Topography presumably controls the distribution of variable source areas evident from intermittent nature of the streams in the studied catchment which very likely continuously expand and contract during hydrological events. The spatial extension of variable source areas in the catchment characterized by well developed and incised stream channel network could on one hand control rapid formation of rainfall runoff and on the other washout of the nitrate accumulated in the forest soils.

5. Conclusion

We believe that presented high-frequency tracing of hydrological events and associated water chemistry dynamics would improve the understanding of the role of specific rainfall events on the total annual budgets of nutrient releases from the forested Padež stream catchment. Furthermore, through the continuous measurements of water chemistry parameters we have obtained an insight into the capacity of forested catchments to control water chemistry constituent exports and investigate the relations between the forest hydrological and biogeochemical export mechanisms over the nitrate which is transported from the catchment by the stream.

High frequency measurements of streamwater nitrate concentrations in the Padež stream throughout different hydrological and seasonal biogeochemical settings disclosed a highly variable but at the same time a strong linkage between the hydrological and biogeochemical controls of the nitrate export from the spatial perspective of a catchment. The role of specific hydrological events proved to be important for predicting the rates of nitrate flushing whose size in the soil pool is large, although it can vary considerably. Consequently, the predicted rates of N cycling and nitrate losses from the catchment can alter greatly from year to year due to the climatic variations and their relative effect on plant growth and soil processes of mineralization and nitrification.

Acknowledgement
We gratefully acknowledge M. Padežnik, A. Vidmar and T. Kralj for field support.

References

[1] Burt T P, Heathwaite A L and Trudgill S T 1993 *Nitrate: Processes, Patterns and Management,* (Chichester: John Wiley and Sons) p 444
[2] Goodale C L, Lajhta K, Nadelhoffer K J, Boyer, E W and Jaworski N A 2002 Forest nitrogen sinks in large eastern U. S. watersheds: estimates from forest inventory and an ecosystem model *Biogeochemistry* **57/58** 239–66
[3] Kirchner J W, Feng X, Neal C and Robson A J 2004 The fine structure of water–quality dynamics: the (high-frequency) wave of the future *Hydrol. Process.* **18** 1353–59
[4] Quinn P 2004 Scale appropriate modelling: representing cause and effect relationships in nitrate pollution at the catchment scale for the purpose of catchment scale planning *J. Hydrol.* **291** 197–217
[5] Cirmo C P and McDonnell J J 1997 Linking the hydrologic and biogeochemical controls of nitrogen transport in near-stream zones of temperate-forested catchments: a review *J. Hydrol.* **199** 88–120
[6] Andersson L and Lepisto H 1998 Links between runoff generation, climate and nitrate–N leaching from forested catchments. *Water Air Soil Poll* **105** 227–37
[7] McHale M R, McDonnell J J, Mitchell M J and Cirmo C P 2002 A field-based study of soil water and groundwater nitrate release in an Adirondack forested watershed. *Water Resour. Res.* **38** 1031–47
[8] Rusjan S, Brilly M and Mikoš M 2008 Flushing of nitrate from a forested watershed: An insight into hydrological nitrate mobilization mechanisms through seasonal high-frequency stream nitrate dynamics. *J. Hydrol.* **354** 187-202
[9] CUAHSI 2006 Advancing Biogeochemical Research in the Field Hydrological Sciences: The CUAHSI Hydrological Measurement Facility – Biogeochemical Component Consortium of Universities for the Advancement of Hydrologic Science p. 25
[10] Hornberger G M, Bencala K E and McKnight D M 1994 Hydrological controls on dissolved organic carbon during snowmelt in the Snake River near Montezuma, Colorado *Biogeochemistry* **25** 147–65
[11] Creed I F and Band L E 1998 Export of nitrogen from catchments within a temperate forest: evidence for a unifying mechanism regulated by variable source area dynamics. *Water Resour. Res.* **34** 3105–20

IOP Conf. Series: Earth and Environmental Science **4** (2008) 012050 doi:10.1088/1755-1307/4/1/012050

Measuring and modelling forest transpiration

M Šír[1], J Čermák[2], N Nadezhdina[2], J Pražák[3] and M Tesař[1]

[1] Institute of Hydrodynamics, v.v.i., Academy of Sciences of the Czech Republic, Pod Paťankou 30/5, 166 12 Praha, Czech Republic
[2] Institute of Forest Botany, Dendrology and Geobiocenology, Mendel University of Agriculture and Forestry, Zemědělská 3, 613 00 Brno, Czech Republic
[3] Institute of Thermomechanics, v.v.i., Academy of Sciences of the Czech Republic, Dolejškova 5, 182 00 Praha, Czech Republic

E-mail: milo_sir@yahoo.com

Abstract. Two transpiration models were tested, one with evaporation control, and another without evaporation control. The principle of the model with evaporation control is that the physical mechanism of transpiration is evaporation, which is actively controlled by plants. The supposed mechanism is: part of the heat (heat equivalent to the energy absorbed from solar radiation) that would cause heating of the plant above 25 °C is dissipated by evaporation. The model has four physical parameters, which are in principle measurable independently. The model without evaporation control is based on the assumption that evaporation dissipates a constant fraction of the incoming solar radiation. In this model, no physical mechanism of transpiration is given. The latter model needs only one parameter, which can be found by means of parameter search. Both models were tested by comparing their results with transpiration measured in the floodplain forest growing along the bank of the Dyje River close to Pohansko (Moravia, Czech Republic). Both models produced a typical difference of about 1.5 mm d^{-1} between measured and calculated daily transpiration totals. We can therefore say that both models are equally valid.

1. Introduction

The aim of this study was to evaluate the validity of two transpiration models based on two different approaches. The first supposes that plants have the ability to control evaporation in order to optimise their temperature [1]. The second is based on the assumption that plants use some constant fraction of the global radiation above the forest canopy for evaporation. The direct measurement of field transpiration in floodplain forests was used as a reference.

2. Experimental site

The experimental part of the study was performed in the floodplain forest growing along the bank of the Dyje River close to Pohansko (about 10 km east of the town of Břeclav, southernmost Moravia). The region is one of the warmest in the Czech Republic. The long-term mean summer temperature is 17 °C.

Floodplain forest is a very specific ecosystem, usually growing in several hundred-metre-wide belts along rivers, mostly surrounded by arable land where cereals and other agricultural crops are grown. They usually have a non-limiting soil water supply (although sometimes soil air supply is limiting, when water remains locally after floods during summer months), and fertile soils that are well supplied by nutrients during regular floods and often by dust blown from surrounding fields. Highly productive broadleaf stands have an upper canopy composed mostly of *Quercus*, *Fraxinus*, *Tilia*, *Ulmus*, *Carpinus*, *Alnus* and *Populus*, an understorey composed of numerous shrubs (e.g. *Cornus*, *Crataegus*, and others) and a rather dense herb layer. Long-term ecosystem-level research has been going on in the region since the late sixties (see comprehensive volumes [2]) along with tree and stand-level water relation studies [3].

The experimental stand of sample trees was about 50 to 100 m from the river bank. An approximately 2 to 3 m deep layer of heavy soils of quaternary origin lies on an approximately 8 m deep layer of sandy gravel of tertiary origin. The altitude is 176 m a. s. l. The underground water table was within the heavy soil layer on most days of the 1998 growing season, water rose above the soil surface during spring floods about four times a year, and precipitation was lower than potential evapotranspiration from summer until late September.

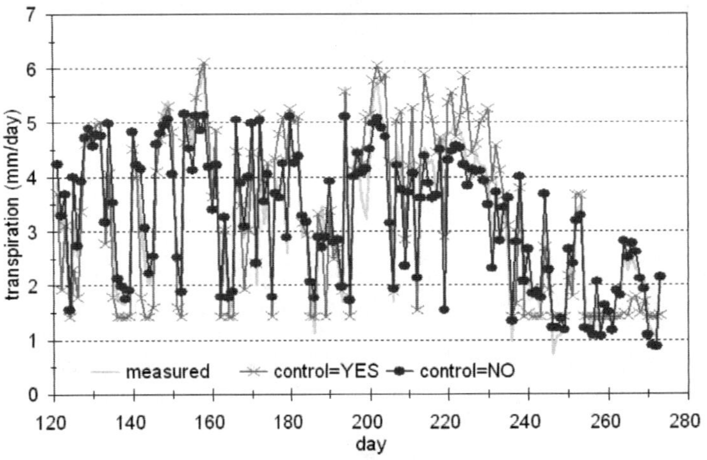

Figure 1. Seasonal course of daily transpiration totals measured directly and those calculated using models a) with evaporation control (YES), b) with no evaporation control (NO), at the experimental site near Pohansko from May 1 to September 30 of 1998.

3. Methods

Daily totals of transpiration were estimated on the basis of sap flow measurements. Sap flow was measured in tree trunks using the trunk tissue heat balance method [4]. Two measuring points (consisting of five electrodes and eight compensating thermocouples, insulated by 2 cm-thick polyurethane foam and 0.5 mm aluminium tin) were installed at breast height on opposite sides of the trunks. Data were recorded by multi-channel data loggers at 15 minute intervals. Altogether 9 sample trees of pedunculate oak (*Quercus robur* L.) were selected from the closed canopy experimental stand for measurement according to the total quintiles, so that each tree represented the same fraction of the basal area. Sample trees ranged in diameter at breast height from 40 to 80 cm and heights ranged from 30 to 40 m. The method gave actual data of daily sap flow in the range 0.05 to 0.6 $m^3\,d^{-1}$ for individual trees, which was scaled-up to the stand level on the basis of the sap flow to basal area relationship [5], with an approximate relative error of about ±10 to 12%. Two transpiration models were used.

Model with evaporation control: The principle of the model is that the physical mechanism of transpiration is evaporation, which is actively controlled by plants. The supposed mechanism is: part of the heat (heat equivalent to the energy absorbed from solar radiation) that would cause heating of the plant above the optimum temperature is dissipated by evaporation [1]. It is supposed that plants are heated by absorbed solar radiation. They are also heated or cooled by the surrounding air, where the flux q of the heat exchange between plant and air is given by equation $q = \alpha.\Delta T$, where α is the coefficient of heat transfer and ΔT is the plant/air temperature difference. The model has four physical parameters, which are in principle measurable independently: optimum temperature, albedo, coefficient of heat transfer, and effective thickness of plant cover [1].

Model with no evaporation control: Transpiration is calculated as the amount of water evaporated when actually using a fixed fraction of the heat equivalent of solar radiation above the forest canopy. In this model, no physical mechanism of transpiration is given. No evaporation control is considered, nor the heating/cooling of plants by the surrounding air or the absorbance/reflection of solar radiation. The model has only one parameter, which can by found by means of parameter search.

Table 1. Seasonal totals (May 1 to September 30 of 1998) obtained from two transpiration models.

	kWh m^{-2}	%
Global radiation income	720	100
Amount applied for transpiration (= latent heat)	320	45
Amount heating plants and air (=sensible heat)	400	55
Transpiration	500 mm	3.3 mm d^{-1}

4. Results

The hourly sums of global radiation and hourly air temperature in the period of May 1 to September 30 of 1998 were used for calculations. The parameters of the model with evaporation control were set: optimum temperature T_0 = 25 °C, albedo a = 0.25, coefficient of heat transfer α = 15 W m^{-2} K^{-1}, effective thickness of plant cover l = 0.001 mm. The parameter of the model with no evaporation control was found to be 0.435. Measured and calculated transpiration data are shown in Figures 1, 2 and 3 as daily sums and cumulative seasonal sums. Seasonal totals are given in Table 1.

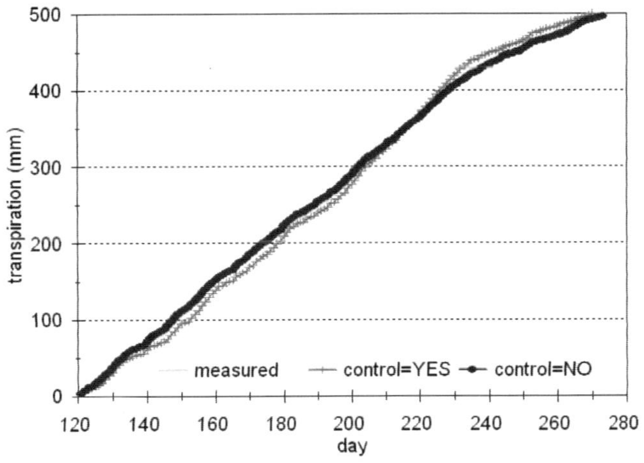

Figure 2. Cumulative seasonal course of daily values of transpiration measured directly and those calculated using models a) with evaporation control (YES), b) with no evaporation control (NO), at the experimental site near Pohansko from May 1 to September 30 of 1998.

5. Discussion

Average daily transpiration 3.3 mm d^{-1} and seasonal total transpiration 500 mm are reliable in the climatic conditions in Pohansko. Transpiration was not limited by water shortage, which means that computed values correspond to potential transpiration. The seasonal sum of sensible heat 400 kWh m^{-2} is very close to the average seasonal value of 408 kWh m^{-2} estimated in the watershed Liz (Šumava Mts., altitude 828 to 1074 m a. s. l., spruce forest) over the period 1983 to 1994 [6].

6. Conclusions

The differences between measured and calculated daily totals were rather small (Figures 1, 2) and non-systematic (Figure 3). The uncertainty included in the measured and calculated data is much higher than the typical difference of about 1.5 mm d^{-1} between measured and calculated daily transpiration totals (Figure 3). We can therefore say that both models are equally valid, based on the similarity of the results. Results presented in this paper indicate that transpiration of the floodplain forest (which is very close to potential transpiration, when there is practically no soil water shortage) is sufficiently modelled by the model with no evaporation control, parameterised by a single constant.

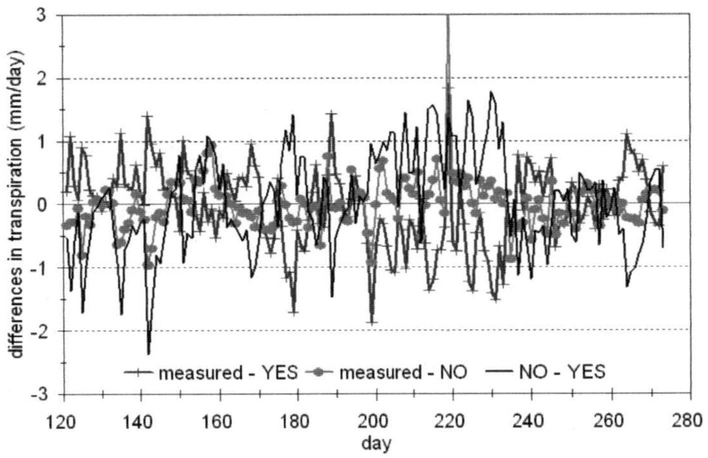

Figure 3. Differences in daily total transpiration values between measured and calculated data, when using the model with (YES) and without (NO) evaporation control.

Acknowledgements

This research was supported by the Academy of Sciences of the Czech Republic (Institutional Research Plan AV0Z20600510 and Project No. 1QS200420562), Grant Agency of the Czech Republic (Projects No. 205/06/0375, 205/08/1174 and 526/08/1016), Ministry of Education, Youth and Sports (Project No. MEB 0808114) and Ministry of the Environment of the Czech Republic (Project No. SP/1a6/151/07).

References

[1] Pražák J, Šír M and Tesař M 1994 Estimation of plant transpiration from meteorological data under conditions of sufficient soil moisture *J. Hydrol.* **162** 409–27

Pražák J, Šír M and Tesař M 1996 Parameters determining plant transpiration under conditions of sufficient soil moisture *J. Hydrol.* **183** 425–31

[2] Penka M, Vyskot M, Klimo E and Vašíček F 1985 *Floodplain Forest Ecosystem: I. Before Water Management Measures* (Praha: Academia) p 466

Penka M, Vyskot M, Klimo E and Vašíček F 1991 *Floodplain Forest Ecosystem: II. After Water Management Measures* (Praha:, Academia/Elsevier) p 630

[3] Čermák J, Úlehla J, Kučera J and Penka M 1982 Sap flow rate and transpiration dynamics in the full-grown oak (*Quercus robur* L.) in floodplain forest exposed to seasonal floods as related to potential evapotranspiration and tree dimensions *Biologia Plantarum* **24** 446–60

Čermák J, Kučera J, Prax A, Bednářová E, Tatarinov F and Nadyezhdin V 2001 Long-term course of transpiration in a floodplain forest in southern Moravia associated with changes of underground water table *Ekologia* **20** Suppl. 1 92–115

Čermák J and Prax A 2001 Water balance of the floodplain forests in southern Moravia considering rooted and root-free compartments under contrasting water supply and its ecological consequences *Ann. Sci. For.* **58** 1–12

[4] Čermák J, Deml M and Penka M 1973 A new method of sap flow rate determination in trees *Biologia Plantarum* **15** 171–8

[5] Čermák J, Kučera J and Nadezhdina N 2004 Sap flow measurements with some thermodynamic methods, flow integration within trees and scaling up from sample trees to entire forest stands *Trees, Structure and Function* **18** 529–46

[6] Šír M, Tesař M and Pražák J 1997 Estimations of seasonal potential transpiration *Landschaftsökologie und Umweltforschung* **25** 281–4

IOP Conf. Series: Earth and Environmental Science **4** (2008) 012051 doi:10.1088/1755-1307/4/1/012051

The experimental watersheds in Slovenia

M Šraj, S Rusjan, S Petan, A Vidmar, M Mikoš, L Globevnik and M Brilly

University of Ljubljana, Faculty of Civil and Geodetic Engineering
Jamova 2, SI-1000 Ljubljana, Slovenia

E-mail: msraj@fgg.uni-lj.si

Abstract. Experimental watersheds are critical to the advancement of hydrological science. By setting up three experimental watersheds, Slovenia also obtained its grounds for further development of the science and discipline. In the Dragonja experimental watershed the studies are focused on the afforestation of the watershed in a mediterranean climate, on the Reka river the water balance in a partly karstic area is examined, and on the case of the Glinščica stream the implications of the urban environment are studied. We have obtained valuable experience and tested new measuring equipment on all three experimental watersheds. Measurements and analysis on the experimental watersheds improved the current understanding of hydrological processes. They resulted in several PhD Theses, Master Theses and scientific articles. At the sane time the experimental watersheds provide support to the teaching and studying process.

1. Introduction

The experimental watersheds of the Dragonja, Reka and Gradaščica with Glinščica rivers are nowdays critical to the advancement of hydrological science, not only in Slovenia, but also on international level. Those watersheds have been equiped with modern measuring equipment for precise measurements of precipitation, intercepted precipitation, discharges, erosion and water quality. Thus contemporary experimental base was established used for scientific research and at the same time provided support to the teaching and studying process.

The Dragonja river basin is situated on the border between Slovenia and Croatia (Figure 1). Due to political and social circumstances and poorly managed agriculture in the past, the catchment area has been depopulated in the past decades, leaving the forests to develop in the abandoned agricultural fields. The changes in land use changed the hydrology of the catchment in a great deal. The Dragonja watershed was chosen as an experimental watershed, since it is interesting because of the intensive natural reforestation in the last decades, which has caused a decrease in minimal and maximal flows. At the same time no noticeable climate (precipitation and temperature) changes were observed. Precise measurements in the last few years were the basis for cooperative scientific work between the Vrije Universiteit from Amsterdam and the University of Ljubljana, which resulted in several PhD Theses, Master Theses and scientific articles. At the same time the experimental watershed provided support to the teaching and studying process. Until now, many students from Amsterdam (VUA), Ljubljana (UL) and Freiburg (UF) have performed their field work on that experimental watershed.

The Reka River is the widest known sinking stream of a classical karst area, and it has been studied since antiquity. The river sinks into the Škocijan Cave system, which was proclaimed by UNESCO as a World Heritage Site in 1986. Then the groundwater stream flows to the karst springs of the Timav, and drains into the Adriatic Sea in the Trieste Bay (Figure 1). Slovenian IHP Committee selected the

IOP Conf. Series: Earth and Environmental Science **4** (2008) 012051 doi:10.1088/1755-1307/4/1/012051

Reka River Basin for an experimental basin. The river has a drainage area of 422 km^2 and a mean discharge of 8.26 m^3/s. In the seventies, the Reka River was one of most polluted rivers in Slovenia. During floods in 1999 and 2000, experimental measurements of velocity, water stage, suspended sediment transport, chemical parameters and toxicity tests were conducted.

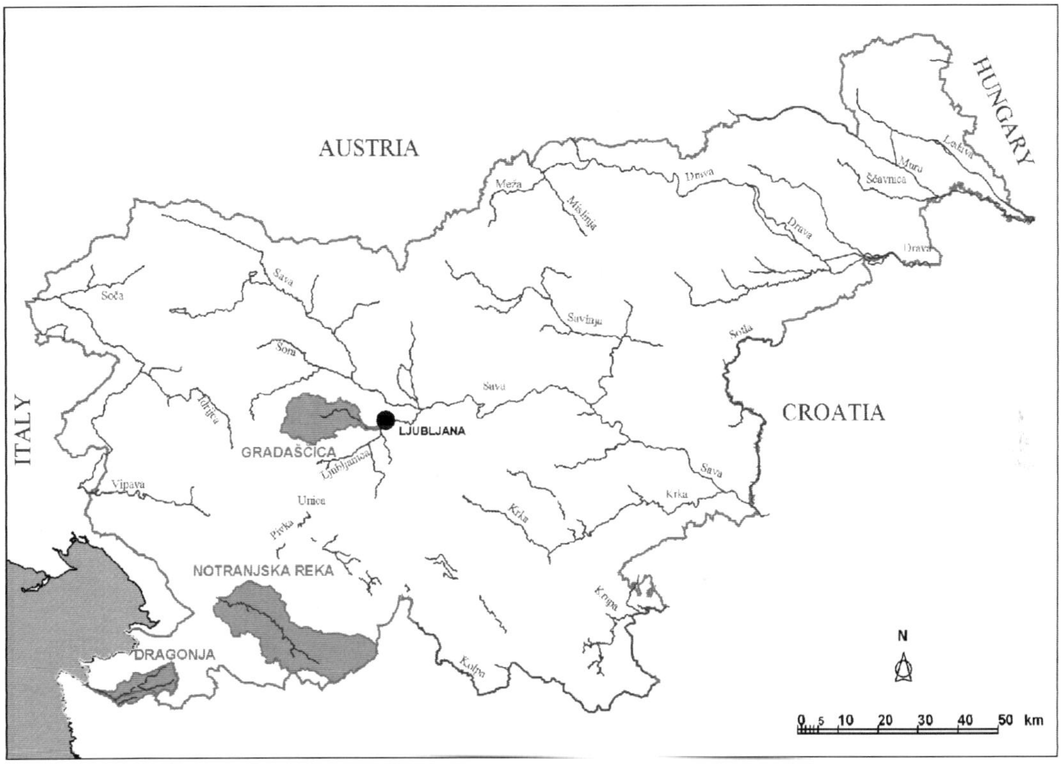

Figure 1. The location of the experimental watersheds in Slovenia.

In urban areas the concept of the watershed becomes more complex and difficult to define because the natural topography has been disturbed; the water may be drained through storm drains and in some cases it may be diverted by drains into other basins. Due to past regulations, urban watercourses are frequently considered only as an architectural element or as conduits for sewer and rainfall. In the area of Ljubljana watercourses Ljubljanica, Gradaščica and the Glinščica are regulated in sterile channel cross sections with high concrete banks which separate the water bodies from surrounding urban areas (Figure 1). The precipitation watershed area of the Glinščica comprises 17.4 km^2. The Glinščica watershed study site is equipped with weather stations, rainfall stations, two water stations with Doppler velocity meters and sample points for water quality measurement by multipurpose probe. The multipurpose probe measures dissolved oxygene and nitrates concentration, pH, specific conductivity, depth, total dissolved solids and temperature. Measurements showed how vulnerable to pollution is the Glinščica creek due to typical urban flow regime with extremely low water flows and water content especially in the summer.

2. The Dragonja river experimental watershed

2.1. The study area

The Dragonja river basin is situated in SW of Slovenia, in the northern part of the Istrian Peninsula (Figure 1). The basin area is hilly region, between 150 and 450 m above sea level. The area of the watershed is 90.5 km^2. In the past the area was mainly inhabited in the hilly ridge tops. In the 1960s and 1970s the region depopulated and this led to the abandonment of agricultural land use. This caused natural reforestation of the land. The forest area has increased from 25 % (in 1953) to more than 60 % (today). The area is geologically composed of Eocene flysch. Soil is mostly carbonate rendzina. The Dragonja watershed belongs to the submediterranean climate region with mean annual temperatures around 14 °C and the annual precipitation of 1000 mm (coast) to 1300 mm (the upper watershed area). Snow is extremely rare. The changes in land-use changed the hydrology of the catchment at great deal.

2.2. Measurments and analysis

The Dragonja river watershed was chosen as an experimental watershed in 1999 [1]. The measurements on the watershed started in autumn 2000 with the aim to define the impact of the natural reforestation on the water balance of the watershed. Measurements included rainfall and meteorological measurements, discharge measurements, forest hydrological cycle components measurements [2] and soil erosion measurements (Figure 2). Rainfall was measured on several locations (Labor (2x), Marezige, Kocjančiči, Sirči, Rokava, Boršt, Kubed and Stara Vala). There were also two meteorological stations in Boršt and Kubed that measured incoming and outgoing radiation, temperature, relative humidity, wind velocity, direction and precipitation. The discharges (velocity) of the Dragonja river were measured on several locations with current meter and automatically with ultrasonic Doppler instrument near the outflow of the river.

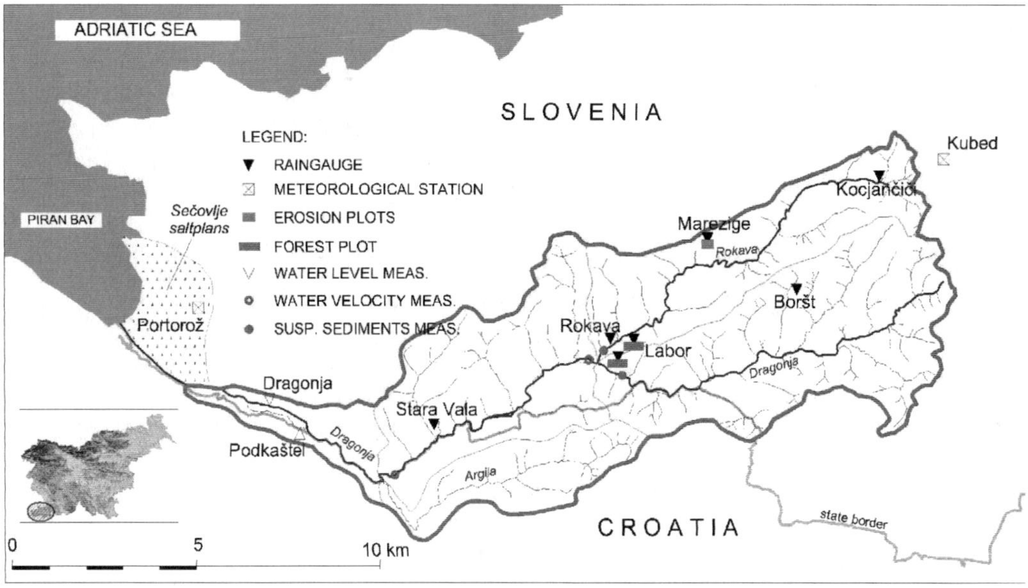

Figure 2. The Dragonja river basin with monitoring sites.

Two study plots in the forest were established in co-operation with Vrije University from Amsterdam where the forest hydrological cycle was analysed [3]. One is located on the north-facing slope (1419 m^2) and the other on the south-facing slope (615 m^2) (Figure 2). Precipitation above the canopy, throughfall and stemflow on each plot were measured (Figure 3). Rainfall above the canopy was measured with a tipping bucket rain gauges and with totalisators (manual gauges) for control. Throughfall was measured with two steel gutters in combination with ten manual gauges in each plot, which were emptied and moved randomly every time. Stemflow was measured on two most typical species in each plot. Litter was collected regularly in 10 baskets for LAI (Leaf Area Index) estimation. For the same purpose and at the same points, hemispherical photographs of canopies were taken and three series of measurements of photosynthetically active radiation (PAR) were made. The specific leaf area (SLA) was estimated for the main species for need of estimating the leaf area index.

Figure 3. Precipitation measurements above the canopy (left), throughfall measurements (right, above) and stemflow measurements (right, below).

For the needs of studying the processes of soil erosion and sediment transport in the Dragonja experimental catchment discharge of the Dragonja river was measured [4] and sediment samples (ISCO) in the Rokava catchment (the biggest tributary in the upper part of the Dragonja catchment) were taken [5]. Also Interril soil erosion measurements were made (Figure 4). Eight 1 m^2 erosion plots were set up. The locations with different land use types: on bare soil in a young olive grove (2), in an overgrown meadow (2) and in the forest (4) were chosen. The erosion plots in the forest were placed on soil with two different slopes, 8° and 21°. The surface runoff from each of the erosion plots was collected in reservoirs. Measurements were performed after every erosive event. The volume of the collected surface runoff was determined first, and then samples of water and detached soil mixture from the reservoirs were taken. The samples were dried at 105°C in the laboratory, where the concentration of undissolved particles was determined.

Figure 4. Erosion plots in the olive grove on bare soil and the tipping bucket rain gauge (left), and the erosion plot in the forest (right).

Part of the research in the Dragonja experimental watershed deals with the development of a new synthesis method of assessing the hydromorphological status of river corridors based on state analysis of the Dragonja river. The bases of the new method were time effectiveness of data gathering and accuracy of the method. Prior to the practical research implementation, a concept of a rapid transect data gathering was designed. Additionally, an extensive hydromorphological record sheet was elaborated, including a combined list of hydromorphological variables.

2.3. Results and discussion

According to the results of one year precise measurements and analysis of individual components of the forest hydrological cycle: precipitation above the canopy, throughfall and stemflow on two forest plots facing south and north, the average values of the throughfall amounted to 67.1 and 71.5 % of the associated rainfall for the south and north research plots, respectively, while the average stemflow fraction was 4.5 and 2.9 %, respectively. The difference is a consequence of the different structure and characteristics of forests on both slopes. On the basis of the measurements, water balance equation and models, it was calculated that the average annual interception losses amount to 28.4 and 25.4 % for the south and north slopes, respectively [6]. The values are relatively high compared with other similar studies, which are usually between 15 and 25 %. Such high loss of the deciduous forest of the Dragonja watershed is considerable and could be the reason for the reduction of surface water runoff. Regression analyses of precipitation measured on individual rain gauges showed good agreement with a correlation coefficient over 0.95. Regression analyses of the measured throughfall with fixed and roving gauges gave expected higher values of roving gauges with very high correlation coefficients.

The measurements were extended in 2005 when eight 1 m^2 erosion plots for interrill soil erosion measurements were set up on locations with different land use types: on bare soil in a young olive grove, in an overgrown meadow and in the forest. In 2006, 33 erosive events were registered, contributing 875 mm of rainfall, while 43 events in 2007 contributed 939 mm of rainfall. The total annual rainfall erosivity factor R in 2006 amounted to 2090 MJ/ha mm/h generating 8.43 kg/m2 of soil loss on the bare soil. The total R factor for 2007 was approximately 75 % higher than in 2006, while the annual soil loss on bare soil was higher by only 25 %. Few major erosive events were responsible for the greater part of the washed soil: the 3 most intensive erosive events in 2006 meant 55 % of the annual R factor, while for the same portion of the annual R factor in 2007 5 erosive events were needed. The interrill soil erosion in the forest is strongly connected to the vegetation period when the most intensive erosive events occur: the portion of the forest soil loss in the summer period, when the vegetation cover is nicely developed, is much lower than the soil loss on the bare soil [7].

3. The Reka river experimental watershed

3.1. The study area

The Reka river has a drainage area of 422 km^2 and a mean discharge of 8.26 m^3/s. It is the widest known sinking stream of a classical karst area, and it has been studied since antiquity. The river sinks into the Škocijan Cave system, which was proclaimed by UNESCO as a World Heritage Site in 1986. Then the groundwater stream flows to the karst springs of the Timav, and drains into the Adriatic Sea in the Trieste Bay. The stream density is about 1.696 km/km^2. Slovenian IHP Committee selected the Reka River Basin for an experimental basin. In the seventies, the Reka River was one of most polluted rivers in Slovenia. The Reka River Basin is situated on the Brkini syncline Eocene flysch rocks surrounded by a large karstic region [8].

3.2. Measurements and results

There is a town Ilirska Bistrica, plus a lot of small villages, along the valley. The chemical and woodwork industry heavily polluted the river, which was one of most polluted streams in Slovenia. The main polluters were the fiberboard factory and the organic acids factory in Ilirska Bistrica. The BOD_5 was between 100 and 200 mg/l during the mean flows in 1969 and 1979. Industrial water treatment was unsuccessful, and the Reka River valley was an unpleasant place with a specific smell caused by the anaerobic processes in the river body, especially during the low flows, with only few hundred liters per second of discharge. Pollution prevention work started in the seventies. A wastewater treatment plant and two reservoirs for low flow recharge were constructed. In the nineties, the industrial production decreased and the pollution diminished.

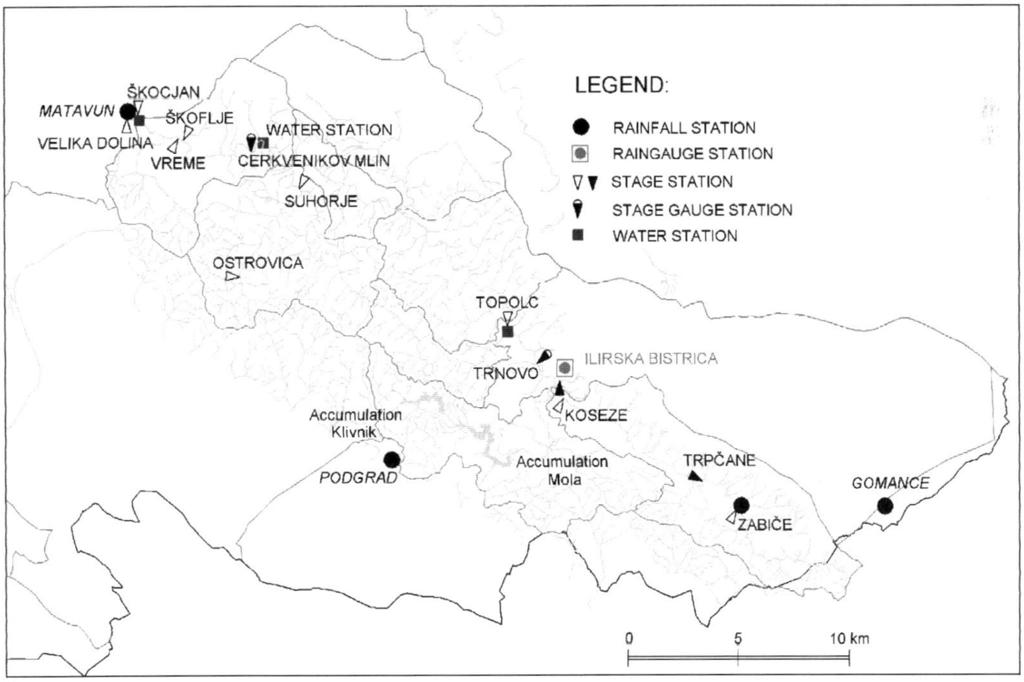

Figure 5. The Reka river gauging stations and measuring sites.

IOP Conf. Series: Earth and Environmental Science **4** (2008) 012051 doi:10.1088/1755-1307/4/1/012051

There were three hydrological gauging stations in the basin (Figure 5). Two of them, Cerkvenikov mlin and Trnovo, are equipped with a recorder. The others are only staff gauges, and some of them have been abandoned. The Cerkvenikov mlin hydrological station was established in 1951. A large fluctuation in discharges can be observed there. The lowest discharges are in the summer time, when the natural discharge is below 200 l/s.

During floods in 1999 and 2000, experimental measurements of velocity, water stage, suspended sediment transport, chemical parameters and toxicity tests were conducted.

The measurement of velocity and discharge took place at the Cerkvenikov mlin state gauge station (Figure 5). Three instruments were used to measure the stream velocity: Valeport self recording current meter, Starflow ultrasonic velocity meter and ISCO 6700 compatible area-velocity module (Figure 6 and Figure 7).

Figure 6. Valeport self recording current meter and Starflow ultrasonic velocity meter (left), and ISCO 6700 compatible area-velocity module (right).

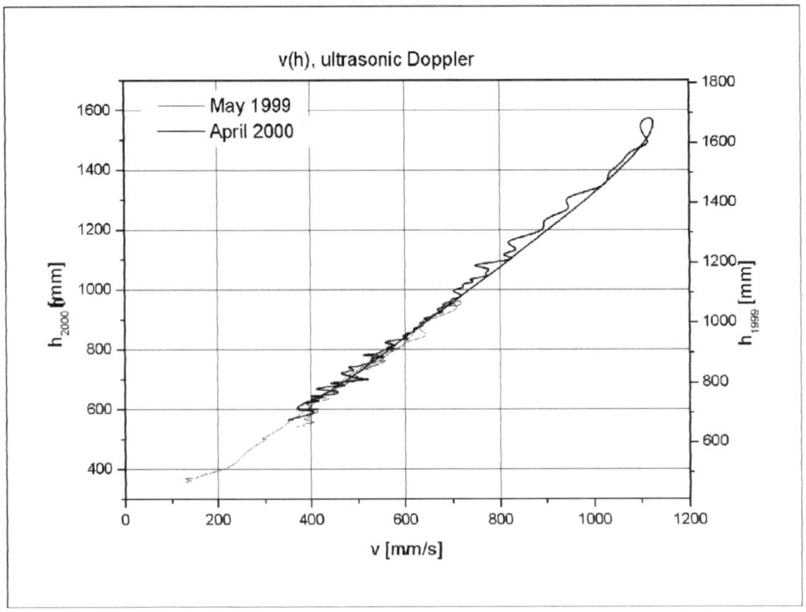

Figure 7. Stage - velocity curve for two events.

The suspended solids concentration were measured with following instruments: Hach SS6 turbidimeter, Partech IR40 infrared sensor and ISCO 3700 sampler (Figure 8). The turbidimeter was not operating all the time, due to problems with pumping.

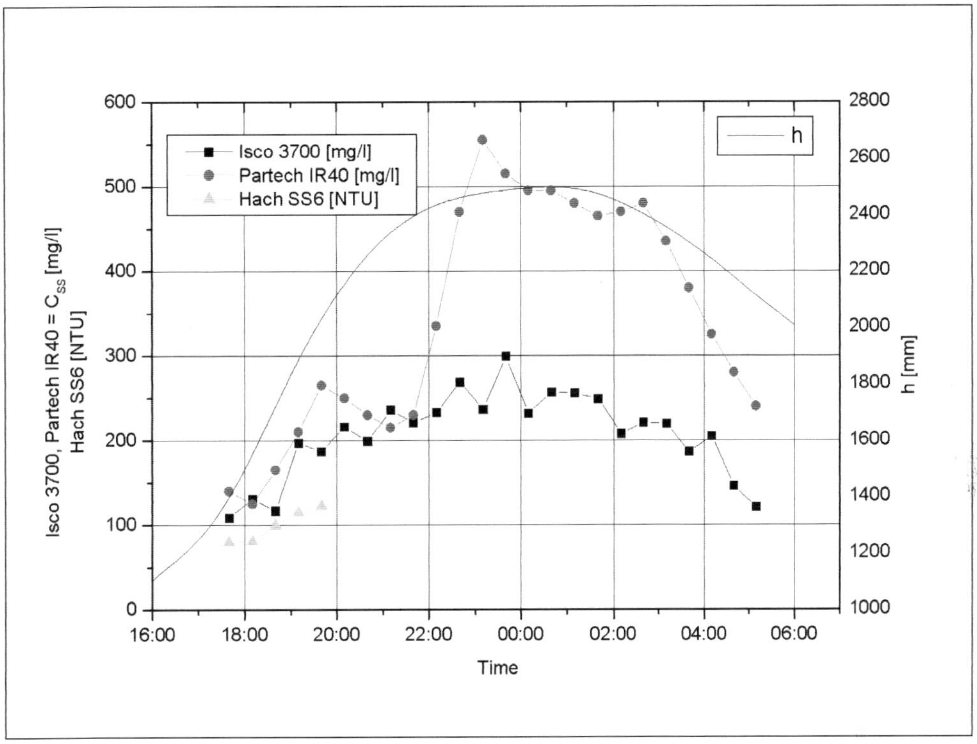

Figure 8. Measured suspended sediment on 29-30 March 2000.

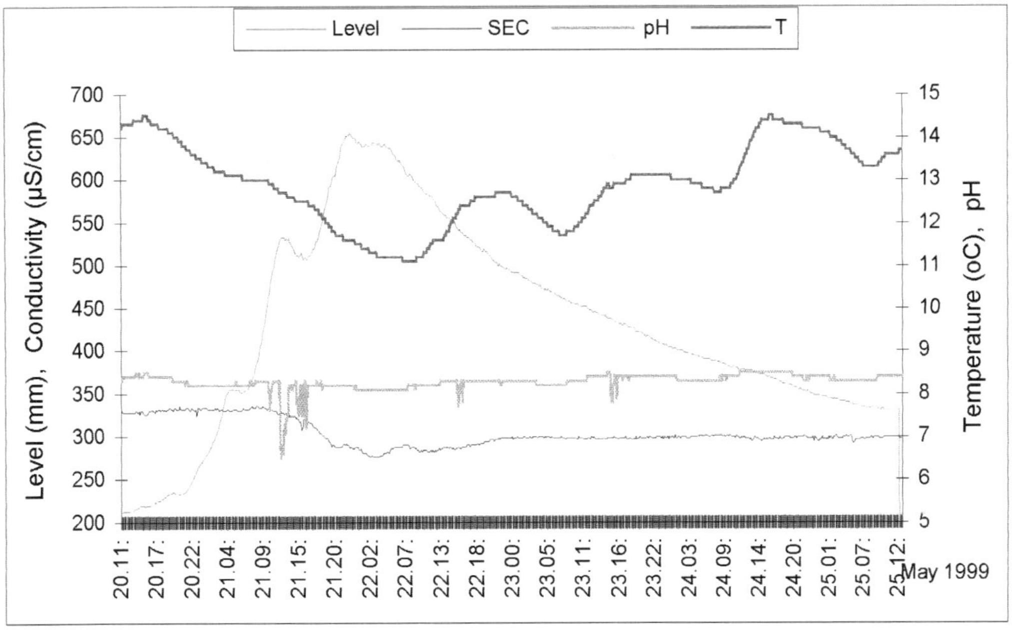

Figure 9. Water quality measurement in May 1999.

Water quality parameters: temperature, pH, conductivity (SEC) and dissolved oxygen (DO) of the Reka River were measured at different hydrological conditions with a Sound YSI 600 and the portable ISCO 6700 auto-sampler (Figure 9).

Three microbiotes were used to test toxicity of damp staining water and toxicity of river Reka. Samples from river Reka were taken at flood wave as well as before and after it. The aim of selecting three microbiotests was to find out if there were any differences in their sensitivity. Namely, the conventional *Daphnia* toxicity test is legally adopted for testing the toxicity of samples from rivers during the regular annual biomonitoring of surface water quality [9].

4. The Gradaščica – Glinščica experimental watershed

4.1. The study area

The Gradascica River basin spreads in the transitional area from the Dinaric into Alpine region in the central part of Slovenia. The headwater section flows through the varied mountain relief of the Dolomites, and is carved with numerous ravines and valleys. The Gradaščica river watershed comprises an area of 154.4 km^2, which reaches far into the Polhov Gradec mountains (Figure 10).

Steep slopes, fairly high altitudes and abundance of precipitation (average yearly quantity from 1600 to 1700 mm) result in a quick rise in the water level of the Gradaščica. The plain area of the Ljubljana basin widens on the eastern part of the watershed. At the Bokalce dam, the Gradaščica splits into two water bodies, the Mestna Gradaščica and the Mali Graben, which flow into the Ljubljanica River. The Bokalce dam controls the discharge to Mestna Gradaščica stream and only about 10 % of discharge of the Gradascica River diverts to the Mestna Gradaščica stream. The Mali Graben carries in total about 90 % of discharge of the Gradascica River.

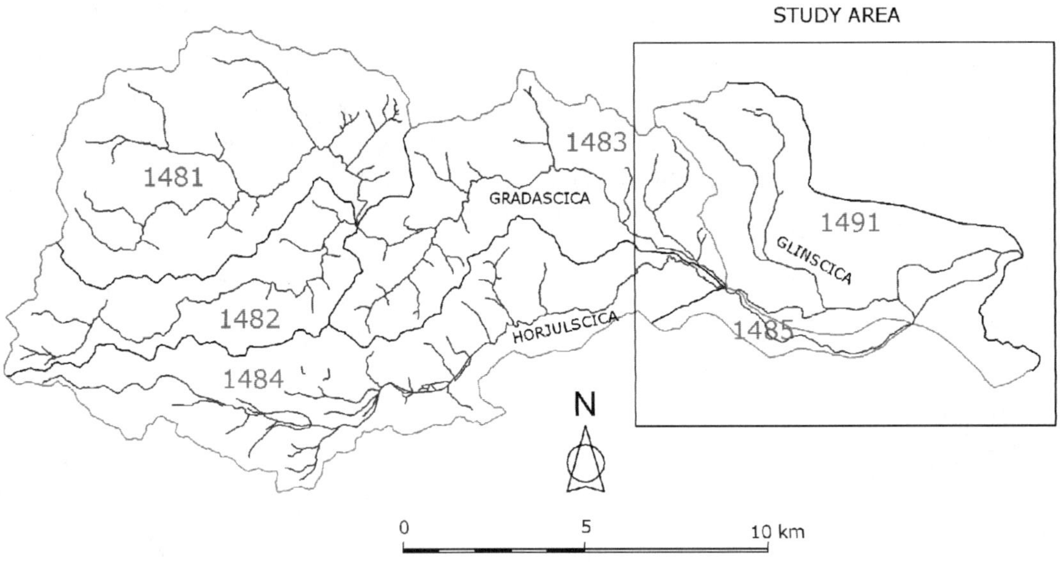

Figure 10. The Gradaščica river experimental watershed.

Glinscica (1491) is the tributary of the Mestna Gradascica stream (Figure 10). The area is close to the city centre and is heavily urbanised. The stream was heavily modified and regulated by concrete blocks (Figure 11) [10]. The Mestna Gradascica also divides into the upper and lower part related to confluence with the Glinscica. The Glinščica watercourse has its source under the north-eastern slopes of Toško čelo (590 m altitude) and passes into the plain area of the Ljubljana Plain, which is of poor permeability and flows into the Gradaščica (293 m altitude), presenting the southernmost point of the basin. The topography of the basin is comprised of a hilly area to the east and to the west and a plain area that spreads out in the southern part. Vast urban areas extend mostly in the eastern and southern parts of the catchment. The precipitation watershed area of the Glinščica comprises 17.4 km^2. The position of the runoff within the urban area is determined by the removal of rainfall water by way of a sewage system, therefore the orographic barrier fails to coincide with the Glinščica drainage. The total drainage area of the Glinščica up to its outlet into the Gradaščica is somewhat bigger and comprises 19.3 km^2 of the catchment area.

Figure 11. The Glinščica watercourse channel in Ljubljana.

4.2. Measurements and results

The measuring equipment includes a one-dimensional ultrasonic Doppler instrument, 2D/3D handheld Doppler velocimeter and water quality multiprobe. One-dimensional Doppler instruments, which are placed at the bottom of streams, continuously record the water level, measure water velocity and temperature. The multi-purpose instrument, designed for in-situ and flow-through applications, measures parameters simultaneously. It can be used for remote or attended monitoring of fresh, salt, or polluted water; in both, surface and groundwater. The multiple parameters include: nitrate, temperature, conductivity, depth, dissolved oxygen, Oxidation Reduction Potential (ORP) and pH. Measurements by multi-purpose probe were carried out in different seasons in approximately 40 measuring cross sections along the streams Glinsica, Mestna Gradascica, Mali Graben and Przanec from July 2003 to July 2004 (Figure 12). The continuous measurements of fluctuations of parameters in streams were also derived by the probe for 24 hours at 15-minute sampling rate in 3 cross sections equipped by the Doppler instruments. Six rain gauge stations were used for rainfall measurement (Figure 12). Measurements showed how vulnerable to pollution is the Glinščica creek due to typical urban flow regime with extremely low water flows and water content especially in the summer. The measurements have been conducted from July 2003 [11].

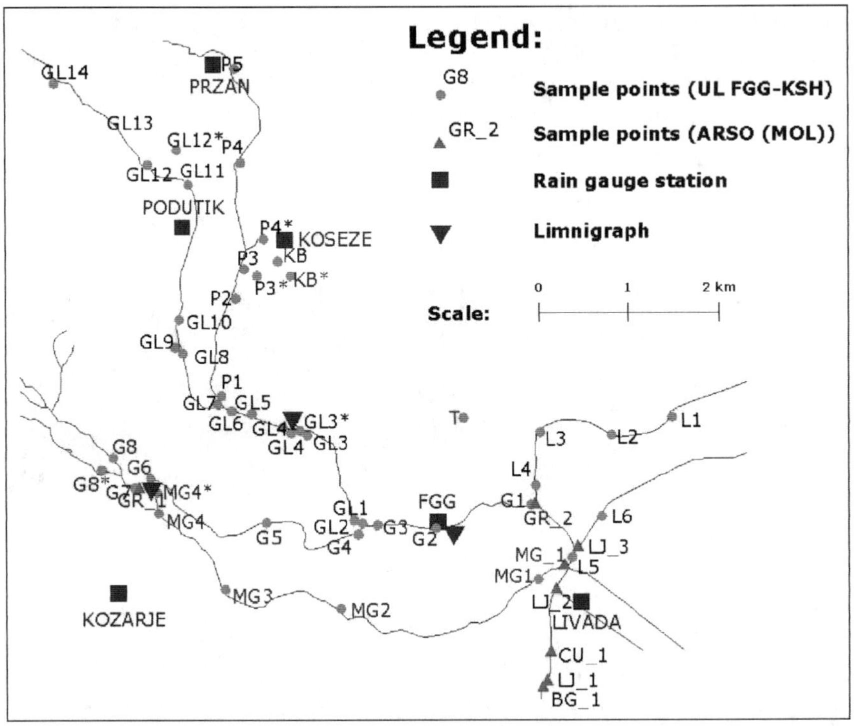

Figure 12. Measuring sites in the study area.

4.3. Results and discussion

The measurements of parameters by the multi-purpose instrument were performed in four two-day campaigns in July, October 2003, and January, March 2004. The number of measuring points was reduced during each campaign because of the rather small scale of the studied area. Among the parameters measured, the measurements of dissolved oxygen and pH value are discussed in detail.

Dissolved oxygen is a basic requirement for healthy aquatic ecosystems. The water in the streams was saturated with dissolved oxygen for more than 70 % of the time. The water was over-saturated in the summer time. The over-saturation with dissolved oxygen was higher in heavily modified concrete channels of the Mestna Gradascica and Glinscica streams than in the rehabilitated channel of the Mali Graben River. The concrete bottom and banks of the Mestna Gradascica and Glinscica streams are intensively overgrown by algae. As the concrete channel is mainly unshaded, the photosynthetic activity is very intense. This is reflected on the high concentration of dissolved oxygen, which is oversaturated during the daytime in the summer.

The pH of the stream water reflects the chemical characteristics of precipitation and land surface. Levels of dissolved carbon dioxide in stream water reflect the balance between uptake of carbon dioxide by aquatic organisms (for photosynthesis) and release of carbon dioxide (by respiration). On the other hand, the concentration of dissolved oxygen is increased by the photosynthetic activity during the daytime and the dissolved oxygen can be depleted during the night by the respiration of water organisms.

The measurements were logged automatically at 15-minute intervals. The relation between the variations of pH and dissolved oxygen saturation is evident (Figure 13). Increased values of pH and dissolved oxygen saturation during the daytime are result of diurnal photosynthetic activity in the water body. The temporal fluctuations of pH and dissolved oxygen saturation are also related to the increased water discharge during a small flood wave produced by 11 mm rainfall.

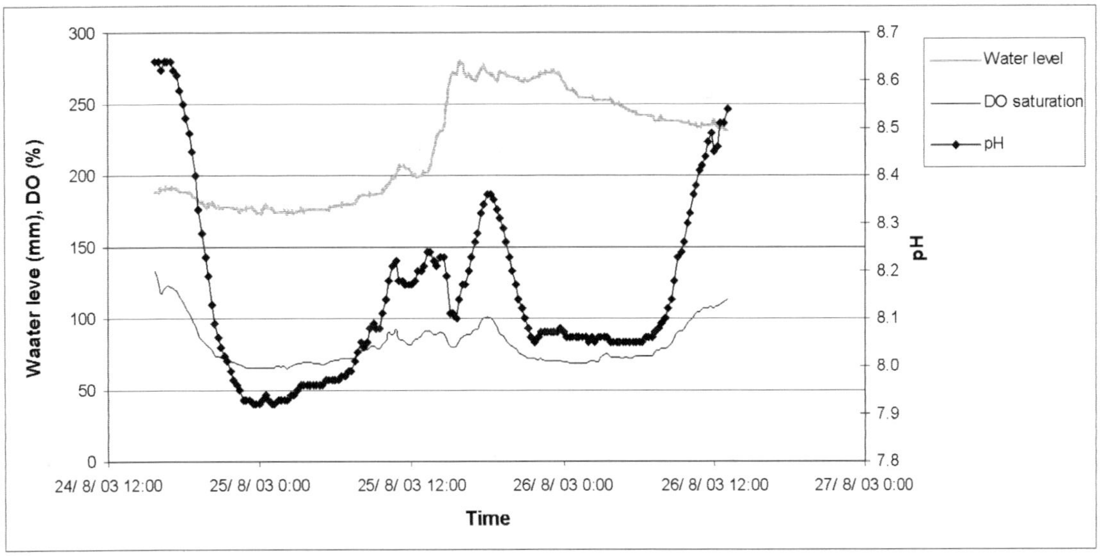

Figure 13. Continuous measurements of water level, pH and dissolved oxygen saturation in the Mestna Gradascica River.

5. Conclusions

The measurements on the Dragonja river basin started in autumn 2000 with the aim to define the impact of the natural reforestation on the water balance of the watershed. According to the result of one-year measurements, the average value of the precipitation throughfall amounted to 67.1 and 71.5 % of the associated rainfall for the south and north research plots, respectively, while the average stemflow fraction was 4.5 and 2.9 %, respectively. The difference is a consequence of the different structure and characteristics of forests on both slopes. The average annual interception losses amount to 28.4 and 25.4 % for the south and north slopes, respectively. The values are relatively high compared with other similar studies. Such high loss of the deciduous forest of the Dragonja watershed is considerable and could be the reason for the reduction of surface water runoff. The conclusions of the research will improve the current understanding of hydrological processes, conditions and interaction between water, soil, deciduous forest and climate in the representative watershed of the Dragonja river.

The process of urbanisation has been accompanied by intensive changes of hydrological characteristics of watersheds. The concept of the watershed in urban areas has become more complex and difficult to define because the natural topography is disturbed in the process. The results of the monitoring on the Glinščica watershed demonstrate a wide range of information that can be gained from continuous measurements of basic water quality parameters in connection with analysis of hydrological regime of the urban environment. One of the major advantages of this type of monitoring is the possibility to quantify water quality extremes that cannot be detected by traditional spot measurements or sampling. This leads to a better understanding of dynamic short term variability in stream water quality and a new insight concerning watershed and in-river processes. The understanding of the processes calls for many field measurements, modelling, data mining and analyses. The use of monitoring is the first step of research, which would help understand the system and identify the key factors for modelling.

IOP Conf. Series: Earth and Environmental Science **4** (2008) 012051 doi:10.1088/1755-1307/4/1/012051

References

[1] Globevnik L 2001 *An integrated approach towards water management in catchments: an integrated analysis of time and spatial components of water regimes as the basis of modern procedures of modelling and planning of water management solutions, land use, and environmental protection: the Dragonja catchment* (Ljubljana: University of Ljubljana) PhD Thesis p 176 (in Slovenian)

[2] Van der Tol C, Dolman A J, Waterloo M J, Raspor K 2007 Topography induced spatial variations in diurnal cycles of assimilation and latent heat of Mediterranean forest *Biogeosciences* **4** 137–154

[3] Šraj M 2003 *Modeling and measuring of rainfall interception* (Ljubljana: University of Ljubljana) PhD Thesis p 236 (in Slovenian)

[4] Keestra S D, van Huissteden J, Vandenberghe J, Van Dam O, de Gier J, Pleizier I D 2005 Evolution of the morphology of the river Dragonja (SW Slovenia) due to land-use changes *Geomorphology* **69** 191–207

[5] Petkovšek G, Mikoš M 2004 Estimating the R factor from daily rainfall data in the sub-Mediterranean climate of southwest Slovenia *Hydrol. Sci. J.* **49**(5) 869–877

[6] Šraj M, Brilly M, Mikoš M 2008 Rainfall interception by two deciduous Mediterranean forests of contrasting stature in Slovenia *Agric. for. meteorol.* **148**/1 121–134

[7] Petan S, Zorn M, Mikoš M 2008 Measurements of interrill soil erosion under different land use in Slovene Istria *Geophysical Research Abstracts: Proc. EGU General Assembly (13–18 April 2008, Vienna, Austria)*

[8] Brilly M, Mikoš M, Petkovšek G, Šraj M, Kogovšek J, Drobne D, Štravs L 2002 The experimental monitoring of water regime in the Reka river *Acta Carsologica* **31**/1 65–74

[9] Brilly M, Mikoš M, Petkovšek G, Šraj M, Kogovšek J, Drobne D 2001 The Reka river monitoring and toxicity tests *The application of ecohydrology to water resources development & management (16–18 September 2001, Venice, Italy, UNESCO)*

[10] Rusjan S, Fazarinc R, Mikoš M 2003 River rehabilitation of urban watercourses on the example of the Glinščica river in Ljubljana *Acta hydrotech.*, **21**/ 34 1–22

[11] Brilly M, Rusjan S, Vidmar A 2006 Monitoring the impact of urbanisation on the Glinscica stream *Phys. chem. earth* **31**/17 1089–96

Institute of Physics
Dirac House, Temple Back
Bristol BS1 6BE UK